# Global Warming and Forest

# Global Warming and Forest

**Dr. M.P. Singh,** *Ph.D., FMA*
*University Professor and Chairman*
*Department of Forest Sciences*
*Birsa Agricultural University,*
*Ranchi – 834 006*
*(Jharkhand)*

2011
# DAYA PUBLISHING HOUSE®
Delhi - 110 002

| | | |
|---|---|---|
| *Published by* | : | **Daya Publishing House**<br>**A Division of**<br>**Astral International Pvt. Ltd.**<br>**– ISO 9001:2008 Certified Company –**<br>4760-61/23, Ansari Road, Darya Ganj<br>New Delhi-110 002<br>Ph. 011-43549197, 23278134<br>E-mail: info@astralint.com<br>Website: www.astralint.com |
| *Laser Typesetting* | : | **Classic Computer Services**<br>Delhi - 110 035 |
| *Printed at* | : | **Chawla Offset Printers**<br>Delhi - 110 052 |

PRINTED IN INDIA

# Preface

Human responses to climate, at the individual national and global labels, have been explored from both natural science and social science perspective. Examination of recent events from global warming to advances in satellite and radar has enhanced current understanding.

The scientific consensus that human are warming the world is based on three sources of information. One of which attracted much attention recently. For a decade, Paleoclimatologists have combed through temperature records locked in various things ranging from ancient tree rings to ice cores. Yet they failed to find a natural warming in the past 1000 years as strong as that of the past century suggesting that humans and their green house gases were behind the recent warming, as did computer studies of warming patterns and the tend of 20th century warming. Consideration of new sorts of measurements and application of a different analytical technique to the data has led to the conclusion that even the surprisingly dynamic climate system may not have produced a natural warming as large as that of the past century. According to P. Jones of the University of East Anglia, the past few decades are still the warmest of the past 1000 years. Temperature records recovered from tree rings and other proxies generally agree that no time in the past millennium has been as warm as recent decades.

It is not often possible to correctly predict the Earth's Climatic Patterns. Storms, drought and floods have affected humanity throughout history. However, records from the past two centuries point to a trend underlying such fluctuations the Earth has been gradually getting warmer. The closely related issues of green house has emissions, global warming and climate change have already ascented to the top of the environmental agenda. Human activities have increased atmospheric concentrations of greenhouse gases so much that the world is committed to future warming ever and above any natural temperature increases that may be occurring. Global warming can alter the global climate patterns which in turn, will raise sea levels and disrupt both terrestrial and oceanic ecosystems. Changes in temperature, weather conditions or soil moisture could severely reduce the productivity of marginal lands. Flood supplies and water resources have already been stretched to the limits in some parts of the world. These could be affected. Low-lying coastal regions could be severely flooded, creating social and economic problems. In the wast-hit regions, many people may be driven from their homes. In the coming few decades we are going to witness the first substantial effects of climate change with changes to the hydrologic cycle including more frequent extreme and high-impact weather events. The time is not too for when oil will be a rare and expensive product: we have probably less than half a century in which to develop and alternative, more efficient energy system.

Environmental questions will need to be addressed in much broader context than before, because these are closely related to development issues. Population will grow unevenly between regions with a shifting population balance between poor and rich countries.

In writing this book, many objective have been to introduce the students of global warming, global cooling, nitrogen deposition and the impact of energy use, and to help scientists as far as possible in solving many pressing issues by elucidating how human activities will affect the goods and services provided by the global environment. Scientists must develop the programmatic tools that will be needed in the next decade to address the broader. Earth system questions posed by society as for instance, how could interdisciplinary approaches be successfully implemented to create partnerships between different scientific communities, and how to deal with problems arising from regional and cultural difference.

For their prosperity, as most countries depend on a stable climate, they must give top priority to addressing the serious issue of global warming.

I have depended heavily in writing this book on useful informative literature, periodical and by many leading researchers all over the world for which I am greatly indebted to them.

*M.P. Singh*

# Contents

# Chapter 1
# General Introduction

## Climate System: An Overview

On Earth, Weather and Climate profoundly influence life everyday uumans experiences them. As both weather and climate are essential for health, food production, and well-being, not only weather but also human–induced climate change can be a matter of some concern. In fact, human activities are already influencing the climate. Understanding, detecting and predicting the human influence on climate can be possible only when we understand, the system that determines the climate of the Earth as also the processes that lead to climate change.

Around us the term 'weather' means the fluctuating state of the atmosphere, characterized by the temperature, wind, precipitation, clouds, and other weather elements. This weather is created from rapidly developing and decaying weather systems such as mid-latitude low and high-pressure systems with their associated frontal zones, showers, and tropical cyclones. Weather can be predicted over a period of hours only, while synoptic scale cyclones may be predicted over a period of some days to about a week. Beyond a week or two, individual weather systems cannot be predicted. The term 'climate' refers to the average weather in terms of the mean and its variability over a certain time-span and a certain area. Climate varies from place to place and depends on latitude, distance from sea, Vegetation, and presence or absence of mountains, or other geographical factor. It varies also in time; from season to season, year-to-year, decade-to-decade, or even on much longer timescales. It is the statistically significant variations of the mean state of the climate or of its variability, typically persisting for decades or longer, which are referred to as 'climate change'.

By external forcing climate variations and change caused may be predictable, particularly on the larger, continental and global, spatial scales. Since human activities, such as the emission of greenhouse gases or land-use change, do result in external forcing, the large-scale aspects of human-induced climate change are also predicted on a short-term basis. What we cannot predict confidently is population change, economic change, technological development, and other pertinent characteristics of future human activity. This means that one has to depend on carefully constructed scenarios of human behaviour and make climate projections on the basis of such scenarios. Knowledge of weather and climate has commonly focused on those variables which affect daily life most directly: average, maximum and minimum temperature, wind near the surface of the Earth, precipitation, humidity, clouds, and solar radiation. These variables can be observed hourly by many weather stations. However, the growth, movement, and decay of weather systems really depend also on the vertical structure of the atmosphere, the influence of the underlying land and sea and certain other factors not directly experienced by human beings. Climate is determined by interactions of the atmospheric circulation with the large-scale ocean currents and the land with its features such as albedo, vegetation, and moisture content. The planetary climate I actually determined by factor that influence the radiative balance, for example, the atmospheric composition, solar radiation, or volcanic eruptions. To understand the Earth's climate and its variations and to possibly predict the changes of the climate caused by human activities, none of the many factor and components that determine the climate can be ignored.

## The Climate System

To an interactive system the term 'climate system' refers consisting of five major components: the atmosphere, the hydrosphere, the cryosphere, the land surface, and the biosphere. These components are forced or influenced by various external-forcing mechanisms, the most important of which is the Sun (Figure 1.1). In addition, the direct effect of human activities on the climate system is treated as an external forcing.

## Atmosphere

Unstable and rapidly changing the most part of the system is the atmosphere whose ever-changing composition is crucial. The Earth's dry atmosphere is composed mainly of nitrogen ($N_2$ 78.1 per cent volume mixing ratio), oxygen ($O_2$ 20.9 per cent), and argon (Ar, 0.92 per cent).

These gases interact in only a limited manner with the incoming solar radiation. They do not interact with the infrared radiation emitted by the Earth. Yet, there are some trace gases, such as carbon dioxide ($CO_2$), methane ($CH_4$), nitrous oxide ($N_2O$), and ozone ($O_3$), which do absorb and emit infrared radiation. These latter are called greenhouse gases (GHS's). Their total volume-mixing ratio in dry air is less than 0.1 per cent by volume and yet they play an essential role in the Earth's energy budget. The atmosphere also contains water vapour, which represents a natural GHG. Its volume-mixing ratio is highly variable, but is typically around 1 per cent. These GHGs absorb the infrared radiation emitted by the Earth and emit infrared radiation up-and downward, thereby raising the temperature near the Earth's surface. Furthermore, water vapour, carbon dioxide, and ozone also absorb solar short-wave radiation.

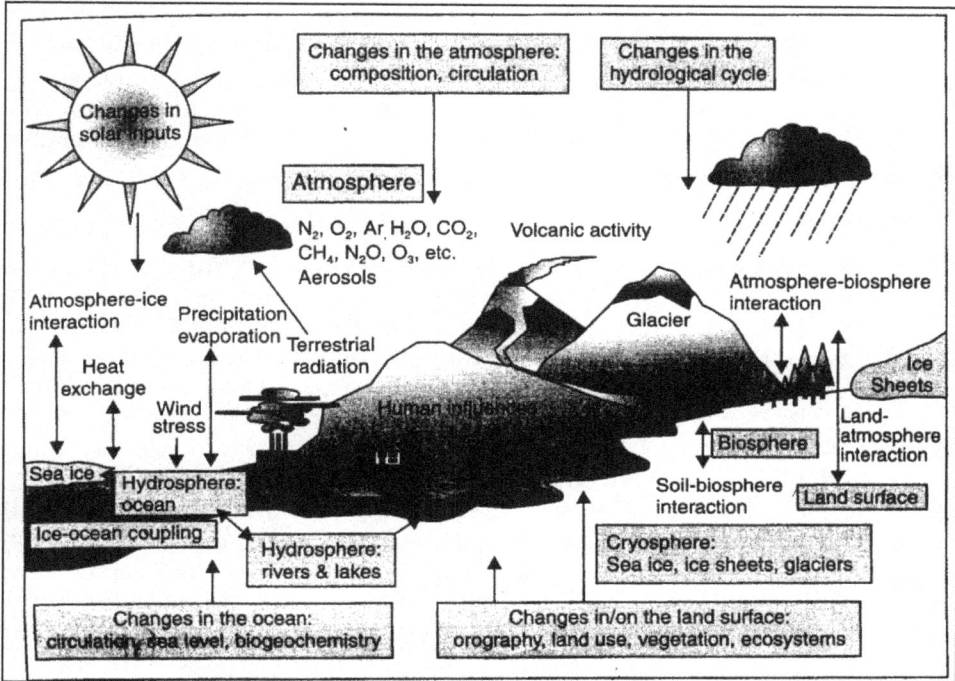

**Figure 1.1: The global climate system the components their processes and interactions and some aspects that may change (after Baede *et al.*, 2001)**

In the atmosphere the distribution of ozone as also its role in the Earth's energy budget is unique. In the lower part of the atmosphere, the troposphere, and lower stratosphere, ozone ($O_3$) acts as a CHG. Higher up in the stratosphere, there is a natural layer of high $O_3$ concentration. This layer absorbs solar ultraviolet (UV) radiation. Therefore, this so-called ozone layer has a critical role in the statosphere's radiative balance; while at the same time, it filters out this potentially damaging form of radiation

Besides the above gases, the atmosphere contains some solid and liquid particles (aerosols) and clouds, which interact with the incoming and outgoing radiations in a highly complex and spatially variable way. The most variable component of the atmosphere is water in its various forms, for example, vapour, cloud droplets, and ice crystals. Water vapour is the strongest know GHG. Indeed, it is central to the climate and its variability and change (Baede *et al.*, 2001).

## Hydrosphere

Hydrosphere is that part of the climate system which comprises all liquid surface and subterranean water, both fresh water (*e.g.*, rivers, lakes and aquifers) and saline water of the oceans and seas. Fresh water runoff from the land, which returns to the oceans through rivers, affects the ocean's composition and circulation. The oceans cover about 70 per cent of the Earth's surface, store and transport a large amount of energy, and dissolve and store large amounts of carbon dioxide. Their circulation is driven by the wind and by density differences created by salinity and thermal gradients (the so-called thermohaline circulation): oceanic circulation is slower than the atmospheric circulation. Chiefly due to the large thermal inertia of the oceans, they modulate (weaken) vast and strong temperature changes and function not only as a regulator of the Earth's climate but also as a source of natural climate variability, especially on long timescales.

## Cryosphere

The ice sheets, continental glaciers, snowfields, sea ice, and permafrost includes cryosphere. Its importance to the climate system accrues from its high reflectivity (albedo) for solar radiation, its low thermal conductivity, its large thermal inertia, and its critical role in driving deep ocean water circulation. A lot of water, variations in their volume can potentially create sea level variations since the ice sheet store.

## Land Surface

From the Sun at the land surface, vegetation and soils influence how energy coming is returned to the atmosphere. Some are returned as long-wave (infrared) radiation; they heat the atmosphere as the land surface warms. Some aid evaporation of water, either in the soil or in plant leaves and so bring water back into the atmosphere. As the evaporation of soil moisture requires energy, soil moisture strongly influences the surface temperature.

## Biosphere

On the atmospheric composition both marine and terrestrial biospheres have a substantial impact. The biota (the combined animal and plant life in an area) influences the uptake and release of GHG's. Through photosynthesis, all plants (especially forests) store significant amounts of carbon from carbon dioxide. Indeed, the biosphere has a central role in the carbon cycle, as well as in the budgets of many other gases, such as methane and nitrous oxide. Other biospheric emissions are the so-called volatile organic compounds (VOC) which often have important effects on atmospheric chemistry, on aerosol formation and hence on climate. With the storage of carbon and the exchange of trace gases being influenced by climate, there are feedbacks between climate change and atmosphere concentrations of trace gases.

On a wide range of spatial and temporal scale, diverse physical, chemical, and biological interaction takes place among the different components of the climate system. These enhance the complexity of the system. The various components of the complex climate system are all connected by fluxes of mass, heat and momentum-all subsystems are open and interrelated (Baede *et al.*, 2001). One typical example is the atmosphere and the ocean being strongly coupled and exchanging, inter alia, water vapour and heat through evaporation and part of the hydrological cycle and leading to condensation, cloud formation, precipitation and runoff; the interaction also supplies energy to weather systems. In addition, precipitation influences salinity, its distribution, and the thermohaline circulation. Atmosphere and oceans also exchange carbon dioxide, besides other gases, and maintain a balance by dissolving it in cold polar water, which sinks into the deep ocean, also by outgassing in relatively warm upwelling water near the equator.

# Variations in Natural Climate

The ultimate source of energy radiation from the Sun is that drives the climate system. About half of the radiation is in the visible short-wave part of the electromagnetic spectrum. The remaining half lies mostly in the near-infrared part, with some in the ultraviolet. Each square metre of the Earth's spherical surface outside the atmosphere receives an average throughout the year of 342 Watts of solar radiation, about 31 per cent of which becomes instantly reflected back into space by cloud, by the atmosphere, and by the Earth's surface. The remaining 235Wm$^{-2}$ is partly absorbed by the atmosphere, but 168 Wm$^{-2}$ warms the Earth's land and the oceans. From the Earth's surface, that heat is returned to the atmosphere, partly as infrared, partly as sensible heat, and as water vapour, which releases it heat upon condensing higher up in the atmosphere. This exchange of energy between surface and atmosphere maintains a global mean temperature near the surface of 14°C, decreasing rapidly with height and reaching a mean temperature of –58°C at the top of the troposphere (Baede *et al.*, 2001).

Between incoming solar radiation and the outgoing radiation emitted by the climate system, a stable climate requires a proper balance. The climate system itself must radiate an average of 235Wm$^{-2}$ back into space (Figure 2). Any physical object radiates energy of an amount and at wavelengths that are typical for the temperature of the object-at higher temperatures, more energy is radiated at shorter wavelength. For the Earth to radiate 235 Wm$^{-2}$, it should radiate at an effective emission temperature of –19°C with typical wavelengths in the infrared part of the spectrum. This happens to be 33°C less than the average temperature of 14°C at the Earth's surface. These facts can only be understood by considering the radiative properties of the atmosphere in the infrared part of the spectrum, *i.e.*, the greenhouse effect.

## Greenhouse Effect: Natural

In the atmosphere several trace gases found absorb and emit infrared radiation. These GHG's absorb infrared radiation, emitted by the Earth's surface, the atmosphere and clouds, except in the 'atmospheric window' which is a transparent part of the spectrum (Figure 2). They emit in turn infrared radiation in all directions including downward to the Earth's surface. GHG's trap heat within the atmosphere, a phenomenon called the natural greenhouse effect. The net result: an upward transfer of infrared radiation from warmer levels near the Earth's surface to colder levels high above. The infrared radiation is effectively radiated back into space from an altitude with a temperature of, on average with the incoming, radiation temperature of, on average, –19°C, in balance, whereas the Earth's surface is kept at a much higher temperature of an average 14°C.

This effective emission temperature of 19°C correspond in mid-latitudes with a height of approximately 5 km. It is essential for the greenhouse effect that the temperature of the lower atmosphere is not constant (isotherm) but decreases with height. The natural greenhouse effect forms the part of the energy balance of the Earth (Figure 1.2) (Baede *et al.*, 2001).

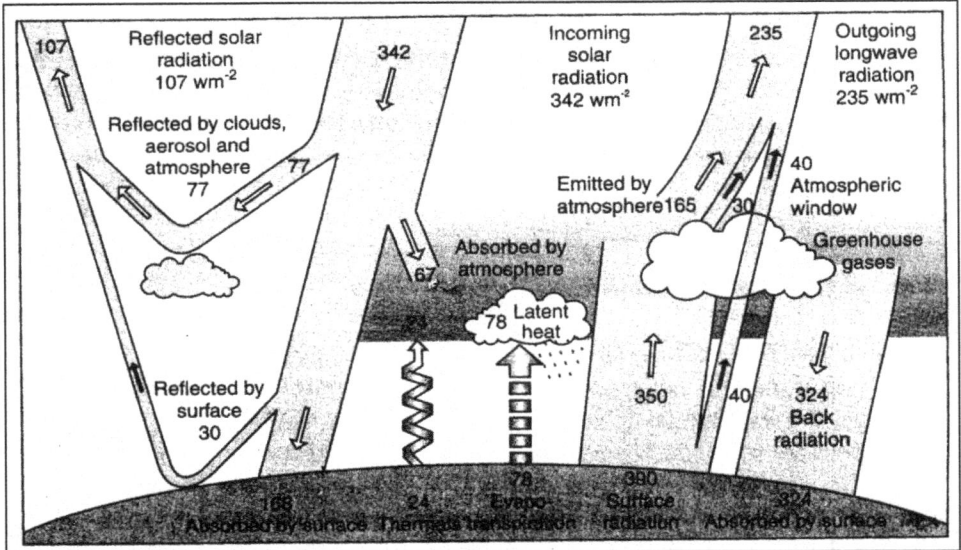

**Figure 1.2: Annual and global mean energy balance of the Earth. Of the incoming solar radiation, 49 per cent (168 Wm⁻²) is absorbed by the surface. That heat is returned to the atmosphere as sensible heat, as evapotranspiration (latent heat) and as thermal infrared radiation. Most of this radiation is absorbed by the atmosphere, which in turn emits radiation both up and down. The radiation lost to space come from cloud tops and atmospheric regions much colder than the surface. This causes a greenhouse effect.**

The Earth's energy balance clouds also affect. Particularly in the natural greenhouse effect, absorbing and emitting infrared radiation and contributing to warming of the Earth's surface. On the other hand, being bright reflectors of solar radiation, most clouds cool the climate system. In fact, the net average effect of the Earth's cloud cover in the present climate is a slight cooling: the reflection of radiation more than compensates for the greenhouse effect of clouds. However, this effect is highly variable and depends on cloud height, type, and optical properties of clouds.

By volcanic eruption when external forcing, such as the solar radiation or the huge amounts of aerosols are thrown into the atmosphere, and which may vary on very different timescale, causing negative or positive natural variations in the radiative forcing, the climate system reacts to store the balance. While any positive radiative forcing warms the Earth's surface on average, a negative one cools it.

## Variability of Natural Climatic

Both from radiative forcing and from internal interaction climate variations can result between components of the climate system. In the case of variations in the external forcing, the response time of the various components of the climate system can be very different. With regard to the atmosphere, the response time of the troposphere is only days to weeks, whereas the stratosphere equilibrates on a scale of a few months. The oceans have a much longer response time, from decades to centuries or even millennia. The response time of the strongly coupled surface-troposphere system is quite low as compared with that of the stratosphere. The oceans chiefly determine it. The biosphere tends to respond fast, for example, to droughts, but rather slowly to imposed change. In fact, the system may respond to variations in the external forcing on a wide range of spatial and temporal scales. On the climate the impact of solar variations exemplifies such externally induced climate variations.

Even in the absence of any changes in external forcing, the climate can very naturally because it is a system of components having very different response times and non-linear interactions; so the components can never be in equilibrium and tend to vary constantly. A good example of such internal climate variation is the *El Nino-Southern Oscillation* (ENSO) resulting from the interaction between the atmosphere and ocean in the tropical Pacific (Baede *et al.*, 2001).

The climate response to both the internal variability of the climate system and external forcing becomes more complicated by feedbacks and non-linear responses of the various components. A process is termed as 'feedback' when the result of the process affects its origin thereby intensifying (positive feedback) or reducing (negative feedback) the original effect. A good example of positive feedback is that of the water vapour in which the atmospheric content of the vapour increases as the Earth warms. Water vapour being a strong GHG, this increase in turn may amplify the warming. A strong, very basic negative feedback is radiative damping-an increase in temperature strongly increases the amount of emitted infrared radiation, which limits and controls the original temperature increase (Baede *et al.*, 2001).

Physical feedbacks involving physical climate processes differ from biogeochemical feedbacks usually involving coupled biological, geological, and chemical processes. The highly complex interaction between clods and the radiative balance exemplifies a physical feedback. A good example of a biogeochemical feedback is the interaction between the oceans. Understanding this feedback enhances an understanding of the carbon cycle.

Several processes and interaction are non-linear, meaning that there is no simple proportional relation between cause and effect. A complex, non-linear system tends to show chaotic behavior, *i.e.*, the behavior of the system critically depends on very small changes of the initial conditions. But this does not imply that the behavior of non-linear chaotic systems cannot be predicted. A good example is the daily weather whose predictability is usually limited to period of at most two weeks.

## Variability in Global and Hemispheric

It appears that the Northern Hemisphere climate of the past thousand years was characterized by an irregular but steady cooling, followed by a strong warming during the 20th century. Temperatures were fairly warm during the 11th to 13th centuries and relatively cool during the 16th to 19th centuries. The warmth of the late 20th century seems to have been unprecedented during the millennium.

The data from the southern Hemisphere are scanty but point to temperature changes in past centuries quaite different from those in the Northern Hemisphere, the only obvious similarity is the strong warming during the 20th century.

## Regional Patterns

Regional or local climate is usually more highly variable than climate on a hemispheric or global scale because regional or local variations in one region are compensated for by opposite variations elsewhere. The spatial structure of climate variability on seasonal and longer timescales is such that it occurs mainly in preferred large-scale and geographically anchored spatial patterns-patterns that result from interactions between the atmospheric circulation and the land and ocean surfaces. However, although geographically anchored, their amplitude some time changes in time as, for instance, the heat exchange with the underlying ocean changes.

The best-know example is the quasi-periodically varying ENSO phenomenon, caused by atmosphere-ocean interaction in the tropical Pacific. The resulting *El-Nino* and *La Nina* events exert a global impact on weather and climate. Another good example is the North Atlantic Oscillation (NAO) that strongly influences the climate of Europe and a part of Asia. This pattern has opposing variations of barometric pressure near Iceland and near the Azores. On average, a westerly current, between the Icelandic low-pressure area and the Azores high-pressure area, moves cyclones toward Europe. The pressure difference between Iceland and the Azores fluctuates on timescales of days to decades, and sometimes I reversed. The NAO variability affects the regional climate variability in Europe, especially during winter.

## Extreme Events

Climate is associated with a certain probability distribution of weather events. Weather events with values very different form the main (such as heat waves, drought and floods) are less likely to occur. Those events, which are least likely to occur in a statistical sense, are termed 'extreme events'. For instance, extreme weather in one region (*e.g.* a heat wave) can be quite normal in another. Small changes in climate occasionally have a strong impact on the probability distribution of weather events in space and time, as on the intensity of extremes.

# Variations in Human-induced

Like other living organism, while humans, have always influenced their environment, it is only since the mid-18th century that human activities started extending to a much larger continental or even global scale. The combustion of fossil fuels for industrial or domestic usage, and biomass burning, produce GHGs and aerosol that affect the atmospheric composition. The emission of chlorofluorocarbons (CFCs) and other chlorine and bromine compounds not only affects the radiative forcing but has also led to a strong depletion of the stratospheric ozone layer. Land-use change, due to urbanisation, afforestation and farming, affects the physical and biological properties of the Earth's surface. These effects change the radiative forcing and can potentially affect regional and global climate.

Since the industrial Revolution, the atmospheric concentration of various GHGs has increased. The amount of carbon dioxide, for instance, has increased by at least 30 per cent and is still increasing at high rate of, on average, 0.4 per cent per year, chiefly due to the combustion of fossil fuels and deforestation. This increase is anthropogenic. The concentration of other natural radiatively active atmospheric components, for example, methane and nitrous oxide, is also increasing due to farming, industrial and other activities. The concentrations of the nitrogen oxides (NO and $NO_2$) and of carbon monoxide are also rising. These gases are not GHGs, but they do have a role in the atmospheric chemistry and have increased tropospheric ozone (a GHG) by 40 per cent since preindutrial time. Furthermore, $NO_2$ is a good absorber of visible solar radiation. CFCs and some other halogen compound are not found naturally in the atmosphere; they have been introduced by human activities. They not only deplete the stratospheric ozone layer but also are strong GHGs. Their greenhouse effects are partly compensated for by the depletion of the ozone layer, which causes a negative forcing of the surface-troposphere system. All these gases, except tropospheric ozone and its precursors, have quite long atmospheric lifetimes and consequently have become well mixed throughout the atmosphere (Baede *et al.*, 2001).

Various industrial, energy-related, and land-use activities of human beings increase the amounts of aerosols in the atmosphere, in the form of mineral dust, sulphates, nitrates and soot. Unlike CFCs, their atmospheric lifetime is short as they are removed by rain. This means that their concentrations are highest near their sources and very substantially regionally, with global consequences.

The increases in GHG concentration and aerosol content in the atmosphere produce some change in the radiative forcing to which the climate system must react to restore the radiative balance (Baede *et al.*, 2001).

## Effect Enhancement in Greenhouse

The increased atmospheric concentration of GHGs enhances the absorption and emission of infrared radiation. The atmosphere becomes more opaque so that the altitude from which the Earth's radiation is effectively emitted into space becomes higher. As the temperature is lower at higher altitudes, less energy is emitted; this causes a positive radiative forcing. This effect is termed the enhanced greenhouse effect.

## Aerosols Effect

The effect of the increasing aerosols on the radiative forcing is highly complex. Part of the incoming solar radiation is scattered back into space, causing a negative radiative forcing which may offset the enhanced greenhouse effect. However, due to the short atmospheric lifetime of aerosols, the radiative forcing is highly inhomogeneous in space and in time and so complicates their effect on the non-linear climate system. Soot can absorb solar radiation directly, leading to local heating of the atmosphere, or absorb and emit infrared radiation directly, leading to local heating of the atmosphere, or absorb and emit infrared radiation, adding to the enhanced greenhouse effect.

Aerosols can also affect the number, density, and size of cloud droplets which changes the amount and optical properties of clouds, hence their reflection and absorption. This could also impact precipitation.

## Change of Land-use

This may result from various human activities such a changes in farming practices and irrigation, deforestation, reforestation and forestation and also from urbanization or traffic. Land-use change in turn changes the physical and biological properties of the land surface and hence the climate system. Land-use change on the present scale has the potential to contribute significantly to changing the local, regional, or even global climate and has a strong impact on the carbon cycle. Some physical processes and feedbacks caused by land-use change, that may impact the climate, are changes in albedo and surface roughness, and the exchange between land and atmosphere of water vapour and GHGs. Land-use change also affects the climate system through biological processes and feedbacks involving the terrestrial vegetation, which produces changes in the sources and sinks of carbon in its various forms.

Urbanization as a land-use change can affect the local wind climate through influencing the surface roughness. It also creates a local climate that is warmer than the surrounding countryside because some heat is released by densely populated human settlements. The outgoing long-wave radiation can be modified through interception by tall buildings.

## Response in Climate

The increase in atmospheric GHG and aerosol concentrations, as also land-use change, produces a radiative forcing or affect processes and feedbacks in the climate system. The climatic response to these human-induced forcing is complicated by the feedbacks, by the strong non-linearity of many processes, and because the various coupled components of the climate system differ greatly in response time to perturbations. While an increase of atmospheric GHG concentrations leads to an average increase of the temperature of the surface-troposphere system, the response of the stratosphere is very different. The stratosphere is characterized by a radiative balance between absorption of solar radiation, primarily by ozone, and emission of infrared radiation, chiefly by carbon dioxide. Any increase in the carbon dioxide concentration, therefore, increases the emission and cools the stratosphere.

# Past and Future: Climate Change

In all civilisations changes in global and regional climates have always played a key role. Entire civilisations have collapsed after prolonged periods of drought. The recent devastating hurricane in the United states wreaked widespread havoc, highlighting the fact that societies need robust infrastructures such as buildings constructed to withstand high-speed winds, good road, dependable transportation, and early warning system to deal with extreme weather conditions. These measures strongly depend on accurate predictions of regional climate change regardless of whether it is a result of natural variability or is caused by GHG emissions.

Barnett *et al.* (2005) underscored the potential impacts of regional climate changes caused by anthropogenic GHG emissions, especially the effects of these changes on water availability. One popular model of the interaction between scientific research and policy making describes an almost linear path whereby an environmental problem is identified, and Research and Development (R&D) is used to investigate the causes, effects and potential solutions (Patrinos and Bamzai, 2005). This information is provided to policy makers who may make appropriate changes to legislation or technology. The best practical example of tackling a problem by this linear route is the discovery of stratospheric ozone layer depletion, which led to the Montreal Protocol, which banned CFCs that harm the ozone layer. However, the usual mechanism is often not so clear-cut or well-defined-repeated interactions can occur between R&D and policymaking. This may be the case for regional climate variability. Patrinos and Bamzai mentioned two examples of multi-level R&D that could in due course provide robust solution for regional climate problems.

One example describes governmental R&D priorities for water availability and usage. The ability to measure, monitor, and forecast the state of national and global freshwater supplies is certainly a problem of national importance. A good research strategy needs to be developed to understand the processes that control water availability and quality; also to gather baseline information and develop monitoring systems for ensuring adequate future supplies. Untapped water resources could well come from water conservation, water re-use, desalination, and aquifer storage. Thee are very often not considered a resource, but should be (Patrinos and Bamzai, 2005).

The second example of multi-level R&D is the important steps taken in response to recent widespread droughts in the united state. The droughts led governor in the concerned states to endorse a drought early-warning system which could provide such water users as farmers, tribes, business owners, wildlife managers and decision makers at all levels of government with valuable information that enables them to assess risks in real time, and allow wise decisions to be made ahead of a drought.

There is no doubt that the risks to humans and property can be minimized by developing repeatedly during the Late Pleistocene. Rapid variations in the atmospheric concentrations of carbon dioxide, $CH_4$ and $N_2O$ occurred in lockstep with temperature, implying a causative relationship. Abrupt changes in the character of thermohaline circulation in the sea, particularly in the North Atlantic, generated swings in temperature across the Northern Hemisphere. Shifts in hydrologic balances in continental interiors were marked, producing severe drought in some region (Pedersen, 2000). In contrast, the climate of the Holocene was probably relatively stable and contributed to the flourishing of human societies. Many of the past variation in climate that have recently been discovered occurred abruptly, within decades or less.

Studies of proxies have revealed that temperatures declined from AD 1000 until the late 19th century. Followed by warming at a rate that is unprecedented in the record, leading to temperatures in the late 20th century that were unique in the context of the entire millennium. The available palaeorecords show that in geological terms, about, natural changes are common, and we can expect them in the future. Given additional human-induced forcing of climate, we may expect enhanced societal distress as climate changes are superimposed on increasingly vulnerable population. Mitigative actions to limit such forcing are thus urgently warranted.

In Figure 1.3 shows the year-by-year values of the global average surface temperature of our planer as compiled at the British Meteorological Office from land and ocean stations in 1996. The most recent year, 1995, was the warmest in the 130 years of record. The smoothed, dark line is a 21 year running average.

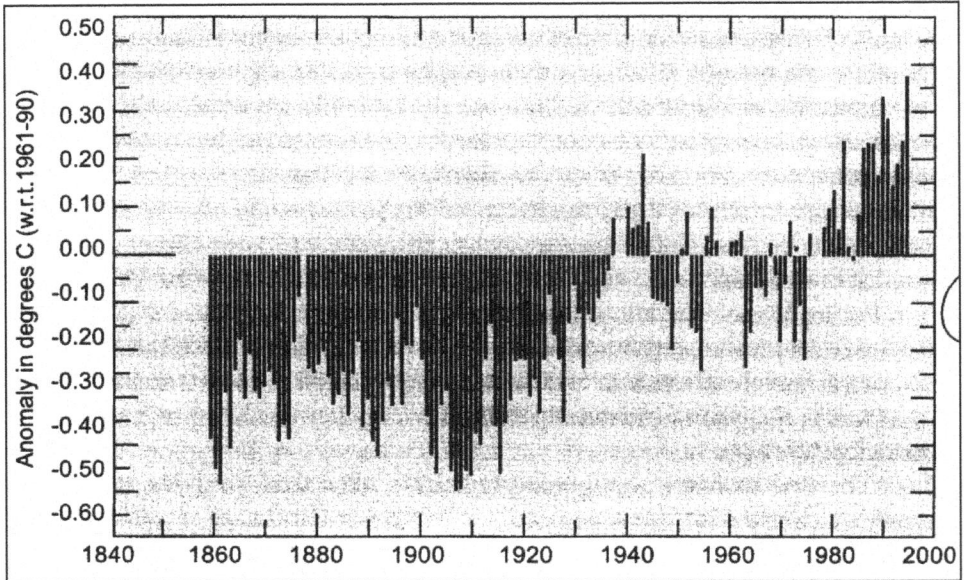

**Figure 1.3: Combined land air and sea surface temperatures of the period 1860-1995 (relative to 1961-90 average).**

The Earth's climate and atmospheric composition have regularly waxed and waned through the glacial-interglacial cycles. Ice core and other palaeorecords provide a fascinating window on the metabolism of the Earth over hundreds of thousands of years. No record is more intriguing than the rhythmic 'breathing' of the planet as revealed in the Vostok ice core records of temperature and carbon dioxide and methane concentrations (Petit *et al.*, 1999) cyclic variations of relatively long cold (glacial) periods were interrupted by shorter warm (interglacial) periods. The atmospheric carbon dioxide concentration varied from 180-200 ppm during the glacial periods to 265-280 during the interglacials.

In the pattern the palaeorecords show other interesting details:

1. During the glacial terminations, the increase in atmospheric carbon dioxide is in phase with southern hemisphere warming; melting of the northern hemisphere ice caps lags by thousands of years;

2. The strong coupling between temperature and atmospheric carbon dioxide suggest that the latter is probably the primary amplifier of climate change during glacial terminations; and

3. The periodicity of the cycles shows a strong correspondence to the cyclic variations in the Earth's orbit (Steffen, 2000).

It seems that the precise nature of the upper and lower limits of atmospheric carbon dioxide concentration points to strong control mechanisms-both terrestrial and oceanic biological processes are critical elements of the control loop.

Biogeochemical interactions between land and ocean transfer control from one to the other on a periodic basis. The lower level of about 180 ppm for atmospheric carbon dioxide represents something of am 'ecoystem/biome compensation point.' Below that level, system lose almost as much carbon through respiration as they can take up through photosynthesis in the cold, dry carbon dioxide-depleted climate. This has implications for the transfer of nutrients between land and ocean. At the upper limit (280 ppm), the solubility-driven flux of carbon dioxide from the ocean to atmosphere is balanced by the uptake of carbon dioxide by the terrestrial and oceanic biota (Steffen, 2000).

Gradually, as the climate warms and carbon dioxide concentration increases, the increasing activity of the terrestrial biosphere accelerates the mobilisation of elements such as phosphorus (P), silicon (Si), and iron (Fe) from the gosphere through enhanced root activity. Theses elements eventually leak from the terrestrial biosphere into rivers and to the coastal ocean. Over millennia, these nutrients become entrained into the oceanic circulation; and, in areas of upwelling, stimulate oceanic net primary production and increase the drawdown of carbon dioxide from the atmosphere. The increasing biotic uptake of carbon dioxide in both oceans and land eventually matches the solubility-driven outgassing of carbon dioxide and the system reaches a balance at an atmospheric concentration of carbon dioxide of about 280 ppm.

However, the strong activity of the terrestrial biosphere has already sown the seeds of its own destruction. The interglacial balance is precarious. The vigour of terrestrial and marine biological uptake is overtaking the outgassing form the oceans. This triggers a set of feedbacks-initial cooling increasing solubility of carbon dioxide, increasing sea ice and further cooling-which drive the system towards the glaciated state. Although the terrestrial biosphere is taking up less carbon dioxide, it also releases phosphorus that was tied up in internal cycling. This P finds its way into the ocean and stimulates productivity there.

As cooling and drying proceed, terrestrial biotic activity declines, and the flux of nutrients from the land into the ocean eventually stop. Oceanic uptake of carbon dioxide nutrients from the land into the ocean eventually stop. Oceanic uptake of carbon dioxide then starts to decline as well, and the system eventually reaches and bounces along its 'floor' at 180 ppm carbon dioxide, effectively controlled by a quiescent terrestrial biosphere until the next cyclic change in the Earth's orbit jolts it back into the glacial termination phase and the whole intertwined loop of forgings and feedbacks starts over (Steffen, 2000).

It is interesting to put the very recent human perturbations to Earth system in the context of the above highly regular pattern. The current concentration of atmospheric carbon dioxide of about 380 ppm is well above the above the upper control limit of the Earth's recent past; there is no evidence of a state or 'domain of attraction' above 280 ppm. The rate of increase of atmospheric carbon dioxide is about two orders of magnitude higher than that during the glacial terminations. The rates of increase of mean global temperature over the past several decades appear to be without precedent in the recent past (Mann *et al.*, 1998).

Some major global environmental issues are:

1. Climate change resulting from human-induced enhancement of the greenhouse effect;
2. Depletion of the stratospheric ozone layer;
3. Disruptive seasonal and interannual variation in temperature and precipitation, such as the *El Nino*-Southern Oscillation (ENSO); and
4. Large-scale changes in land use and land cover.

The climatic, ecological, and biogeochemical records of the Earth can help advance our understanding of anthropogenic effects on the Earth system. Historically, the Earth's geography, climate, and ecosystems have suffered dramatic changes and fluctuations on timescales varying from less than a decade to millennia.

Although global average conditions changes slowly, intramural to decadal variations do occur at regional (subcontinental) scales, driven by changes in sea surface temperature and ocean circulation patterns, volcanic eruptions, soil moisture anomalies, and variations in the amount of solar radiation reaching the Earth. Table 1.1 lists some major factors involved in global climate change.

**Table 1.1: In global climate change Major factors involved (after Jones, 1997).**

| Period of Operation | Factors | Process and Effects |
|---|---|---|
| Years to millennia | Planetary orbits | Gravitational pulls affect solar activity, cosmic radiation and volcanic activity. Radiation level and atmospheric capacity/reflectivity are affected |
| Decades | Sun's Internal processes | Affect solar output of electromagnetic and particulate (solar wind) radiation |
| 20,000 to 100,000 years | Earth's orbit | Distance from the Sun, angle of polar axis and seasonability affect distribution and intensity of solar radiation. |
| Hours to decades | Earth's atmospheric composition | Transmission, radiation and reflection of electromagnetic radiation are affected |
| Days to millennia | Earth's surface properties | Radiation and heat balance, evaporation, precipitation and wind movement are affected |
| Years to millennia | Earth's core and mantle | Movements (plant tectonics) affect volcanic emission of dust and gas; land-sea distribution; mountain building. Fluctuations in geomagnetic fields affect cosmic radiation distribution/intensity, ocean currents. |

Since the advent of the industrial era, atmospheric abundances of greenhouse gases (GHGs) have increased because of human activity, causing an increase in the radiative forcing of the Earth's atmosphere. Some part of this increase may have been counter-balanced by the cooling associated with simultaneous increase in atmospheric sulphate and carbonaceous aerosols or by the effects of stratospheric ozone depletion. The quantitative regional and global impacts of these changes on the Earth's climate are strongly region-dependent, and there are large uncertainties.

In land use and land cover global changes and the resulting losses of productivity of terrestrial ecosystem are taking place. Land-use change often depletes localised resources of common value to social system. Transformation of land from native forest cover to agriculture, grasslands, and urban development can change the natural water cycle, resulting in the degradation of previously productive land and the loss of biodiversity.

Though the biggest challenge for human civilization today is poverty, climate change is the biggest environmental challenge because no part of the world is immune to it. It is paradoxical that despite increasing evidence of climate change, many governments are not convinced enough to agree on the issue of reducing emissions of GHGs. A recently released Shanghai conference report stated that global warming over the last century could not be explained without human influence. A new assessment by the Intergovernmental Panel on Climate change (IPCC) has projected a marked increase in global warming by 1.4 to 5.8° C over the 21$^{st}$ century. Progress in reducing emissions is slow because while the problem is environmental, the response is economic. Economists dominate the negotiations and seek either to shift the 'costs' elsewhere or at least to defer them. The negotiations are not seen as a response to a common threat but as sharing of burdens. Climate change is the biggest environmental issue, but the global environment is not so high on political agenda of several countries.

One of the hottest issues is that of carbon sinks- the questions of how much can be allowed to developed countries to offset their emissions by the uptake of carbon in forests and farmland. This issue is of great interest both to developed and developing countries.

Historically, much of the responsibility for the increase in emissions of GHGs lies with the countries that were industrialized earliest. These countries have already agreed to cut emissions first. The 'polluter-pays-principle' requires them to take measures that will result in lower emissions. However, these measures involve costs and there lies the 'payment' factor. The fact is that we have failed in integrating ecological awareness into economic decision-making largely because of the low clout of environment ministries and preoccupation of business with the returns of capital. A judicious combination of regulation, incentives, and corporate responsibility can tilt the economy towards environmental rationality.

Emissions are rising and economies are growing. India's emission of carbon dioxide have been rising during the past few years. The longer we take to address this problem, the bigger will it become. One argument says: "Act now, it will cost less"; another says: "Delay it and technology will be cheaper". The first option is certainly better. It is not just about a technological fix but also about a change in habits, perceptions, and consumption patterns. One cannot start too early. We have the power, and must act now. A strange gulf between the great importance of earth's climate and the low level of public interest in it warrants concrete actions to ensure that earth's climate future gets more attention than it has got. The general public is greatly concerned at headlines about natural calamities such as tornadoes, killer heat waves, typhoons and floods, but is generally less concerned about equally serious but less spectacular effects of long-term climatic changes such as global warming.

The fundamentals of climate and how it is being changed are straightforward: As humans add GHGs like carbon dioxide and methane to the atmosphere, they form a blanket that intercepts infrared radiation as it leaves the Earth. This 'greenhouse effect' has been well understood for over a century. Modeling work tracking average global temperature over its fluctuations during the past 10 centuries reveals that it followed natural events (such as volcanic eruptions and variations in solar flux) fairly well until the 20th century where after it started rising rapidly, associated with an increase in atmospheric carbon dioxide from its preindustrial level of 280 parts per million (ppm) to the present level of about 380 ppm. This value is still rising as we continue business as usual. Much of the present warming trend is attributed to human activity.

A recent scenario-building exercise (see Kennedy, 2004) suggested a sudden breakdown in the North Atlantic circulation, producing a dramatic regional cooling. We do not know whether global warming will continue to increase steadily or possibly cross the threshold of some nonlinear process.

Scientists disagree about what the future may hold. The highly sophisticated general circulation models used in the world's major centers are producing results that generally agree, but there is debate on the possibility of altered relationships between oceans and atmosphere, the role of clouds and aerosols, the influence of changes in the Earth's ability to reflect light, and the regional distribution of climate effects. These disagreements confuse the public even more-because the scientists do not agree, the issue tends to be ignored. However, the issue should not be ignored. The models project that a doubling of the atmospheric concentration of carbon dioxide from reindustrial levels (probable by this century's end) would increase average global temperature by between 2°C and 4°C, and they predict an increase in the average frequency of unusually severe weather events. In addition, the modest increases already observed in this century are changing the rhythms of life on earth. While the effects of global warming have been most marked in the Arctic, where dramatic glacial retreats and changes in the reflectivity of the land have occurred, even at low latitudes, mountain glaciers have shrunk so much that the photogenic snowcap of Mount Kilimanjaro in Kenya will vanish by 2020 (Kennedy, 2004). Some plants and the organisms that depend on them have already changed their schedules in several parts of the world, advancing their flowering and breeding times at a rate of about 5 days per decade. Sea levels have risen 10 to 20 and some parts of the northern United States, groups of icebergs were episodically launched into the North Atlantic. The melting of this freshwater ice, which also freshened ocean surface water, most probably changed the strength of the oceanic thermohaline circulation (Clark *et al.*, 2002), thereby producing abrupt climate changes (Clarke *et al.*, 2003). Large volumes of glacial meltwater resulting from the deglaciation of North America also influenced the circulation of the North Atlantic. Some 8,500 years ago, the Lauarentide Ice Sheet, which at its maximum formed a 3-km-thick dome over Hudson bay, was disintegrating rapidly. Shortly before its disintegration, the glacial Lake Agassiz had become a superlake. It eventually released its waters to the Hudson Bay.

Although modern analogues to subglacial outburst floods from Lake Agassiz are found in Iceland and elsewhere, in terms of the released water volume, the flood from superlake Agassiz is by far the largest known glacial outburst of the past 100,000 years. A physical model of subglacial outburst flooding (see Clarke *et al.,* 2003) reveals that the maximum discharge of the flood was 5 to 10 Sv (Sverdrups) and that it lasted less than a year.

There are several gaps in our knowledge of the 8,200- year cold event. Further studies of how ice sheet margins, oceans, and vegetation zones might have been affected can enhance understanding of the series of responses that followed the initial outburst. Geological and geophysical studies in the Hudson Bay region and the Labrador Sea could determine where the water release occurred and whether it took place as a single or multiple events (Clarke *et al.,* 2003).

The well-publicised impacts of events such as *El Nino* have generated an unequalled public awareness of how climate affects the quality of life and environment. This unequal awareness has in turn increased the demand for accurate climatological information. This kind of information has become available in *The Encyclopaedia of World Climatology,* edited by Oliver (2005). This encyclopaedia covers most subfields of climatology, supplies information on climates in major continental areas, and explains the intricacies of climatic processes. It provides a lucid explanation of current knowledge and research directions in modern climatology. It emphasizes climatological developments that have evolved over the past two decades. The relationship between climatology and both physical and social sciences is fully explored, as also the significance of climate for our future well-being.

## Ecology and Climate

Both natural and anthropogenic global changes are closely associated with major ecological disturbances. These issues have been commonly studies by means of scale-up, scale-down, and scale-up investigations with embedded scale-down components. None of these approaches by themselves can provide the most reliable ecological assessments. A fourth research paradigm, called strategic cyclical scaling (SCS), seems to be more effective. It involves continuous cycling between large and small-scale studies, so offering better understanding of the behaviour of complex environmental centimeters in the past century. Without any doubt, out climate future is crucial and it certainly needs more attention than it has gotten so far.

The looming concern about the magnitude and rate of future climate change warrants a proper understanding of the mechanisms underlying past abrupt climate changes. An extreme cold event that happended 8,200 years ago may be the most amendable to detailed examination because it happens to be the most recent such event (Clarke *et al.,* 2003).

The ice-core record from Greenland shows that the abrupt cooling 8,200 years ago was the largest climate excursion of the past 10,000 years (Dansgaard *et al.,* 1993). The mean temperature dropped by about 5°C for about 200 years, accumulation of snow declined greatly, precipitation of chemical impurities increased, and forest fires became more frequent., this event affected much of the Northern Hemisphere. It was probably triggered by the sudden release of fresh water from a huge, glacier-dammed lake, which appeared during the deglaciation of North America (Barber *et al.,* 1999).

As Clark *et al.* (1999) suggested, changes in the extent of the ice sheets that once covered much of North America directly influenced the freshwater balance of the North Atlantic; they are implicated in many abrupt climate events of the past 100,000 years. During the last Ice Age, when a few kilometers-thick ice sheets covered most of Canada systems and allowing more reliable forecast capabilities for analyzing the ecological consequences of global changes (Root and Schneider, 1995).

Increasing human populations are modifying atmospheric composition, water quality, and land surfaces, and introducing a variety of novel chemicals into the environment. Human have transported species beyond their natural boundaries. When such changes take place on a globle scale (*e.g.*, climate change caused by an enhanced greenhouse effect), or regionally but with high enough frequency to be global in scope (*e.g.* habitat fragmentations), they are termed 'global changes' (Woodward, 1992; Kareiva *et al.*, 1993; Root and Schneider, 1995).

The ecological implications of any global change cannot be predicted easily because the rates of human-induced change are often much faster than those related to natural causes, which limits the reliable application of historic analogues (Woodwell, 1990). The scales at which different research disciplines operate make interdisciplinary connections difficult and require devising methods that can bridge scale gaps (ehleringer and field, 1993). Many disciplines need to be integrated. Uncertainties exist in virtually every aspect of the analyses.

## Paradigm of scale-Up

The 'scale-up' or 'bottom-up' paradigm is very commonly attempted in most natural science studies. Observations made at small scales are used to determine possible mechanistic associations or 'laws of nature' that are then extrapolated to predict larger scale responses. Some of the conspicuous features observable at smaller scales, however, may not reveal dominant processes that generate large-scale patterns.

Extrapolating inferences from small-scale experiments to more complex or larger scale environmental systems depends on whether external large-scale processes can be easily observed at smaller scales. For example, forest ecosystem models driven by global warming scenarios with doubled concentrations of carbon dioxide project dramatic alterations in the current geographic pattern of global biomes (woodward, 1992, pp. 93-116). Extending these models to account for the direct effects of doubled carbon dioxide on water use due to decreased stomatal conductance, however, relies on results extrapolated from single-plant studies to the scale of whole forests. This kind of scale-up drastically changes the percentage of area predicted to experience biome change.

One important limitation of the scale-up paradigm is that it cannot always predict the behaviour of complex systems.

## Paradigm of scale-Down

In this 'scale-down' or 'top-down' paradigm, observed large-scale patterns are correlated with other large-scale patterns to reveal possible causal relations. The drawback is that the discovered associations may be statistical artifacts that do not reflect the causal mechanisms needed for reliable forecasting.

Scale-down techniques have long been used to delineate biogeographic boundaries of biomes (area having similar species and climates). For instance, the Holdridge life-zone classification distinguishes biomes by means of two predictors: temperature and precipitation. The static nature of such an approach has attracted some criticisms (Root and Schneider, 1995).

## Techniques of Scale Transition

Those who build scale-up models must use scale-down parameterizations to treat unresolved phenomena. However, the scale typically addressed by climate models and ecological observations differ considerably. This has prompted some ecologists to increase the number of large-scale ecological studies-and some climatologists to shrink the grid size of climate models. Actually, both are required, along with techniques to bridge the scale gaps.

## Paradigm of Strategic cyclical Scaling

In the strategic cyclical scaling (SCS) paradigm, scale-down and scale-up approaches are cyclically applied and strategically designed to tackle practical problems. Large-scale associations are used to focus small-scale investigations with a view to ensuring that tested causal mechanisms generate the large-scale relations. According to Root and Schneider (1995), this allows more reliable forecasts of the consequences of global change disturbances. The SCS is not merely a two-step process; rather it is a continuous cycling between strategically designed large and small-scale studies, with each successive investigation building on previous insights obtained from all scales. The use of the SCS has given good results in the study of birds in relation to changes in climate variables and may prove equally profitable in studies on reptiles.

## Integrated Assessment

The overall process for studying global change problems is termed integrated assessment. It refers to the 'end-to-end' assessment of physical, biological, and social causes of global changes and their implications for environment and society. It involves coupling scenarios of population, land use, affluence, and technology changes over time to biogeochemical models driving climate models, in turn driving ecological models, all used to provide global change scenarios for economic models. It is possible to investigate policy options for mitigation and adaptation by employing integrated assessment techniques (see Alcamo, 1994). The SCS approach forms an integral part of the Earth systems' science subcomponent of this assessment process. An integrated assessment framework enables decision-makes to formulate and select better policy choices. But to conduct proper integrated assessment will require addressing issues across many scales and disciplines as well as necessitate the elimination of constraints imposed by inflexible traditions at existing institutions.

It is no simple task to disentangle the consequences of climate variation at an ecosystem level. No general ecological theory adequately elucidates the functioning of marine ecosystems. But ecologists certainly can track patterns and processes. Stenseth *et al.* (2004) discussed many examples that bring out apparent links between climate and marine ecology. They also elaborated concepts, models, and statistical and simulation techniques for quantifying species interactions. Their monograph examines the argument that large-scale climate indices, such as the North Atlantic Oscillation (NAO), may serve as simplified proxies that capture the essence of the overall physical variability better than the complex of local observational details. As nature happens to be intricate, the effects of climate fluctuations on ecology may be nonlinear, act with time lags that are difficult to detect, and may have both direct effect on life history traits and indirect effects through the food web.

Cod (fish) is a key predator species having central role in the ecosystem dynamics of the North Atlantic. Stenseth *et al.* (2004) showed how a given NAO pattern could correspond to varying levels of cod recruitment (good or poor) in different areas, which yield apparently contradictory results. Comparative analysis of the factors (*e.g.* air and water temperatures, ice cover, and winds) that appear to control cod recruitment at an ecosystem level among the different environmental settings of the Barents Sea, the North Sea, and Canadian waters offers a good approach to unraveling this paradox.

Recently, the world ocean has been subdivided into sixty-four large marine ecosystems (www.edu.uri.edu/lme), and there is need to intensify research in these ecosystems. Stenseth *et al.* (2004) have suggested a comprehensive approach to appraising ecological interactions between exploited and no exploited species in the context of climate change.

## Between Climate and Vegetation Interactions

Reciprocal interactions take place between climate and vegetation or global distribution of plants. Whereas such distribution is mostly controlled by climate, the vegetation cover not only changes the albedo of the Earth's surface but also exchanges carbon dioxide, oxygen, and water vapour with the atmosphere, thereby influencing climate. This climate-vegetation interaction has been going on since the land surface was colonized by vascular plants around 400 million years ago. However, even before this greening of the land, several feedback pathways allowed biotic systems to influence their abiotic environment. The gas and water vapour exchange is regulated primarily by the stomata on leaf surfaces and their response to the boundary layer meteorological condition. The growth habit of the plant and the spatial and seasonal distribution of leaf cover also affect the exchange of materials between plants and the environment, and hence the energy balance between the land surface and the atmosphere.

The fact that the atmospheric carbon dioxide concentration is rising and the appreciation of its effect on global climate have stimulated interest in vegetation-climatic-atmosphere interactions. Many laboratories and field works have addressed plant-atmosphere interactions at the individual plant and community level. The desire to understand the global impact of these processes has in turn inspired the creation of computer models, simulating the functioning of the Earth's climate, oceans and terrestrial biosphere with a view not only to simulating the full range of atmospheric, climatic and bio-geochemical processes but also to generate a basis for predicting the impact of anthropogenic perturbations (Beerling *et al.*, 1998).

The proportion of atmospheric oxygen and carbon dioxide has undergone significant changes through geological history, as also has the pattern of global climate. The past record of the vegetation climate-atmosphere feedback system is an important means of testing the validity of the ocean, climate, and biosphere models. But this does not mean that ancient climates can be invoked as analogues for a future 'greenhouse' world- a future climate response will have a significant non-equilibrium component quite different from the long-term average picture of past warm periods.

## Climate Change, Fire Risk and Drought Risk

Climatologists are predicting that the risk of rural fires will increase with climate change over the rest of the 21st century. Currently, there are about 3,000 rural fires a year, burning, for instance, about 7,000 ha of land in New Zealand. Dangerous fire weather results from a combination of strong winds, high temperatures, low humidity and seasonal drought.

The drier, windier weather that we may expect with climate change can lead to easier ignition, faster fire spread, more burned areas, and increased fire suppression costs and damage. Increased frequency of drought will lead to longer fire seasons, increased fire intensities, increased resource requirements, and more difficult fire suppression. The frequency of thunderstorms with attendant lightning will increase although some of these risks maybe offset by increased rainfall.

In many countries, drought is a major natural hazard and costs enormously at the farm gate alone. A severe drought can be defined as the rare 1-in- 20 year drought. Under a 'low-medium' scenario, by the 2080s, severe droughts are likely to occur at least twice as often in many drought-prone eastern areas of New Zealand and some other countries.

Under a 'medium-high' scenario, in drought-prone arid regions, pasture could start to dry out about a month earlier in spring by late this century (see www.Climatechange.govt.nz/resources/reports/drought-risk-May05).

## Effect: Wind on Climate

Wind happens to be the faster growing nonfossil source of primary energy in view of the fact that global wind-power capacity is growing by approximately 8GW yr$^{-1}$ (AWEA, 2004). The cost of electricity from wind power is now approximately 40 dollars per MWh$^{-1}$ at the best sites, but costs are declining rapidly (Keith *et al.*, 2004).

In the face of various plans to curb the carbon dioxide emissions in many countries, wind power has a major role in global energy supply. Its widespread use can modify local as well as global climate by extracting kinetic energy and altering turbulent transport in the atmospheric boundary layer. However, although the local environmental impacts of wind power have been explored to some extent, the climatic impact of wind turbines has not been attempted.

Although wind power is a renewable resource (Gipe, 1995), the rate of its renewal is fairly low. The yearly average horizontal flux of kinetic energy at the 100-m hub heights of large wind turbines can be more than 1kWm$^{-2}$ (see Keith *et al.*, 2004). These large power fluxes make it possible to economically extract wind power, but a collection of wind turbines cannot extract this power arbitrarily because turbines interfere with their neighbors by slowing local winds. Most of the kinetic energy that drives wind turbines comes from the generation of available potential energy at planetary scales, which drive winds throughout the atmosphere. Within the atmospheric boundary layer, turbulent mixing carries momentum downward to the surface, converting kinetic energy to heat. The downward flux of kinetic energy averages approximately $1.5W.m^{-2}$ over the global land surface (Pleixoto and Oort, 1992). The power that can be extracted by wind-turbine arrays is limited by this downward flux of kinetic energy.

Though the generation and dissipation of kinetic energy is a minor (-0.3 per cent) component of global energy fluxes, the winds do mediate quite large energy fluxes by transporting heat and moisture. Indeed, alteration of kinetic energy fluxes exerts much greater climatic effects than alteration of radioactive fluxes by an equal magnitude (Peixoto and Oort, 1992; Keith *et al.*, 2004). Keith *et al.*, attempted climate simulations that address the possible climatic impacts of wind power at regional to global scales by using two general circulation models and several parameterizations of the interaction of wind turbines with the boundary layer.

Very large amounts of wind power can produce measurable climatic change at continental scales. Large scale effects are observed, but wind power exerts a negligible effect on global-mean surface temperature; it would deliver enormous global benefits by reducing emissions of carbon dioxide and air pollutants. The results may enable a comparison between the climate impacts due to wind power and the reduction in climatic impacts achieved by the substitution of wind for fossil fuels.

According to Keith *et al.*, the direct climatic changes due to wind power may be beneficial in view of the likelihood that they reduce, rather than increase, aggregate climate impacts.

## Human Role in Earth System Changes/ International Trade in Virtual Water

In the changes humans play a leading role currently occurring in the Earth system, whether it be the hydrosphere, the biosphere, or in the modeling of these complex interactions. Hoekstra (2003) introduced the novel concept of 'virtual water trade' and discussed how world trade of water-intensive products contributes to changes in regional water systems. People in one country can affect the hydrological system in a remote national through contributing to changes in the global climate system. Local emissions of GHGs contribute to the predicted change of the global climate, so affecting temperature, evaporation and precipitation patterns elsewhere. Countries affect water systems in other parts of the world through another mechanism. There is a direct link between the demand for water-intensive products (notably crops) in countries such as Japan and the Netherlands, and the water used for production of export goods in countries such as the United States (US),and Brazil. The water used for producing export goods for the global market significantly contributes to the change of regional water systems (Hoekstra, 2003).

Japanese consumers impose some stress on water resources in the US by contributing to the mining of aquifers, emptying of rivers and increased evaporation in North America. Likewise, Dutch consumers significantly contribute to the water demand in Brazil. The issue here is: how significant are these teleconnections in the global water system via the mechanism of global trade? Hoekstra (2003a) discussed how the impact of global trade on water systems are already occurring today.

Producing goods and services generally requires water. That water, which is used in the production process of an agricultural or industrial products, is termed the 'virtual water' contained in the product. For example, in order to produce 1 kg of grain, we need 1 to 2 m³ of water. Producing 1 kg of cheese requires 5 m³ of water on average. If country A exports a water-intensive product to country B, A is exporting water in virtual form. In this way, some countries support the water needs of other countries. Trade of real water between water-rich and water-scarce regions is generally impossible due to the large distances and associated costs, but trade in water intensive products (virtual water trade) is quite realistic.

The virtual water flows between nations can be assessed by multiplying international trade volumes (ton/yr) by their associated virtual water content (m³/ton). Trade data can be had from the United National Statistics Division in New York. The virtual water content of crops has been estimated per crop and per country on the basis of various Food and Agriculture Oroganisation (FAO) databases (CropWat, ClimWat, FAOSTAT).

Table 1.2 shows the estimated global average virtual water content for some products. There are significant differences between countries, mainly relating to differences in climate conditions; but in the case of livestock products, these also relate to differences in animal diets indifferent countries.

The global virtual water trade is estimated to be $1+10^{12}$ m³/yr in the period 1995-1999, of which 67 per cent relates to international trade of crops, 23 per cent to trade of livestock and livestock products and 10 per cent to trade of industrial products (Hoekstra, 2003). For comparison: the global water withdrawal for agriculture (water use of irrigation) in the same period was about 2,500 Gm³/yr. taking into account the use of rainwater by crops as well, the total water use by crops in the world has been estimated at 5,400 Gm³/yr (Hoekstra, 2003). The total water use in the world for domestic and industrial purposes is estimated at 1,200 Gm³/yr. This implies that 15 per cent of the water used in the world for human purposes is not used for domestic consumption but for export (in virtual form).

**Table 1.2: Virtual water content of a few selected products in m³/ton (after Chapagain and Hoekstra, 2003; Hoekstra and Hung, 2003).**

| Products | Water Content (in m³/ton) |
|----------|---------------------------|
| Potato | 160 |
| Maize | 450 |
| Milk | 900 |
| Wheat | 1200 |
| Soybean | 2300 |
| Rice | 2700 |
| Poultry | 2800 |
| Eggs | 4700 |
| Cheese | 5300 |
| Pork | 5900 |
| Beef | 16000 |

Different nations do not have comparable shares in global virtual water trade. Major virtual water exporters are the US, Canada, Australia, Argentina and Thailand. Countries with a large net import of virtual water are Japan. Sri Lanka, Italy, South Korea and the Netherlands.

On the basis of the estimated global virtual water trade flows, draft national virtual water trade balances may be calculated by adding all virtual water imports and subtracting all virtual water exports.

The virtual water concept facilitates analysis of the impacts of consumption patterns on water use. Per country, the cumulative virtual water content of all goods and services consumed by the individuals of the country has been calculated (Hoekstra and Hung, 2002) and represents the 'water footprint' of a nation, analogous to the ecological footprint (Wackernagel and Rees, 1996). The water footprint of a nation equals the use of domestic water resources, minus the virtual water export flows, plus the virtual water import flows.

Figure 1.4 exemplifies how the real water balance and the virtual water balance of a country link to each other. The total use of domestic water resources in China is

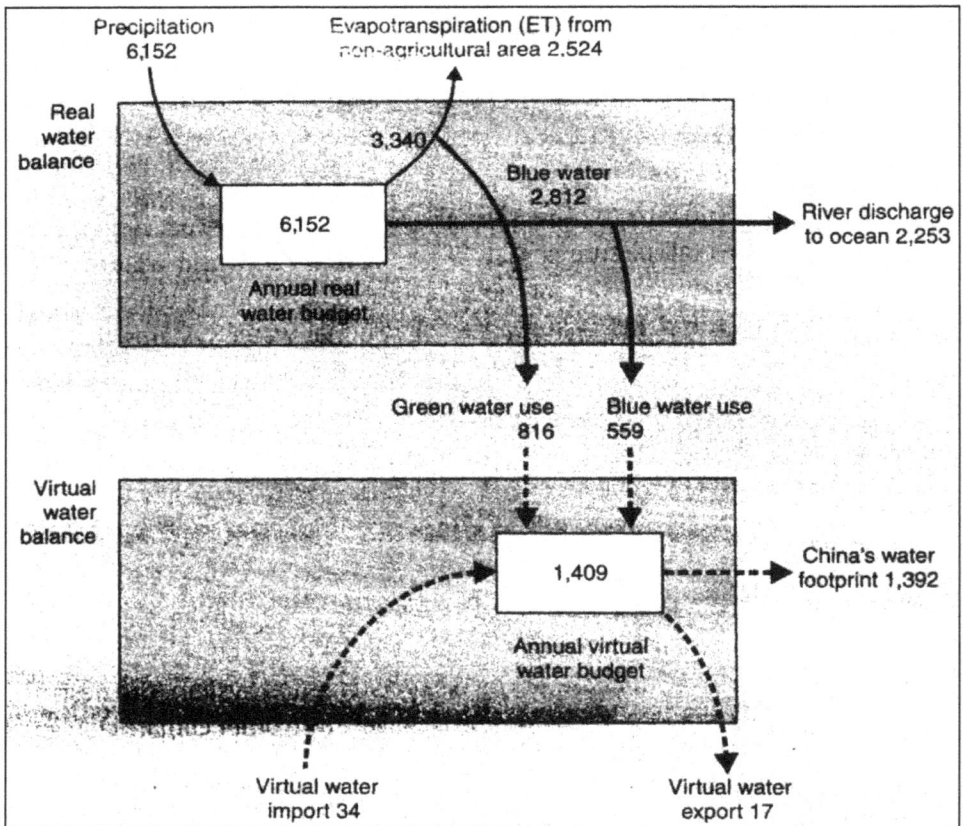

**Figure 1.4: The real and the virtual water balance of China in 1999 (data in Gm³/yr) (after Hoekstra, 2003).**

1,375 billion m³/yr (59 per cent 'green' water, 41 per cent 'blue' water), of which 1.2 per cent is used for export. The water footprint in China is 1,392 billion m3/yr, pressing on domestic resources for 97.6 per cent and on foreign water resources for the remaining 2.4 per cent. This is expected to increase in the future. Some countries, for example, the US, Canada, Australia, Argentina and Thailand, have net export of virtual water and do not depend on foreign water resources. An extreme example at the other end of the spectrum is Jordan, which relies on foreign water resources to satisfy 60 to 90 per cent of its domestic water need. The overall picture is that 15 per cent of the world water use is not for meeting domestic demands, but for meeting foreign demands. In other words, roughly speaking, 15 per cent of the disturbances of regional water systems that have been widely reported are linked to demands for water-intensive products in other parts of the world. In the face of the growing globalization, these teleconnections are likely to become increasingly important in the coming years.

## In Peril: The World

The world must not ignore the peril of the increasing pace of climate change. Unchecked resource use can potentially cause cataclysmic changes such as the breakup and melting of the polar ice caps, the collapse of the ocean circulation system, the destruction of some big cities by tornadoes, and widespread havoc from other climate changes. Many scientists have warned against the unbridled excesses of a petroleum-based economy. While some recent movies have tended to exaggerate the impact of global climate change (such as those outlined above), Speth (2004) has presented a careful and thoughtful assessment of the current and future state of the planet. We should not rest by concentrating mostly on local environmental concerns while ignoring the more important global-scale ones. There is little doubt that the Earth's environment has been deteriorating at an increasing pace despite local and national legislation and international agreements. The international climate convention has not significantly protected climate. The biodiversity convention is not proving very effective in protecting biodiversity. The desertification convention is not preventing desertification. The convention on the Law of the Sea has not proved strong enough to protect fisheries except only marginally. It appears that international efforts to help the environment are simply not working (Speth, 2004).

According to Speth, the following are some major causes of environmental decline:

1. The population is too large; the Earth's biodiversity (resources in economic terms) has been raped and pillaged;
2. The use of toxic pesticides has skyrocketed (nearly $2.7 \times 10^6$ metric tons per annum);
3. Use of fossil fuels as well as deforestation are enhancing the concentration of atmospheric carbon dioxide, which increases the temperature at the Earth's surface; and
4. The bulk of the world's nitrogen and water have been appropriated, and wasted, by industrialization.

Indeed, the environmental situation today is much worse than it was 25 years ago. To address this sad state of affairs, Speth (2004) proposed an 8-point plan of action or linked transitions that can redefine and redirect growth. These eight transitions are:

1. Stabilize and shrink the global population;
2. Eliminate poverty;
3. Modernize industry and agriculture through environmentally friendly technologies;
4. Develop environmentally honest prices in all markets;
5. Reduce consumption to sustainable levels;
6. Make universal environmental education compulsory;
7. Govern responsibly; and
8. Change our consciousness.

## Sustainable Development: Climate Change

Sustainable development forms a conspicuous element of climate-change policy discussions at the global level, under Agenda 21 and the various Conventions resulting from the UNCED-1992. It is defined as the development that meets the needs of the present without compromising the ability of future generations to meet their own needs. Sustainable development integrates economic, social and environmental issues. It does not preclude the use of exhaustible natural resources but requires that any use be appropriately offset. However, sustainable development cannot take place without significant economic growth in the developing countries (Goldemberg *et al.*, 1996).

Three crucial components that promote sustainable development are economic growth, social equity and environmental sustainability. Politicians in developing countries tend to see a trade-off between economic growth and environmental sustainability, but environmental conservation for sustainability of natural resources is not a luxury- it is a necessity when considering long-term economic growth and development, particularly in the least developed countries. The decline and degradation of natural resources such as land, soil, forests, biodiversity and groundwater, resulting from current unsustainable use patterns can increase due to climate change in the next coming ten decades. Africa, South Asia and some parts of Latin America are already suffering from severe land degradation and freshwater scarcity. The following are some sustainable development strategies that can help to mitigate climate change:

☆ Adoption of cost-effective energy-efficient technologies in electricity generation, transmission, distribution, and end-use can lower costs and bring down pollution level besides reduction of GHG emissions.

☆ Shifting cost-effective renewable enhances sustainable energy supply, reduces local pollution and GHC emissions.

☆ Adoption of forest conservation, reforestation, afforestation, and sustainable forest management practices help in conservation of biodiversity, watershed protection, rural employment generation, increased incomes to forest dwellers and carbon sink enhancement.

☆ Efficient, fast and reliable public transport systems such as metro-railways reduce urban congestion, local pollution and GHG emissions.

☆ Rational energy pricing based on long-run- marginal-cost principle can trip the balance in favour of renewable, increase the spread of energy-efficient and renewable-energy technologies, and the economic viability of utility companies, and ultimately lead to GHG emission reduction.

## A Possible Scorching Future

Work done on thousands of personal computers has confirmed that a fairly strong heating of the globe in the coming centuries cannot be ruled out–in fact, it could get even hotter than the previous worst case had it. However, no modeler can yet predict that such an extreme scenario is any less likely than the moderately strong warming that most climate scientists expect. This extreme perspective differs from the perspective that emerged a few years ago (in summer 2004)- many climatologists and computer modelers at that time were inclined to opt for a middle ground (Science, 13 August 2004, p. 932). Three different kinds of studies, *viz.*, the latest expert-designed climate models, perturbations of a single model, and studies of past climate, considered the issue of how sensitively the climate will respond to increasing GHGs. All pointed to a moderately strong sensitivity: When GHG's are doubled; warming would be about 2°C to 4°C. A low, nearly harmless climate sensitivity much less than 2°C appeared quite unlikely; likewise, an extreme one of more than 6°C to 7°C, although possible, was much less likely than the consensus range (Kerr, 2005). There is an urgent need for the major industrial countries to join with China and India to address the problem of climate change together. The long-term impact on the environment, even of a 2°C rise in average global temperatures, can be devastating. A 14-member independent panel of scientists and policy makers has produced a report (www.americanprogress.org/climate) that seeks to find common ground between those nations, which have ratified the 1997 Kyoto Protocol, and those (*e.g.*, the US and Australia) that have not. The Report suggested a global effort to set up a cap and trade system for emission that would extend belong the Kyoto framework that expires in the year 2012 and a shirt in agricultural subsidies from food crops to biofuels. It wants wealthy nations to help developing countries in controlling their emissions as their economies grow. Failure to mobilize in the face of this threat can approve extremely costly. Current global temperatures are around 0.8°C above reindustrialize levels, and a 2001 report by the IPCC projected a rise of 1.4° to 5.8° by 2100. Recent analyses conclude that even higher temperatures cannot be ruled out (Kintisch, 2005). Even a 2°C rise could have catastrophic effects on Earth, such as the decimation of coral reefs worldwide, the melting of the West Antarctic Ice Sheet, and the severe degradation and disruption of various ecosystems. Indeed, there exists a possibility that temperatures could exceed the 2°C goal before emissions controls are enforced and bring down the temperature by the year 2100. In response to the report's suggestions, the US is likely to develop clean coal technologies and go in for vehicles that are more efficient.

# Chapter 2
# System of Earth

## General Description

An ever-changing planet is the earth. Around the Sun its orbit varies, continents drift, mountains are driven upwards and erode, animal and plant species evolve, and terrestrial and marine ecosystems change. In the past large changes have generally occurred as the result of natural forces beyond human influence or control. But human beings have now become powerful agents of environmental change on global, regional, and local scales on the environment. With an increasing world pollution, an expanding global economy, and the development of new technologies, the human impact is sure to increase in the future.

Since the mid-1800s, it is now clear that increased use of fossil fuels is changing the composition of the Earth's atmosphere and that this change is exerting a warming influence on the global climate. Global-scale variations in climate alter regional patterns of rainfall and temperature. Changes is land cover such as the conversion of forest to pasture in the tropics, and changes in land use such as increases in fertilizer applications to croplands worldwide have contributed to changes in atmospheric composition and may also contribute to climate change on both regional and global scales. Emissions of chlorofluorocarbons (CFCs) and other chlorine- and bromine containing gases have led to depletion of the stratospheric ozone in both the Southern and Northern Hemispheres that will last for decades and cause adverse impacts on human health and ecological systems.

In the Earth system all these and other changes are having profound consequences. Changers in any single component of the earth system affect the entire system. For example, changes in the tropical Pacific Ocean not only affect weather in North America but also in other parts of the world.

The complexity of the earth system and the many feedbacks among its components make understanding and predicting climatic and environmental change and all of its ramifications a very difficult challenge indeed. The *El Nino*-Southern Oscillation (ENSO) cycle happens to be the most prominent year-to-year climate fluctuation affecting global climate. It originates through coupled ocean-atmosphere interactions in the tropical Pacific. It affects the rest of the globe through oceanic and atmospheric teleconnections. Warm phases (*El Nino*) and cold phases (*La Nina*) recur with a periodicity of about 3 to 7 years. The most recent *El Nino* in 1997-98 was the strongest of the 20th century; it was followed by cold *La Nina* conditions from 1998 to early 2001.

As a single the earth system functions self-regulating system made up of physical, chemical, biological and human components. The interactions and feedbacks among these components are complex and show multi-scale temporal and spatial variability. Knowledge of the natural dynamics of the Earth system provides a sound basis for evaluating the effects and consequences of human driven change.

Human activities are substantially influencing the Earth's environment, for instance through greenhouse gas emissions, and climate change, anthropogenic changes to earth's land surface, oceans, coasts and atmosphere and to biological diversity, the water cycle and biogeochemical cycles can be clearly identified beyond natural variability. They may even equal some of the great forces of natural in their extent and impact. Many are accelerating.

Global change is real and is taking is taking place now. To understand global change in terms of a simple cause effect paradigm it is not possible. Human-driven changes produce many effects that cascade through the earth system and interact with each other and with local and regional scale changes in multidimensional patterns which cannot be predicted easily.

The interactions between environmental change and human societies have a long, complex history extending over many milalennia, but they have changed strongly in the last century. Human activities are now so pervasive and profound that they are altering the earth in ways that threaten the very life support system upon which humans depend. Steffen (2004) reviewed what is known about the earth system and the impact of changes caused by humans. He discussed the consequences of these changes with respect to the stability of the earth system and the well-being of humankind and showed future paths towards Earth system science in support of global sustainability.

The Earth system dynamics have characteristic critical thresholds and abrupt changes. Human activities could inadvertently trigger such changes with strong consequences for the Earth's environment and inhabitants. They can potentially switch the earth system to alternative modes of operation, some of which may well be irreversible and less hospitable to humans and other life.

The Earth systems has moved in terms of some key environmental parameters, we beyond the range of the natural variability over the last half million years at least. The nature of changes now taking place simultaneously in the earth system, the magnitudes and rates of change are unprecedented (Larigauderie, 2002).

About the Earth system although much has been learn, from satellite observation, we still lack a good operational global observing system for the ocean interior, biosphere, the cryosphere and the lithosphere.

Continuous global observation of the three-dimensional structure of the atmosphere oceans, biosphere, cryosphere and lithosphere can lead to producing validated numerical models of the Earth system. These models will help scientists to extract the predictable portion of natural variability or change, to be used for different forecasting purposes and to support decision making for intelligent Earth system management.

The World Climate Research Programme (WCRP) is engaged in designing of such an observing system for climate parameters and validated coupled climate models (encompassing atmosphere, ocean and land surface, including vegetation). Such an observing system will incorporate:

☆ Suitable microwave sensors on satellites for observing radiation matter interactions from the top of the atmosphere through the clouds to the surface.

☆ Reference sites for high quality in situ measurements and surface-based remote sensing to obtain globally validated satellite-based time series of physical, chemical and biological parameters.

Data generated from using the above system is expected to form the basis for testing of global and regional coupled models that can help predict 'signals' within the climate system such as those associated with *El Nino*-Southern Oscillation (ENSO), soil moisture, sea ice, extra tropical upper ocean temperature/salinity and other related phenomena that occur at different timescales ranging from weeks up to interannual.

## Forcings of Climate

Human activities have already become so pervasive that they are exerting significant changes in the atmosphere and at the land surface-change that perturb the Earth's natural fluxes of solar and infrared (heat) radiations. These human-induced changes, often called 'enhanced radiative forcings', lead to changes in temperature, precipitation and other climatic variables.

## On Solar Radiation: Aerosol Effects

Very fine (microscopic) particles in the atmosphere (called aerosols) reflect some solar radiation back to space, thus exerting a cooling influence. However, a relatively new finding is that regional increases in short-lived aerosols resulting from human activities can suffice to alter the Earth's radiation balance. Emissions of sulphur dioxide from coal combustion, and of other gases from biomass burning, cause the atmosphere over and downwind of major industrial regions and regions of tropical deforestation to reflect some of the incoming solar radiation back to space. This exerts a regionally distributed cooling influence.

In many industrialized nations, the particle loading from fuel burning is being rigorously controlled so as to lessen human health impacts, acid rain, and visibility problems. However, as aerosol concentrations are reduced, the global warming they have masked becomes apparent.

## The *El Nino*-Southern Oscillation

The ability to make seasonal-to-interannual forecasts, particularly for tropical and subtropical regions has emerged from an improved understanding of the irregular cycling of the *El Nino*-Southern Oscillation (ENSO). The ENSO phenomenon involves the warming and cooling of large areas in the tropical Pacific Ocean, with strong associated shifts in atmospheric pressure and rainfall in other regions.

The ENSO phenomenon is the most prominent mode of climate variability as it affects weather and climate in large parts of the world (Diaz and Markgraf, 2000). In the early 1940s, unusually high values of total ozone were observed over several sites in europe, and exceptional climatic conditions were registered at the Earth's surface. A prolonged *El Nino* occurred in 1939-42, suggesting a possible relation between *El Nino*. European climate and the northern stratosphere (Labitzke and van Loon, 1999). Using historical observations and reconstruction techniques. Bronnimann *et al.* (2004) analysed the anomalous state of the troposphere and stratosphere in the Northern Hemisphere from 1940 to 1942 that occurred during a strong and long lasting *El Nino* event. Exceptionally low surface temperatures in Europe and the north Pacific Ocean coincided with high temperatures in Alaska.

In the lower stratosphere, there were high temperatures over northern Eurasia and the north Pacific Ocean, and a weak polar vortex. Also, there were frequent stratospheric warmings and high column ozone at Arctic and mid-latitude sites. By comparing historical data for the period 1940-42 with more recent data and a 650-year climate model simulation. Bronnimann *et al.* (2004) concluded that the observed anomalies constitute a recurring extreme state of the global troposphere-stratosphere system in northern winter that is related to strong *El Nino* events.

The work of Bronnimann *et al.* (2004) suggests that the global climate anomaly in 1940 to 1942 constitutes a key period for our understanding of large scale climate variability and global *El Nino* effects. The tropical Pacific Ocean is in a neutral state (no *El Nino* or *La Nina*), but equatorial Pacific sea surface temperatures (SSTs) continue to be above average. Although this positive anomaly is often a sign of *El Nino*, most forecast models indicate that neutral conditions will continue until the end of 2005.

## Circulation Cells of ENSO and Atmospheric

Bjerknes (1969) was the first to visualize a close relation among the Southern oscillation, east-west SST contrast in the equatorial Pacific Ocean, and the thermally driven zonal Walker circulation. Since then, the Walker circulation ahs been recognized to be associated with the interannual phenomenon of ENSO (*e.g.*, Webster and change, 1988). However, the manner of evolution of the walker circulation cell during ENSO has not been well studied, because of a lack of observational data. The atmosphere also has meridional circulation cells: the Hadley cell and the ferrel cell (*e.g.* trenberth *et al.*, 2000). Little is known about how these atmospheric meridional cells vary during the evolution of ENSO. Some recent data have made possible the study of the atmospheric circulation associated with ENSO (Wang, 2002).

## Atmospheric Cell of Mean and Interannual

Atmospheric circulation cells are identified by atmospheric vertical motion and the divergent component of the wind; and Wang has summarized the mean state and anomaly of the equatorial zonal walker cell, the tropical meridional Hadley cell, the extratropical meridional ferrel cell, and the MZC (mid-latitude zonal cell) (Figure 2.1). The mean Walker circulation cell is the air ascending in the equatorial western pacific, flowing eastward in the upper troposphere, sinking in the equatorial eastern Pacific and returning towards the equatorial western Pacific in the lower troposphere (Figure 2.1a). The western Pacific has a single mean Hadley cell, with the air rising in the tropical region, flowing poleward in the upper troposphere, and returning to the tropics in the lower troposphere.

In the eastern Pacific, the tropical circulation has two meridional cells with moist air rising in the intertropical convergence zone (ITCZ), then diverging northward and southward in the upper troposphere and descending over the regions of the subtropical high and the equatorial eastern Pacific cold tongue. The extratropics show a classical Ferrel cell, with upward motion in the mid-latitudes and downward motion in the southern extratropics (Wang, 2002).

As per the mean MZC, the air rises in the central North Pacific, diverges eastward and westward in the upper troposphere, descends over regions of the west coast of North America and the east coast of Asia, then flows back to the central North Pacific in the lower troposphere. As in the Walker and Hadley cells, the MZC is identified by the divergent wind and the pressure vertical velocity. If we also consider the rotational wind, the MZC is identified by the divergent wind and the pressure vertical velocity. If we also consider the rotational wind, the MZC is not a closed cell (Wang, 2002a).

The atmospheric cells also very with the interannual phenomenon of ENSO (Figure 2.1b). During the mature phase of *El Nino*, both the walker cell and the MZC become weak. The anomalous Hadley cell in the eastern Pacific during the mature phase of *El Nino* shows the air rising in the tropical region, flowing northward into e upper troposphere, descending in the mid-latitude, and returning to the tropics in the lower troposphere. The anomalous Hadley cell in the Western Pacific has an opposite rotation as that of the anomalous Hadley cell in the eastern Pacific (Wang. 2002 C).

The SST anomalies in the tropical North Arlantic (TNA) also are affected by the Pacific *El Nino*. Since the mature phase of the Pacific *El Nino* occurs around December of the Pacific *El Nino* year, the TNA SST anomalies tend to peak around subsequent May of the Pacific *El Nino* year. The Pacific *El Nino* affects TNA northeast (NE) trade winds that reduce latent heat flux and then increase the TNA SST. But how does the Pacific *El Nino* affect the TNA trade winds? It appears that the Walker and Hadley circulations link the Pacific *El Nino* with the TNA.

The anomalous walker and Hadley cells during the mature phase of *El Nino* (Wang, 2002b; Wang and Enfield, 2002) are summarized in Figure 2.2. The air anomalously ascends in the far equatorial eastern Pacific, diverges eastward in the upper troposphere, and then descends over the equatorial Atlantic. Associated with

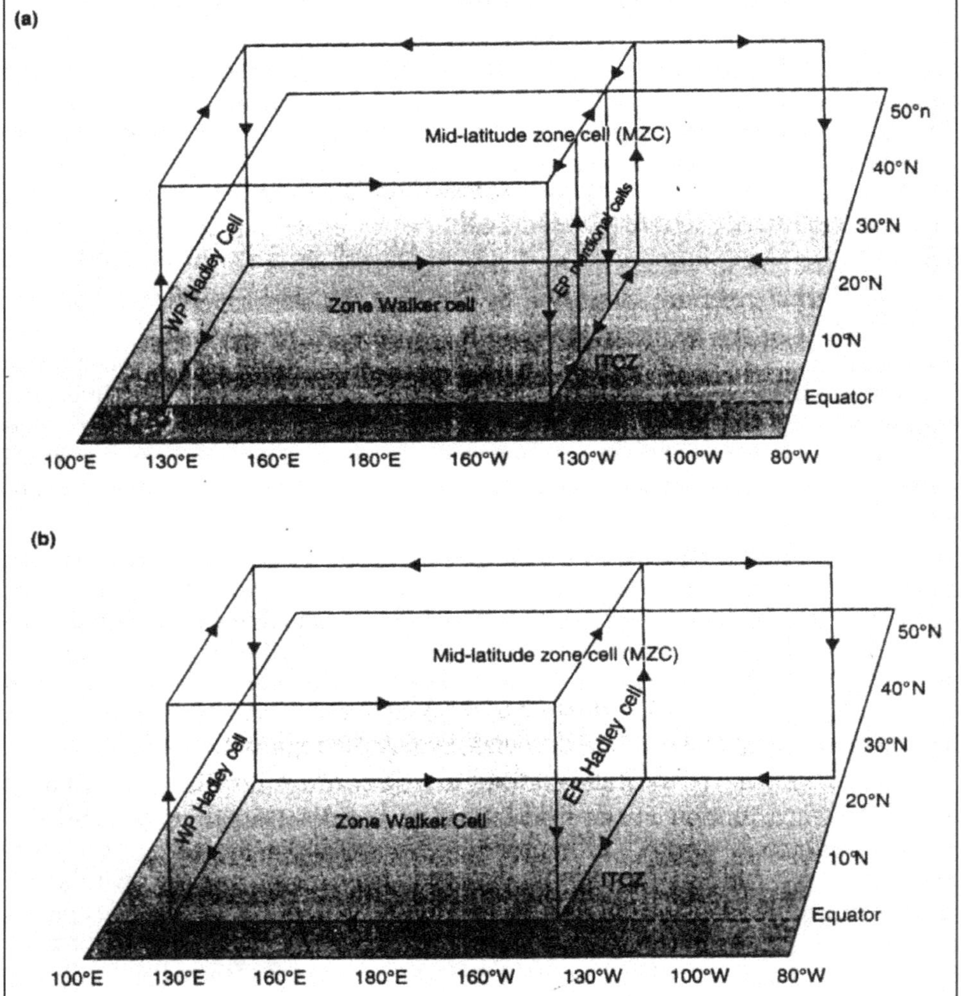

Figure 2.1: Schematic diagrams summarizing (a) mean state of atmospheric circulation cells and (b) anomalous atmospheric cells during the mature pahse of *El Nino*. Shown are the equatorial zonal Walker cell, the merdional Hadley cell in the western Pacific (WP), the meridional Hadley and ferrel cells in the eastern Pacific (EP),and the mid-latitude zonal cell (MZC)(after Wang, 2002a).

this anomalous Atlantic Walker circulation is an anomalous Hadley circulation cell in the Atlantic. The Hadley cell shows anomalous ascending motion in the region of the subtropical high. The anomalous ascending motion weakens subsidence in the region of the subtropical high and the associated NE trade winds over it southern limb in the TNA region. The weaker NE trades reduce evaporation, leading to warm SST anomalies over the TNA in the subsequent spring of the Pacific *El Nino* year.

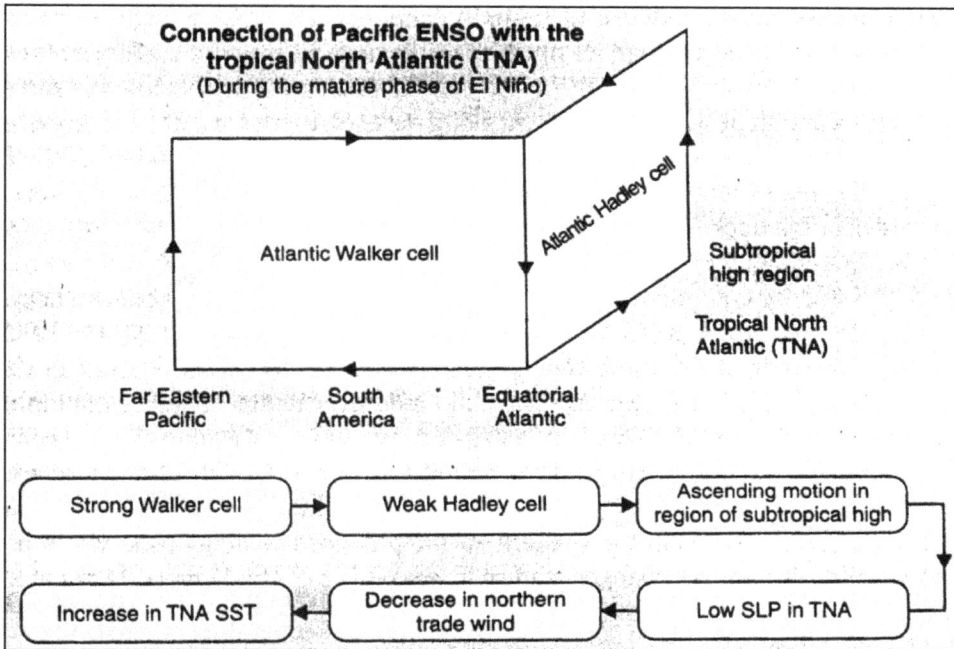

**Figure 2.2: Schemewise sketches showing linkage of the Pacific *El Nino* with the Tropical North Atlantic (TNA) (after Wang, 2002b).**

When the warm TNA SST anomalies are established in spring, the TNA region shows anomalous ascending motion (Wang, 2002b).

The ENSO Observing System, developed over the period 1985-94 (Mc Phaden *et al.*, 2001) consists of satellite and in situ components. The latter includes drifting and moored boys, ships of opportunity XBT and meteorological measurements and island and coastal tide gauge stations. These observation have recently been augmented with profiling floats (Mc Phaden, 2002).

In the moored buoy component of the ENSO Observing System, moorings utilize compatible sensor suites, sampling strategies, and data processing procedures. Near real-time and delayed mode data are managed and distributed as a unified and integrated data set via the World Wide Web (www) from mirror sites in the US and Japan, and are also distributed in real-time via the Global telecommunications System.

## In Early 2002 Ocean Warming

The basin scale development of warm conditions in the tropical Pacific in early 2002 has been captured with considerable detail. In April 2002, sea surface temperatures (SSTS) were close to their seasonal highs and slightly warmer than normal (by about 0.5° C) near the equator between 160°E-160°W and east of 110°W. Near normal easterly trade winds prevailed throughout much of the eastern and central equatorial Pacific, though west of the date line weak westerly anomalies were found. Below the surface, the thermocline sloped down to the west along the equator east of 165°E in response to easterly trade wind forcing. However, the upper part of the thermocline was slightly depressed in the eastern and central Pacific as indicated by the warm 1 to 2°C subsurface anomalies in the upper 50 to 150m (Mc Phaden, 2002).

The evolution of these weak warm conditions may be seen in 5-day averaged zonal wind, SST, and 20°C depth anomalies along the equator for the past 2 years. Cool *La Nina* conditions, which were established in the aftermath of the 1997-98 *El Nino*, extended into early 2001. These cool conditions were associated with stronger than normal trade winds in the western basin and reduced cloudiness and rainfall in the central Pacific. Then, in mid-2001, SSTs began to warm near the date line in keeping with an increase in westerly wind burst activity in the western Pacific. These westerlies were associated with certain active phases originating in the equatorial Indian Ocean. Intensified surface winds in the western Pacific produced cooling of the warm pool west of 160°E. These SST changes lead to a weakening of the large-scale zonal SST gradient, and to an eastward shift in deep atmospheric convection along the equator to near the data line.

In some respects these above events are similar to what occurred during the onset of the 1997-98 *El Nino* in early 1997. But conditions in April of each year, for example, indicate that zonal wind, SST, and subsurface temperature anomalies were significantly larger in April 1997 than in April 2002 for the region encompassed by the moored byoy array. In general, warm phase ENSO-like conditions were more pronounced in April 1997 than in April 2002. trends toward ENSO warm phase conditions were also stronger between January and April 1997 than for the same four month period in 2002. Moreover, the development of warm conditions appears to be waning at present rather than amplifying as in April 1997. by the end of April, anomalously cool SSTs re-emerged in the equatorial cold tongue.

It appears that heat content near the equator rebounded from low values observed in 1998-99 at the height of the recent *La Nina* to higher levels in 2000-01 creating conditions conducive to development of an *El Nino* in 2002. Large-scale ocean atmosphere interactions in the Pacific are probably being mediated by intraseasonal timescale fluctuations, consistent with the notion that the ENSO cycle may be thought of as a damped or weakly unstable oscillator (Moore and Kleeman, 1999; Fedorov and Philander, 2000).

Several statistical and dynamical ENSO forecast models have predicted that the ocean warming observed in the early 2002 will continue for the next two to three seasons. But other forecast models indicate persistent neutral conditions, or diminishing warm anomaly amplitudes over the same period. Anomalous ocean warming has already changed weather patterns in some parts of the Pacific basis. Unusually heavy rains in Ecuador and Peru in March 2002, for example, fell in association with warmer than normal eastern Pacific SSTs. These rains led to significant flooding, crops losses, and fatalities. While such regional climate anomalies may not necessarily be the harbinger of more widespread global climatic disruptions, developments so far underscore the importance of monitoring, understanding and predicting anomalous oceanic and atmospheric conditions in the tropical Pacific, whether or not they achieve the status of *El Nino* (or *La Nina*) (McPhaden, 2002).

As the most important tropical the ENSO phenomenon has long been recognized ocean-atmosphere coupled phenomenon. More recently, the Indian Ocean Dipole (IOD) phenomenon has emerged as another important manifestation of the tropical air-sea interaction (Saji *et al.*, 1999; Webster *et al.*, 1999; Behera *et al.*, 1999; Murtugudde *et al.*, 2000.; Rao *et al.*, 2002; Vinayachandran *et al.*, 2002). The impact of the IOD is not limited to the equatorial Indian Ocean; it influences the Southern Oscillation, the Indian summer monsoon rainfall, and even the summer climate condition in Asia (see Yamagata *et al.*, 2002). The robust nature of the IOD as an ocean atmosphere coupled phenomenon is simulated successfully by using high-resolution coupled General Circulation Models (GCMs) that also resolve ENSO in the Pacific (Lizuka *et al.*, 2000).

## Ocean Dipole of The Indian

Yamagata *et al.* (2002) showed that the Indian Ocean Dipole (IOD) is a physical mode involving dynamics of the tropical Indian Ocean. The IOD concept has raised a new possibility to reappraise the old problems of the relation between the Indian summer monsoon rainfall and *El Nino* from a new angle, and to strengthen the predictability of climate variations originating in the tropics.

## In the Indian Ocean Coupled Mode Inherent

As it appears that the IOD really exists in the Indian Ocean, the pertinent question is whether or not IOD events in the Indian Ocean are independent of ENSO events in the Pacific. According to Yamagata *et al.* (2002), the IOD events occur as a part of ENSO events (see Allan *et al.*, 2001; Baquero–Bernal and Latif, 2002). However, the non-orthogonality of two time series does not necessarily mean that the two phenomena are always connected in a physical space.

## On Climate Influence

Since the indices of ENSO and the IOD are non-orthogonal, it is also possible that the Pacific ENSO itself may be affected by the IOD. Using partial correlation analysis, Yamagata *et al.* (2002) succeeded in revealing such an inverse influence; they also suggested that some IOD events precede some ENSO events. The influence of the IOD is not just confined to the tropical region, but extends to the whole globe (see Yamagata *et al.*, 2002), even though because of the non-orthogonality of the IOD and ENSO indices, the IOD influences have not been properly appreciated.

Using a partial correlation analysis, Yamagata *et al.* (2002) demonstrated that the enhancement of the East African rain is dominated by the positive IOD rather than EI Nino. Interestingly, the positive IOD and the warm episode of ENSO have opposite influence in the Far East including Japan and Korea; positive (negative) IOD events give raise to warm and dry (cold and wet) summer owing to enhancement of downdraft in the troposphere. This has been clearly recorded during the IOD events in 1961, 1967, 1977 and 1994, as against the simultaneous occurrence of the positive IOD and the *El Nino* during 1997. The Indian summer monsoon rainfall (ISMR) is enhanced (decreased) during positive (negative) IOD events; the recent weakening of ENSO ISMR relation may be interpreted in terms of frequent occurrence of the positive IOD in recent years (Ashok *et al.*, 2001).

In the Southern Hemisphere, the IOD impact is notable in the South Western Australia and Brazil because of propagation of planetary waves in the winter hemisphere; positive (negative) IOD events cause warm and dry (cold and wet) conditions (see Yamagata *et al.*, 2002).

As discussed, the IOD may evolve without the external forcing from the Pacific ENSO though it does interact with the Pacific phenomenon on some occasions, possibly through the atmospheric bridge and partly via the oceanic through flow around the Australian continent. The strong seasonal phase-locking of the interannual IOD events may be related to the latter.

However, it is important to appreciate first the unique nature of the IOD as the air-sea coupled phenomenon unique to the tropical Indian Ocean and then to clarify ways of interaction with other important phenomena such as the ENSO events.

## Lithosphere/Pedosphere

The solid outer section of the Earth is called the lithosphere. The huge mass of the lithosphere is not generally vulnerable to change by humans, but it serves as a source of raw materials (coal, oil, natural gas, ore, sand and ground=water) and is also used as a disposal site for wastes of all kinds.

The lithosphere's outer zones of contact to the other spheres (*viz.*, soils and sediments), on the other hand, represent a sensitive area of great significance for living organisms and can be greatly changed by humans.

Soils cover large parts of the ice-free surface of the continents like a thin skin. In a zone of thickness ranging from a few continents like a thin skin. In a zone of thickness ranging from a few centimeters to several metres, the lithosphere, hydrosphere, atmosphere and biosphere form the pedosphere. In this way, soils represent structural and functional elements of terrestrial ecosystems.

Sedimens are the biotically active zones in aquatic areas corresponding to soils. They have great importance in biogeochemical cycles.

## Soils and Sediments: Significance

The significance of soils and sediments for plants, animals, microorganisms and humans as well as for the balance of energy, water and elements can be summarized on the basis of three principal functions: *viz*, habitat, regulation and utilization.

### Habitat

Soils are the habitat and basis of life for a diversity of plants, animals and microorganisms, on whose metabolism the regulations function and production function of soils is based. Soil organisms mediate synthesis, conversion and decomposition of substances in the soil and they influence the stability of ecosystems by decomposing toxic substances, delivering substances for growth and generating a flowing balance between processes of synthesis and decomposition. Soils are the basis for the primary production of terrestrial systems and thus the basis for existence of human societies as well.

## Regulation

The regulation includes transport, transformation and accumulation of substances. Soils facilitate the exchange of substances between the hydrosphere and atmosphere as well as neighbouring ecosystems. The regulation function comprises all biotic and biotic internal processes in the soil which are triggered by material inputs and non-material influences.

## Utilization

Soils are the centerpiece of agricultural and forestry production (production function); this depends on their capacity of supplying plants with water and nutrients and serving as their rhizosphere and the producing biomass that is usable for people.

Like all assets that are impaired by human activity in their functions or sustainability, soils are in need of effective protection. Some soil degradation occurs naturally and slowly. But human activities (*e.g.*, forest clearing, ploughing up grassland, draining wetlands or irrigating dry areas, the mechanization and application of chemicals in agriculture and forestry as well as the overexploitation of fields, pasture land and forests) can greatly hasten soil degradation.

### Local-level Changes

Local-level causes of change also include the increasing crop monocultures up to large scale cultivation of genetically uniform types, which reduced biodiversity. This not only influences the communities of organisms themselves but also the regulation and habitat functions of soils and so the long-term usability and conservation of these resources. Soil destruction and pollution result from deposition or output of toxic substances, surface sealing and fragmentation of areas by means of roads and settlements. These forms of harmful environmental impact are frequently connected with the intensification of industrial activities and traffic and therefore, play a great role in industrial national and urban agglomerations.

### Regional-level Changes

Emissions of acidifiers, nutrients and toxic substances which cross borders are some of the causes at the regional level. Increasing urbanization and the related geographical decoupling of food production and consumption also harm the environment. Hydraulic engineering measures, such as regulation of rivers, construction of dams, lowering or raising of the ground water table and construction of irrigation channels and drainage, affect the water balance of soils, resulting in adverse environmental impact leading to soil degradation.

### Global-level Change

In soils, global-level changes are caused by climatic changes. Altered temperatures and precipitation can directly affect the soil by speeding up or slowing down conversion and transport processes. An indirect effect can occur via the vegetation cover by virtue of changes in the cover and in biomass production. The anthropogenic spread of alien species can also drastically alter the soil. Examples are the global spread of the eucalyptus tree and the introduction of a fluke (animal) that out competes earthworms in Ireland and England. There is an increasing trend towards such processes, which occur even beyond natural barriers due to the greater mobility of people.

## Observing System: The Earth

The US National Aeronautics Space Administration (NASA) has been trying to understand global climate change through its Earth Observing System (EOS). On 15th July, 2004, a spacecraft Aura bristling with instrument to measure the earth's atmospheric chemistry soared into orbit. Three large EOS platforms have been launched since 1999; about a dozen smaller US satellites are also monitoring everything from the world's ice sheets to solar radiation.

The Earth scientists have for long thought of EOS as a means to collect huge amounts of data for help in understanding the mysteries of the complex global climate system. That vision has led to plans that called for NASA to build and launch three massive platforms; terra, aqua, and Aura that would gather simultaneous data on a host of ground, ocean and atmosphere parameters. Terra is about the size of a school bus. It examines land surface changes, atmospheric aerosols, global cloud cover, and ocean temperatures. Aqua measures stratosphere temperatures and earth's thermal radiation budget, among other parameters. Aura's focus is on atmospheric chemistry.

The scientific output has been enormous. While 17 terabytes of data were delivered in 1999, EOS may well deliver over a thousand terabytes in 2005. Hopefully, in a few years, the data may help scientists to produce superior climate models based on enhanced understanding of how the land surfaces, oceans and atmosphere interact (Lawler, 2004). However, some critics feel that the NASA satellite and data system have not delivered as a coordinated system providing long-term coverage. Although satellites have given a better view of global systems than that obtained from in situ measurements taken on ocean buoys or balloons, they do have some shortcomings-orbits decay and satellites drift. Satellites also have their limits-they cannot tell us much about the ocean depths or what's happening under the Antarctic ice sheets.

Climate and weather are not the same; for climate systems, much longer-time data are needed than for weather monitoring. Weather work requires high-resolution images without the absolute accuracy and stability that climate researchers need. A weather forecaster need not store data; for climate researchers, an organized and accurate long-term database is indispensable. Some EOS spacecraft are rolled in orbit, so they can spot the moon and use it to calibrate delicate climate instruments-this kind of manoeuvre might be too dangerous for an operational satellite critical for national weather forecasting (Lawler, 2004). Unfortunately, most climate research has been fragmented. There has been criticism that the great majority of it is old-fashioned Earth science in disguise. Scientists, instead of working on a problem such as how clouds interact with radiation, aerosols, and general planet circulation, very often just extend previous work on cloud physics. Climate research is too focused on small-scale issues. Effective and meaningful contribution to understanding the climate system can only come from sustained work in several disciplines ranging from the solid earth to upper atmosphere to weather, climate ecosystems, and oceanography. Although in recent years, astronomers, solar system researchers, and solar physicists have reached consensus on long-term plans and priorities for their respective fields, reconciling the many and competing desires of climate researchers is very difficult.

One critical question is how to create and deploy a climate observing system that can provide consistent and accurate data. A climate observation and data system has been proposed (Lawler, 2004) that would tie together all the world's environmental satellites, along with in situ data, a global telecommunication network, comprehensive models of the land, ocean, and atmosphere.

## Between Climate Change and Biosphere: Feedback

At the global level, there is a strong feedback between climate change and the biosphere. Changes in vegetation structure affect the physical properties of the land surface, such as its albedo (the percentage of sunlight it reflects), surface roughness and canopy conductance to water vapour. Changes in vegetation structure and function also influence the exchange of carbon dioxide, methane and nitrous oxide, between the atmosphere and the biospohere. In this way plants can influence the percentage of incoming sunlight absorbed at the surface or the atmospheric concentration of GHGs and hence the rate of magnitude of climate change.

Positive feedback from the biosphere may possibly exacerbate global warming. Models of Earth's climate 6,000 years ago suggest that variations in the Earth's orbit by themselves could not have produced the high annual average surface temperatures deduced from paleoclimate analyses. Instead, orbital forcing could have been exaggerated by subsequent vegetation responses (Foley *et al.*, 1994). Because boreal forests absorb much more solar radiation than tundra does, poleward shifts in the location of the forest/tundra boundary during a period of warming can amplify climate changes by as much as 50 per cent.

Under some circumstances, the carbon cycle may act as a second feedback loop. If during warming the rate of forest dieback on the southern edge of a forest range (in the Northern Hemisphere) were to be higher than the rate of forest expansion on its northern edge, there would occur a net release of carbon to the atmosphere, exacerbating the greenhouse effect responsible for the climate change in the first place. What happens depends on the poleward migration rate. transient carbon releases owing to this and related mechanisms might be substantial enough during predicted warming that the biosphere might be substantial enough during predicted warming that the biosphere might no longer be the sink for carbon dioxide than what it is today (Smith and Shugart, 1993).

We still cannot predict how the interaction of dispersal modes and landscape patterning will affect the response to climate change. The two major plant dispersal modes likely to be important in large-scale migrations of functional types are wind and vertebrate dispersal. Wind dispersal is especially common among shade-intolerant species such as grasses and many forest trees and these species often are highly clumped. Species dispersed by vertebrate animals-including most tropical trees-may be either shade-tolerant or shade-intolerant but are often adapted to fill

**Box 1: Feedback's (after Ridgwell, 2002)**

Different components of the Earth system can be connected in two different ways:

☆ With a positive correlation (*i.e.* an increase in the state of one component causes an increase in a second or a decrease in the state of one component causes a decrease in a second component);

☆ With a negative correlation (*i.e.* an increase in the state of one component causes a increase ina second or vice-versa.

small gaps or patches of disturbed ground, growing as scattered individuals in highly diverse communities. The best approach might be to gradually narrow the range of possible outcomes by analyzing specific regions and vegetation types. For example, we may analyse the effect of climate change on Europe, which has many tree species with low dispersal rates and a highly fragmented landscape. This approach, together with that of evaluating various scenarios of climate change, offers the best hope of avoiding surprises in biosphere dynamics.

The marine iron cycle is a central component of a climatic sub-system having the necessary properties for feedback. This is because the atmospheric carbon dioxide and, therefore, climate may respond to dust due to iron fertilization, with dust supply in turn depending on global climate (Ridgwell, 2002). Feedbacks are crucial to the behavior of the earth system and its response to both natural and anthropogenic perturbations. The marine iron cycle feedback will probably play a role in the future climate response to anthropogenic change-as it has already probably done so during past glacial periods when dust appears to have been highly sensitive to small changes in climate. Open ocean iron 'fertilization' experiments carried out in the equatorial Pacific, the Southern Ocean and North Pacific, have shown that insufficient iron availability limits phytoplankton growth in the ocean.

An important source of this iron to the biota of the open ocean is via the deposition of mineral aerosol (dust.). Records of past dust deposition suggest that during the last Ice Age the Aeolian flux of iron to the surface ocean must have been 2 to 3 times higher than at present. At that time, mixing ratios of carbon dioxide in the atmospohere (around 190 ppm) were also much lower. This correspondence led Martin (1990) to propose the glacial iron hypothesis in which low atmospheric carbon dioxide is explained as a result of enhanced iron fertilization of the biota. Since carbon dioxide in the atmosphere exerts radiative forcing on climate, climate is , therefore, also sensitive to changes in the Aeolian iron supply to the ocean.

Dust transport out too the open ocean is generally more efficient under a less vigorous hydrological cycle. Carbon dioxide in the atmosphere also directly stimulates vegetation growth. That climate (and $CO_2$) probably exert a strong control on dust is revealed by analysis and study of ice cores which suggest that enhanced dust concentrations are associated with cold dry glacial periods.

Taking these two different linkages in the Earth system together; if changes in dust flux affect atmospheric carbon dioxide (and therefore climate) and dust fluxes are in turn responsive to global climate, a positive 'feedback' loop is formed (Ridgwell and Watson, 2002; Ridgwell, 2002). In this feedback system, any cooling of global climate will tend to increase dust available and transport efficiency which in turn could decrease atmospheric carbon dioxide (through iron fertilization of the biota), causing additional climate cooling and so further enhanced dust supply. Likewise, the loop could operate in the reverse direction and so amplify a warming of climate. If a path of successive connections can be traced from any given component back to itself, a closed or 'feedback' loop is formed (Figure 2.3). An even number (including zero) of negatively correlated connections counted around the loop gives a positive feedback, which will act to amplify an initial perturbation in the state of any component within this loop. Conversely, an odd number of negative correlations gives a negative

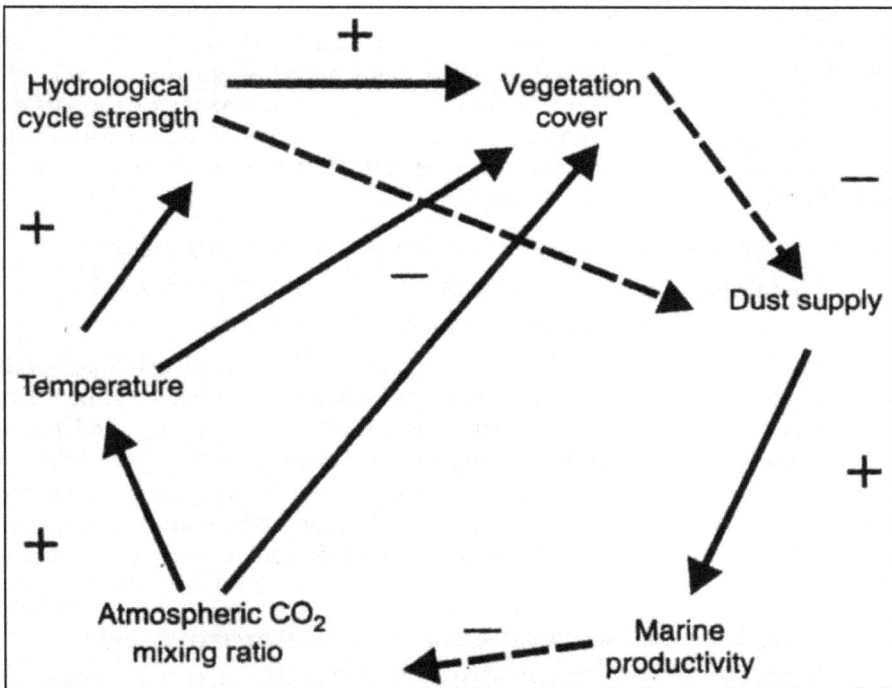

Figure 2.3: Some of the potential feedbacks involving dust and iron fertilization in the climate system. Positively correlated connections are shown in solid lines and negative ones in broken lines. Four main (positive) feedback loops exist in this system, each having a total of two negatively correlated connections within the loop;

1. Dust supply → productivity → $CO_2$ temperature → hydrological cycle → vegetation → dust supply
2. Dust supply → productivity → $CO_2$ → temperature → hydrological cycle → dust supply
3. Dust supply → productivity → $CO_2$ → vegetation → dust supply
4. Dust supply → productivity → $CO_2$ → temperature → vegetation → dust supply

feedback, which will tend to dampen any perturbation, thus stabilizing the system (Ridgwell, 2002).

It is conceivable that operation of the above feedback could generate two distinct glacial states in the earth system, one of 'high-carbon dioxide low-dust' and the other of 'low-carbon dioxide high-dust' (Ridgwell, 2002).

Climate change can alter ecosystems and thereby trigger feedback effects that can either enhance or retard it (Harte, 2002). Such feedbacks are especially likely in montane and high-latitude ecosystems where soils are carbon-rich, sharp transitions in ecosystem community structure prevail as a result of topographic variability, vegetation is sensitive to climatic variables such as snowmelt date and length of growing season, and climate change is expected to be quite strong due to snow-albedo feedback. According to Harte (2002), climate warming in montane regions will have important ecological consequences, and will trigger climate-ecosystem feedback of at least local to landscape scale. Identifying and quantifying the role of feedback systems in amplifying (or suppressing) climatic change can enhance our understanding of how the climate system operated during the glacial-interglacial cycles of the past few million years and of how it may respond in the future to continuing GHG emissions (Ridgwell, 2002).

## Land-atmosphere Exchange of Reactive and Long-lived Compounds: Key Interactions and Feedbacks in the Earth System:

The land-atmosphere exchange processes of a variety of substances are tightly coupled, highly sensitive to climate change, and contribute to climate forcing through their effects on troposphere chemistry and radiative flux. Long-lived gaseous compounds, such as carbon dioxide, methane, and nitrous oxide, as well as reactive volatile organic compounds and nitrogen oxides, are linked in the geochemical cycles of carbon and nitrogen. Of particular interest is the interaction among production, transport, transformation and deposition, with a focus on biology, physics and chemistry.

## Feedbacks Between Land Biota, Aerosols, and Atmospheric Composition in the Climate System

The interaction of biogenic and anthropogenic aerosol particles with the climate system, and coupling of biological and hydrological processes with atmospheric reactions controls the self-cleansing mechanism of the atmosphere, particularly in the tropics. Direct emissions of natural and anthropogenic aerosols, as well as secondary aerosol formation and the production of cloud condensation nuclei should be investigated to enhance the understanding of the direct and indirect effects of aerosols (including dust, biomass smoke and biogenic particles) on radiative flux and cloud-precipitation processes.

Surface-atmosphere exchange processes are important in determining the concentration of hydroxyl radical-the main oxidant determining the rate of chemical removal of compounds from the atmosphere. Changes in land use and cover directly or indirectly affect the oxidizing capacity of the atmosphere and surface removal processes. Hence, surface-atmosphere exchanges, as well as mixing and transport, play a key role in regulating chemical transformations.

Changes in gas-phase chemistry also affect aerosol formation and growth processes. Vegetation promotes the formation of aerosols by oxidation of emitted VOCs and the formation of clouds through the combined control it exerts on evaporation. In turn, aerosols and clouds affect the amount of light received by vegetation (Global change Newsletter, Dec. 2005).

## As a Sustainable Spaceship: The Earth

In a spaceship, the internal environment has to be kept within certain tolerable limits for humans and the right mix of gases has to be maintained in the spaceship's atmosphere; also the spaceship needs to have an effective plumbing system that deals with astronaut wastes and also provides water for their drinking and washing. For feeding the astronauts, sufficient supplies of food of the right kind have to be provided as determined by the number of astronauts on board; indeed all astronaut needs and requirements are contingent upon the actual number of astronauts in a spaceship. Builders of spaceships must ensure that there is no overcrowding.

Despite many spaceships having been built, engineers have not been able to build a spaceship that it self-sustaining; in other words, a spaceship that does not need to receive supplies from the Earth. The hard fact is that the only known unique sustainable spaceship is the planet Earth.

## Sustainable Development

In the recent past, the world has been moving in unsustainable directions, which led to the emergence of several methodological challenges:

1. Spanning the range of spatial scales;
2. Accounting for temporal inertia and urgency;
3. Dealing with functional complexity and multiple stresses on human and environmental systems;
4. Recognizing the wide range of outlooks;
5. Linking such themes and issues as poverty, ecosystem functions, and climate;
6. Understanding and reflecting deep uncertainty;
7. Accounting for human choice and behaviour;
8. Incorporating surprise, critical thresholds, and abrupt change;
9. Effectively combining qualitative and quantitative analyses; and
10. Linking with policy development and action through stakeholder participation (Swart *et al.*, 2002).

As sustainable development is connected to climate change, global collaboration on climate change should be approached on multiple levels through local and national development programmes. Current analyses of climate change policy have been driven mostly by concerns about climate changes; related ancillary benefits from energy efficiency, such as reduced health impacts of local air pollution, may also be significant albeit only of secondary importance-they may reduce the total costs of compliance with climate change commitments.

Energy initiatives and other climate-favouring activities emerge as side-benefits of sound development programmes. Price reform, agricultural soil protection, sustainable forestry, energy sector restructuring all undertaken without any reference to climate change mitigation or adaptation-have had substantial effects on the growth rates of GHG's emissions.

Contributions by developing countries to the managements of the risks of climate change should be seed not as a burden of legal commitments to be avoided, but as a side-benefit of should and internationally supported development. Development problems should be solved not only in the most climate-friendly way but also most sustainable development friendly way.

This leads to an alternative strategy for establishing cooperations between developing and developed nation: to ground analyses and implementation programmes in development objectives and to work out from the foundation to climate policy in the context of sustainability (Figure 2.4).

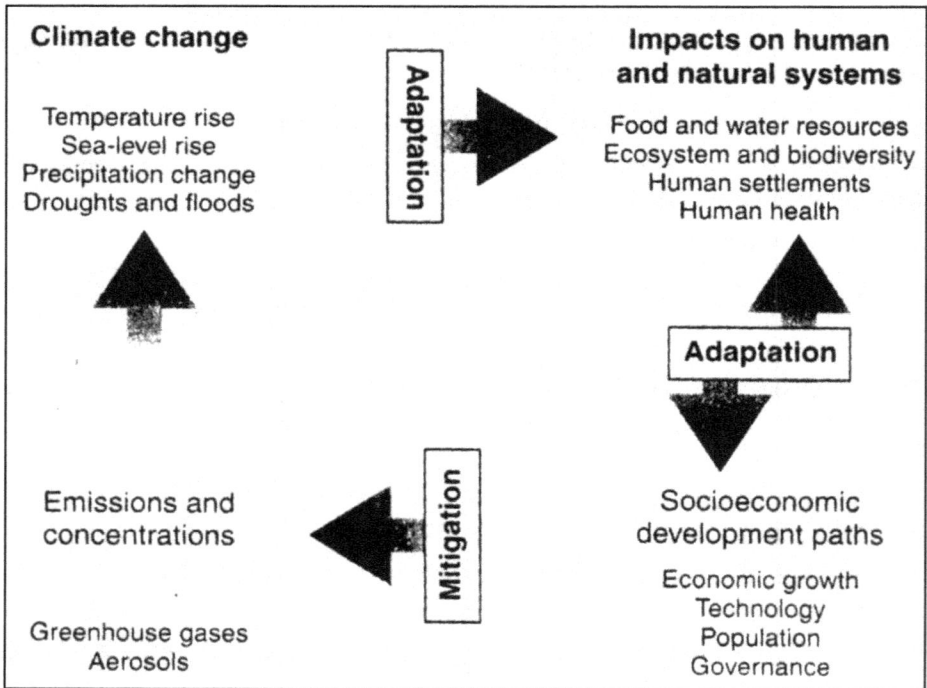

**Figure 2.4: An integrated framework for climate change. Simplified sketch of an integrated framework for considering the relationship between climate change and sustainable development.**

Sustainability science will have to go beyond the incremental responses to perturbation that are dominating much research on the dynamics of combined socio-ecological systems.

The evolution of methods that can adequately and rigorously capture uncertainty, the capacity for system discontinuity, and the normative content of sustainability problems generates the need for a rich and urgent research agenda. Participatory scenario development can be a good option for addressing some of the above challenges. Indeed, this has already been used in the work of the Intergovernmental Panel on Climate Change (Nakicenovic and Swart, 2000). Such an exercise has been successful in bridging gaps between science and policy by engaging a wide range of experts and stakeholders in a systematic exploration of various global futures.

## The Nature of the Earth's Life Support Systems

Life on Earth and the systems supporting it are confined to a fairly thin 'skir' the biosphere-that envelops a huge, inert, lifeless mass. This biosphere extends some tens of kilometers into the atmosphere, some kilometers below the Earth's surface (depending on position) and some tends of kilometers below the ocean's surface (depending on position). The total diameter of the Earth is approximately 13,000 km.

Two fundamental processes are at work within the biosphere: firstly the constant input of energy from the Sun (solar radiation) and, secondly, the cycling of elements needed to maintain life. Solar radiation arrives in large quantities, but most is reflected; the part that is not reflected penetrates the atmosphere and provides the energy (after harvesting by plants) that maintains life; it also powers oceanic and atmospheric events. The cycling of essential elements ensures that the finite amount of these remains continuously available (Williams, 2004).

The above description of the two processes should not convey that they are simple in reality, the events involved are extremely complex. The complexity is borne only by the huge diversity of plants and animals and their interactions with each other and with the inanimate world.

Besides the above, another fact is the presence of a protective, ozone 'shield' in the stratosphere. This shield protects life at the surface from excessive amounts of harmful, ultraviolet-B radiation. It was in the second half of the 20th century that humans could for the first time see the Earth as a sphere suspended in space. This sight made them acquire a new perspective of their home planet. The view from space conferred a sense of unity, with oceans, deserts, forests, and even some signs of human activity all being part of one Earth system. It was also at this time that the image of the 'Spaceship Earth' entered the public mind-for all but the last 100-200 years of human history, humanity was clearly only a passenger on spaceship Earth. But now, humankind has come out of its passenger seat and it attempting to take control of the ship. This is a dangerous courses of action, because we still do not know clearly how the spaceship responds to perturbations, how the control are wired, an what all the indicators signaling change really mean.

The best illustration of the central role of the biosphere in controlling conditions on spaceship Earth emerges by looking at the chemical composition of the atmosphere, where major and minor gases show a composition very far from chemical equilibrium-a condition that could not persist long without life's influence. The biota and their geophysical and geochemical environments are linked through strong and complex feedbacks. There are strong linkages between biological activity, atmospheric constituents and climate, and human activity is perturbing the natural feedbacks.

## Biota, Aerosols, Clouds and Climate

Charlson *et al.* (1987) proposed a abiosphere/climate feedback loop in which marine phytoplankton emits a volatile sulphur containing substance [dimethylsulphide (DMS)] that escapes from seawater into the atmosphere. There DMS becomes oxidized to sulphate aerosol particles that serve as cloud condensation nuclei (CCN). Increased CCN numbers proliferate the number of cloud droplets, making clouds brighter, and reduce the amount of sunlight absorbed by the Earth. This enhanced albedo cools the Earth, changing the living conditions for plankton, and so their rate of DMs emission. Recently, in this DMS/CCN/clouds/climate feedback, the possibility of climate effects from the addition of anthropogenic CCN has entered the limelight (Andreae, 2002).

Three different mechanisms for CCN/clouds/climate interactions modify cloud properties to amplify the effects of aerosol particles on climate, and are called 'indirect effects' in contrast to the 'direct effect' that is due to the interaction of sunlight with the aerosol particles themselves.

First, the addition of anthropogenically produced aerosol particles originating mostly from biomass burning and from the oxidation of sulphur dioxide from fossil-fuel burning increase cloud droplet numbers and so increase cloud albedo. This 'first indirect aerosol effect', may well be the largest climate forcing due to anthropogenic aerosol emissions. It appears to be much more marked over the ocean than over land because of the type of clouds occurring there.

Second, since the amount of water available for droplet formation in a given cloud is more or less fixed, producing more droplets in a cloud means that the droplets are smaller. This implies some consequences for the production of raindrops since for a cloud droplet to become large enough to be able to drop out of a cloud and turn into rain, it must gather other droplets- a process that only occurs when the largest drops in a cloud can sink rapidly enough to collect smaller droplets in their path, and hence to keep growing and falling faster until they rain out.

But when there are no droplets bigger than about 15mm, this coalescence process fails and the cloud is left to dissipate without rainfall (Rosenfeld, 2000). This prolongs the lifetime of the cloud, thereby increasing the total amount of cloud present in the atmosphere, again increasing planetary albedo and cooling the planet the 'second indirect effect'.

All the aerosol effects mentioned above reduce solar radiation at the earth's surface, which reduces ground heating and evaporation, and thus influences the hydrological cycle (Lohmann, 2002), also exerting a negative climate forcing.

A third effect occurs, if an 'overseeded' cloud acquires enough convective potential energy to ascend high enough for ice particles to form. Then, alternative rain production mechanisms involving ice take over, and rainfall occurs from such polluted clouds. Such a shift of rain formation from lower level water clouds to higher-level ice clouds has the following consequences: rainfall intensity increase, lightning is enhanced, additional energy from the freezing of water is released, and the level of conversion of latent to sensible heat is shifted upwards (Andreae, 2002).

This rainfall production and the associated energy conversion processes affect large scale climate dynamics.

In the pristine state, both marine and terrestrial biota seem to regulate CCN concentrations to remain at fairly low values. Increased CCN over previously pristine continental regions also have climatic effects. A particularly powerful perturbation results from the third indirect effect, when CCN are added in the tropics, because this region is the engine of global circulation.

Thus, by providing most of the CCN to the natural atmosphere, the biosphere strongly influences cloud radiative and microphysical properties, and thereby both climate and the hydrological cycle. This natural regulation mechanism is now being overwhelmed by anthropogenic emissions (Andreae, 2002).

## Biota and the Self-cleaning Atmosphere

Large amounts of chemical compounds are emitted into the atmosphere, the largest share being methane and other hydrocarbons from biogenic sources, with a combined flux of over 1000 million tons per year (Prather *et al.*, 2001). Were it not for an extremely rapid and effective self cleaning mechanism of the atmosphere, these and other compounds would soon accumulate to unacceptable concentrations. Hydrocarbons, being very weakly soluble in water, cannot be removed effectively by being incorporated in rain. Their elimination requires some oxidative reaction steps to be transformed into water soluble compounds, either polar organics or carbon dioxide. The most important first reaction step is that with the hydroxyl radical (OH), aptly called the 'detergent' of the atmosphere (Crutzen, 1995). This is a short-lived, very reactive molecule. It is produced from the photo dissociation of ozone to dioxygen and an oxygen atom, and the subsequent reaction of this oxygen atom with water vapour ($H_2O$). Because of the high levels of ultraviolet radiation and water vapour in the tropics, the OH concentrations are highest in the tropics, and most of the oxidation of methane, carbon monoxide, and other trace gases occurs in the 'Great Tropical Reactor' the region of high OH in the tropical troposphere. This fact shows clearly that the tropical region has a key role in regulating physical climate and also in maintaining the chemical composition of the atmosphere.

Besides the presence of water and ultraviolet light, the abundance of $O_3$ and the relative amounts of hydrocarbons and nitrogen oxides ($NO_x \rightarrow NO + NO_2$) also have crucial, inter-related roles. Whereas at very low levels of $NO_x$, hydrocarbon oxidation removes $O_3$ and consumes OH, at higher $NO_x$ levels more $O_3$ and reactive radicals are formed. Under pristine conditions, the biosphere is the dominant source of both hydrocarbons and $NO_x$ in the lower and midtroposphere and their relative amounts emitted are such that $NO_x$ concentration are low and so the troposphere is in a low-ozone state.

In the rainforest regions of the tropics, this happens through a tight interaction of biological, chemical and physical processes, which promote turnover of nitrogen but prevent this nutrient from escaping easily into the atmosphere. Some NO is formed during the nitrogen turnover which accompanies the decomposition of organic matter in soils; a part of this nitric oxide (NO) diffuses into the air layers over the soil where it reacts with ozone to form $NO_2$ which becomes deposited on plant surfaces in the forest canopy and becomes available for plant growth.

The conversion of rainforest to grass and crop lands (deforestation) and the resulting removal of the tree canopy disrupt this tight $NO_x$ recycling system. Biomass burning for deforestation and land management supplies additional $NO_x$ and hydrocarbons to the regional atmosphere leading to a transition from a low-ozone state of the Great Tropical Reactor to a high-ozone photochemical smog situation (Andreae, 2002). This change in gas phase chemistry has some consequences for aerosol production. Under natural conditions, the aerosol yield from the photo-oxidation of terpenes is fairly low. But at higher $O_3$ concentrations, more low-volatility compounds are formed, which can then form aerosol particles (Kanakidou *et al.*, 2000). Tropical deforestation and land-use change enhances aerosol loading in three ways by biomass burning, by emissions from fossil fuel combustion in vehicles and power plants, and by increasing the aerosol yield from the oxidation of biogenic hydrocarbons. Elevated aerosol levels change cloud dynamics and increase lighting frequency. The changed microphysics in polluted clouds reduces their power of cleaning the atmosphere, and enhanced lightning results in increased $NO_x$ formation-both factors that increase ozone production and pollution of the upper troposphere (Andreae, 2002).

Pervasive human activities influence all the earth's compartments and processes. In particular, the atmospheric concentration of carbon dioxide has resin from 280 ppm (parts per million) in 1750 to 367 ppm in 1999 (IPCC, 2001). The effects of this increase are very likely responsible for the global increase in temperature seen over the past half a century. Effects of both the increase in carbon dioxide concentration and global warming on ecosystems have started manifesting. These may influence organisms directly by acting either on the physiology (*e.g.* photosynthesis) or on the species seasonal cycle. They may also affect biological systems indirectly by modifying biotic factors, in turn affecting the spatial distribution of species (Beaugrand *et al.*, 2002).

Solar radiation reaching the ocean's surface stimulates evaporation; more water is evaporated from the ocean than is precipitated back, with the balance being precipitated over land, which results in the formation of lakes, rivers and groundwater. Ultimately, all water precipitated over land re-enters the ocean. Lakes, rivers and groundwater are the sources of water for drinking, irrigation, waste transport and other purposes. Collectively, they are the earth's plumbing system.

Solar radiation that is fixed by plants in photosynthesis serves as the energy to support life, *i.e.* the earth's feeding system. The Earth's biological diversity is also involved in this feeding. It seems to regulate and protect the system; that is as a security system-the maintenance of the Earth's biodiversity is crucial in maintaining the health (vitality) of the Earth's ecosystems. The extinction of species may be compared with the loss of rivets that hold an aeroplane together (Williams, 2004).

The carbon, oxygen, and nitrogen cycle are important in maintaining the right mix of atmospheric gases they have a vital role in the Earth's respiratory system. Carbon also has a role in climate control: the more the carbon dioxide present in the atmosphere, the more the solar radiation that is retained. So the carbon cycle is also important in Earth's climate control systems. The cycles of other essential elements are no less significant. Phosphorus (P) forms part of the molecules that release energy in all living organisms, although it is needed in only small amounts. This is also true for almost all other essential elements (some of which, in excessive amounts, can even be toxic).

## To the Life Support Systems: Extent of Damage

There is no doubt that humans have substantially damaged the earth's life support systems. Most of the earth's lakes, reservoirs and rivers-its plumbing system-have been abused and misused, and are in crisis. Siltation, salinisation, exotic introduction, overfertilisation, acidification, river diversion, damming, drainage and poisoning are some of the more notable impacts. The overall natural character of inland waters has changed significantly, adversely and, in some cases, irreparably.

As for the Earth's feeding system, the human population has crossed six billion, of which many millions are below starvation level. In many regions, the 'carrying capacity' of the land to support a given human density has already been exceeded. To maintain the present population, vast areas of natural habitat have been grossly changed by agriculture, pastoral activity, and irrigation in dry regions. The demands on certain essential elements (particularly phosphorus and nitrogen in forms accessible to plants) have risen, with significant disturbances to the natural nutrient cycles. The Earth's feeding system is either approaching, or has already exceeded, its ability to provide adequate food for the global human population (Williams, 2004).

The destruction of large areas of natural habitat to grow food, and for many other human activities, has contributed to a significant loss in the Earth's biodiversity-unfortunate, as it spells loss of potentially useful pharmaceutical products; even more unfortunate in terms of the loss of mechanisms that might regulate and protect the Earth's feeding system from unpredictable outbreaks of parasites or other adverse biological and non-biological impacts upon it; in other words, lead to a decrease in the efficacy of the Earth's security system. As, in general, no species is without some significance within its ecosystem, on an evolutionary scale, the Earth's biodiversity reflects the spaceship's genetic intelligence encoded over almost four billion years of evolution. Any decrease in biodiversity increases the risk of damage to the Earth's feeding system.

Although it appears that humans have not yet affected the atmospheric mix of gases to a point which places the Earth's respiratory system at risk, still in many large cities, smog (mixture of various oxides of nitrogen and small particles) has created or aggravated human health problems.

In fact, the incidence of emphysema, asthma and other human respiratory diseases have increased greatly in many countries in recent years.

Human activities have already exerted strong impacts upon the carbon cycle in terms of the emission of large amounts of carbon dioxide into the atmosphere. Motor vehicles and industrial activities are the primary sources. In the atmosphere, the carbon dioxide traps heat escaping from the earth, so increasing the Earth's overall temperature. The latest prediction from an intergovernmental committee of experts is that temperatures will rise from to 5°C within the 21$^{st}$ century depending on the region. The chief effect of this rise in tropical regions will be to increase evaporation rates and, thus, cloud cover and so impact the global hydrological cycle. A major, initial effect will be to decrease ocean density and melt polar ice, leading to a rise in ocean levels. This will impose further risk to millions of humans already at risk from sea flooding as, for example, in Bangaldesh, the Netherlands and many Pacific islands.

Besides carbon dioxide, some other gases released into the atmosphere by human activities sulphur dioxide ($SO_2$), nitrous oxides, methane ($CH_4$) and various CFCs. All these have produced negative environmental effects; sulphur dioxide to acid rain, nitrous oxides to smog, and methane to enhancing the effects of carbon dioxide as a GHG (Kumar and Hader, 1999).

Nitrous oxides contribute both to greenhouse effect and ozone depletion. CFCs decrease ozone concentration in the upper atmosphere (the stratosphere) thus allowing increased amounts of ultraviolet-B radiation to reach the earth. Excessive ultraviolet-B radiation adversely affects living organisms. While in Australia and New Zealand attempts are being made to restrict human exposure to ultraviolet-β radiation, nothing can be done anywhere to restrict the exposure of the natural environment. Serious damage has already occurred to the Earth's protective shield.

The spaceship Earth is already overcrowded and is likely to be more so in the coming decades. Its life support systems do not recognize any national boundaries-and if they fail, they fail for all countries. Also, it takes the failure of only one life support system-the most vulnerable to damage-for the sustainability of the whole spaceship to fail. Quite probably, this may manifest as dramatic increase in the worldwide incidence of old and new viral and bacterial diseases. An immediate focus on conservation of our life support systems and of biological diversity is needed.

## In Amazonia: Large-scale Biosphere-atmosphere Experiment

The Amazon contains the largest extent of tropical forest on earth. However, the Amazon is rapidly changing due to human activities. Large-scale biosphere-Atmosphere (LBA) experiment has recently been started as one of the largest coordinated scientific endeavours in the humid tropics, involving 80 closely-linked research groups and around 600 scientists from South and North America. Europe and Japan. Besides increasing our understanding of the importance of amazonia for the planet, the LBA is expected to enhance the scientific understanding needed to guide the sustainable use of the Amazonian forests.

The Amazon Basin accounts for a large proportion of the planet's animal and plant species. However, over the past 25 years, rapid developments have led to the deforestation of over 500,000 sq km in Brazil alone. A small number of field studies carried out over the last 15 years showed local changes in the water, energy, carbon and nutrient cycling and atmospheric composition caused by deforestation and biomass burning (Nobre *et al.*, 2001).

The LBA project was started in 1998. Its research strategy is two pronged, the first is to carry out a collection of process-based studies at small scales where models and remote sensing are used to scale up to the basin scale. The second study focuses along two ecological transects with ecoclimatic and land use intensity gradients across the basin. The LBA studies are organized around seven themes (physical climate, atmospheric chemistry, carbon storage and exchange, biogeochemistry, hydrology and surface water chemistry, land use and land cover change, and the human dimension of Amazonian development).

Preliminary results indicate a picture of an intrinsically coupled rain-producing system where different processes interact at different space and timescales and where the underlying surface, either forest or deforested, actively participates in several mechanisms of cloud formation in the Amazonia (Silva Dias *et al.*, 2001).

From an Earth system point of view, it is impossible to decouple the atmospheric properties and composition from the Amazonian ecosystem as a whole. The atmospheric composition in Amazonia and its physical and chemical properties are very much regulated by the underlying Amazone tropical rain forest. The forest has a great deal of importance in regulating the concentration of key trace gases such as methane, carbon monoxide, hydrocarbons volatile organic compounds (VOC) and many others. The forest is responsible for the emission of significant amounts of natural biogenic aerosol particles, which provide most of the CCN that are key to the Amazonian hydrological cycle. The aerosol particles carry and distribute essential plant nutrients such as phosphorus. The direct radioactive effects from the layer formed by aerosol particles plus the indirect effects of clouds strongly impact the atmospheric radiation balance in Amazonia. The powerful convention in Amazonia makes global transport of trace gases and aerosol particles plus water vapour very efficient (Artaxo *et al.*, 2001).

# Chapter 3
# Change in Climate

## General Description

The statistical description of weather and its anomalies is climate. It changes continuously, since both the interactions among the climate system components (atmosphere, ocean, biosphere, cryosphere,and lithosphere) and the irradiance of the Sun vary on all timescales up to billions of years. On timescales which are of most interest for mankind, *i.e.*, up to centuries, the major factors which cause climate variability are: changing composition of the atmosphere; spectral solar irradiance variations; systematic changes of ocean-atmosphere-land interaction, if resulting in major changes of ocean circulation; volcanic eruptions and land cover change modifying the surface energy budget. Two of these influencing factors, namely changing atmospheric conditions and land cover change are now largely anthropogenic and so the climate discussion mostly involves anthropogenic influence. At different periods, different influencing factors have dominated. Since the complex interactions are only partly understood, it is difficult to attribute observed climate parameter changes or trends to particular causes (Grassl, 1999; IPCC, 1995, 1996).

The following tools needs: The complex and difficult task of climate change detection.

☆ Continuous global observations at least at the Earth's surface and in the atmosphere.

☆ Known history of external forcing factors.

☆ Tested coupled ocean-atmosphere-land (climate) models.

☆ Statistical techniques to detect changing anomaly patterns.

Mankind has been altering the land surface for millennia, and thus has changed the surface energy budget appreciably at least on regional scales. At present, about 40 per cent of the area of the continents not covered by land ice is used for agriculture and settlements. The changes in surface albedo (ratio of reflected to incoming solar irradiances), evapotranspiration, and surface roughness change local climate, for example through the urban heat island effect but whether or not global climate also has been affected is not very clear (Grassl, 1999).

A recent survey of ancient levels of lakes, rivers and sea, and changes in stalagmites and sediments has revealed a strong correlation between climate change and rise and fall of civilizations in the Middle East (Issar and Zohar, 2004). Warm periods were characterized by aridisation, economic crisis and mass migration. Cold periods brought abundant rain, prosperity and settlement. Climate change appears to have been the decisive factor in the origins of the cradle of civilization.

Some significant environmental problems facing the world today include shortages of clean and accessible freshwater, degradation of terrestrial and aquatic ecosystems, increases in soil erosion, loss of biodiversity, changes in the chemistry of the atmosphere (such as increasing ozone in the troposphere and its depletion in the stratosphere), declines in fisheries, and the ongoing changes in climate. These changes superimpose over and above the stresses imposed by the natural variability of a dynamic planet; they interest significantly with the effects of poverty, disease, and malnutrition. In fact, the changes are profound and mirror those in the human-nature relationship. They cascade through the Earth's environment in unpredictable ways and call for societies to develop a range of creative response and adaptation strategies. If this does not happen, there is every possibility that the earth itself may be driven into a different state and one that would be less hospitable to humans and other forms of life.

Global climate change is of interest because of the probability that it may affect the ease with which humans make a living, and perhaps, the carrying capacity of the planet for humans and other species. There exists the distinct possibility that human activities will cause global climate change because our choices can then affect outcomes.

With global environmental change now occupying an increasingly central place in human affairs, a vigorous international debate has ensued about the nature and severity of global change and its implications for lifestyles. Science can help societies in developing and communicating the essential knowledge base that may be instrumental in deciding on how to respond to global change.

The Earth behaves as a system in which the oceans, atmosphere and land, plus it living and non-living parts are all connected. As human activities are responsible for some parts of global change, successful attempts have been made to separate natural and anthropogenically induced variability in the Earth system. Global change is an outstanding environmental issue facing humankind today. Some major research findings in the last decade are outlined below (Steffen and Tyson, 2001).

The Earth is a system that life itself helps to control. Biological processes interact strongly with physical and chemical processes to create the planetary environment.

---

**Box 1: Global change is more than climate change**

The phrase the Earth system refers to the interacting physical, chemical, biological and human processes that transport and transform materials and energy and so provide the conditions necessary for life on the planet. Climate refers to the aggregation of components of weather-precipitation, temperature, cloudiness, for example-but the climate system includes processes involving ocean, land and sea ice in addition to the atmosphere. The Earth system encompasses the climate system, and many changes in the Earth system functioning directly involve changes in climate. However, the Earth system includes other components and processes, biophysical and human, important for its functioning. Some Earth System changes, natural or human-driven, can have significant consequences without involving any changes in climate. Global change should not be confused with climate change- it is significantly more than climate change (Steffen and Tyson, 2001).

---

Global change is much more than climate change. It is real, it is happening now and it is accelerating. Human-induced changes are clearly identifiable belong natural variability and are equal to some of the great forces of nature in their extent and impact. The human enterprise drives multiple, interacting effects that cascade through the Earth system in complex ways.

The Earth's dynamics are charactrerised by critical thresholds and abrupt changes. Human activities could inadvertently trigger changes with catastrophic consequences for the Earth system. The Earth is currently operating in a no-analogue state. In terms of key environmental parameters, the system has recently moved well outside the range of the natural variability shown over at least the last half million years. The nature of changes now occurring simultaneously in the earth system, their magnitudes and rates of change are unprecedented.

The above scientific results warrant an urgent need for ethics of global stewardship and strategies for the earth system management. The 'business-as-usual' way of dealing with the earth has to be replaced- as soon as possible-by deliberate strategies of good management. The Earth itself is a single system, within which the biosphere is an active, essential component (Steffen and Tyson, 2001). It is the interaction of biological processes with physical and chemical processes that creates conductive environment for living organisms on Earth. Life itself plays the crucial role in the functioning of the Earth system and in helping to control it.

Terrestrial biota constitute an important component in Earth system functioning. The type of vegetation present on the land surface influences the amount of water transpired back to the atmosphere and the absorption or reflection of the Sun's radiation. The vegetation's rooting patterns and activity control not only carbon and water storage but also fluxes between the land and the atmosphere. The biological diversity of terrestrial ecosystem functioning in the face of a changing environment. Indeed, variability and change are realities of the Earth system.

## Linkages

One important universal feature of the planetary machinery is the various linkages that connect processes in one region to consequences in others, thousands of kilometers away. This is best illustrated by atmospheric and oceanic circulation which transports heat from the tropics to the poles.

The atmospheric transport of materials is usually considered only in the context of air pollution, but it also has a major role in natural biogeochemical cycles by linking land and ocean processes across long distances.

Anthropogenic waste heat does not have direct consequences for global climate. However, if the concentrations of long-lived gases which influence the transfer of solar and terrestrial radiation in the atmosphere change, a long-term change will be caused in the principal climate forcing factor, the planetary radiation budget. This is mainly achieved through emissions of carbon dioxide, methane, and nitrogen oxide ($N_2O$, nitrous oxide) caused by the burning of oil, coal and natural gas and by agriculture. Emissions of short-lived gases may also influence climate if they become chemically transformed into either aerosol particles, scattering and absorbing mainly solar and to a lesser extent, terrestrial radiation, or radioactively more active gases such as troposphere ozone (Grassl, 1999).

If all parameters in the atmosphere remain constant at mean observed values except for the concentration of a certain radioactively active gas, a net radiation flux change will result. This net flux change is a measure of the stimulus for climate change. Such a stimulus immediately provokes the climate system to change its troposphere and stratospheric structures, which are held constant in the instantaneous radioactive forcing calculations.

The five most important radioactively active gases of the atmosphere (Table 3.1) share one common property, *viz.*, thermal infrared or terrestrial radiation is absorbed much more strongly than solar radiation. Thus, these gases shield the Earth's surface from a strong direct loss of heat to space. They keep the Earth's surface warmer by at least 30°C compared to the atmosphere without them. Increasing the concentration of such a gas will warm up the surface and the lower atmosphere until as much energy is radiated to space as is absorbed from the down welling solar radiation. The debatable issue is no longer whether the observed concentration increase of these gases stimulates a mean global warming but how the water cycle (water vapour, clouds, snow and ice surfaces, ocean circulation) feeds back, and what regional effects will occur (Grassl, 1999). Figure 3.1 shows that the radioactive forcing resulting from the addition of long-lived greenhouse gases (GHGs) since 1850 already amounts to +2.5 Wm$^{-2}$ and is more effective by a factor of 100 than anthropogenic waste heat flux into the atmosphere. The carbon dioxide share is roughly 60 per cent, next comes methane with less than 20 per cent, closely followed by halocarbons with about 15 per cent and the rest due to nitrous oxide.

To predict climate change requires observational constraints on the current climate state, knowledge of the way the coupled air-ocean-ice-Earth-life system behaves, and information on changing forcing such as solar variability. Studies of past climate are also required, to focus model-building efforts on climate components that are likely to change and to allow testing of the ability of models to predict time-evolution of the system (Alley *et al.*, 1999).

**Table 3.1: Lifetimes, concentrations, growth rates, and (direct) radioactive forcing sonly for radioactively active atmospheric gases. The Wm² column refers to radioactive forcing since the reindustrialize period, while the W m²/ppbv column is accurate only for small changes about the current atmospheric composition; in particular, $CO_2$, $CH_4$ and $N_2O$ concentration changes since reindustrialize times are too large for linearity to be assumed (after Grassl, 1999).**

| Species | Lifetime Years | Uncer Tainty in Lifetime % | Concentration ppbV1992 | Reindustrialize | Current Growth ppvc/yr | Radioactive Forcing Rate $wm^{-2}/ppbv$ | Radio-active Forcing, $Wm^{-2}$ |
|---------|----------------|----------------------------|------------------------|-----------------|------------------------|------------------------------------------|--------------------------------|
| *Natural anthropogenically influenced gases* | | | | | | | |
| Carbon dioxide ($CO_2$) | Variable | – | 356000 | 278000 | 1600 | $108 \times 10^{-6}$ | 1.56 |
| Methane ($CH_4$)* | 12.2 | 25 | 1714 | 700 | 8 | $3.7 \times 10^{-4}$ | 0.47 |
| Nitrous oxide ($N_2O$) | 120 | – | 311 | 275 | 0.8 | $3.7 \times 10^{-3}$ | 0.14 |
| Chloroform ($CHCl_3$) | 0.51 | 300 | *0.012 | – | *0 | 0.017 | |
| Methylene chloride ($CH_2Cl_2$) | 0.46 | 200 | *0.030 | – | *0 | 0.03 | |
| *Gases phased out before 2000 under Montreol Protocol and its amendments* | | | | | | | |
| CFC-11 ($CCl_3F$) | 50 | 10 | 0.268 | 0 | +0.000** | 0.22 | 0.06 |
| CFC-12 ($CCl_2F_2$) | 102 | – | 0.503 | 0 | +0.007** | 0.28 | 0.14 |
| CFC-113 ($CCl_2FCClF_2$) | 85 | – | 0.082 | 0 | 0.0001 | 0.28 | 0.02 |
| CFC-114 ($CClF_2CClF_2$) | 300 | – | 0.020 | 0 | – | 0.32 | 0.007 |

*: Methane increases have been calculated to cause increases in tropospheric ozone and stratospheric $H_2O$: these indirect effects, about 25 per cent of direct effect, are not included in radioactive forcing given here.

**: Gas with rapidly changing growth rate over past decade: trend since 1992 reported.

It is possible to determine ages of sediments by counting of annual layers (in favourable ice cores, tree rings, corals, cave formation, and ocean and lake sediments), by radiometric techniques or by correlation techniques to other well-dated record. Good agreements among independent results provide confidence.

The Earth's climate is driven chiefly by the differential input of solar energy across various latitudes and the response of the atmosphere, oceans and land masses to it. No aspects of the global environment, physical or human, remains untouched by climate. Hence, any significant, rapid change in mean weather patterns can affect the integrity of these environments and cause irreversible damage.

Unlike the past, when climate change was externally driven and fairly slowly, current change is mostly driven by human activities-specifically, the emission of carbon dioxide and other trace gases through the burning of fossil fuels, cement manufacturing, deforestation and certain land-use practices. Because emission rates of these gases exceed their rate of sequestration and uptake by the oceans and

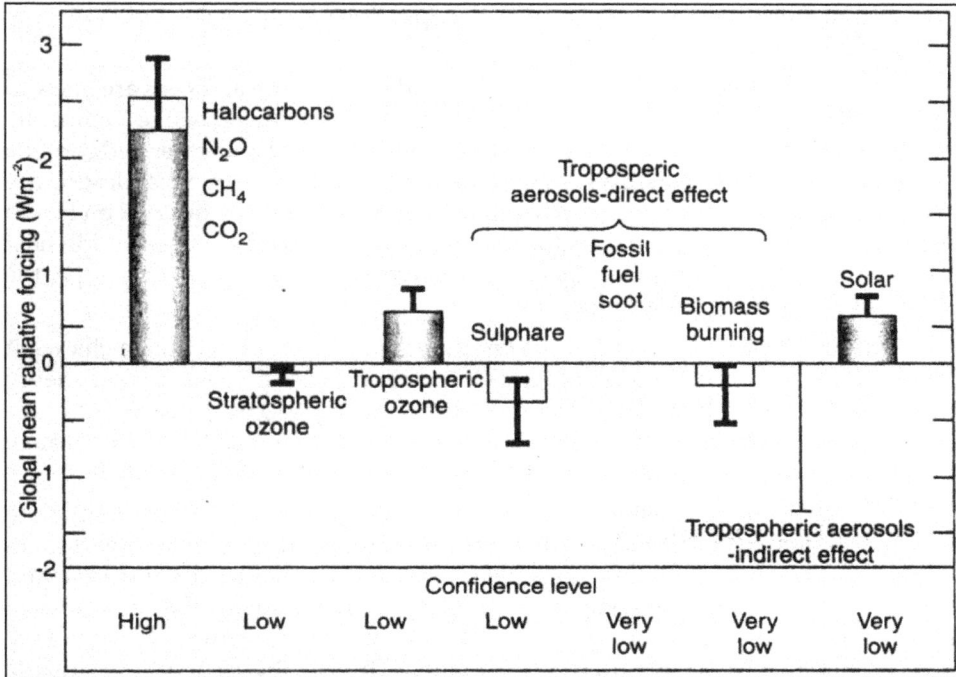

**Figure 3.1: Estimates of globally and annually averaged anthropogenic radioactive forcing due to changes in concentrations of GHGs and aerosols from reindustrialize times to present day and to natural changes in solar output from 1850 to present. The heights of the bars indicate mid-range estimates of the forcing. The error bars show an estimate of the uncertainty range. The contributions of individual gases to the direct greenhouse forcing are indicated on the first bar. The indirect greenhouse forcing associated with the depletion of stratospheric ozone and the increased concentration of tropospheric ozone are shown in the second and third bars respectively. The direct contributions of individual troposphere aerosol components are grouped into the next set of three bars. The indirect aerosol effect, arising from the induced change in cloud properties, is shown next. The final bar shows the estimate of the changes in radioactive forcing due to variations in solar output. The forcing associated with stratospheric aerosols resulting from volcanic eruptions is not shown, as it is highly variable over this time period (after IPCC, 1996).**

biosphere, their concentration in the atmosphere has increased rapidly since the beginning of the industrial age, and it is threatening to double by the first half of the $21^{st}$ century. The projected consequence is a warmer earth (Houghton *et al.*, 1996).

According to Houghton *et al.*, GHG's have grown by between 0.25 and 5 per cent annually, dependent on the type, since the beginning of the industrial age, and human activities have been responsible for almost all the increase. Coupled atmosphere-ocean general-circulation models (GCMs) project a 'best estimate' global mean temperature rise of 2°C by the year 2100. Sulphate aerosols should reduce the magnitude of the warming. Sea level should rise about 50cm which is a slightly lower estimate than the 66cm projected in 1990. Precipitation should increase globally, but regional changes are uncertain (Nkemdirim, 1997).

Models described in Watson *et al.* (1996) predict that tropical forests are likely to be more affected by land-use changes than by the projected climate change. Temperate forests may not be much affected. Some losses may occur in the southern edge of the boreal forests. Overall, models suggest a net loss of forest cover. Most deserts are likely to become more extreme as a result of climate change. The outlook on health includes potential increases in vector organisms of infectious diseases and in heat waves. Agricultural yields may increase or decrease depending on crop type, the water balance and the geographical location.

Both the physical and the human environment can adapt to climate change if the change evolves slowly technological innovations, use of low-emission energy sources and sustainable land-use practices can help mitigating the impacts of climate change. A wise course of action is to slow down the rate and magnitude of change through behaviors modification and reduced population growth (Nkemdirim, 1997).

Climate can cause considerable changes in living conditions. Many temperate and high-latitude countries would be winners and many tropical nations and small islands will be losers. Bruce *et al.* (1996) stressed the need for environmental reconciliation based on the so-called 'no regrets' option- namely, that investments designed to curb or reduce the growth of GHG's should have positive socio-economic effects whether or not climatic change occurs. Though the issue of sustainable development is important, but there is need to differentiate between weak sustainability and strong sustainability. The former requires that any depletion of natural capital be offset either by human produced capital or by substitution of another form of natural capital, such as renewable assets, in place of nonrenewable ones. Strong sustainability demands that some natural capital, being irreplaceable, must be preserved. Bruce *et al.,* seem to favour the latter course in light of the fact that there are no close substitutes for the atmosphere and the climate it produces.

Whereas climatologists and oceanographers have focused on the changes in our physical environment-changes in the climate, the oceans, and the chemistry of the air we breathe-environmental biologists have addressed issues of conservation and the extinction of species. In reality, both these broad concerns are intertwined and mutually dependent, and past changes in biodiversity have both responded to and caused changes in the Earth's environment. Understanding of global environmental change can help in creating scenarios of biodiversity in the next century. Combining physical earth science with conservation biology is a starting point for regional assessments on all scales.

## Change: Drivers

Global change is comprised of a wide range of changes in the global environment caused by human activities. Many of the changes are accelerating and interact with each other and with other environmental changes at local and regional scales (see Reid and Miller, 1989). Over the past two centuries, both the human population and the economic wealth of the world have grown rapidly leading to increased resource consumption in agriculture and food production, industrial development, international commerce, energy production and urbanization. All these enterprises transform the earth system. Over six billion humans share basic human needs, such as the demand for water, food, shelter, community health and employment.

The manner in which these needs are met is a critical determinant of the environmental consequences at all scales (Table 3.2).

**Table 3.2: Proximate and underlying drivers of human transformation of earth (after Steffen and Tyson, 2001).**

| | Proximate Driver* | Underlying Driver |
|---|---|---|
| Land | Clearing (cutting forest, +burning), agricultural practices (*e.g.* tillage, fertilization, irrigation, pest control high yielding crops), abandonment. | Demand for food (+dietary preference), recreation, recreation. Other ecosystem goods and services. |
| Atmosphere | Fossil fuel burning, land use change (*e.g.* agricultural practices), biomass burning, industrial technology. | Demand for mobility, consumer products, food |
| Water | Dams, impoundments, waste disposal techniques, management practices | Demand for water (direct human use), food (Irrigation), consumer products (water usage in industrial processes). |
| Coastal/Marine | Land-cover conversion, groundwater removal, fishing intensity and technique, coastal building patterns, sewage treatment technology, urbanization. | Demand for recreation, lifestyle, food, employment |
| Biodiversity | Clearing of forest/natural ecosystems; introduction of alien species. | Demand for food, safety, comfort, landscape amenity. |

*: Proximate drivers are the immediate human activities that drive a particular environmental change underlying drivers are related to the fundamental needs and desires of individuals and groups. Proximate and underlaying drivers are the end point in linked sequence with such intermediate linkages as markets, institutions, infrastructure, policy, political systems, cultural values.

Many discussions of global climate change have focused on increasing surface temperatures, while changes in the water cycles such as precipitation, evaporation and river discharge have received much less attention. Water has far-reaching effects on the Earth's climate. The natural greenhouse effect is mostly caused by water vapour; the distribution of vegetation types is sensitive to the local water balance; and regional climate patterns are largely influenced by ocean currents (Stocker and Raible, 2005).

Modeling of the many aspects of the water cycle facilitates reliable assessment of the changes resulting from rising levels of GHG's in the atmosphere. Wu *et al.* (2005) investigated changes in the freshwater balance of the high northern latitudes, using a climate model which connectors the influences of the oceans, atmosphere and land surface on climate, and that has been used to show the contribution of rising GHGs to warming during the 20th century (Stott *et al.*, 2000). This modeling showed large seasonal cycles and interannual variability in the global climate over the past 140 years. A comparison between modeling simulations with and without GHGs revealed that the increased Arctic river discharge is associated with warming induced by rising concentrations of GHG's. Such trends in water balance have also been seen in the latest data from observations of the atmosphere (Simmons and Gibson, 2000), particularly in the mid-to-high latitudes of the Northern Hemisphere. It appears that an accelerated hydrological cycle in the atmosphere expresses itself as increasing river discharge in the circum- Arctic regions and consequently in a freshening of the northern North Atlantic (Curry *et al.*, 2003).

Wu *et al.* (2005) extended their simulations to cover the entire 21$^{st}$ century, assuring the increased levels of GHG's predicted by two standard emissions scenarios (Houghton *et al.*, 2001). In these simulations, the speeding up of freshwater delivery to the Arctic, which started in the last three decades of the 20$^{th}$ century, continues and by about 2020 the discharge shows a rise above the upper end of the simulated range of variability. These results support predictions that fundamental environmental changes will occur in the high latitudes of the Northern Hemisphere (Corell *et al.*, 2004). This warming will melt the permafrost, alter the seasonal snow cover, increase the warming, and further speed up the water cycle (Stocker and Raible, 2005).

As per the model of Wu *et al.* (2005), freshwater delivery to the high latitudes is likely to increase in both hemispheres. But the rate of change will be higher in the Northern Hemisphere, which will warm more rapidly. This will lead to a net transfer of fresh water from the Southern to the Northern Hemisphere. Should such a transfer occur for long enough, it would change the properties of the major water masses in the world's oceans and hence alter the balance of deep-water formation in the far north and far south (Stocker *et al.*, 1992). It is , in fact, deep-water formation at high latitudes by which warm water, flowing from low latitudes, cools, becomes denser and sinks. The return flow at depth completes the 'meridional overturning circulation'- a major phenomenon in oceanic circulation (Stocker and Raible, 2005).

Delivery of more freshwater to the Arctic will further reduce salinity in areas where deep-water formation occurs, so decreasing water density. This might weaken the meriodional overturning circulation in the North Atlantic. Further and more rapid warming will increase the vulnerability of this circulation system (Stocker and Schmittner, 1997) and may even lead to a permanent circulation change in the North Atlantic (Knutti and Stocker, 2001).

## Variability: Climate

The Earth's climate, ecosystems and human activities vary greatly in both space and time. Past changes in climate, atmospheric composition and land use have affected the Earth's surface differently at various locations. Pointers to future changes suggest that the pattern in every sphere will continue to change. Reconstructions of past environments and mapping of today's environments enhance understanding of the processes that drive the Earth system. It is becoming possible to make reliable model simulations of future change.

Maps can be drawn from ground-based and satellite-derived data, conceptual and numerical models, census and additional relevant databases. Several global change-related data compilations and directories are already available. Some maps can be so developed that past conditions may be compared visually with the present, and also with future environmental conditions predicted on the basis of current models and forcing scenarios (Sahagian and Cramer, 2002).

Over the past decade, scientific research has greatly advanced our understanding of global change. The growing understanding that the current and future state of the Earth system is intimately linked to human activities, and the increasing societal concern about the implications of global environmental change, underline the need for and importance of these scientific efforts.

## To Interannual Variations: Seasonal

The weather fluctuat4es day-to-day, even hour-to-hour. Meteorologists understand enough about atmospheric behaviour to be able to make useful predictions about the specific conditions of the atmosphere up to almost a week in advance. Climate means the general patterns of weather in a region averaged over seasons, years, and decades. Climate also shows natural variations and fluctuations. Thus, monthly and seasonally averaged temperature, precipitation, sunshine, cloudiness, and wind very from year to year and decade to decade. Over the past decade, climatologists have improved their ability to predict rainfall and temperatures up to a year in advance. Even though they are still experimental, these seasonal-to-interannual forecasts are now being made with confidence and are used by agricultural and water resource planners in some parts of the world, particularly in the tropic, to adjust planting schedules, crop selections, and water released from dams in order to reduce the economic and social impacts of droughts and floods.

Our picture of global decadal to centennial climate variations is hazy. Lack of past data is the primary reason for our poor knowledge of decadal climate variability. The *El Nino* phenomenon varies on an interannual, not a decadal timescale. To understand decadal change, we clearly need information on longer timescales than our instrumental records can provide.

Proxies are the only source of records long enough to understand variations on timescales of decades and longer. By going to the South Pacific, where even now instrumental data are almost nonexistent, Linsley *et al.* (2000) used the Sr/Ca ratio in corals as a true proxy thermometer, unliked $^{18}O$, which responds to both temperature and salinity. Their temperature record is persuasive. Generating many Sr/Ca measurements relatively quickly and cheaply has become possible now.

Most of the variance in the climate record on timescales of less than a century has been captured in a few preferred modes of climate variability. For example, the North Atlantic Oscillation (NAO), usually defined by means of the sea-level pressure difference between Iceland and the Azores, has variability at several periods; investigators have singled out the biennial and decadal bands. The NAO correlates with climate variations in Europe, the Middle East, and the eastern part of North America. The Pacific Decadal Oscillation (PDO), defined in terms of sea surface temperature (SST) in the North Pacific, is related to similar climate variations over North America (see Cane and Evans, 2000; Markgraf, 2001). As regards decadal variability, unstable ocean-atmosphere interactions at mid-latitudes have been suggested, with the decadal timescale determined by the oceanic gyre circulation. The available evidence is that the atmospheric response to mid-latitude SST perturbations is too weak to sustain long-term modes. In fact, the climate system varies at every possible frequency from the seasonal to the millennial and beyond.

Both the PDO and the NAO extend into the tropics. Observed variability I the tropical ocean seems to be accounted for by low-latitude winds. Conceivably, the same coupled ocean-atmosphere physics that generates the interannual *El Nino* cycle may generate variability at periods longer than interannual. The *El Nino* example demonstrates that variations I the tropics can generate variations at all latitudes.

In recent years, the following changes relating to climate have been recorded:

1. Global mean air temperature at 2m height has increased by abut 0.5°C in the 20th century: both hemispheres show similar warming, with the Northern Hemisphere experiencing stronger variability.

2. The stratosphere has cooled especially at high latitudes during recent decades as a result of both enhanced GHG concentrations and ozone depletion in the lower stratosphere.

3. Mountain glaciers have been retreating in most extrapolar areas: glaciers with measured mass budgets have lost 0.3 m per year on average in the last decade.

4. Mean sea level has risen by 10 to 25cm (best estimate is 18cm) in the 20th century, mainly because of thermal expansion of seas water and glacier melting.

5. Night time temperatures over continents have risen more strongly than daytime temperatures, leading to a reduction of the daily temperature amplitude by about 0.6°C in the 20th century.

6. Precipitation has increased in northern mid-latitudes and high latitudes. Subtropical precipitation in the Northern Hemisphere has decreased by about 10 per cent since 1970 (Grassl, 1999).

## Global Climate Change: Simulating

In 1998, a collaborative global change study involving scientists from the Stanford University and the Carnegie Institution was launched with a view to looking at four main ways in which the global environment is changing. In this manipulative experiment, grassland ecosystems are exposed to four different components of global change:

1. Elevated carbon dioxide.
2. Warming,
3. Increased water inputs, and
4. Nitrogen deposition.

The treatments incljde two levels of each of the factors (enhanced versus ambient) and all possible combinations of the four factors, for a total of 16 treatments.

Three main questions have motivated the design. First, to know when effects of global change factors are additive. This is important because most experiments address the global change factors individually, which involves a risk of misinterpretation if the factors interact in unexpected ways.

Second, how much of the ecosystem response comes from changes in the composition of the community of plants, microbes and animals in the plots, versus changes in the physiology,biochemistry and morphology of species present at the beginning of the experiment. Most experiments run for a duration that is shorter than the longevity of the dominant plants, and therefore have limited access to the components of the response driven through changes in species composition.

Third, to identify the routes through which each of the global change factors affects primary production, carbon storage, nutrient dynamics, the wear budget, and species composition. One of the most interesting dimensions of global change effected on ecosystems is that the responses are rarely direct. It is not uncommon for an ecosystem response of interest to be separated from the direct response by three or even more layers of indirect responses.

Virtually, nothing is known about how ecosystems of different complexity will respond to global warming. Microcosm permit experimental control over species composition and raters of environmental change. By using microcosm experiments, Petchey *et al.* (1999) have shown that extinction risk in warming environments depends on trophic position but remains unaffected by biodiversity. Warmed communities disproportionately lose top predators and herbivores, and become increasingly dominate by autotrophy and bacterivores.

Changes in the relative distribution of organisms among tropically defined functional groups lead to differences in ecosystem function beyond those expected from temperature-dependent physiological rates. Diverse communities retain more species than depauperate ones, as predicted by the insurance hypothesis, which suggests that high biodiversity buffers against the effects of environmental variation because tolerant species are more likely to be found (Naeem and Li,1997; Yachi and Loreau, 1999).

Studies of single trophic levels have shown that warming can affect the distribution and abundance of species, but complex responses generated in entire food webs greatly complicate inferences based on single functional groups (Patchey *et al.*, 1999).

## Climatology: Urban

All over the world, people are clustering together in towns and cities. With increasing development of agriculture and commerce, the population in rural areas declines while more people migrate to organ commercial centers. The urban environment attracts many people. Good knowledge of the urban environment, in all its aspects, to keep pace with the growing urban population is among the most urgent of humankind's needs today.

# Contrasts: Urban-Rural

Modern urban areas differ physically in five important ways from rural areas. The differences lie in:

1. Surface materials,
2. Shapes of surfaces,
3. Heat sources,
4. Moisture sources, and
5. Air quality.

Cities are made up of more granite-like materials, with larger thermal capacities, than the countryside. The three-dimensional nature of the city, with its many buildings and multiple-level surfaces, presents a complex geometry to the atmosphere. The city has major heat sources, such as industrial and domestic heating and automobiles, not found in such concentration I the countryside. The rural areas, on the other hand, are sources of moisture relative to the city. The comparative absence of vegetation and the efficient removal of precipitation in the city highlight this contrast. Finally, the wastes of industry, domestic activity, and modern transportation all contribute to changes in the quantity and quality of materials suspended in the urban atmosphere.

For a city with clean air and with wind flowing from the country into the city, the urban maximum temperature usually exceeds the rural maximum by a small amount. Under the same air quality conditions, the relative warmth of the city increases at night. This increase would be present even if the wind were not calm, and it is due primarily to the increased fuel consumption and the recovery of heat stored by day tin streets and buildings. On the other hand, when the urban air is dirty and reflects more sunlight from airborne contaminants, the temperature contrast is reduced during the day, as compared with the situation when air is clean. At night, in contrast, the temperature contrast is increased because of the partial closing of the radiative 'window' to the exit of long wave energy from the city's canopy. When cooler air is not being adverted into the city, the temperature contrast increases.

# The 'Heat Island'

Isotherms in the vicinity of any town or city generally show a pattern like the one for London in Figure 3.2, where temperatures are highest in the most densely built up portions. The appearance of the isothermal pattern has prompted the term 'urban heat island' for the pattern and its variants.

Whereas, in some cities, topographical effects on climate may contribute to an apparent heat island, the island can be shown to be man-made-the direct result of the existence of the city itself. Two kinds of evidence support this view. First, in areas of topographic simplicity, such as London, the heat island shows up very clearly. Second, the intensity of the contrast- the 'height' of the island-increases as the city's population grows (Mitchell, 1961).

In general, it is the magnitude of the physical contrasts between city and country, as affected by several variables, which determines the climatic contrasts. A very

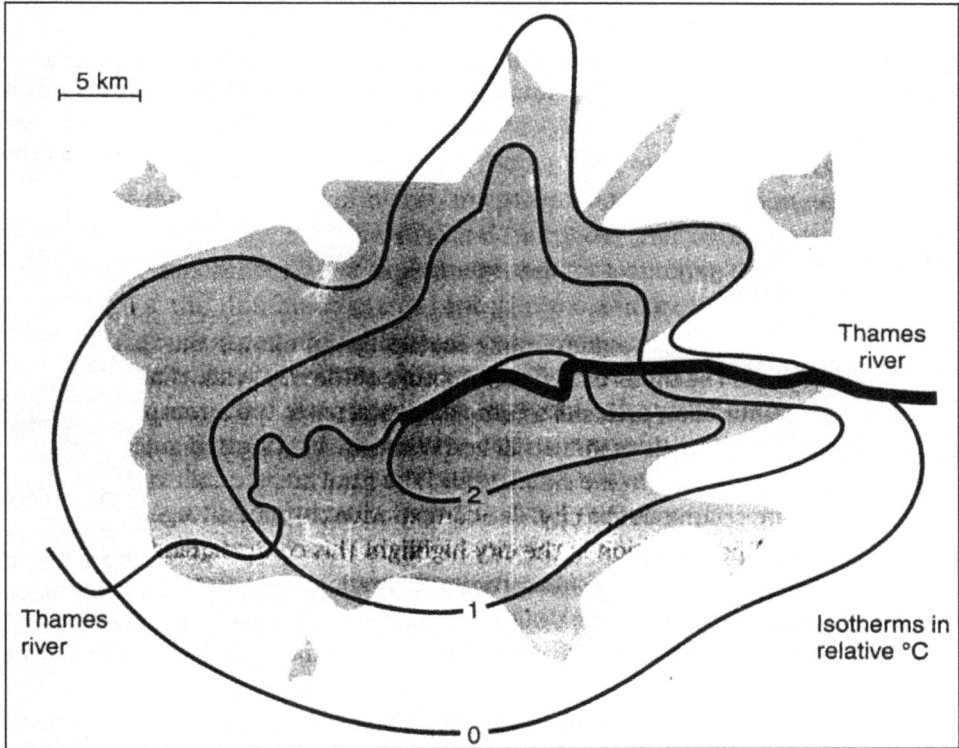

**Figure 3.2: Representative isotherms (in relative °C) illustrating the 'heat island' effect of a city (after Chandler, 1967).**

populous city may have such as suburban 'sprawl' that it blends very gradually into its rural surroundings and the horizontal temperature gradients become inconspicuous. Thus even a small city can show up pa distinct heat island if it contrasts physically with its immediate environments. As regards the effects of regional weather for temperature contrast, it matters little what the weather map looks like, so long as the requisite conditions of clear skies and calm winds are met locally.

Some countries are distributing practical implementation tools that enable local governments to use various measures to reduce the urban heat island effect.

Dark rooftops and paved surface absorb heat and increase the temperature or urban centers by a few degrees Celsius. Higher temperatures can increase the demand for electricity and greatly increase urban air pollution by encouraging ozone formation. Lightening of rooftops and other surfaces reduces temperature and significantly decreases electricity use and ozone formation. This improves air quality improvement and reduces GHG emissions.

# Global Change and Mountain Regions

Mountain regions occupy about one-fourth of the Earth's surface. They are home to approximately one-tenth of the world's population and provide goods and services such as water, forest products, refugia for biodiversity, and storage of carbon and soil nutrients to humanity. The United Nations (UN) declared the year 2002 as the 'International Year of Mountains'.

Recognising the sensitivity of mountain environments and the consequences that changes in these environments might have for humanity, several regional, national and even global research initiatives are focusing on mountain areas.

Globalisation processes and related cumulative and systemic environmental changes can potentially threaten the future ability of mountain regions to provide goods and services that are often taken for granted. The strong altitudinal gradients in mountain regions provide excellent opportunities to detect and analyse global change processes and phenomena, because:

1. Meteorological, hydrological, cryospheric and ecological conditions change strongly over relatively short distances. This enhances biodiversity and characteristic sequences of ecosystems and cryospheric systems are found along mountain slopes. The boundaries between these systems (*e.g.* ecotones, snowline, and glacier boundaries) tend to be climatically sensitive and may experience rapid shifts due to environmental change and thus can be used as indicators;

2. The higher areas of mountain ranges are not commonly affected by direct human activities; they serve as locations where the environmental impacts of climate change alone, including changes in atmospheric chemistry, can be studied;

3. Mountain regions are distributed all over the globle, from the equator almost to the poles and from oceanic to highly continental climates. This global distribution provides unique opportunities to carry out comparative regional studies from widely separated parts of the globe and to analyse the regional differentiation of environmental change processes (Reasoner *et al.*, 2001).

In keeping with the changing environmental conditions along mountain slopes, socioeconomic conditions, land-use and land-management practices, resources exploitation and the appeal of mountain regions for tourism also change. Unsustainable management practices often worsen the living conditions, making people migrate elsewhere; some mountain areas become depopulated, whereas others become over-populated. Such processes have several strong and mostly adverse side effects.

The important role of global change issues in mountain regions is widely acknowledge since mountain environments are essential to the survival of the global ecosystem and many of them are suffering from accelerated soil erosion, landslides, and rapid loss of habitat and genetic diversity. The traditional view that mountains represent pristine systems isolated from human impact and only marginally connected to economic, political and cultural centres of influence is no longer valid. The extensive, widespread retreat of alpine glaciers brings out the high impact of global climate change at high elevations and the consequences for lowland agriculture, hydroelectric power, mitigation of natural hazards and ecotourism. Similarly, greater physical, administrative and market integration of mountain and upland agriculture with mainstream systems have basically changed local resource management strategies leading to resource use intensification and overexploitation. Threatened by the increasingly global scale of both systemic (impact environments at global scale) and cumulative (operate at local scale but are becoming globally pervasive) human impacts, several mountain systems are developing along a trajectory towards critical regions-places where high rates of environmental change are seen in fragile and sensitive ecosystems coupled with economies strongly dependent on local environmental resources and limit response capability.

Globalisation also introduces change in mountain environments as reflected in demographic changes, the incorporation of mountain economic into extra-regional economics, the increasing influence of urban processes and perspectives and increases in consumption.

In view of the fragility of mountain environments, the complex network of both physical and socioeconomic factors that may impact these environment and the substantial consequences that changes in mountain regions may have on humankind, there is a clear need for an integrated approach to addressing these issues, For example, human-water interactions will become critical in the near future. Some mountains in arid and semiarid regions provide over 80 per cent of the water resources to the surrounding lowlands for irrigation, drinking water, industry and domestic use. Knowing that the least 60 per cent of current freshwater resources is being used for food production, and that the complex issue of food security is likely to become quite important in the 21$^{st}$ century, effective water management strategies will be required to tackle a broad range of issues and consequences, will require a focus on mountain regions, and will require inputs from both physical and social sciences. According to Reasoner *et al.* (2001), mountain regions provide unique and valuable settings in which to study the specific facets of environmental changes, their regional consequences, and resource management strategies to adapt to and mitigate these consequences. However, most previous works have not been structured to facilitate a clear understanding of the interactions between climate, land surface processes, and human activities, taking into account the specific conditions in mountain environments. This deficiency needs to be addressed.

The relationship between humanity and the planet Earth must change. We are approaching crossroads that require significant choices to be made. Business as usual is no longer an acceptable option. Global change research in mountain regions will become increasingly important in the coming decades. A new system of global environmental science is required that can help integrate across disciplines, environment and development issues and the natural and social sciences.

# Himalayan Glaciers

The Himalayas which, in a way, form a roof over part of the world, are experiencing the impacts of climate change. With the climate warming up, melting glaciers threaten the livelihoods of millions of people in India, Nepal and China. All along the mountain range, local communities have at least one thing in common-their way of life is threatened by changes to their environment. Already in Nepal, rising temperatures are swelling glacial lakes to burst. Across the mountains in Tibet, herdsmen have to struggle to feed their livestock on an increasingly deteriorating landscape. The locals have started blaming global warming for many of their troubles. Billions of people depend on water from the Himalayas whose snow peaks feed the flow of several important rivers.

Hydrological models discussed by Barnett *et al.* (2005) point to the far-reaching effects the climate change may have in the region. The WWF (2005) referred to the possibility of vanishing glaciers and declining water supplies. Some future disasters could include floods, droughts, land erosion, biodiversity loss, and changes in rainfall and the annual monsoon (Cyranoski, 2005).

Climate change is particularly critical for cattle herdsmen in Tibet and sharp mountain guides in Nepal, Agrawal (2005) and others have shown that about a score of glacial lakes are at risk of bursting in Nepal, and even more in Bhutan. The lakes form naturally from the melt water of glaciers, but when climate change forces the glaciers to melt very fast, the excess water floods into the lakes, Earthquakes, landslides or slope instability can collapse many natural dams.

Melting glaciers convincingly show the impact of climate change. Glaciers and lakes in the source region for the Yellow River have been shrinking recently, threatening a crucial water resource for millions of Chinese. The affected regions are experiencing vegetation loss of 3 to 10 per cent each year. If the trend of rising temperature continues, a decrease in water availability of 20 to 40 per cent is predicted over the next 50 to 100 years, and a fall in total agricultural output including wheat, rice and corn cross of 10 per cent by 2030-50 (Cyranoski, 2005).

# Land Cover and Land Use

Human activities have modified or changed all except a few of the Earth's landscapes. These changes alter vegetation, change the capacities of landscapes to cleanse water and air, affect how animals, plants, and ecosystems can migrate, and alter biological diversity.

The current pattern of global land cover most often reflects past and present land use. Since different land uses and vegetation types (forests, farmlands, grasslands, urban developments, etc.) differ in their capacities to absorb and store carbon, proper monitoring of land cover and land use are critical.

The larger patterns of land cover can be observed and monitored from space. From historical achieves, including the last 20 years of satellite data, a quantitative assessment of landscape and land-use change is being built up.

One high priority area is to combine knowledge of species habitat requirements with measurements of actual land cover, derived from satellite remote sensing, aircraft, and field surveys, to assess the likelihood that sufficient habitat will remain for broad assemblages of species with similar requirements. This efforts depends on the results of basic biological research, land-use and planning information and simple modeling to provide direct guidance to managers and policymaker faced with difficult tradeoffs over the uses of land.

## Hot Spots of Land-use Change

The Earth's vegetation cover has traditionally been regarded as a passive component of the climate system. On the other hand, many studies carried out in the last decade point towards a key role which land surface and biospheric processes can have in weather and climate at local, regional and global scales. The role of the land surface ranges from purely physical influences (*e.g.* the aerodynamic drag on the atmosphere, role of soil characteristics in controlling soil moisture and runoff) to some major biological influences (*e.g.*, leaf stomata response to environmental changes and biogeochemical cycles). This points to the need for a more general definition of a climate system- one that encompasses not only the biotic world (atmosphere, hydrosphere, cryospohere and pedosphere) but also the living biosphere. There appears to be a significant input from land surface and biogeochemical processes and their feedbacks in determining our present and future climate (Kabat *et al.*, 2000).

Theoretical considerations about the Earth's climate system suggest that large-scale (regional) land-cover changes, particularly in the tropics, should have remote climatic effects. In view of the fact that the three major tropical convective heating centers are associated with the land surfaces in Africa, amazonia, and the maritime region of South-East Asia, changes in vegetation cover in threes regions could affect the structure, strength and positioning of convective storms. Even small changes in the magnitude and spatial pattern of tropical convection may then change the magnitude and pattern of upper-level tropical outflow which feeds the h higher altitude zonal jet, therefore affecting regions for beyond the actual hot spots of land-use change. Land-cover changes which result in changes in tropical convection appear to affect weather and climate remotely both in the tropics and at high latitudes, analogous to the remote affects attributed to the opposing phases of *El Nino*= Southern Oscillation (ENSO). Also, changes in tropical and mid-latitude vegetation cover play some roles in tropical monsoon circulations, which have global effects and interactions extending far from the tropics.

According to Kabat *et al.* (2000), adding the human and socioeconomical dimension as a part of an already very advanced globalization process may only strengthen the importance of regional to global land use change climate feebacks.

## Asian Monsoon and Land-use Change in South-East Asia

More than half of the world population lives under the asian monsoon climate. The life, agriculture and economy of the people living in this region depend upon the monsoon rain water as a part of a huge water cycle system. Although the asian monsoon system has been traditionally described as a seasonally changing atmospheric circulation, induced simply by the heating contrast between the land mass and the ocean, the seasonal monsoon cycle, as well as its interannual variability, are attributed to highly nonlinear processes in an interactive, land surface-atmosphere-ocean system. The Asian monsoon system is tightly coupled with the ENSO in the Pacific Ocean.

The variability and long-term change of the Asian monsoon may be sensitive both to natural forcings as well as to humana-ainduced changers of land use and atmospheric composition (*e.g.*, deforestation and fossil-fuel emissions).

During the last 3,000 years, the changes in land use over East Asia have been one of the largest in the world. Past and current anthropogenic land-cover changes have brought about significant variations of the land surface dynamic parameters, such as surface albedo, surface roughness, vegetation leaf area index and vegetation fractional coverage. These variations generated changes in surface energy and water balance, which is turn impacted upon atmospheric circulation over the region and beyond.

Sometimes, there is a weakening of the summer monsoon as a result of destruction of the natural vegetation in parts of East Asia. The regional-to-continental land-use-atmosphere feedbacks are a strong part of the Asian monsoon system dynamics. The effects of land use changes in South-East Asia are very likely to propagate for beyond the region itself, and can, therefore, become a significant factor in the Earth system dynamics (Kabat *et al.*, 2000).

## Terrestrial Human-Environment Interactions

Over the last decade, considerable changes have taken place in our understanding of the patterns and drivers of land-use change. These changes were previously seen in a simplistic manner-as recent, local, spatially homogeneous alterations of previously pristine land (mostly forest) to agriculture, largely caused by population pressures. A wide variety of different landscape types have been modified by humans for millennia in a heterogeneous way-driving forces are not only global and regional but also local, the they include socio-economic and political as well as population and biogeochemical triggers (Ojima *et al.*, 2002). The impacts of these changes are more far-reaching than believed hitherto, and are affecting biodiversity, albedo and human health as well as water and biogeochemical cycles. Current terrestrial carbon sinks are not permanent features; rather they are likely to stop growing, probably diminish, and perhaps even disappear, some time in the 21st century. This finding has a strong bearing on policies and targets aimed at stabilizing the accumulation of GHG's. Recent progress is aiding development of the mechanisms underlying projections of food production under climate and atmospheric change over the next hundred years, and of how changes in biodiversity may alter ecosystem functioning and services critical to human well being (Ojima *et al.*, 2002).

The next decade of global change research on terrestrial systems is going to adopt the 'human-environment' paradigm which implies consideration of the tight coupling between the functioning of ecological systems and the dynamics of human societies as the key factor involved in the response of terrestrial systems to global change and in their feedbacks to the atmosphere and the oceans. The focus will be on the diversity of ecological and human systems on the land surface and on critical vertical and horizontal flows, not only within terrestrial ecosystems but also across to the atmosphere and ocean. Figure 3.3 illustrates the three major nested issues, representing progressive steps towards urgently needed knowledge for this purpose. Some pertinent issues are:

1. What drivers and dynamics of variability and change are involved in terrestrial human-environment (T-H-E) systems

2. How is the provision of environmental goods and services affected by changes in terrestrial human-environment system?

3. What are the characteristics and dynamic soft vulnerability in terrestrial human-environment systems ?

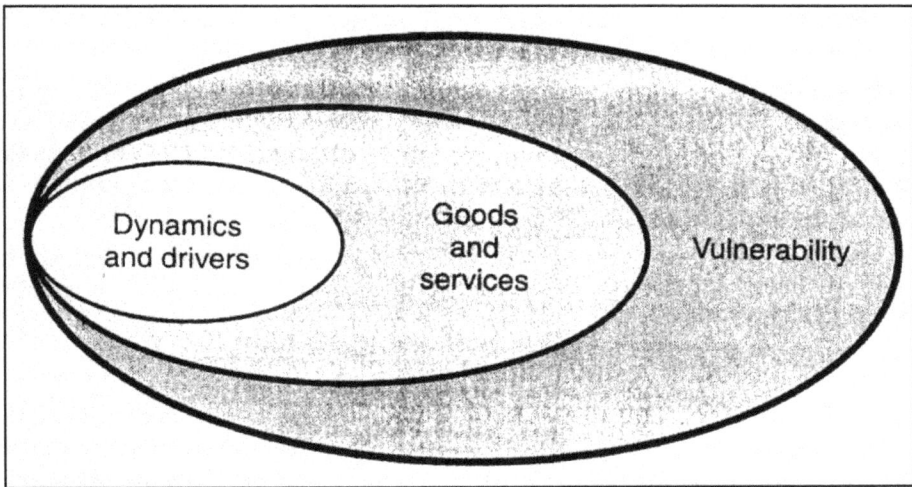

**Figure 3.3: Three nested issues that should be addressed to understand and evaluate terrestrial-human-environment systems (after Ojima *et al.*, 2002).**

The major need is to develop appropriate tools and knowledge to detect environmental change; to attribute the causes and underlying mechanisms of impacts and feedbacks, and to develop the capabilities to predict future trajectories of provision of ecosystem goods and services and of resulting vulnerability of T-H-E systems. These objectives can be achieved by critical studies of the main processes in terrestrial human-environments, including biogeochemical cycling (*e.g.* nitrous oxide fluxes to the atmosphere, nutrient loading on river systems), the response and effects of biodiversity (*e.g.*, effects of functional diversity on ecosystem productivity), biophysical phenomena (*e.g.*, landscape effects on mesoscale climate, soil erosion and changes in sediment delivery to the coastal zone), and human demographic and socio-economic dynamics (Ojima *et al.*, 2002).

# Land Atmosphere Interaction

Such atmospheric processes as physical climate variability and chemical deposition act as major constraints on both natural and anthropogenic terrestrial biogeochemical cycles. For instance, interannual variability in carbon uptake at ecosystem, regional and global scales is strongly affected by variation in climate and by the related feedbacks to physiology and productivity. The deposition of gases and aerosols from the atmosphere also influences ecosystem functioning by acting either as fertilizers or as toxic substances (Kabat *et al.*, 2002).

Many of the processes taking place on land cannot be considered independently from what is happening in the atmosphere. Atmospheric chemical and physical properties themselves are directly determined by the terrestrial biosphere, including humans. The emerging perspective within the Earth system science community is to treat the land-atmosphere interface as a system, defined by intricately linked processes and dependencies that have important implications for the functioning of the Earth as a whole (Kabat *et al.*, 2002).

The chief determinants of the coupled atmosphere-terrestrial biosphere system are energy partitioning, cloud processes, and aerosols and trace gas exchanges. For instance, land cover type constraints the regional build-up of clouds and precipitation patterns (Pielke *et al.*, 1997) at the landscape scale and has even been implicated in the changing dynamics of the Asian Monsoon System (Tyson *et al.*, 2001).

The vegetation cover can affect climate via atmospheric aerosols. Indeed, the vegetation itself probably plays a key role in influencing regional precipitation patterns.

# General Impacts

If GHG concentrations continue to rise, global warming would have wide-ranging effects on the environment and human society. These effects on individual countries and societies maybe assessed by estimating how the climate is likely to change at the local or regional level, and how this change might affect both the natural and the human environment on which people depend for their survival. The United Nations Environment Programme (UNEP) (1993) reviewed these effects and possible impacts of climate changes both on the natural environmental and on human society, taking as a deference standard the Intergovernmental Panel on Climate Change (IPCC) Working Group II assumptions: an increase in GHGs equivalent to a doubling of pre-industrial atmospheric carbon dioxide by 2025-2050 for a business as usual scenario; a corresponding increase in global mean temperature from 1.5 to 4.5°C; and a sea-level rise from 0.3 to 0.5 m by the year 2050 and of about 1 m by 2100 (see IPCC, 1990; Jager and Ferguson, 1991; Houghton *et al.*, 1992). The likely impacts of climate change were analysed first for the natural environment:

☆ Ice and snow;

☆ Oceans and coasts;

☆ The hydrological cycle; and

☆ Ecosystems and vegetation.

Secondly, the possible effects of climate change on human society were assessed, in terms of:

★ Water resources;

★ Food and agriculture;

★ Coastal dwellers;

★ Economic activity; and

★ Human settlements and health.

Of course, impacts will occur in more complicated ways, and impacts on one sector will affect those on other sectors.

## Impact on Natural Resources

Climate change is likely to exert considerable stress on natural resources throughout Asia. This continent has over 60 per cent of the world's population; its natural resources already are under stress. Most sectors in Asia show weak resilience to climate change. Many countries depend socio economically on such natural resources as water, forests, grassland, rangeland, and fisheries. The climate change sensitivity of some vulnerable sectors in Asia and the impacts of these limits are shown in Table 3.3.

**Table 3.3: Sensitivity of Selected Asian Regions to Climate Change (after WMO/ IPCC, 2001).**

| Climatic Elements and Seed Level Rise | Vulnerable Region | Primary Change | Impacts | |
|---|---|---|---|---|
| | | | Primary | Secondary |
| 0.5-2°C(10-to-45 rise) | Bangladesh Sundarbans | Inundation of about 15 per cent (~750km²) increase in salinity | Loss of plant species Loss of wildlife | Economic loss cm sea-level Exacerbated insecurity and loss of employment |
| 4°C (+10 per cent rainfall) | Siberian permafrosts | Reduction in continuous permafrost shift in southern limit of Siberian permafrost by ~ 100-200 km northward | Change in rock strength Change in bearing capacity Change in compressibility of frozen rocks Thermal erosion | Effects on construction Industries Effects on construction Effects on mining industry Effect on agricultural development |
| >3°C (>+20 per cent rainfall) | Water resources in Kazakhstan | Change in runoff | Increase in winter floods decrease in summer flows | Risk to life and property Summer water stress |
| ~ 2°C (-5 to 10 per cent rainfall; 45-cm sea-level rise) | Bangladesh lowlands | About 23-39 per cent increase in extent of inundation | Change in flood depth category Change in monsoon rice cropping pattern | Risk to life and property Increased health problems Reduction in rice yield |

The region's vulnerability to climate change is presented in Table 3.4 for selected categories of regions/issues.

Food insecurity poses the primary concern for Asia. Crop production and aquaculture would be threatened by thermal and water stresses, sea-level rise, increased flooding and strong winds associated with intense tropical cyclones. Areas in mid- and high latitudes will experience increases in crop yield, but yields in lower latitudes generally will decrease. A longer duration of the summer season should lead to a northward shift of the agroecosystem boundary in boreal Asia and favour an overall increase in agriculture productivity. Climatic variability and change also will affect scheduling of the cropping season, as well as the duration of the growing period of the crop. In India, acute water shortages combined with thermal stress may adversely affect wheat and, more severely, rice productivity even under the positive effects of elevated carbon dioxide in the future. Crop diseases such as wheat scab, rice blast, and sheath and calm blight of rice also are likely to become more widespread in temperate and tropical regions of Asia, should he climate become warmer and wetter. Adaptation measures to rescue the negative effects of climatic variability can involve changing the cropping calendar to take advantage of the wet period and to avoid the extreme typhoons and strong winds during the growing season.

Asia leads the world in aquaculture, producing over 75 per cent of all farmed fish, shrimp, and shellfish. Many wild stocks are under stress as a result of overexploitation, trawling on sea-bottom habitats, coastal development and pollution from land-based activities. Marine productivity is greatly affected by plankton shift, such as season shifting of sardine in the Sea of Japan, in response to temperature changes induced during ENSO. Storm surges and cyclonic conditions routinely lash the coastline and add heavier sediment loads to coastal waters. Effective conservation and sustainable management of marine and inland fisheries are needed at the regional level so that living aquatic resources can continue to meet regional and national nutritional needs (see Kumar, 2002).

Besides Asia, in several Latin American countries, such as Argentina, Brazil, Chile, Mexico and Uruguay, many studies based on GCMs and crop models have projected decreased yields for numerous crops (*e.g.* maize, wheat, barley, grapes) even when the direct effects of carbon dioxide fertilization and implementation of moderate adaptation measures at the farm level are duly considered. Predicted increases in temperature will reduce crops yields in these countries by shortening the crop cycle.

Freshwater availability is highly vulnerable to anticipated climate change. Surface runoff increases during winter and summer periods would be more pronounced in boreal Asia. Countries in which water use is more than 20 per cent of total potential water resources available will experience severe water stress in drought conditions. Surface runoff may decrease greatly in arid and semi-arid Asia under projected climate change scenarios.

Clinical change may not only change streamflow volume but also the temporal distribution of stream flows throughout the year. Water is going to be a scarce commodity in many south and south-east Asian countries, particularly where

**Table 3.4: Vulnerability of key sectors to climate change impacts for selected subregions in Asia. Asterisks point to general confidence-level rankings (after WMO/IPCC, 2001).**

| Regions | Food and Fiber | Biodiversity | Water Resources | Coastal Ecosystems | Human Health | Settlements |
|---|---|---|---|---|---|---|
| Boreal Asia | Slightly resilient**** | High vulnerable*** | Slightly resilient*** | Slightly resilient** | Moderately vulnerable** | Slightly or not vulnerable*** |
| Arid and semi-arid Asia-CA | Highly vulnerable**** | Highly vulnerable** | Highly vulnerable**** | Moderately vulnerable** | Moderately vulnerable*** | Moderately vulnerable*** |
| Tibetan Plateau | Slightly or not vulnerable** | Highly vulnerable*** | Moderately vulnerably** | Not applicable | No information | No information |
| Temperate Asia | Highly vulnerable**** | Moderately vulnerable*** | Highly vulnerable**** | Highly vulnerable**** | Highly vulnerable*** | Highly vulnerable*** |
| Tropical Asia-SA | Highly vulnerable**** | Highly vulnerable*** | Highly vulnerable**** | Highly vulnerable**** | Moderately vulnerable*** | Highly vulnerable**** |
| South-East Asia | Highly vulnerable**** | Highly vulnerable*** | Highly vulnerable**** | Highly vulnerable**** | Moderately vulnerable*** | Highly vulnerable*** |

reservoir facilities to store water for irrigation are insufficient. Growing populations and concentration of populations in urban areas are bound to exert strong pressures on water availability and water quality.

Tropical cyclones and storms are taking a heavy toll on life and property both in India and Bangladesh. Any increase in the intensity of cyclones combined with sea-level rise would spell more loss of life and property in low-lying coastal areas in cyclone-prone countries (Table 3.5).

**Table 3.5: Potential land loss and population exposed in Asian countries for selected magnitudes of sea-level rise, assuming no adaptation (after WMO/IPCC, 2001).**

| Country | Sea-level Rise (cm) | Potential Land Loss | | Population Exposed | |
|---------|---------------------|------------|------------|------------|------------|
|         |                     | $(km^{-2})$ | (per cent) | (million) | (per cent) |
| Bangladesh | 45 | 15668 | 10.9 | 5.5 | 5.0 |
|            | 100 | 29846 | 20.7 | 14.8 | 13.5 |
| India | 100 | 5763 | 0.4 | 7.1 | 0.8 |
| Indonesia | 60 | 34000 | 1.9 | 2.0 | 1.1 |
| Japan | 50 | 1412 | 0.4 | 2.9 | 2.3 |
| Malaysia | 100 | 7000 | 2.1 | >0.05 | >0.3 |
| Pakistan | 20 | 1700 | 0.2 | n.a. | n.a. |
| Vietnam | 100 | 40000 | 12.1 | 17.1 | 23.1 |

# Streamflow and Water Availability in Changing Climate

Water availability on the continents is crucial for human health, economic activity, ecosystem function and geophysical processes (vorosmarty *et al.*, 2000; Shiklomanov and Rodda, 2003; Reiter *et al.*, 2004; Mooney *et al.*, 2005). The saturation vapour pressure of water in air is very sensitive to temperature, so that perturbations in the global water cycle can accompany climate warming (Allen and Ingram, 2002). Regional patterns of warming-induced changes in surface hydro climate are not only very complex but also much less certain than those in temperature, especially in light of the regional increases and decreases that are expected in precipitation and runoff. Streamflow constitutes a temporally lagged, spatial integral of runoff over a river basin. When averaged over many years, runoff generally equals the difference between precipitation and evapotranspiration and, hence the convergence of horizontal atmospheric water flux (Milly *et al.*, 2005). From a resource perspective, runoff serves as an index of sustainable water availability. But streamflow is highly vulnerable to anthropogenic disturbances, which sometimes generate spurious (no climatic) changes; the most significant of these disturbances can be associated with the diversion of water for the irrigation of cropland.

Milly *et al.* (2005) showed that an ensemble of 12 climate models exhibits qualitative and statistically significant skill in simulating observed regional patterns of twentieth century multidecadal changes in streamflow. These models project 10 to 40 per cent increases in runoff in eastern equatorial Africa, high-latitude North

America and Eurasia, and 10 to 30 per cent decrease din runoff in southern Africa, southern Europe, the Middle east and mid-latitude western North America by the year 2050. These changes is sustainable water availability would have significant regional-scale consequences not only for economies but also for ecosystems.

On the basis of their modeling work and data analysis. Milly *et al.* (2005) suggested that a significant part of the twentieth-century hydrocliamtic change was externally forced, that larger changes should be expected in the coming decades, and that current climate models can help to characterize future changes. For the future, it may be wise to include projections of forced hydroclimatic change as factors in assessments of water availability, thereby facilitating their consideration not only in water management by also in economic and ecological assessments and planning (Milly *et al.*, 2005).

## Impacts of Human Activities

Human activities can alter the global atmosphere is many ways. The atmospheric composition is changing and the consequences of these changes are quite serious. Three major issues in this context are:

1. Stratospheric ozone depletion;
2. The greenhouse effect; and
3. The global spared of air pollution

The above-mentioned global issues are interlinked. The growth of the global population and the subsequent increased demand for energy, food and water is one specific cause of these problems. Stratospheric ozone depletion and the greenhouse effect are linked because chlorofluorocarbons, which play a major role in the destruction of stratospheric ozone, are also potent GHGs. Moreover, the stratospheric cooling expected as a result of the greenhouse effect will alter stratospheric chemistry and so ozone depletion rates. Other causal links between the global atmospheric issues include, for instance, that fossil fuel combustion not only adds GHGs to the atmosphere, but also emits other pollutants which can be transported to long distances in the atmosphere, giving rise to transboundary pollution.

Many of the concern about rising concentrations of atmospheric carbon dioxide relate to how climate may change. But there may also be some direct biological effects. In terrestrial ecosystems, extra atmospheric carbon dioxide can have a fertilizing effect through increased photosynthesis. But such as increase seems unlikely in most marine systems because most algae use bicarbonate ions rather than carbon dioxide as a photosynthetic substrate. There have, however, been indications of a negative biological response in the marine environment. Calcification rates of reef-building corals and coralline algae are depressed by increased levels of carbon dioxide (Riebesell *et al.*, 2000). Riebesell *et al.*, studied the influence of carbon dioxide on the calcification of coccolithophorid algae that are widely distributed in the surface layer of coastal waters and the open ocean. The calcification rate of two species of coccolithophorids maintained in culture was lower at carbon dioxide levels predicted for the year 2100 than at reindustrialize levels. Natural communities sampled in the

North Pacific also showed a similar response. Coccolithophorids may be the world's most important producers of calcium carbonate ($CaCO_3$). They form low-magnesium calcite shells and their distribution extends into sub polar waters (Gattuso and Buddemeier, 2000).

The reef-building corals and coralline algae are sensitive to high carbon dioxide levels and are botton-dwelling, found usually in tropical and subtropical regions. They typically form skeletons of aragonite or high-magnesium calcite. Together, the coccolithophorids, reef-building corals and coralline algae are responsible for well over half of the world's calcium carbonate production. Given their widespread distribution, it is likely that the carbonate biogeochemistry, and possibly the ecology, of the entire world ocean will be altered by rising $CO_2$ levels.

Atmospheric carbon dioxide equilibrates rapidly with the surface layer of the ocean, where most additional carbon dioxide combines with carbonate ions:

$$CO_2 + CO_3^{2-} + H_2O \longrightarrow 2HCO_3^-$$

This decreases the concentration of $CO_3^{2-}$, a building block of calcium carbonate.

Decreased calcification in response to increased concentration of carbon dioxide has now been reported across phylogenetically distant groups that precipitate different types of calcium carbonate crystals (low-magnesium calcite, high-magnesium calcite and aragonite), either through internal or external calcification. All of the calcifying organisms shown to have a strong saturation-state response are fighter plants or animals containing photosynthetic symbionts.

Rising carbon dioxide levels have two antagonistic effects on its fluxes mediated by calcification. The amount of carbon dioxide generated by calcification will decrease as a result of the decreased rate of calcium carbonate precipitation. But increased carbon dioxide concentration also shifts the seawater carbonate equilibria with the result that more carbon dioxide is released per mole of calcium carbonate precipitated. These two effects could be balanced in coral reefs, but the response of pelagic calcification could lead to increased carbon dioxide storage capacity in the upper ocean. Calcium carbonate compensation is its dissolution that maintains a balance between inputs from land and burial in deep-ocean sediments.

The Information of calcareous skeletons by marine planktonicorganisms and their subsequent sinking to depth generates a continuous rain of calcium carbonate to the deep ocean and underlying sediments. This process helps to regulate marine carbon cycling and ocean-atmosphere carbon dioxide exchange. The present rise in atmospheric carbon dioxide levels causes significant changes in surface ocean pH and carbonate chemistry. Such changes slow down calcification in corals and coralline macro algae, but the majority of marine calcification occurs in plank tonic organisms.

Riebesell *et al.* (2000) reported reduced calcite production at increased carbon dioxide concentrations in unialgal cultures of two dominant marine-calcifying phytoplankton species, *Emiliania huxleyi* and *Gephyrocapsa oceanica*. This was accompanaied by an increased proportion of abnormal coccoliths and incomplete coccospheres. Diminished calcification led to a reduction in the ratio of calcite precipitation to organic matter production. According to Riebesell *et al.* (2000), the progressive increase in atmospheric carbon dioxide concentrations may slow down the production of calcium carbonate in the surface ocean. As the process of calcification releases carbon dioxide to the atmosphere, the response observed by them could potentially act as a negative feedback on atmospheric carbon dioxide levels.

*Human activities*–mainly burning of fossil fuels and changes in land cover-change the concentration of atmospheric constituents of the surface that absorb or scatter radiant energy. Most of the observed warming over the last five decades appears to have been due to the increase in GHG concentrations. Future climate changes may include more warming, changes in precipitation patterns and amounts, sea-level rise, and changes in the frequency and intensity of some extreme events.

Many Earth systems that sustain human societies are quite sensitive to climate. Impacts are expected in ocean circulation; sea level; the water cycle; carbon and nutrient cycles; air quality; the productivity and structure of natural ecosystem; the productivity of agricultural, grazing, and timber lands; and the geographic distribution, behaviour, abundance, and survival of plant and animal species, including vectors and hosts of human disease. Change in these systems in response to climate change, as well as direct effects of climate change on humans would affect human well-being, positively and negatively.

Welfare is impacted through changes in supplies of, and demand for, water, food, energy, and other goods derived from these systems; changes in incomes; changes in loss of property and lives from extreme climate phenomena; and changes in human health. Climate change will affect sustainable development in different parts of the world and many further increase existing inequalities. Impacts will vary in distribution across people, places, and times, thereby raising important issues about equity. Indeed the stakes are quite high.

On the other hand, the risks associated with climate change cannot be definitely established as they are a function of the probability and magnitude of different types of impacts. Possible impacts include those threatening irreversible damage to or loss of some systems within the next century; modest impacts to which systems may readily adapt; and impacts that could prove beneficial for some systems (WMO/ IPCC, 2001).

Figure 3.4 illustrates the scope of the climate change assessment. Human activities that change the climate expose natural and human systems to a changed set of stresses or stimuli. Those systems which happen to be sensitive to these stimuli are affected by the changes and so can trigger autonomous, or expected, adaptations. These adaptations reshape the residual or net impacts of climate change.

Policy responses to impacts already perceived or in anticipation of potential future impacts can assume the form of planned adaptations to mitigate adverse effects or enhance beneficial ones. Policy responses also can involve actions to mitigate climate change through GHG emission reductions and enhancement of sinks.

## Methods and Tools of the Assessment

Climate change impacts, adaptations, and vulnerability are assessed by means of a large variety of methods and tools which have improved detection of climate

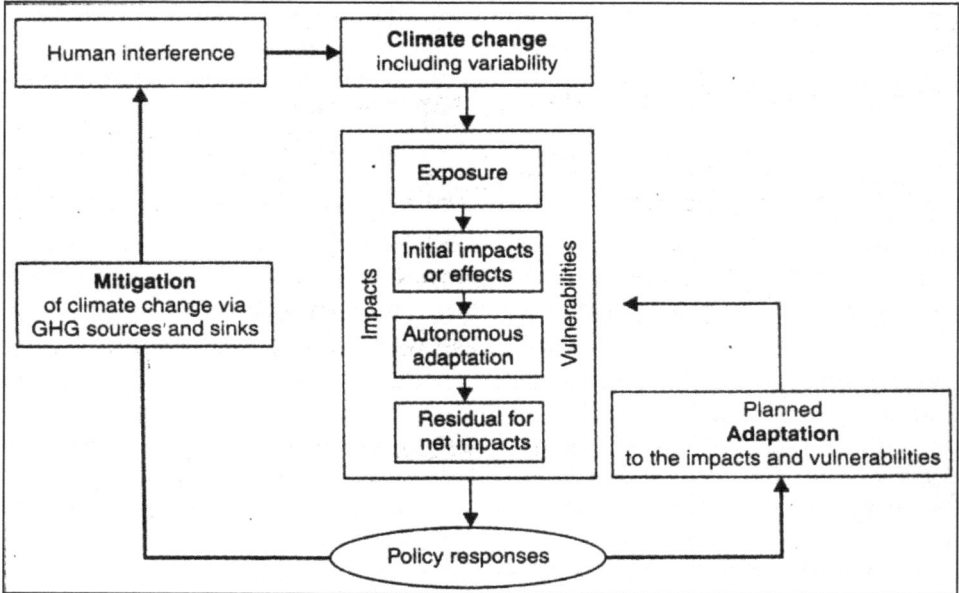

**Figure 3.4: Policies/responses to climate change assessment (after WMO/IPC, 2001).**

change in biotic and physical systems and generated many new findings. Some tools effectively address the human dimensions of climate as both causes and consequences of change and can deal directly with cross-sectoral issues concerning vulnerability, adaptation, and decision making. Methods and tools are being increasingly employed for costing and valuing effects, treating uncertainties, integrating effects across sectors and regions, and applying appropriate frame works for evaluating adaptive capacity.

## Using Indicator Species or Systems for Detecting Responses to Climate Change

Indicator species (*e.g.* butterflies,m penguins, frogs, sea anemones) may be used to detect responses to climate change and to infer its more general impacts on natural systems (*e.g.* in native meadows, coastal Antarctica, tropical cloud forest, and the Pacific rocky inertial, respectively).

## Integrated Assessment

The internal assessment is an interdisciplinary process that combines, interprets, and communicates knowledge from diverse scientific disciplines form the natural and social sciences for investigating and understanding causal relationships within and between complicated systems. Such assessments employ computer-aided modeling, scenario analyses, simulation gaming and participatory integrated assessments as well as qualitative assessments based on experience and expertise.

## Scenarios of Future Change

A scenario represents a coherent, internally consistent, and plausible description of a possible future state of the world. Scenarios offer alternative view of future conditions that appear likely to influence a given systems or activity. Climate scenarios describe the forcing factor of focal interest, whereas no climatic scenarios provide the socioeconomic and environmental context within which climate forcing operates.

## Nonclimatic Scenarios

Nonclimatic scenarios describing future socioeconomic, land-use, and environmental changes are crucial for characterizing the sensitivity of systems to climate change, their vulnerability, and the capacity for adaptation. Socioeconomic scenarios have been used more frequently for projecting GHG emissions than for assessing climate vulnerability and adaptive capacity. These scenarios usually identify several different topic or domains, for example, population or economic activity, as well as background factors such as the structure of governance, social values, and patterns of technological change; they make it possible to create baseline socioeconomic vulnerability, preclimate change; determine climate change impacts; and assess post-adaptation vulnerability.

## Land-use and Land-cover Change Scenarios

Land-use change and land-cover change (LUC-LCC) involve several processes central to the estimation for climate change and its impacts. The LUC-LCC influences carbon fluxes and GHG emissions, which directly change atmospheric composition and radioactive forcing properties. LUC-LCC modifies land-surface characteristics and, indirectly, climatic processes. Land-cover modification and conversion sometimes change ecosystem properties and vulnerability of ecosystems to climate change. Some strategies for mitigating GHG emissions involve land cover and changed land-use practices. Many different LUC-LCC scenarios have been created; most do not address climate change issues explicitly, but they do focus on other issues (*e.g.*, food security and carbon cycling).

## Environmental Scenarios

The environmental scenarios deal with changes in environmental factors other than climate that will occur in the future regardless of climate change. As these factors could modify the impacts of future climate change, scenarios are needed to think of possible future environmental conditions such as atmospheric composition (*e.g.* carbon dioxide, tropospheric ozone, acidifying compounds, and ultraviolet-B radiation); water availability, use, and quality and marine pollution.

## Sea-level Rise Scenarios

Sea-level rise scenarios are needed to asses a variety of threats to human settlements, natural ecosystems, and landscape in the coastal zone. Sea-level rise scenarios with reference to movements of the local land surface arouse the greatest interest for impact and adaptation assessments.

## Climate Scenarios

Three chief types of climate scenarios aid impact assessments; incremental, analogue, and climate model-based. Incremental scenarios involve simple adjustments of the baseline climate according to anticipated future changes; they help in testing system sensitivity to climate. Since they involve arbitrary adjustments, they are usually not realistic meteorologically. Analogues of changed climate from the past record or from other regions are fairly difficult to identify and are seldom applied, but they sometimes offer useful insights into impacts of climate conditions outside the present-day range.

The commonest scenarios use outputs from general circulation models (GCMs) and usually are developed by adjusting a baseline climate (typically based on regional observations of climate over a reference period (*e.g.*, 1961-1990) by the absolute or proportional change between the simulated present and future climates. The vast majority of scenarios represent changes in mean climate, but some recent ones also have included changes in variability and extreme weather events, which may mean important impacts for some systems.

## Scenarios of the 21st Century

In the year 2000, the IPCC completed a Special Report on Emissions scenarios (SRES). These scenarios consider the period 1990 to 2100 and include several socioeconomic assumptions (*e.g.* global population and gross domestic product). Their implications for other aspects of global change also have been assessed (Table 3.6).

## Human Diseases

Many diseases occur during certain seasons or erupt from unseasonal floods, drought or other changes in weather. Emerging concerns about global warming, accompanied by greater climate variability, have in recent years prompted many studies of diseases fluctuations related to short-term or interannual climate oscillations (*e.g.* from weather extreme driven by *El Nino*). Yet, there have been doubts as to whether or not there has been any documented change in human disease trends in response to long-term climate change, since warming has already occurred over the last century (Folland and Karl, 2001; Easterling *et al.*, 1997).

In the year 2002, Rodo *et al.*, found a strong relationship between progressively stronger *El Nino* events and cholera prevalence in Bangladesh, spanning a 70-year period. This study seems to represent the first evidence that warming trends over the last century are affecting human disease (Patz, 2002). Rodo *et al.* (2002) found that the association of cholera incidence in the earlier half of the century (1893-1940) is weak and uncorrelated with ENSO; whereas late in the century (1980-2001)., the relationship is strong and consistent with ENSO. Past climate change, therefore, may have already affected cholera trends in the region through intensified ENSO events (Patz *et al.*, 2002).

According to the IPCC, since the late 1950s global average surface temperature has increased by 0.6°C, snow cover and ice extent have diminished and ocean heat content has increased. Also, sea level has risen on average by 10 to 20 cm during the past century. The Indian ocean has progressively warmed since 1960 (Levitus *et al.*, 2000) and the subcontinent has warmed by 2 to 3°C over the last century. There has been an increase over time in the frequency and amplitude in the ENSO. As the rate of change in climate is faster now than in any period in the last thousand years, the finding of progressively more intense ENSO in the region is highly pertinent to health professionals.

**Table 3.6: The SRES scenarios and their implications for atmospheric composition, climatic and sea level. Values of population, GDP and per capita income ratio (a measure of regional equity) are those applied in integrated assessment models used to estimate emissions (after WMO/IPCC, 2001).**

| Date Global Population $(10^{12}$ (ppm)$^d$ | | Global GDP Temperature (ppm)$^e$ | Per Capita Income Ratio$^c$ Change $(OC)^f$ | Ground-level $O_3$ Sea-level Rise | $CO_2$ Concentration (Billions)$^a$ | Concentration | Global USS $(yr^{-1})^b$ |
|---|---|---|---|---|---|---|---|
| 1990 | 5.3 | 21 | 16.1 | - | 354 | 0 | 0 |
| 2000 | 6.1-6.2 | 25.28 | 12.3-14.2 | 40 | 367 | 0.2 | 2 |
| 2050 | 8.4-11.3 | 59-187 | 2.4-8.2 | ~60 | 463-623 | 0.8-2.6 | 5-32 |
| 2100 | 7.0-15.1 | 197-550 | 1.4-6.3 | >70 | 478-1099 | 1.4-5.8 | 9-88 |

a   values for 2000 show range across the six illustrative SRES emissions scenarios; values for 2050 and 2100 show range across all 40 SRES scenarios.

b   see footnote a; gross domestic product (trillion 1990 US$ yr$^{-1}$).

c   see footnote a; ratio of developed countries and economies-in-transition (annex 1) to developing countries (non-Annex 1).

d   model estimates for industrialized countinents of northern hemisphere assuming emission for 2000, 2060 and 2100 from the A1F and A2 illustrative sires emissions scenarios at high and of the sires range.

e   observed 1999 values for 1990, 2050, and 2100 are from simple model runs across the range of 35 fully quantified sires emissions scenarios and accounting for uncertainties in carbon cycle feedback related to climate sensitivity.

f   Change in global mean annual temperature relative to 1990 averaged across simple climate model runs emulating results of seven AOGCMs with an average climate sensitivity of 2.8C for the range of 35 fully quantified sres emissions scenarios.

g   based on global mean temperature changes but also accounting for uncertainties in model parameters for land ice, permafrost and sediment deposition.

Seasonality in disease incidence often point stop an association with weather factors. Epidemics of meningococcal meningitis in sub Saharan Africa consistently erupt during the hot dry season and weaken or vanish soon after the onset of the rainy season. Mosquito-borne diseases, such as dengue fever, show strong seasonal patterns, transmission being highest in the months of heavy rainfall and humidity. Enteric diseases also show significant seasonal fluctuations (Patz, 2002).

Colwell (1996) reported a connection between sea surface temperature (SST), marine ecology, and human cholera in Bangladesh. Copepods (or zooplankton) which feed on algae, serve as reservoirs for *Vibrio cholerae* and other enteric pathogens. This observation may explain why cholera follows seasonal warming of SST that can enhance plankton blooms. *Vibrio* species are influenced by temperature and salinity which, along with SST, are consistent with the role played by sea surface height (Lobitz *et al.*, 2000).

# Effects of Natural and Human Systems on Climate Change and Vulnerability

## Hydrology and Water Resources

The impact of climate change on stream flow and groundwater recharge varies regionally and between different climate scenarios, largely following projected changes in precipitation. One consistent projection across many climate change scenarios is that annual mean stream flow in high latitudes and in South-East Asia will increase, but in Central Asia, the area around the Mediterranean, southern Africa, and Australia will decrease. The retreat of most glaciers may accelerate, and many small glaciers would disappear. In general, the projected changes in average annual runoff are weaker than impacts based solely on temperature change because precipitation changes show grater variation between scenarios. At the catchment scale, the effect of a given change in climate varies with physical properties and vegetation of catchments, and may be in addition to land-cover changes.

Approximately 1.8 billion people (about one-third of the world's population) are living in water-stressed countries (defined as using more that 20 per cent of their renewable water supply). This number may shoot up to 5 billion by # the year 2025, depending on the rate of population growth. The projected climate change may further decrease the streamflow and groundwater recharge in many of these water-stressed countries-for instance, in central Asia and southern Africa-but may increase it in some others.

Although demand for water has been increasing due to population growth and economic development, it is falling in some countries because of greater efficiency of use. While climate change may not markedly affect municipal and industrial water demands, it could substantially affect irrigation withdrawals, which depend on how increases in evaporation are offset or exaggerated by changes in precipitation. Higher temperatures, hence higher crop evaporative demand, point towards increased irrigation demands. The strongest vulnerabilities are expected to be in unmanaged water systems and systems that are currently stressed or unsustainably managed due to policies that discourage efficient water use and protection of water quality, inadequate watershed management, failure to manage variable aster supply and demand, or lack of sound professional guidance. In unmanaged systems, the capacity or ability to buffer the effects of hydrologic variability on water quality and supply is usually lacking. In unsustainably managed systems, water and land uses add stresses that heighten vulnerability to climate change (see WMO/IPCC, 2001).

## Agriculture and Food Security

There is a wide variation in crop yield responses to climate change, depending upon species and cultivar; soil properties; pests, and pathogens; the direct effects of carbon dioxide on plants, and interactions among carbon dioxide, air temperature, water stress, mineral nutrition, air quality, and adaptive responses. Although increased carbon dioxide concentration sometimes stimulates crop growth and yield, this benefit does not always overcome the bad effects of excessive heat and drought. Some costs will certainly be involved in coping with climate-induced yield losses and adaptation of livestock production systems. There agronomic and husbandry adaptation options may include adjustments to planting dates, fertilization rates, irrigation applications, cultivar traits, and selection of animal species.

If autonomous agronomic adaptations are also considered, crop modeling assessments indicate that climate change can elicit generally positive responses at less that a few degrees Celsius warming and generally adverse responses for more than a few degrees Celsius mid-latitude crop yields. Yields of some crops in tropical areas would decline generally with even very small increases in temperature, because such crops are already near their maximum temperature tolerance and dryland/rainfed agriculture predominates. Where there is also substantial decrease in rainfall, tropical crop yields would suffer more. With autonomous agronomic adaptation, crop yields in the tropics tend to be less adversely affected by climate change than without adaptation, but they still tend to remain below level estimated with current climate. Global mean annual temperature increases of a few degrees Celsius or greater may push up food prices due to a slowing in the expansion of global food supply relative to growth in global food demand.

## Terrestrial and Freshwater Ecosystems

Vegetation modeling studies have shown the potential for significant disruption of ecosystems under climate change. Migration of ecosystems or biomes as discrete units may not take place; but at a given site, species composition and dominance will change. The results of these changes will lag behind the changes in climate by years to decades to centuries. Distributions, population sizes, population density, and behaviour of wildlife will be affected directly by changes in global or regional climate and indirectly through changes in vegetation. Climate change will lead to poleward movement of the boundaries of freshwater fish distributions along with loss of habitat for cold-and cool-water fishes and gain in habitat for warm-water fishes (WMO/IPCC, 2001). In fact, many species and populations are already at high risk, and could be exposed to greater risk by the synergy between climate change rendering portions of current habitat unsuitable for many species and land use change fragmenting habitats thereby creating barriers to species migration. Without appropriate management, these pressures will cause some species currently classified as 'critically endangered' to become extinct and the majority of those labeled 'endangered or vulnerable' to become rarer, and thereby closer to extinction, in the 21st century. Some suitable strategies for minimizing risks to species are: (*a*) establishment of refuges, parks, and reserves with corridors that allow migration of species, and (*b*) use of captive breeding and translocation. But these options may suffer from limitations imposed by high costs. Terrestrial ecosystems are generally storing increasing amounts of carbon, through some small productivity gains; it appears that the terrestrial uptake may be due more to change in uses and management of land than to the direct effects of elevated carbon dioxide and climate.

## Human Health

A number of vector-and water-borne infectious diseases are sensitive to changes in climatic conditions. Under various climate change scenarios, it appears that there would be a net increase in the geographic range of potential transmission of malaria and dengue-two common vector-borne infections. Even within their current ranges, these and other infectious diseases may increase in incidence and seasonality. In all cases, however, actual disease occurrence is greatly influenced by local environmental conditions, socioeconomic circumstances, and public health infrastructure. Projected climate change is likely to be accompanied by an increase in heat waves, and exacerbated by increased humidity and urban air pollution. The impact would be greatest in urban populations, affecting particularly the elderly, sick and those lacking access to air-conditioning. Any increase in flooding is going to aggravate the risk of diarrhoeal and respiratory diseases and, in developing countries, hunger and malnutrition. Cyclones could produce devastating impacts in densely settled populations with inadequate resources.

## Human Settlements, Energy and Industry

Human settlements are affected by climate change in any of the following three main ways:

1. The economic sectors that support the settlement are affected from changes in resource productivity or in market demand for the goods and services produced there.
2. Physical infrastructure (*e.g.* energy transmission and distribution systems), buildings, transportation systems, and industries (*e.g.* agroindustry, tourism, and construction) may be directly affected.
3. Populations are affected by extreme weather, changes in health status, or migration.

Flooding and landslides strongly affect human settlements; these changes are driven by increases in rainfall intensity and, in coastal areas, sea-level rise. Riverine and coastal settlements are particularly at risk. Urban flooding poses problems wherever storm drains, water supply and waste management systems are not up to the mark.

Rapid urbanization in low-lying coastal areas everywhere increases population densities and the value of human-made assets exposed to coastal climatic extremes such as tropical cyclones. Settlements that are not economically well diversified, and those where a high percentage of income comes from climate-sensitive primary resource industries (agriculture, forestry and fisheries) are more vulnerable than more diversified settlements. Industrial, transportation, and commercial infrastructure are vulnerable to the same hazards as settlement infrastructure. Energy demand is likely to increase for space cooling and decrease for space heating, but the net effect is scenario and location-dependent. Some energy production and distribution systems may suffer from impacts that can reduce supplies or systems reliability, whereas some other energy systems may benefit.

## Aspects of the Adaptive Capacity, Vulnerability, and Key Concerns

Some aspects of the adaptive capacity, vulnerability, and key concerns for some regions are listed below (see WMO/IPCC, 2001).

# Asia

☆ Adaptive capacity of human systems is low and vulnerability is high in the developing(Asian) countries; the developed (Asian) countries are more able to adapt and less vulnerable.

☆ Extreme events have increased in temperate and tropical Asia, including floods, droughts, forests fires, and tropical cyclones.

☆ Decreases in agricultural productivity and aquaculture due to thermal and water stress, sea-level rise, floods and droughts, and tropical cyclones would adversely affect food security in arid, tropical, and temperate Asia; agriculture would expand and become productive in northern areas.

☆ Human health would be threatened by possible increased exposure to vector borne infectious diseases and heat stress in some parts.

☆ Sea-level rise and an increase in the intensity of tropical cyclones would displace several millions of people in low-lying temperate and tropical coastal areas; increased intensity of rainfall would increase flood risks in temperate and tropical Asia.

☆ Climate change would increase energy demand, decrease tourism attraction and influence transportation in some Asian regions.

☆ Climate change would strengthen threats to biodiversity due to land-use and land-cover change and population pressure. Sea-level rise would put ecological security at risk, including mangroves and coral reefs.

# Europe

☆ Adaptive capacity is generally high for human systems; southern Europe is more vulnerable than other parts of Europe.

☆ Summer runoff, water availability, and soil moisture may decrease in southern Europe, and would increase the gap between the north and drought-prone south; increases are likely in winter in the north and south.

☆ Half of alpine glaciers and large permafrost areas could disappear by end of the 21st century.

☆ River flood hazard will increase substantially with implications for human settlement, industry, tourism, agriculture, and coastal natural habitats.

☆ Northern Europe agriculture will be broadly benefited, but productivity will decrease in southern and eastern Europe.

☆ Upward and northward shift of biotic zones will take place. Loss of important habitats (wetlands, tundra, isolated habitats) could threaten some species.

☆ Higher temperatures and heatwaves may change traditional summer tourist destinations, and less reliable snow conditions may impact adversely on winter tourism.

# North America

☆ Adaptive capacity of human systems is generally high and vulnerability low, but come communities (*e.g.* indigenous peoples and those dependent on climate-sensitive resources) are more vulnerable; social, economic and demographic trends are changing vulnerabilities in subregions.

☆ Some crops would benefit from modest warming accompanied by increasing carbon dioxide, but effects would vary among crops and regions, including declines due to drought in some areas of Canada's Prairies and the US Great Plains, potential increased food production in areas of Canada north of current production areas, and increased warm-temperate mixed forest production. However, further warming would be harmful to crop yields.

☆ Unique natural ecosystems such as prairie wetlands, alpine tundra, and cold-water ecosystems will be exposed to risk and effective adaptation is unlikely.

☆ Sea-level rise would enhance coastal erosion, coastal flooding, loss of coastal wetlands and risk from storm surges, particularly in florida and much of the US Atlantic coast.

☆ Vector-borne diseases-including malaria, dengue fever, and Lyme disease-could extend their ranges; exacerbated air quality and heat stress morbidity and mortality would occur; socioeconomic factors and public health measures would largely determine the incidence and extent of health effects.

# Chapter 4

# Greenhouse Effect and Global Warming

## General Description

The amount of some trace gases, notably carbon dioxide ($CO_2$), nitrous oxide ($N_2O$), methane ($CH_4$), chlorofluorocarbons (CFCs) and tropospheric ozone, have been increasing in the atmosphere. These gases are transparent to incoming short-wave radiation but absorb and emit long-wave radiation and so influence the Earth's climate. They are referred to as greenhouse gases (GHGs).

The atmospheric 'greenhouse' works because the sun's energy reaches the Earth as shortwave radiation, passing fairly easily through the atmosphere to warm up land and oceans. But the heated-up Earth sends energy back toward space as long-wave radiation, and this is blocked to a considerable extent by certain atmospheric gases. The most troublesome is carbon dioxide, because it is released in large quantities by the burning of fossil fuel. The proportion of carbon dioxide in the air has risen from 280 parts per million (ppm) early in the 19th century to over 360 ppm today.

Brown (1996) felt that a tame, nicely balanced greenhouse effect, such as the Earth has enjoyed since the last Ice Age, is beneficial since it keeps the planet warmer than it would otherwise be by about 30°C. But everyone agrees that human activities are pushing in the direction of disruptive atmospheric change. Estimates now predict a warming of 1.5 to 4.5°C by the time carbon dioxide concentration has doubled from pre-industrial Revolution levels. This may well mean planetary disaster for the Earth's ecosystems and for human society.

The Earth is an astonishingly complex planet. For instance, whereas cold oceans absorb huge quantities of carbon dioxide, warmed oceans absorb much less. Increased warmth leads to increased evaporation, leading to more clouds, which can diminish heat absorbed from the sun.

The interesting concept of carbonaceous aerosol ahs recently emerged from atmospheric pollution studies, although standard nomenclature and terminology are still unsettled. Gelencser (2004) has described the nature and atmospheric role of carbonaceous aerosol particles.

Robinson (2006) surveyed the contribution of satellite data to the study of the ocean, focusing on the special insights that only satellite data can bring to oceanography. He has reviewed topics ranging from ocean waves to ocean biology, at scales ranging from basins to estuaries. He also explained how satellite data can be used operationally for tasks such as pollution monitoring or oil-spill detection.

Ebbin *et al.* (2005) made a systematic assessment of the international 200 mile exclusive economic zone. To data, 145 states have ratified the Law of the Sea Convention, and most have established the Exclusive Economic Zones (EEZs). Ebbin *et al.,* focused on the specific nature of the EEZ and the construction and evolution of institutions stemming from its introduction and specifically examined developments at local, national and international levels.

Urbanisation has reached unprecedented levels in the estuarine and coastal zone, particularly in the Asia Pacific region where mega-cities and mega-harbours are registering impressive growth. Wolanski (2005) reviewed the different solutions and pitfalls, successes and failures in a large number of ports and harbours in the Asia Pacific Region, and showed how science can provide ecologically sustainable solutions that apply wherever the growth of mega-harbours occurs.

Consider the burning of fossil fuels. The resulting increase in carbon dioxide can trap heat, but more carbon dioxide promotes plant growth, which offsets the greenhouse effect. Some questions that need to be asked are: How readily will the lucky, newly warm areas in high latitudes, for example, Canada's Northwest Territories-accept refuges from lands made uninhabitable by flooding or desertification? Will plant and animal populations migrate as well, and at the same pace, so that ecosystems survive? According to Brown (1996) and Schneider (1996), the prospect of continued global warming is worrisome indeed.

Our knowledge of the global biogeochemical cycles that determine atmospheric concentrations of methane, nitrous oxide and ozone is not up to the mark as a basis for policy decisions on how to reduce or limit the future growth of their concentrations.

Although changes in global climate due to increasing concentration of aerosols in the atmosphere have probably not been significant in the past, the possibility that they may become important in the future, particularly regionally, cannot be excluded. Future changes in aerosol concentrations cannot be projected with any certainty.

An evaluation of results from climate models leads to the conclusion that the increase in global mean equilibrium surface temperature due to increases of carbon dioxide and other GHGs equivalent to a doubling of the atmospheric carbon dioxide concentration is likely to be in the range of 1.5 to 5.5°C. The temperature change due to the changing concentrations of these GHGs up to the present is about one-half of the change calculated for the increase of the atmospheric carbon dioxide alone [about 70 parts per million by volume (ppmv)]. The effect of the other GHGs is equivalent to an additional increase of carbon dioxide by 40 to 50 ppmv.The concentrations of several of these gases have been increasing more rapidly than that of carbon dioxide. If their rates of increase continue unchanged during the next four to five decades, such a scenario would be equivalent to a doubling of atmospheric carbon dioxide concentrations well before the middle of the next century (Figure 4.1). Chlorofluorocarbons would then become the most important gases in addition to carbon dioxide, if no preventive measures are taken. On the other hand, their regulation would be easier to achieve than the limitation or reduction of carbon dioxide emissions.

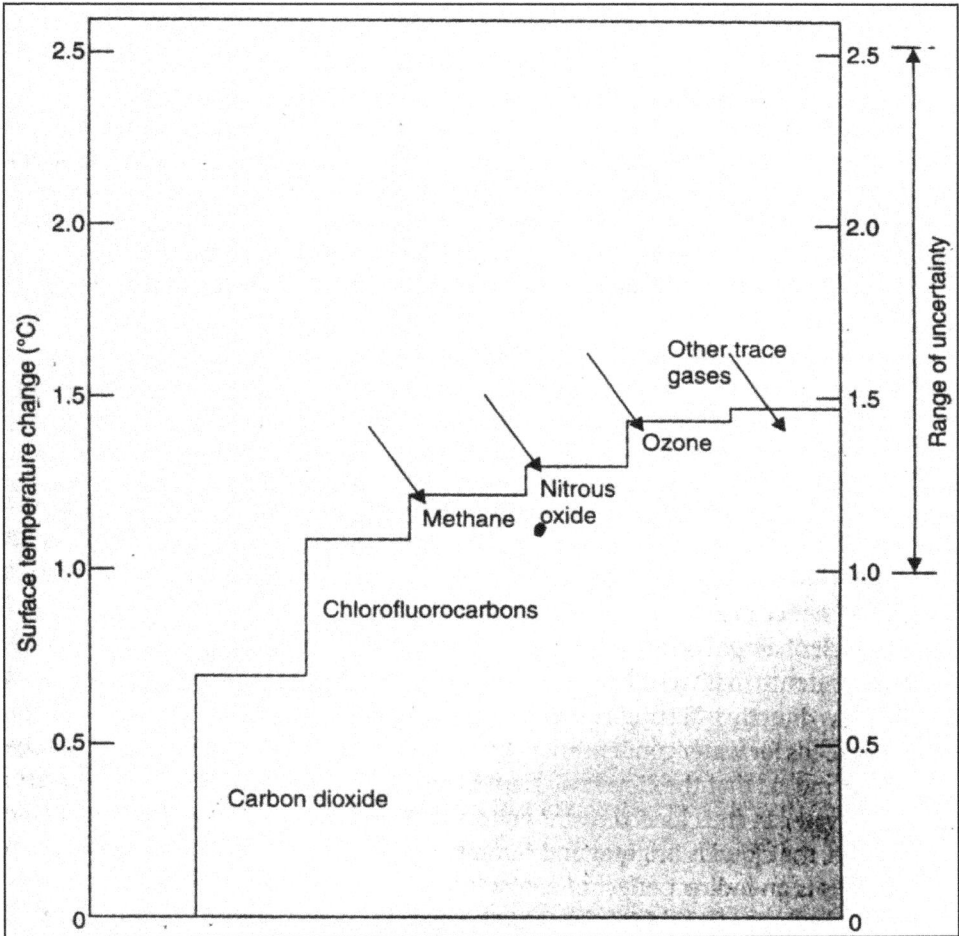

Figure 4.1: Cumulative equilibrium surface temperature warming due to increase in carbon dioxide and other trace gases from AD 1980 to 2030 as computed by a one-dimensional model. Due to feedback mechanisms, expected changes are 0.8 to 2.6 times the values given in this figure (Bolin *et al.,* 1991).

The observed increase in mean temperature during the last century (0.3-0.7°C) cannot be ascribed in a statistically rigorous manner to the increasing concentration of carbon dioxide and other GHGs although such a magnitude lies within the range of predictions (0.3–1.1°C)

The expected change of the global mean temperature due to a doubling of carbon dioxide is of about the same magnitude as the change of global temperature from the last glacial period to the present interglacial.

Continental or regional-scale changes in climate cannot yet be modeled with confidence, except that there are some indications that warming will be enhanced in high latitudes and that summer dryness may become more frequent over the continents at middle latitudes in the Northern Hemisphere.

## Uncertainty

One strong uncertainty in projecting global warming arises from the extremely complex problem of accounting for the effects of small particles in the Earth's atmosphere. Climate projections for the year 2100, compared with today, predict a global average temperature increase ranging from 5.8°C to 1.4°C. The uncertainty arises, in part from aerosols-tiny particles in the atmosphere- and their effect on clouds. Ackerman _et al._ (2004) concluded that the role of these particles in increasing the water content of clouds, and hence cloud reflectivity, is smaller than believed previously.

As a result of air pollution, the mass of small particles in the atmosphere is about 40 per cent greater than it would be in a pristine atmosphere. These particles reflect solar radiation, so directly affecting climate. When clouds are formed, these tiny particles also act as weeds for water condensation. Clouding in a region containing a large number of particles means that the cloud will have higher concentrations of smaller droplets if the liquid water in the cloud remains constant. As the smaller droplets have a higher surface area, the cloud is brighter and can reflect more solar radiation, which cools the climate. This is an indirect effect of aerosols on clouds (Penner, 2004).

Another indirect effect is the change in 'precipitation efficiency' that occurs because smaller droplets are less likely to collide against each other and form precipitation. This second indirect effect increases both the total cloud cover and the total amount of liquid that is held within clouds. The two effects increase the total reflected solar radiation even more. As some climate models are highly sensitive to this change, they predict a total cooling effect that is larger than the warming by GHGs: but the net effect of GHG and aerosol changes over the past hundred years is a warming (Penner, 2004).

## Global Carbon Budget

Although the year-to-year increase in the carbon dioxide concentration of the atmosphere is largely driven by combustion of fossil fuels and deforestation, natural processes such as exchange of carbon dioxide with the ocean, photosynthesis, respiration of living organisms, and decomposition also play a role. These processes, which constitute the Earth's carbon cycle, account for about 95 per cent of the total exchange of carbon dioxide with the atmosphere. In reindustrialize times, the global carbon cycle was quite balanced and the atmospheric carbon dioxide concentration remained stable. At present, approximately two-thirds of the carbon dioxide added to the atmosphere by human activities is sequestered by the natural processes-as bicarbonate in the ocean, wood in the forests, and organic matter in the soil. The remaining one third accumulates in the atmosphere, perturbing the radiative properties of the atmosphere and possibly altering the climate.

Carbon dioxide sequestration by the global carbon cycle is, therefore, an important factor mitigating the potential impact of human activity on climate. The future course of anthropogenic climate change will depend significantly on how these natural components of the Earth's carbon cycle respond in the future. Much effort is now focused on understanding where the carbon is going and in establishing a system that will permit scientists to monitor the dynamics of carbon cycling by the world's oceans and terrestrial ecosystems.

The most powerful approach developed to date is based on a 'top down' analysis of the dynamics of carbon dioxide and other tracers in the atmosphere. Flask samples of air are collected at frequent intervals at many stations all over the world and sent to central laboratories for analysis of oxygen and carbon dioxide concentrations and other trace gases. One critical measurement is the $^{13}C$ and $^{18}C$ isotopes of carbon dioxide. These global measurements need to be supplemented by process-level studies of isotopic effect sat the regional, ecosystem and plant scales. Such data are needed to calibrate and test the models of isotopic fractionation used to interpret the global observations.

A new technology for measurements of carbon dioxide concentration and isotopic composition of carbon dioxide in air samples has been evolved at the Carnegie institution of Washington. It is specifically intended for ecosystem and plant or leaf-scale measurements and it uses very small samples of air (less than 5 ml for a complete analysis) while achieving levels of accuracy and cross-calibration that approach those of the flask networks. In addition, the procedure can be automated. In combination with a mass spectrometer, the new technique allows scientists to calculate the carbon dioxide concentration in a sample.

The various manipulative treatments with added water and added nitrogen fertilizer mimic important global changes: many of the world's ecosystem are already heavily impacted by nitrogen deposition, and climate models almost universally predict that precipitation increases in parallel with temperature. Second, the water and nitrogen treatments provide probes for untangling the direct and indirect effects of the carbon dioxide and warming treatments. For example, carbon dioxide can increase plant growth directly by increasing photosynthesis, but it can also act indirectly by decreasing stomatal conductance and increasing soil moisture. Similarly, whereas warming can accelerate the decomposition of soil organic matter through metabolic stimulation of the soil microorganisms, it also can have the opposite effect by drying the soil.

The analysis following growth showed some patterns that the researchers expected and some they did not. One expected response was the earlier flowering of grasses in the heated treatments. However, the grasses grew the tallest in the treatments with both added water and added nutrients. Elevated carbon dioxide unexpectedly enhanced the success of shrub seedlings, suggesting that it could play a role in transforming the grassland to shrubland. So far, most of the significant responses are in the one-factor and two-factor treatments, rather than in those with three or four factors.

The Mauna Loa (Hawaii) atmospheric carbon dioxide measurements began an 1958 and happen to be the longest continuous record of atmospheric carbon dioxide concentrations available anywhere n the world. The Mauna Loa site is a most appropriate location for measuring undisturbed air because the possible local influences of vegetation or human activities on atmospheric carbon dioxide concentration are minimal there. From 1958 to 2000, the Mauna Loa record shows a 17 per cent increase in the mean annual concentration, from 316 ppmv of dry air to 369 ppmv. The increase in mean annual concentration from 1999 to 2000 was 1.1. ppmv. (The largest single yearly mump in the Mauna Loa record was the 2.9 ppmv increase from 1997 to 1998) (see http://cdiac.ornl.gov/trends/co2/sio-keel.htm).

Since the year 1751, approximately 270 billion tons of carbon have been emitted to the atmosphere from the consumption of fossil fuels and cement production. Half of these quantities were emitted since the mid-1970s. The 1998 fossil-fuel emission estimate for global carbon dioxide emissions-6,608 million metric tons of carbon-represents a 0.3 per cent decline from 1997. This small 1997-1998 decline is the first decline in the global record since a 1.6 per cent decline from 1991 to 1992.

Globally, liquid and solid fuels accounted for about 78 per cent of the emissions from fossil-fuel burning in 1998. Combustion of gas fuels (*e.g.* natural gas) accounted for 18.5 per cent (1220 million metric tons of carbon) of the total emissions from fossil fuels in 1998 and reflects a gradually increasing global utilization of natural gas. Emissions from cement production fell slightly, to 207 million metric tons of carbon, but this represents a 20-fold increase since the 1920s. Emissions from gas flaring for 1998 were estimated to be 47 million metric tons of carbon, well below the levels of the 1970s.

Collectively, emissions from cement production and gas flaring made up 4 per cent of total emissions for 1998.

The average surface air temperature of the globe has warmed approximately 0.5°C since the middle of the 19th century. Also, except for the last few years of the record, the warming varied in extent and magnitude across the globe, and a few areas have cooled since the 19th century (http://cdiac.ornl.gov/trends/emis/em_cont. htm). The Northern Hemisphere has warmed at a rate of 0.6°C/100 years, and the Southern Hemisphere at a rate slightly greater than 0.5°C/100 years. The warming rate for the globe is 0.58°C/100 years. In the global records, the 1990s were much warmer than the years in the rest of the record. In the global record, the ten warmest years have occurred since 1981. In descending order they are 1998, 1995, 1990, 1997, 1991, 1988, 1987, 1981, 1994 and 1989. The mean temperature deviation for the 1990s is 0.42°C.

The mean temperature deviation for the entire record is $-0.02°C$ including the 1990s and $-0.06°C$ without them. The warming in the 1990s is unprecedented (CDIAC Communications, March 2002, p.5).

Models of global climate change predict that increasing levels of atmospheric GHGs will raise average global temperatures and change regional levels of precipitation, also that the incidence of drought will increase with a warming global climate. Forests in those locations where evapotranspiration demand is high and is predicted to increase as temperatures rise, would be particularly vulnerable to declines in annual precipitation. Potential responses of such forests to future drought associated with climate change include a reduction in net primary production and stand water use, alongwith increased mortality of seedlings and saplings (CDIAC Communications, March 2002).

Attempts to 'balance' the global carbon cycle (*i.e.* reconcile the known sources and sinks of carbon) are hindered by two major unknowns: the flux between the atmosphere and the oceans and the flux between the atmosphere and terrestrial ecosystems. To address the latter, several investigators have attempted to estimate the flows of carbon between the atmosphere and both temperate and tropical ecosystems (Houghtor and Hackler, 2002).

This database consists of annual estimates, from 1850 through 1990, of the net flux of carbon between terrestrial ecosystems and the atmosphere resulting from deliberate changes in land cover and land use, especially forest clearing for agriculture and the harvest of wood for wood products or energy. The data are provided on a year-by-year basis for nine regions (North America, South and Central America, Europe, North Africa and the Middle East, Tropical Africa, the former Soviet Union, China, South and South-East Asia, and the Pacific Developed region) and the globe. Some data begin earlier than 1850 (*e.g.* for six regions, areas of different ecosystems are provided for the year 1700) or extend beyond 1990 (*e.g.* fuelwood harvest in South and Southeast Asia, by forest type, is provided through 1995).

The estimated global total net flux of carbon from changes in land use increased from 397 Tg of carbon (1 teragram = $10^{12}$ gram) in 1850 to 2187 Tg, or 2.2 Pg of carbon (1 petagram = $10^{15}$ gram) in 1989 and then decreased slightly to 2103 Tg, or 2.1 Pg of carbon in1990. the global net flux during the period 1850-1990 was 124 Pg of carbon. During this period, the greatest regional flux was from South and Southeast Asia (39 Pg of carbon), while the smallest regional flux was from North Africa and the Middle East (3 Pg of carbon). For the year 1990, the global total net flux was estimated to be 2.1 Pg of carbon (Houghton and Hackler, 2002).

## Soil Carbon

Changes in climate and land use seem to be the main factors in the large-scale loss of carbon from soils in recent years. Globally, soils store some 300 times the amount of carbon now released annually through the burning of fossil fuels. Most of the carbon locked up in soils are insert and remain there. Bellamy *et al.* (2005) reported that soil carbon may be more vulnerable to changing climate and patterns of land use than believed hitherto. This carbon is termed as soil organic carbon (SOC). Organic carbon occurs in the upper layers of mineral soil as humus or above the mineral soil as peat or litter. Soils receive dead organic material (litter) from the plant cover. This litter is decomposed by the soil biota, partly mineralized, and subsequently released to the atmosphere in the form of carbon dioxide and methane, or by leaching into groundwater.

The input-output balance determines whether a soils is accumulating or losing carbon. The SOC contains diverse compounds and is divided into active and passive pools, the latter being more resilient to further degradation and possibly existing in soil for centuries. All factors that reduce biological activity, and that stabilize the SOC by physical protection or binding to calyh silicates or metals, will promote accumulation; those that increase biological activity and destablisation encourage degradation (Schulze and Freibauaer, 2005).

The highly complicated interplay of these factors has a strong bearing on the subtle balance and affects climate change substantially. In humid regions, global warming may increase microbial activity and accelerate the SOC mobilization. In drier areas, the converse may apply. Changing patterns of land use also have significant effects. The SOC losses occur when natural ecosystems are cultivated-because of degradation of soil fertility, intensified soil disturbance and reduced carbon input. If conservation measures are applied to degraded soils, the SOC content, however, can be maintained or even increased.

Bellamay *et al.* (2005) determined the changes in the SOC stock in the top 15 cm of soils in England and Wales during the past 25 years and found an unusually alarming SOC. Extrapolating to the entire United Kingdom (UK), Bellamy *et al.*, estimated annual losses of 13 million tons of carbon, equivalent to 8 per cent of the UK emissions of carbon dioxide in 1990, and as much as the entire UK reduction in carbon dioxide emissions achieved between 1990 and 2002 (12.7 million tons of carbon per year). These losses offset the past technological successes in reducing carbon dioxide emissions. The Kyoto Protocol does not obligate countries to account for changes in the stock of soil carbon. An effective climate policy should require a more comprehensive approach including all major carbon sources and sinks in the biosphere (Schulze and Freibauer, 2005), with emphasis on protecting existing pools of stored carbon (GACGC, 2004).

The work of bellamy *et al.* (2005) provide the first hint that regional climate variation may be contributing to a surprisingly large release of carbon dioxide from soils to the atmosphere.

## Active Carbon Cycle

While most of the Earth's carbon is stored in rocks and sediments, only a small fraction is 'active', *i.e.* circulates in the carbon reservoirs of the oceans, atmosphere and terrestrial and marine biospheres. But this small active fraction is crucial for the maintenance of life on Earth-it is within this pool of carbon that a series of biological, physical and chemical cycles constantly re-distribute the element, effectively transferring energy from sunlight into fuel for the biosphere. These processes make up the 'active carbon cycle' (Lowe, 2000). These cycles are vital for the maintenance of life on Earth.

Since the advent of the industrial and agricultural revolutions over 250 years ago, the flow and distribution of carbon between the various active carbon reservoirs has changed as a result of fairly rapid addition of fossil carbon to the atmosphere in the form of carbon dioxide and other carbon-containing gases. These gases are emitted by processes such as the burning of coal and oil, large-scale clearing of forests, widespread ruminant animal farming and changes in wetlands. It took about 220 years, 1750-1970, for atmospheric carbon dioxide to increase by about 45 ppm. However, it has taken only 30 years since measurements began at Baring Head, New Zealand, for atmospheric carbon dioxide to increase by the same amount. This large increase in growth rate is due to the dramatic escalation in fossil-fuel usage in the second half of the 20th century. This additional carbon exerts significant impacts on some physical and chemical properties of the atmosphere and appears to be the cause of current and projected climate change.

The rate of increase of atmospheric carbon dioxide measured of Baring Head has fluctuated around 1.5 ppm/yr and similar fluctuations are seen at other measurement sites. High growth rates in the late 1980s and low growth in the early 1990s are belived to be related to the effect of climate variations own the active carbon cycle. Gaining a better understanding of theses changes is crucial- it can generate information on how the carbon cycle may be affected by future climate change (Lowe, 2000).

Understanding the changes in the portioning of carbon dioxide between the reservoirs of the active carbon cycle is extremely important because storage of excess carbon dioxide in the deep oceans, for example, could be 'safar' for humanity- that carbon dioxide appears les likely to re-enter the atmosphere over a short time-scale. Conversely, carbon stored in terrestrial ecosystem may be only temporary, as warranted by projections for the degradation of forests as a result of global warming. Furthermore, forests are susceptible to future human intervention through clearing and other land management practices. Fossil-fuel usage adds about 6 billion GTC tons of carbon to the atmosphere each year. Of this, about half is taken up by the oceans and the biosphere. The remainder stays in the atmosphere, leading to an increase in the concentration of atmospheric carbon dioxide.

There is uncertainty as to where carbon dioxide from current fossil fuel emissions is going, and the effects of future climate change increase the uncertainties for future predictions. Some oceanographic studies suggest that the Southern Ocean is already changing in ways that will reduce its uptake of carbon dioxide. Reducing these uncertainties has strong relevance to policy aimed at long-term mitigation strategies for carbon dioxide emissions (Lowe, 2000). However, in spite of the uncertainties in understanding the physical environment and how it may respond, the largest uncertainties are in projections of how people might behave in future.

Recent studies of scenarios for social and economics development and their implications for carbon dioxide emissions point to the possibility that the historical links between population, gross domestic product (GDP) and emissions may not hold good in future. The rate of transfer of technology from developed to developing countries, subsidies in overseas energy markets and the worldwide growth and development of environmental value sare emerging as key factors. Research institutions have a duty to provide unbiased, accurate scientific information to society on human-induced disturbances to the active carbon cycle.

Indeed, understanding the nature and implications of human-induced disturbances to the active carbon cycle will undoubtedly be one of the major scientific challenges of the 21st century.

## Linking the Biophysical and Human Components of the Carbon Cycle

The dynamics of the global carbon cycle have been driven by increasingly complex drivers and controls over time. Before major human intervention, this cycle used to be driven by climate variability and the internal dynamics of the linked land-ocean-atmosphere system. But since some two centuries ago, industrialization and accelerating land-use change increased the complexity of the carbon cycle by adding fossil fuel. Humans now realize that changes in climate and its variability (through changes in the carbon cycle and other GHGs)may significantly influence not only their welfare but also the functioning of the Earth system. The development and implementation of institutions and regimes to manage the global carbon cycle has added new feedbacks in the global carbon cycle (Hibbard *et al.*, 2002). These are outlined below:

1. *Patterns and variability*: The current geographical and temporal distributions of the major stores and fluxes in the global carbon cycle.

2. *Processes, controls and interactions*: the underlying mechanisms and feedbacks that control the dynamics of the carbon cycle, including its interactions with human activities.

3. *Future dynamics of the carbon cycle*: The range of plausible trajectories for the dynamics of the carbon cycle into the future (Hibbard *et al.*, 2002).

## High-Arctic Carbon Sink

One of the best indicators of the health of our planet is the carbon budget of the High Arctic. Maxwell(1992) proposed a climate model that predicted that warming caused by elevated atmospheric carbon dioxide would be strongest and fastest in the Arctic. The most recent regional climate predictions are for an average Arctic temperature increase of 1.7°C by the middle of 21st century, and location-specific temperature increased of 1 to 4°C (see Soegard *et al.*, 2004). Because of strong feedback mechanisms in the Arctic, even moderate temperature changes may be amplified and can result in large environmental responses (Soegaard *et al.*, 2004). Since there are multiple and complex feedbacks, a systems approach, rather than a single process-study, is needed for determination of the High-Arctic carbon budget.

In Northern Scandinavia, during summer, the terrestrial ecosystem constitutes the upper step of the carbon cascade where carbon dioxide fixation takes place nearly 24 hours a day due to the midnight sun. Soegaard *et al.* (2004) recorded summer carbon dioxide fluxes for two contrasting vegetation types: a sedge dominated fen (1996-1999) and a dwarf shrub health (1997, 2000-2004). The temporal variation in carbon dioxide exchange showed carbon dioxide emission in late June when the snow disappears from both vegetation types. Carbon dioxide emission from the fen, however, was greater due to the higher soil carbon content. Combining various aerial carbon sequestration rates with the area extent of the two vegetation types, yielded an estimate of the terrestrial carbon balance for the summer (10g m$^{-2}$ per season) (Soegaard *et al.*, 2000). By applying a soil respiration model, the winter carbon budget was estimated, and summing the winter and summer budgets provided an estimate of the annual budget, which includes the carbon emitted as methane (15 g m$^{-2}$ yr$^{-1}$) from the fens covering 2 to 3 per cent of the area.

Carbon transport occurs through the High Arctic rivers during June-September, with the peak discharge occurring in June-July, associated with snowmelt. Carbon is transported to coastal waters in both dissolved forms and particulate forms [particulate organic carbon (POC)]. The dissolved carbon originates partly from the decomposition of soil organic matter and partly from the dissolution of soil carbonate minerals. Over the summer, the average carbon concentration is usually relatively constant at around 4-5 mg L$^{-1}$ although it can increase substantially after land-slides. POC transport accounts for about one-quarter of the dissolved carbon transport, but it nearly doubles at the time of the maximal biological activity in early August.

The open Arctic seas serve as a year-round carbon sink, pluming atmospheric carbon dioxide into the sea and thereby producing dissolved inorganic carbon (DIC). The uptake rate is chiefly controlled by the carbon dioxide difference across the air-sea interface ($dCO_2$), the presence of see-ice and the atmospheric wind forcing of the surface waters (Soegaard *et al.*, 2004).

## Greenhouse Effect

The greenhouse effect is best described in terms of the annual global average radiative energy budget of the Earth-atmosphere system. The Earth-atmosphere system radiates about 236 Wm$^{-2}$ (long-wave radiation) to space. This amount balances the incoming short-wave radiation from the sun. at a temperature of 15°C, the Earth's surface radiates about 390 Wm$^{-2}$. The reduction of the long-wave radiation to space as a result of the intervening atmosphere is referred to as the greenhouse effect. The most important atmospheric constituents contributing to the greenhouse effect are water vapour, carbon dioxide, and clouds. Without the greenhouse effect, but with the same amount of incoming solar radiation and the same albedo, the global average surface temperature would be about-19°C. With GHGs in the atmosphere, the long-wave radiation from the Earth's surface is partly absorbed and then re-emitted by the gases at the temperature of the air at their level. This has the effect of raisign the effective emitting level to several km above the surface where the temperature is about 30°C lower than near the Earth's surface. So, the Earth's surface temperature can be higher by about this amount (the global average surface temperature is about 15°C). The Earth and atmosphere together emit the same amount of radiation to space as in the absence of greenhouse gases (GHGs).

In recent years, this balance has been disturbed by the addition of GHG's to the atmosphere. An increase in the atmospheric concentration of a GHG's, changes the radioactive energy balance, which increases the temperature of the Earth's surface and lower atmosphere, and decreases the temperature in the upper atmosphere. The greenhouse effect caused by injection of billions of tons of carbon into the atmosphere each year by human activities explains why gases produced by human activity are likely to cause the Earth's average temperature to increase within the lifetimes of most people living today.

Atmospheric concentrations of carbon dioxide have increased by about 25 per cent since coal, oil, and gas became the primary sources of energy to fuel the Industrial Revolution. Carbon dioxide concentrations are currently increasing by about 0.4 per cent each year.

After water vapour, carbon dioxide is the most plentiful and effective GHG's. It occurs naturally but is also produced substantially during the combustion of fossil fuels, particularly coal. When the fuel is burned, its carbon is oxidized to carbon dioxide and released. Carbon dioxide also is released as forests are cleared and the organic matter is burned or allowed to decay. These human activities are injecting almost 6 billion tons of carbon into the atmosphere each year. When this figure is compared with the actual increases in concentrations of carbon dioxide (about 3 billion tons annually), scientists presume that about one-half of the carbon injected into the atmosphere is being absorbed by oceans and plant life and the other half remains in the atmosphere.

Routh estimates can be made for the percent of global warming between 1980 and 2030 arising from each gas and sector, assuming current trends (Table 4.1).

**Table 4.1: Approximate Contribution of Various Greenhouse Gases by Sector (%)**

| Sector | $CO_2$ | CFC | $CH_4$ | $O_3$ | $N_2O$ | Total |
|---|---|---|---|---|---|---|
| Energy | 35 | - | 4 | 6 | 4 | 49 |
| Deforestation | 10 | - | 4 | - | - | 14 |
| Agriculture | 3 | - | 8 | - | 2 | 13 |
| Industry | 2 | 20 | - | 2 | - | 24 |
| Per cent warming by gas | 50 | 20 | 16 | 8 | 6 | 100 |

## Greenhouse Gas Emissions

### Carbon Dioxide

Between 1860 and 1984 a cumulative total of about 183 billion tons of carbon was emitted as a result of fossil-fuel combustion and about 150 billion tons of carbon from deforestation and land-use changes.

The emission for the world as a whole increased from 4.8 billion tons of carbon in 1971 to about 6 billion tons of carbon in 1988.

When burned, various fossil produce different amounts of carbon dioxide for a given release of thermal energy: coal releases about 20 per cent more carbon dioxide than oil, which in turn releases more than natural gas.

Terrestrial ecosystems represent another source of carbon dioxide emissions as a result of human activities. When forests or grasslands are converted into farmland, organic matter is oxide and emitted as carbon dioxide into the atmosphere. It has been estimated that the annual emission from deforestation and land-use changes are in the range of 1.6 billion tons of carbon per year, which is considerably smaller than the annual rate of emissions from fossil-fuel combustion (about 6 billion tons of carbon in 1987).

The chief reservoirs of carbon are the atmosphere, the terrestrial biosphere including soils, the oceans and reserves of fossil fuels. Reservoir sizes and transfers are uncertain. The total amount of carbon stored in the atmosphere is about 725 billion tons, in the biota plus soils about 2,180 billion tons, and in the oceans about 38,400 billion tons. Of the estimated 7 billion tons of carbon that are added to the atmosphere each year as a result of human activities, about one half remains in the atmosphere and the other half goes to the ocean. The magnitude of the natural fluxes of carbon is much greater than that of the fluxes due to human activities.

## Methane

In the middle latitudes of the northern hemisphere, there was an average annual increase of about 1.1 per cent in the concentration of methane in the atmosphere in recent years. Methane is produced by microbial activities during the mineralization of organic carbon under strictly anaerobic conditions, for example in waterlogged soils and in the intestines of herbivorous animals. There is much uncertainty about the natural and anthropogenic sources of methane. The numbers in Table 4.2, however, give the orders of magnitude.

**Table 4.2: Methane production sources.**

| Anthropogenic Source (in millions of tones per year) | | Natural Sources (in millions of tones per year) | |
|---|---|---|---|
| Enteric fermentation (cattle, etc.) | 75±35 | Enteric fermentation (wild animals) | 5±3 |
| Rice paddies | 70±30 | Wetlands | 110±50 |
| Biomass burning | 70±40 | Lakes | 4±2 |
| Natural gas and mining losses | 50±25 | Tundra | 3±2 |
| Solid waste | 30±30 | Oceans | 10±3 |
| | | Termites and other insects | 25±20 |
| | | Others | 40±40 |

## Surprising Source of Methane

Carefully controlled experiments by Dr. frank Keppler of the Max Planck Institute for Nuclear Physics in Germany have resulted in the remarkable recent discovery that terrestrial plants emit methane. A powerful greenhouse gas, methane traps more than 20 times as much heat per molecule as carbon dioxide. Keppler *et al.*, extrapolated their measurements to the whole world. The numbers are uncertain but surprising and staggering: 10 to 30 per cent of the annual total of methane entering the atmosphere could be coming from living plants. This raises the question whether plants are still an important carbon sink. The answer is yes. Conceivably, some of plants' absorption of carbon by photosynthesis greatly outweights their release of methane. The range of estimates puts the global warming effect of methane from plants at 1 to 10 per cent of the positive benefits from their uptake of carbon dioxide. This research appears to explain several previous puzzles, including satellite measurements of inexplicably large plumes of methane above tropical forests, and the current slowing of the global growth rate of atmospheric methane (probably due to tropical deforestation).

In the past, no textbooks have suggested that green plants in oxygenic atmosphere produce methane. Secondly, the current estimates for identified methane sources and sinks are quite inexact, so finding another source does not necessarily 'blow out' the numbers. Conceivably, some of the methane emitted from plant is already being counted but is wrongly attributed to microbial activity in wetlands, including swamps and rice paddies.

## Nitrous Oxide

A recent increase in the atmospheric concentration of nitrous oxide has been observed in the troposphere, although the rate of increase in considerably lower than that of methane. An annual increase of 0.2 to 0.3 per cent has been noted. Nitrous oxide is reduced in the stratosphere and there are no known significant troposphere sinks. The emissions of nitrous oxide into the atmosphere are primarily due to microbial processes in soil and water and are a part of the nitrogen cycle. There is much uncertainty about the emissions due to human activities, especially the magnitude of the emissions from fossil-fuel combustion. Other anthropogenic sources include biomass burning, and cultivated soils, especially with the intensive use of fertilizers.

Nitrous oxide plays an important role in both radioactive forcing and stratospheric ozone depletion, but its global budget remains unresolved, and the natural and magnitude of the sources and sinks continue to be debated. Dore *et al.* (1998) reported that nitrous oxide within the lower-euphotic and upper-aphaotic zone of the North Pacific ocean is depleted in both 15 N and 18 O relative to its tropospheric and deep-ocean composition. These findings are consistent with a near-surface isotopically depleted oceanic nitrous oxide source. It seems that this source, probably produced by bacterial nitrification, contributes significantly to the ocean-atmosphere flux of nitrous oxide in the oligotrophic subtropical North Pacific Ocean. This source may also buffer the isotopic composition of tropospheric nitrous oxide budget. Because dissolved gases in near-surface waters are more readily exchanged with the atmospheric reservoir than those in deep waters, the existence of a quantitatively significant nitrous oxide source at a relatively shallow depth has potentially important implications for the susceptibility of the source and the ocean-atmosphere flux, to climatic influences.

## Implications

Because dissolved gases in near-surface waters are more readily exchanged with the atmospheric reservoir than those in deep waters, the existence of a quantitatively significant nitrous oxide source at a relatively shallow depth has potentially important implications for the susceptibility of the source, and the ocean-atmosphere flux, to climatic influences. The implications are the following:

1. If sea-air nitrous oxide fluxes are dependent not so much on slow diffusion across the thermocline but more upon production and recycling (that is, nitrification) of organic matter above the thermocline, then significant variability in nitrous oxide flux is to be expected at timescales on the order of days to weeks.

2. If the primary source of oceanic nitrous oxide is near the surface rather than in deep waters, variability in wind-driven mixing, such as can be expected in response to climate change, will have a strong effect on the regional and global ocean-atmosphere fluxes of this important trace gas.

3. Significant climatic phenomena such as *El Nino* may produce opposing effects on nitrous oxide flux in different regions of the ocean.

While suppression of upwelling by *El Nino* may reduce nitrous oxide in the surface waters of the equatorial Pacific, ecosystem shifts in the subtropical North Pacific in response to *El Nino* (which favour the dominance of nitrogen-fixing cyanobacteria) would increase nitrification rates in the upper water column, so increasing near-surface nitrous oxide production (see cline *et al.*, 1987).

## Chlorofluorocarbons

Chlorofluorocarbons (CFCs) have been produced for a variety us uses such as solvents, foaming agents, refrigerator fluids and spray can propellants. A rapid increase in the emissions of the most important CFCs (CFC's-11 and CFC-12) until about 1970 changed to a decline after the mid 1970s as a result of restrictions on the use of CFCs introduced by some countries in view of the possible threat of CFCs to stratosphere (near ground level) but are photochemically decomposed in the stratosphere.

## Tropospheric Ozone

The concentration of atmospheric ozone varies considerably both in space and in time as a result of interactions between atmospheric motions and chemical reaction. This concentration increases due to photochemical processes. Methane, carbon monoxide and other hydrocarbons and nitrogen oxides have important roles in this context. Their increasing concentrations and reactions with the hydroxyl (OH) radical are important for the ozone chemistry of the troposphere.

An increase in the atmospheric ozone concentration has clearly taken place at middle and high latitudes of the Northern Hemisphere I the last few decades, particularly during the summer months. The present rate of change is estimated to be 1 to 2 per cent per year.

In the stratosphere, ozone shields the planet from ultraviolet radiation; nearer the ground in the troposphere, the moisture-rich atmospheric layer below the stratosphere, this GHG is produced through reactions involving hydrocarbons and nitrogen oxides, all released through the combustion of fossil fuels in motor vehicles and in industry. Concentrations of tropospheric ozone have been increasing at many places. Tropical forests act as a sink for ozone; their continued destructions significantly affects regional ozone balances.

## Effects of Global Warming

Various uncertainties about the feedbacks within the between the physical, chemical and biological systems make it difficult to evaluate the potential effects of climatic change, but it is possible to develop some scenarios that describe plausible pictures of the future. It is equally difficult to predict the consequences of climatic change, for example, on agriculture, forestry, the energy sector and waste systems; instead scenarios are developed to describe the likely changes.

As about one-half of humanity lives in coastal regions, much attention has been given to potential changes in sea levels as a result of global warming. Global warming induced by GHGs will accelerate sea-level rise, giving an increase of about 20 cm by the year 2030 and 65 cm by the end of the 21$^{st}$ century. These changes would be a result of thermal expansion of the sea water and melting of ice. The rate of increase of sea level could be 3 to 10 times greater than the 0.01 m per decade long-term average observed during the 19$^{th}$ century. The effects of sea-level rise will include erosion of beaches and coastal margins, land use changes, loss of wetlands, changes in frequency and severity of flooding, damage to coastal structures and ports and damage to water management systems.

Agriculture can experience marked changes in the coming decades, some linked to climate and others to fundamental technological, socioeconomic, and environmental changes. Changes associated with global warming are likely to cause intra-regional shifts in agricultural productivity. For all but the most rapid warming, adaptation based on agricultural research should permit general maintenance of total global food supplies with some local disruptions. A rapid rate of warming could lead to more erratic reductions in food availability.

The effects of climatic change on relatively unmanaged ecosystems could be very significant. For example, if the rate of climatic change were very high, major effects on forests in the mid-latitudes may begin early in the 21$^{st}$ century with forest dieback as a result of climatic change beginning between 2000 and 2050. On the other hand, with a rate of temperature change of about 0-1°C/decade, species extinction, reproductive failure and large-scale forest dieback may not occur before the year 2100.

Changes in the tundra and the northern limit of the boreal forests will include both a stimulation of growth and carbon fixation and rapid decay of organic matter. In both the high and middle latitudes, there is the possibility of a large net release of carbon dioxide and methane from soils, tundra, melting permafrost, forest dieback and ecosystem changes. Such release would further enhance the greenhouse effect.

Strategies for responding to climatic change are generally divided into two categories: adaptation strategies adjust the environment or the ways it is used to reduce the consequences of a changing climate; limitation strategies control or stop the growth of GHG concentrations in the atmosphere.

# Importance of Water Vapour

Water vapour is one of the most important greenhouse gases. It also has an important role in stratospheric ozone chemistry. Increases in water vapour promote the formation of polar stratospheric clouds-key prerequisites for ozone depletion. Water vapour is not included in the group of GHG's covered by the Kyoto Protocol because it does not have a large direct anthropogenic source.

Stratospheric water vapour measurements at Boulder, Colorado (USA) for the period 1981-2000 show a statistically significant increase of approximately 10 per cent per decade over altitudes of 15 to 28 km. Approximately half of this increase results from increases in tropospheric methane which oxidizes to water vapour in the stratosphere. The other half of the increase is currently unexplained. These changes have contributed to ozone decreases in the lower stratosphere.

# Methane Fluxes from Landfills

Methane is a fairly long-lived GHG. Its atmospheric concentration has grown from about 700 ppvc in pre-industrial times to over 1700 ppbv today. It is the only long- lived gas that shows chemical feedback effects- increases in atmospheric methane reduce the concentration of the hydroxyl radical (OH),and thus increase the methane lifetime, and also lead to increases in tropospheric ozone.

The global annual input of methane to the atmosphere is estimated to be $535\pm125$ Tg (IPCC, 1995), of which about half is considered to be both anthropogenic and originating from biospheric processes such as anaerobic bacterial fermentation. Decomposition of refuse in municipal landfills is a major component of this biogenic methane. Landfills appear to be the largest anthropogenic source of atmospheric methane in the United States and Europe. Many countries have targeted this as one source that can be controlled by recovery of the $CH_4$ and using it as a fuel, thus potentially providing a way of reducing current GHG emissions (Smith and Bognet, 1997).

In contrast, in developing countries, urban refuse is often disposed off in open dumps, which do not emit much methane even though they do create other environmental problems. As these open dumps may be covered in the future, methane production will increase, and in most cases this will not be recovered for use as fuel but released to the atmosphere.

Landfills typically harbour two contrasting microbial ecosystems: anaerobic methanogenic communities in the refuse, and methanotrophic zones in aerated cover soils. Rates for both methane production and oxidation sometime greatly exceed observed rats for other terrestrial ecosystems. Field flux measurements (net emissions) vary from less than 0.0004 to about 4000 g m$^{-2}$ d$^{-1}$. These net emissions are the result of methane production, oxidation, and gaseous transport processes in the cover soil (Figure 4.2). The figure shows that both methanotrophic oxidation and engineered control systems (pumped gas recovery) may reduce emissions.

Till hitherto 'top-down' approaches have been used to estimate methane fluxes from landfills. The quantities and types of decomposable refuse deposited have been

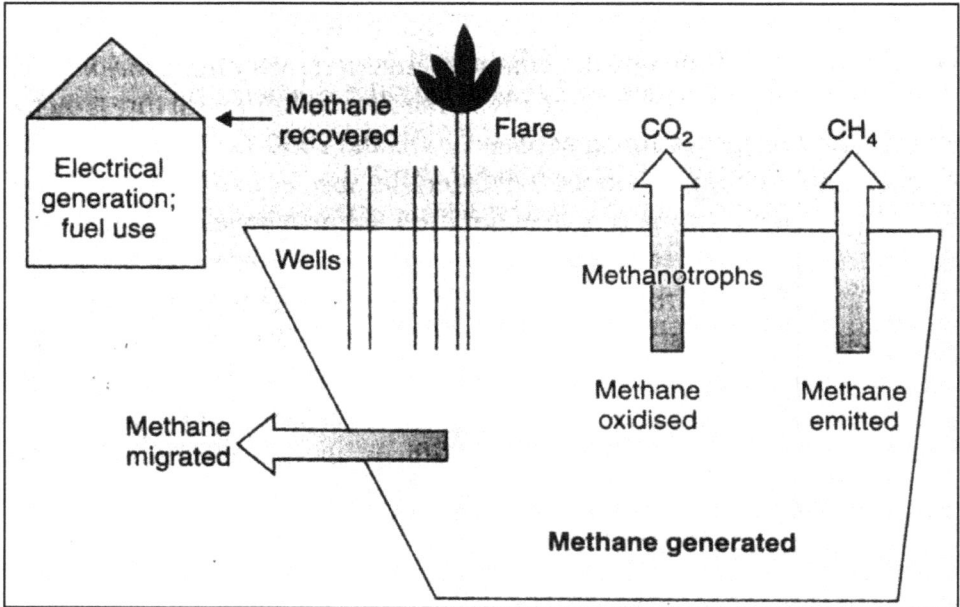

**Figure 4.2: Landfill methane balance (after Smith and Bogner, 1997)**

calculated, and multiplied by assumed rates of methane generation. Such estimates have not considered many factors which affect net emissions.

Emission data can be obtained by chamber, inert tracer, and micro-meteorological methods. As all these methods have certain advantages and disadvantages and are not uniformly applicable to all landfill types, different methods need to be used in combination.

Much work has been done on the effect of methanotrophic methane on net emission versus gross production, and possible isotopic approaches to quantify the relationship. Important variables include soil texture, gas-filled and total porosity, tortuosity, dynamic water content and moisture-holding capacity, clay mineralogy, and nutrient and organic matter content. In landfill soils containing organic matter with a low C/N ratio, methane oxidation can be suppressed because of increased nitrogen turnover. Soil cover design and management practice are also quite important.

Figure 4.3 is a conceptual model for better quantification of country based landfill methane emissions. It is applicable to many developed countries with available solid wastes and landfill management statistics. The primary criterion is wet vs. dry sites (based on the moisture content of the bulk land filled waste); the second tier criterion is the presence or absence of pumped gas recovery, the third criterion is size (small vs. large), the fourth criterion is fractional methane oxidation, and the fifth criterion relates to site construction (above-ground versus below-ground at small sites.).

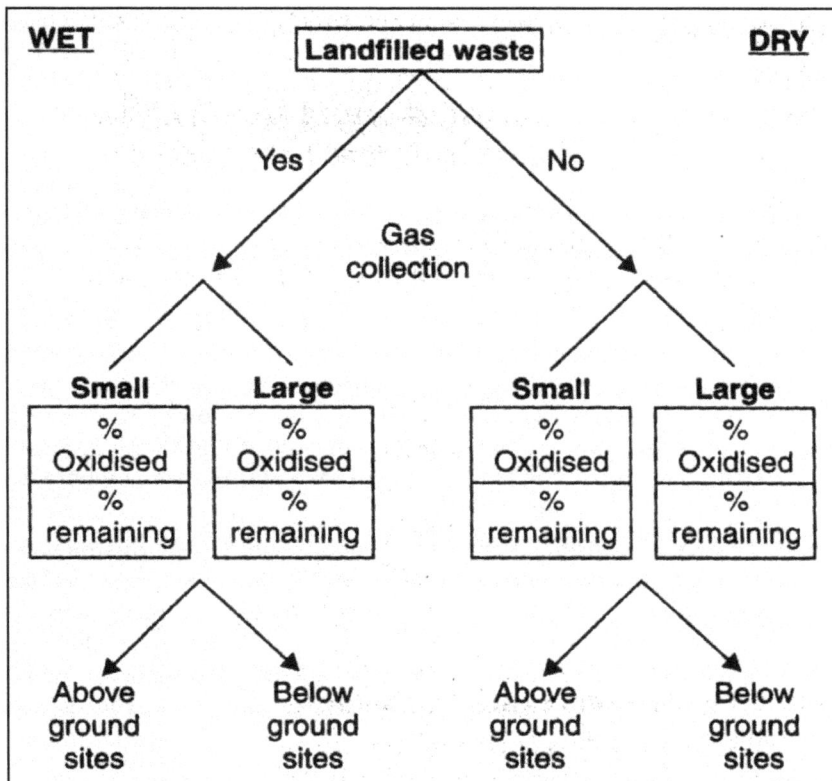

**Figure 4.3: A conceptual model for quantification of landfill methane emissions (after Smith and Bogner, 1997).**

## The Greenhouse Era

That the Earth is warming up is generally accepted now. The pace of global warming has increased in recent times. The warming may be attributed to several causes, for example as a product of sunspots, as acyclical event and more importantly, due to an increase of GHGs in the atmosphere. Besides fossil fuels, land clearing and cultivation release the carbon stored in vegetation (primarily trees) and in the soil into the atmosphere (Blair, 2002).

The increased production of ruminating livestock and retained waste has increased methane production. Methane is over 20 times as potent as carbon dioxide. Ruminants emit methane as a waste product from their digestion processes. Between 4 and 12 per cent of the energy contained in the food of cattle is lost in the form of methane. About 90 per cent of the methane is burped out by the animal, while the other 10 per cent is extruded with the manure. Cattle that can efficiently process grasses and fodder have lower methane emissions. So, faster growing cattle have higher feed efficiencies and lower methane emissions as compared with slower growing cattle of the same weight. Larger cattle usually eat more and emit more methane than smaller cattle. The amount of methane produced is intimately linked to the digestibility of the diet of the animal. Tropical pastures are typically less digestible than temperate pastures. Grains are higher in digestibility than either tropical or temperate grasses, which indicates that cattle in feed-lots have lower methane emissions than similar cattle on grass (Rolfe, 2002).

In nature, gaseous carbon is recovered through the action of photosynthesis in all vegetation and the absorption of carbon into sea water. But, the world's forests have been increasingly felled or cut in search of timbre products or in the process of land clearing.

If all the countries are made responsible for their GHG emissions, some will comply with the obligation better than others. For example, Germany, France and most of the European Union (EU) depend greatly on nuclear power to generate energy. This produces no greenhouse emissions although, for other reasons, it is not universally acceptable. Nevertheless, it does present the EU with a considerable advantage over nations like the United State, Japan and Australia where energy is generated primarily by burning coal (Blair, 2002).

In 1997, the Kyoto agreement proposed targets to which the industralised nations should reduce net greenhouse emissions by 2008-12. The agreed targets are expressed as a percentage of what each such nation's greenhouse emissions were in 1990. The Kyoto Protocol accepted that each nation can achieve its target by whatever means it wished. This can include new forests established for the purpose of carbon recovery and not at the expense of removing existing forests (so-called Kyoto forests) absorbing and retaining gaseous carbon (Blair 2002).

Emitters should be either prosecuted or punitively taxed until they reduce emissions to an acceptable level. The United Kingdom (UK) favours a punitive taxation system to reach its target by the year 2012. to this end it has enacted that a tax is to be paid by emitters unless they effect a fixed reduction of emissions. GHG emissions in the UK are however, still increasing. The EU (which operates as a single unit) is resorting to a rationing approach. It has the benefit of nuclear and gas-generated electricity and a surplus of energy from the former East Germany.

There seems no doubt that the Greenhouse Era will strongly affect the cost of all goods and all services. As the cost of all goods and services will be radically altered, the value of all assets is likely to be altered as well (Blair, 2002).

The United State has not ratified the Kyoto Protocol because of the high cost of meeting the targets. Australia also is unwilling to ratify it. Agriculture contributes significantly to Australia's GHG emissions and is also largely responsible for releases from land clearing, which bring the total contribution of the sector to about one third of national emissions.

Over 65 per cent of the contribution from the agricultural sector (excluding land clearing) comes from livestock, chiefly as methane emissions. Cattle and sheep release methane as a normal by-product of digesting a grass. Methane is also released room pounded manure systems in piggeries, dairies and feedlots. On average, dairy cows emit about 115 kg of methane per head per year in Australia and sheep at least 6 kg.

Climate change will affect pastrol industries. If attempts are to be made to reduce emissions further, the diary industry will have to contribute. It may be cheaper to make reductions in sectors of agriculture than to make equivalent reductions in industry. In this case, the use of carbon offsets or other incentive mechanisms can provide financial incentives to pastoralists to make larger reductions (Rolfe, 2002).

Emissions vary between cattle in three ways. The first relates to the type of feed eaten, for example, tropical grasses, temperate grasses, or feedlot rations. The second relates to the weight of the beast-heavier cattle produce more methane than lighter cattle. The third is related to growth rates, where higher live weight gains mean better feed utilization and lower methane emissions (Rolfe, 2002).

## Carbon, Warming and Human Activities

It appears increasingly unlikely that we will escape with only some mild greenhouse warming. The world has warmed by 0.6°C during the past century. Researchers agree that human activities-mostly burning fossil fuels to produce the GHG carbon dioxide have caused most of that warming. But there is continuing debate on how warm could it get and how bad could the greenhouse threat turn out to be.

Several official assessments of climate science have been quite vague on future warming. Sensitivity, or how much a given increase in atmospheric carbon dioxide will warm the world, falls into the same subjective range: At the low end, a doubling of carbon dioxide- the traditional benchmark- was thought to eventually warm the world by only 1.5°C or less. At the other extreme, temperatures might rise by a worrying 4.5°C or more. but now highly sophisticated climate models have made it possible to estimate statistically the true uncertainty of the models' climate sensitivity. A general consensus for a moderately strong climate sensitivity has emerged (Kerr, 2004). Most of the evidence points to 3°C as the most likely amount of warming for a doubling of carbon dioxide. Such a figure could lead to a dangerous warming by the zcentury's end, when carbon dioxide may have doubled. However, some researchers still feel that the figure of 3°C may not be valid and that it might not exceed 1.5°C.

Much of the warming that has been observed can indeed be attributed to human activities because both the geographical pattern across the Earth's surface and the vertical pattern into the atmosphere of temperature change are what is expected from changes induced by human influences.

### Geographical Pattern

For much of the 20th century, more warming has been occurring in the southern Hemisphere than in the Northern Hemisphere because of the cooling influence of the grater amount of aerosols in the Northern Hemisphere because of the cooling influence of the greater amount of aerosols in the Northern Hemisphere; in the latter part of the 20th century and into the 21st century, more equal warming of the hemispheres has occurred and is expected to continue.

### Vertical Pattern

Cooling is occurring in the stratosphere due to ozone depletion and GHG increases, in contrast to warming in the troposphere (the lower atmosphere and ground surface) from GHG.

The above-mentioned patterns are unlike those expected from naturally varying forcing factors, such as solar radiation and volcanic eruptions. The extent of warming since the 19th century is also in accordance with the range of model estimates of the surface warming that is expected from the increases in GHGs and aerosols that have occurred. Inclusion of a more complete set of human influences, including the cooling influences of aerosols and stratospheric ozone depletions, ahs reconciled observed and predicted changes and clarified that it is essential to consider all important factors than can influence the climate, rather than drawing conclusions or making comparisons based on the estimated change from any single factor.

## Climate Extremes

From model predictions about climate change, it can be inferred that there will be:

☆ Increases in the frequency and intensity of floods, droughts and heat waves.

☆ Changes in the frequency of occurrence of climatic extremes due to increase in GHG emissions.

It is not proper to attribute any particular extreme weather event to climate change, but the trends in their occurrence can be examined. Although detailed worldwide information is not available to explore this question, analyses of data for the US during the 20th century show a consistency between inferences drawn from model predictions and observed responses. Such modeled and observed changes include, for example, an increase in intense rainfall events.

## Terrestrial and Aquatic Ecosystem Feedbacks and Effects

Changes in the atmospheric composition and climate will have a significant influence on terrestrial and aquatic ecosystems. Because these ecosystems also have a major influence back on the atmosphere and climate, understanding their role in potential positive and/or negative feedbacks is critical.

### Types

The feedbacks and effects that the land and aquatic environments exert on climate are of two types:

1. *Biogeochemical*: Changes in the distribution and circulation of chemicals such as carbon- and nitrogen-containing compounds among the atmosphere, oceans, biosphere, and soils, resulting in changes in the atmospheric composition of GHG's.

2. *Physical*: Changes in land-surface properties such as albedo (the ability to reflect light) and roughness (the ability to alter wind speed and direction), resulting in changes in the Earth's energy balance.

# Hosttest Decade

The year 2005 was the second warmest year in the last 125 years, with mean global temperature of 14.5°C. And, in the last one decade, 9 out of the 10 years were the warmest during the past 125 years. This shows that the impact of global warming may become the most crucial environmental issue in the 21st century. The Earth's mean acclimate is determined by a balance in exchanges of energy and gases among the atmosphere, the ocean and the biosphere. According to some experts, the Earth's climate is so complex that small perturbations introduced by human beings cannot change it markedly. In contrast, others argue that, the climate system being nonlinear, small perturbations by human beings can and do cause significant or even substantial changes in the global climate.

As the Eath's climate is modulated by many different factors, sophisticated climate models are required to predict how the climate will evolve in the future. These models incorporated the laws of dynamics and thermodynamics and need robust computational resources for simulating future climate. Reliable predictions are made by these models for parameters such as global mean temperature. For instance, a prediction was made in 1988 (on the basis of climate models) that global mean temperature in 2005 would be 14.45°C; the value that was actually recorded was 14.50°C. This points to the accuracy and trustworthiness of the prediction made by global models.

The surface air temperature in many parts of India increased by 0.5°C during the second half of the 20th century, but in the Himalayas, it increased by 1°C during the same period. This led to the rapid melting of the Himalayan glaciers which, in turn, increased the volumes of glacial lakes. As several such lakes have fairly thin walls of ice or debris, these lakes can potentially cause floods in Nepal, a Bhutan and North India, discharging millions of m³ of water in a few days. The melting of all the Himalayan glaciers can create a serious water crisis in North India.

According to WRI (2001) developing countries accounted for about 37 per cent of cumulative carbon dioxide emissions from industrial sources and land-use change during the period 1900 to 1999 (Figures 4.4 and 4.5), whereas industralised countries accounted for 63 per cent. However, in view of their much higher population and rising economic growth rates, the fossil-fuel carbon dioxide emission from developing countries are expected quite soon to equal, or even exceed, those from the industralised countries. China and India could even equal the USA's greenhouse gas emissions (in 2000) in just two to three decades. If we consider fossil-fuel carbon dioxide emissions alone, in the face of population and economic growth in the coming decades, the contribution of developing countries as a group will quickly overtake the industrialized countries.

# Nitrogen Deposition

With the growing focus on carbon emission in the industrialized world during the past few decades, another global threat that could be even worse, has been sidelined. Today, many people have become aware that carbon dioxide emission cause global warming. But, if some common men or women were to be asked about

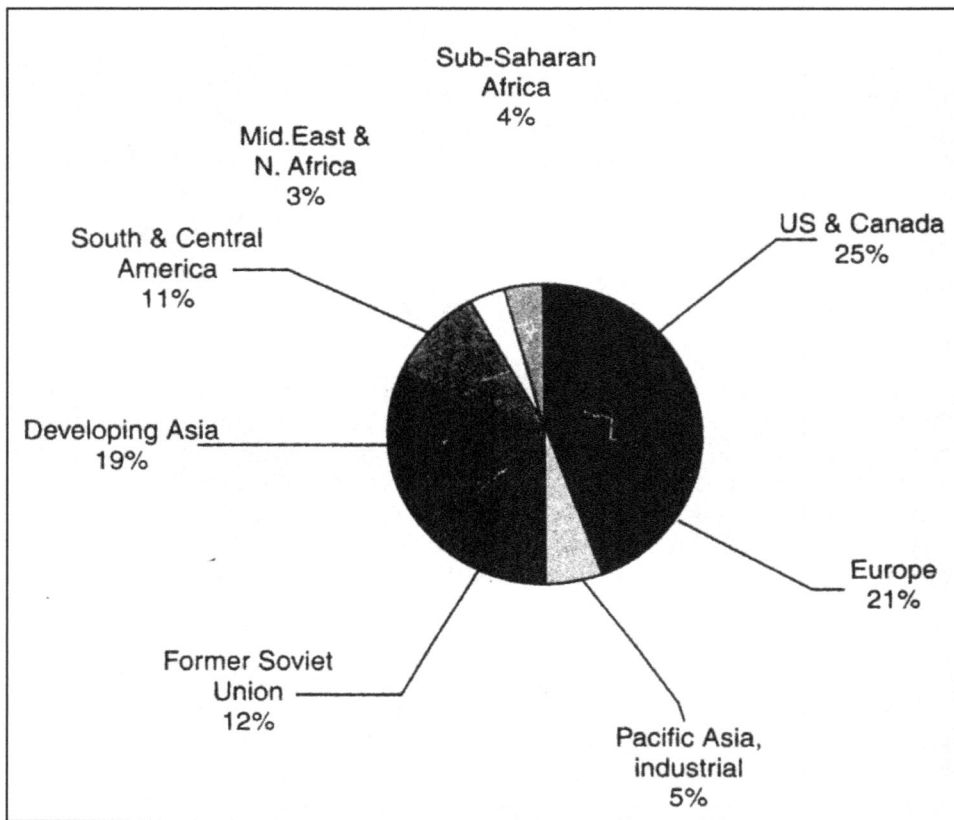

Figure 4.4: Per cent cumulative global carbon dioxide emissions from industrial
sources and land-use change during the 20th century.

nitrogen emission, chances are that they would not know. Nitrogen is becoming a
problem that we cannot ignore. The consequence of a 10 per cent increase in carbon
dioxide has generated an alarming concern, while levels of polluting nitrogen
compounds in the environment have almost doubled and still not attracted any
significant attention. If we continue to ignore them much longer, the consequences
could turn out to be even worse than 'just' global warming. Human health, biodiversity,
ozone levels and global climate are already being affected (Hooper, 2006). Some even
think that long-term, anthropogenic nitrogen could be a greater environmental threat
than anthropogenic carbon.

There has always been a great deal of nitrogen in the environment. Although
about 78 per cent of air by volume is nitrogen gas ($N_2$), until it becomes oxidized or
reduced into a reactive form, it is neighter harmful nor useful to most living things.
For many millennia, the only way to reactive nitrogen was via nitrogen-fixing
microorganisms and lightning. These fixed nitrogen out of the air which then
underwent a series of cyclic changes through soil, plants and animals, back into the
soil, and finally, back into the air (Figure 4.6).

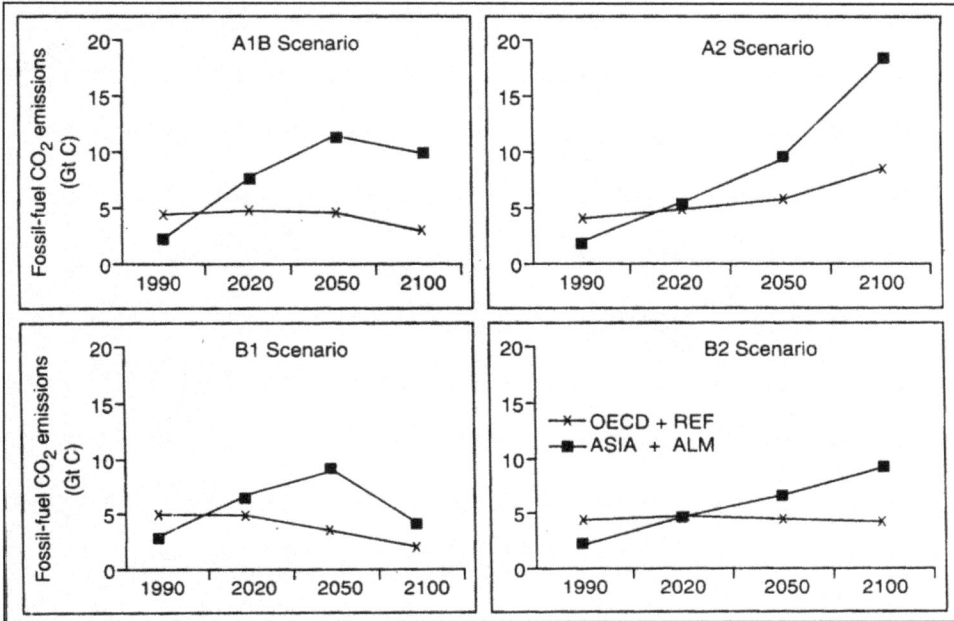

Figure 4.5: Fossil-fuel carbon dioxide emissions (GtC) in industrialized and developing regions under four SRES marker scenarios. OECD90 region includes countries of the Organisation for Economic co-operation and development as a 1990. REF region includes countries undergoing economic reform and groups together in the East and central European countries and the newly independent states of the former Soviet Union. ASIA region includes developing countries in Asia (excluding the Middle East). ALM region stands for the rest of the world and corresponds to developing countries in Africa, Latin America and Middle East. Scenarios A1B, A2, B1 and B2 (after Nakicenovic *et al.*, 2000).

With the advent of the Industrial Revolution and the large-scale burning of fossil fuels, atmospheric nitrogen gas started reacting to form nitrogen oxides ($NO_x$). A century ago, fritz Haber, a German chemist, discovered how to produce ammonia from hydrogen and nitrogen; this later led to the industrial production of nitrogen fertilizer. Levels of reactive nitrogen began rising sharply. Today as much as 70 per cent of reactive nitrogen cycling through the atmosphere, Earth and sea is generated through human activity.

In fact, the nitrogen 'cycle' no longer exists. It is more a cascade (Figure 4.6), or even like a torrent (Hooper, 2006).

While carbon dioxide has been projected into the public mind as the bad guy of global warming, the nitrogen problem continues to be largely ignored partly because the nitrogen story is much more complicated than carbon. In its various chemical forms, nitrogen can have at least seven apparent changes, or 'oxidation states', which can pose a probe to health or the environment. Indeed, they make its passage through the environment the most complex of the major elements (Table 4.3).

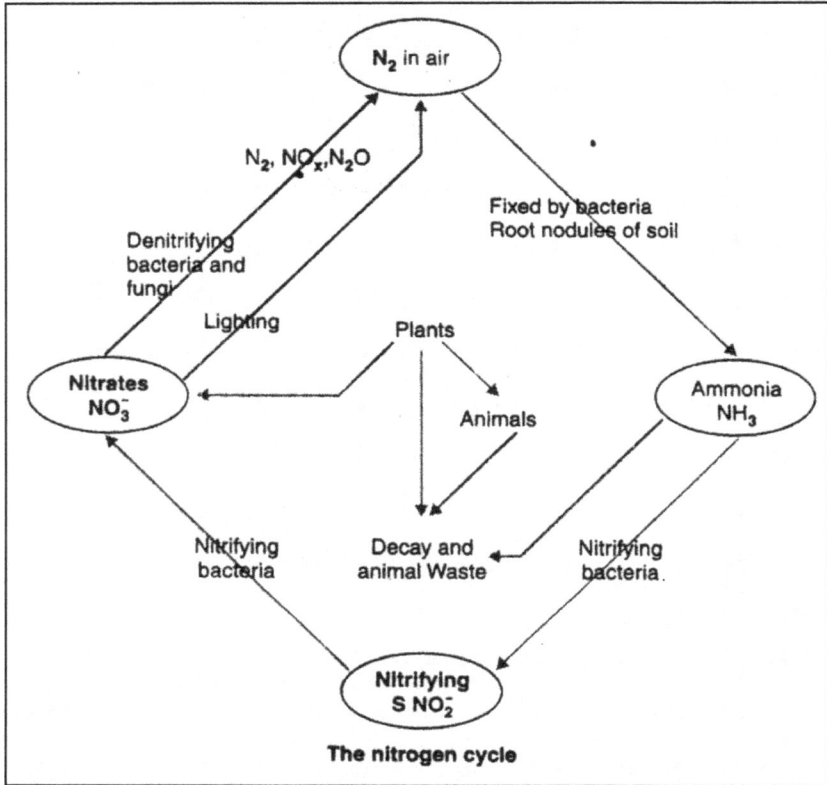

Figure 4.6: Nitrogen in the environment. Pre-industrial: The nitrogen cycle-nitrogen-fixing microbes in the soils or legume root nodules reduce nitrogen gas ($N_2$) to ammonia, which is taken up by plants, or oxidized into nitrites and nitrates by nitrifying bacteria. Lightning strikes split nitrogen molecules, which then combine with oxygen to form nitrates. Finally, nitrogen is cycled back into the atmosphere by denitrifying microbes, which turn nitrates back into $N_2$, nitrous oxide ($N_2O$)and nitrogen oxide ($NO_2$) (after Hooper, 2006).

It is indeed striking that while more nitrogen is falling out of the air (as acid rain and particulates) today than farmers used to add to their fields half a century ago, yet, because this nitrogen input comes from diffuse sources, it is an extremely difficult problem to tackle. One unique property of nitrogen is that the same atom in reactive nitrogen molecule can cause many effects in the atmosphere, in terrestrial and aquatic ecosystems and on human health, as it moves through the environment. The cascade can be stopped only by converting reactive nitrogen back to non-reactive $N_2$-an extremely daunting task.

So much of extra nitrogen in the environment has so overwhelmed the natural cycle that the various denitrifying bacteria just cannot convert the excess back into atmospheric nitrogen gas fast enough. And though agricultural sources of nitrogen cause problems mainly locally, nitrogen compounds from burning fossil fuels can

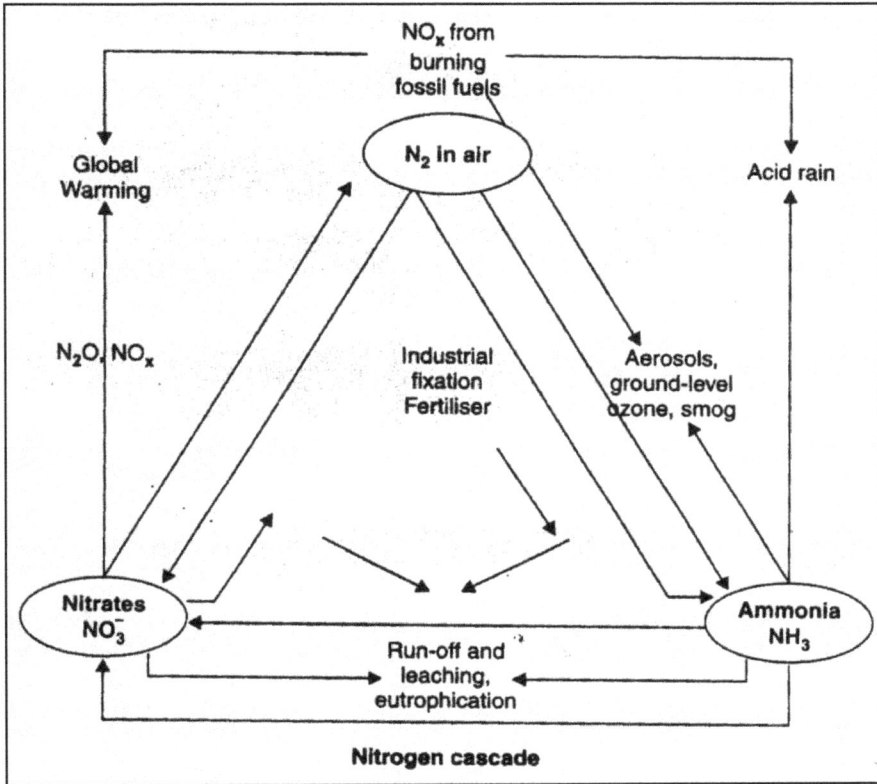

**Figure 4.7: The nitrogen cascade today-the natural nitrogen cycle still operates, but is completely overshadowed. For over a century, fossil fuels have been burnt and nitrogen fixed from the air for fertilizers on an industrial scale. The result is a torrent of reactive nitrogen flowing into the soil, rivers and atmosphere (after Hooper, 2006).**

cross national borders, causing acid rain in neighbouring countries and, even worse, global warming !

Even though the problem is serious, we surprisingly know little about its adverse effects on the environment and health. At high levels there can be an effect on ecosystem, but there may well be some unexpected and unknown effects between low and high levels. It is also not known whether there is a difference between adding small amounts of nitrogen over a long period and adding large amounts very quickly; the effects of different forms of nitrogen on the same ecosystems; and interactions with temperature, carbon and other environmental factors, are also largely unknown.

What we do know, however, is alarming. Nitrogen has a clear connection with human health-there is no such connection with carbon dioxide, according to Elizabeth Holland of the National Centre for Atmospheric Research (NCAR) in Boulder, Colorado. For instance, nitrates present in run-off from agricultural fields can get into drinking water, causing methaemoglobinaemia, in which they are transformed into nitrites in the body, affecting the ability of haemoglobin to carry oxygen.

**Table 4.3: Different forms of one element. Once nitrogen has been fixed into a reactive form, it can be transformed between any of seven oxidation states, six of which cause problems. The only way to address the problem is to turn the nitrogen back into molecular nitrogen gas (after Hooper, 2006).**

| Compound | | Oxidation state | Effect |
|---|---|---|---|
| $NO_3$ | Nitrate ion | +5 | Eutrophication, health problems |
| $HNO_3$ | Nitric acid | +5 | Acid rain, eutrophication |
| $NO_2$ | Nitrogen dioxide | +4 | Smog, acid rain,eutrophication |
| $HNO_2$ | Nitrous acid | +3 | Acid rain, among |
| NO | Nitric oxide | +2 | Smog, acid rain |
| $N_2O$ | Nitrous oxide | +1 | Greenhouse gas, stratospheric ozone destruction |
| $N_2$ | Nitrogen gas | 0 | Inert |
| $NH_3$ $NH_4$ | Ammonia/ammonium | -3 | Smog, eutrophication, aerosols |

Excess nitrogen compounds spell trouble for ecosystems. Livestock farms discharge large quantities of nitrates and ammonia into rivers, lakes and the sea, where they cause havoc on biodiversity and commercial fisheries. This is best seen in the Gulf of Mexico, where nitrates flowing from the Mississippi river have created a 'dead zone'.

The zone, one of 146 worldwide, was first recorded in the 1970s, and appears each year when an excess of nutrients from the Mississippi basin washes into the sea, causing eutrophication and troublesome algal blooms. When the bloom algae die, the bacteria that feed on them actively use up the available oxygen, suffocating anything that cannot move out of the area. Nitrate leaching acidifies freshwater rivers and lakes.

When nitrogen and sulphur dioxides in the atmosphere combine with moisture, nitric, nitrous and sulphuric acids are formed, killing fish and damaging tree roots and leaves (Hooper, 2006). In sunlight, $NO_x$ compounds in the atmosphere react with hydrocarbons, forming ground-level ozone which is a lung irritant and a potential factor in asthma.

## Nitrogen and Climate Change

The biggest puzzle of all seems to be the complicated relationship between nitrogen and climate change. Nitrous oxide ($N_2O$) is found in the atmosphere at concentrations of 311 parts per billion, compared with 360 parts per million for carbon dioxide.

However, molecule for molecule nitrous oxide is a GHG with a least 300 times the global warming potential of carbon dioxide. Some experts have viewed nitrogen as helping to reduce global warming by stimulating plant growth, thereby locking more carbon away, but this benefit is likely to be overshadowed by an increased rate of denitrification in the soil, which pumps nitrogen oxide ($NO_x$) back in the atmosphere (Hooper, 2006).

Recently, an international research project 'NitroEurope' has been launched by the European Union. China and India to cooperatively study the connection between the nitrogen and carbon cycles. Some urgent needs are to find out more about where nitrogen is coming from, and to assess its impact so as to throw light on the effects of the nitrogen cascade. Once these are known, a Kyoto-style agreement will have to be made. In fact, a Kyoto-style document, the Nanjing declaration, ahs already been released in 2004 at the 3rd International Nitrogen Conference. It has been adopted by the European Union and is awaiting more signatures at the United Nations Environment Programme (UNEP).

Nitrogen creates problems primarily because, worldwide, five times as much reactive nitrogen comes from food production as from energy production- it is very difficult to provide plants with the exact amount of nitrogen they need. Any excess simply ads to environmental burden of nitrogen. Animals too use only about 20 per cent of the nitrogen in their feed. The rest is excreted and either runs back into the ground or is given off as fumes (Hooper, 2006).

Precision agriculture may be one appropriate technological solution to address the issue; like fuel injection in car engines-quantities of nitrogenous fertilizer would be matched to the plants' needs. Nitrogen-proofing farms could also help-woodland catches ammonia emissions well, so planting trees around livestock farms can lessen the impact. Another option is to keep livestock inside so that the air can be 'scrubbed' to remove nitrogen compounds before venting it to the atmosphere.

Indeed, in some parts of the US (*e.g.* California), cows are already being subjected to environmental regulations that are otherwise normally restricted to the dirtiest cars. Any farm in the state with more than a thousand dairy cows is required to apply for a license from the Air Resources Board. In the eastern US, the Clean Air Interstate Rules being enforced so as to reduce $NO_x$ emissions by 60 per cent by the year 2015.

In the years to come, it seems certain that agriculture could be asked to manage the emissions of ammonia and other organic compounds from livestock operations. As nitrogen compounds do not respect national boundaries, the above local initiatives will have to be regulated globally.

## Biogeochemical Cycling

One major challenges has been to find out where all of the carbon dioxide being emitted as a result of human activities is going. Natural biological and physical processes, such as photosynthesis and oceanic uptake, can limit the annual increase in the atmospheric loading of carbon dioxide to just under half annual emissions. Together, the land and ocean components of the climate system must be taking up that half of the carbon dioxide that does not remain in the atmosphere.

Much of the excess carbon from human activities that is being removed from the atmosphere appears to be going into terrestrial systems. Land surface processes are removing about 2 billion metric tons more of carbon from the atmosphere each year than they release. This net carbon uptake by land ecosystems is related to several processes, including the regrowth of previously deforested areas and to enhanced storage of carbon by plants due to the elevated carbon dioxide concentration. Several previously deforested areas are showing substantial forest regrowth.

The role of ocean ecosystems in 'pumping' carbon to the deep ocean as dead animals and plants sink to the ocean floor is also important, and would be especially so if the ocean circulation were to be slowed down by global warming. Data collected in a series of ship cruises indicate that about 2 billion metric tons more carbon is being taken up by the oceans than is being released by them through natural processes.

The characteristics of land and ocean surfaces also influence the climate directly. Plants control the amount of evaporation and runoff at the surface, thereby influencing the hydrological cycle. By enhancing or reducing evaporation, soils and surface vegetation have an influence on the near-surface temperature. Shifts in precipitation, whether brought about by seasonal-to-interannual climate fluctuations, volcanic eruptions, or long-term climate change, will cause shifts in temperature and soil moisture. Depending on their magnitude and timing, persistent changes in precipitation are predicted to cause shifts in ecosystems as well as feedbacks affecting climate.

The terrestrial cycles of energy, carbon, water and nutrients are fundamentally connected with climate processes, the global carbon cycle and ecosystem function. Through these connections, the management of ecosystems to provide food, fibre and timbre has impacts at both global and local scales. Local changes are often manifested as shortages (where excess harvesting of a resource in one place or time creates a deficit elsewhere) or as leakages (where water or nutrients leak into a region where they cause harm, as in dryland salinity due to rising water tables or the contamination of surface or underground water by nutrients). It is for these reasons that those concerned with the management of resources are becoming more and more concerned about the ecosystem function as determined by the terrestrial cycles of carbon, water and nutrients, and increasingly interested in the behaviour of these cycles at both small and large scales (Raupach *et al.*, 2001).

Environmental costs (such as nutrient leakages) increase sharply with increasing nutrient inputs. The implication is that a reduction in nutrient inputs will have more benefit through lower leakage than cost through decreased net production.

## Stabilising Atmospheric Carbon dioxide Concentrations

The Kyoto Protocol constitutes a historic, significant step towards the Earth's sustainability. It limits the emissions to the atmosphere of six GHGs for the 30 ratifying countries from the developed world. Stabilising atmospheric carbon dioxide concentrations [a $(CO_2)$] at a level thought to avoid dangerous interference in the climate system poses a strong challenge. There is no consensus as to what atmospheric carbon dioxide will avoid dangerous climatic interference. It depends not only upon the sensitivity of the major Earth System processes to climate change but also to sensitivity to, and capacity to adapt- of different economic, environmental and social sectors. There is no single atmospheric carbon dioxide that we may target, unless a lowest-common-denominator approach is applied (Canadell and Raupach, 2005).

Some experts believe that human societies would be safeguarded from dangerous interference in the climate system by a stabilization of atmospheric carbon dioxide equivalent to a global warming of 2°C. This means atmospheric carbon dioxide of less than 550 ppm. For ready reference, the preindustrial carbon dioxide concentration was 280 ppm, and the current concentration is 378 ppm.

Several possible scenarios covering major possible routes that societies could take in this century, have been developed (IPCC SRES, 2000) that lead to alternative future emission pathways. These scenarios were created on certain assumptions about population and income growth, the cost and availability of current and future energy production and utilization, and other driving elements. The Special Report on Emissions Scenarios (SRES) cover a very broad range of carbon emissions. The Intergovernmental Panel on Climate Change (IPCC) does not indicate a best guess it the most likely future carbon emission scenario. Carbon emissions for the end of this century in the SRES scenarios, range from 3-35 Pg $yr^{-1}$ unconstrained requirements for the extent of change needed to avoid dangerous interference in the climate system, for whatever target we choose for the purpose of evaluating the challenge (Canadell and Raupach, 2005). Some uncertainty emanates from the difficulty of quantifying the impact of major technological improvements on atmospheric carbon dioxide, and understanding the difference we can make by, for example, collectively adopting automobile-hybrid technology, improving household energy efficiency by 50 per cent, or generating two-thirds of our electricity from renewable energies. But none of these major technological changes can materialize in the near future.

The IS92a IPCC scenario is a 'business-as-usual' scenario-one which attempts to highlight what could happen if we do not take specific measures to address the climate change issue, or in other words, what could happen if we let energy markets evolve as they have in the past specific policies to curb carbon dioxide emissions. This scenario does not include any carbon dioxide emission reduction targets, nor any broad policy proposals to bring down deforestation rates. The scenario also assumes business-as-usual in technological development, based on the experience of the last century. Thus, it assumes a decrease in energy intensity by 0.8 per cent annually up until the year 2025, and a 1.0 per cent decrease annually from 2025-2100. More strikingly, the IS92a IPCC also assumes that by the end of this century, 75 cent of power energy will be carbon free, and that energy generated from bio-fuels will provide more energy than the combined global production of oil and gas in 1990 (Edmonds *et al.*, 2004).

A significant change towards renewable and zero-emission energies might suggest that the carbon dioxide stabilization problem would be largely solved by the time we achieve such transformations. In reality, far from it, atmospheric carbon dioxide by the end of this century would be over 700 ppm under IS92a IPCC, *i.e.* about three times the preindustrial level.

To appreciate the strength of the technological challenge involved in limiting atmospheric carbon dioxide to 700 ppm, one can project atmospheric carbon dioxide under a 'freezing' of technology at 1990 levels without efficiency improvements. This scenario provides a reference that illustrates the scale of the advancements already expected to occur. Any attempts to stabilize atmospheric carbon dioxide below 700 ppm will require an even larger effort (Canadell and Raupach, 2005).

The difference in carbon emissions between a given business-as-scenario (*e.g.* IS92a with atmospheric carbon dioxide at about 700 ppm) and a chosen stabilization level (for instance 550 ppm) is termed as the 'energy gap'. The energy gap between IS92a and a 550 ppm stabilization level is a staggering 14 Pg C yr$^{-1}$. This gap can only be bridged by implementing emission reduction policies and clear emission cuts, most likely with high costs involved. For a number of SRES scenarios, the carbon emission gaps by the year 2100 would range from 1 to 25 Pg yr$^{-1}$ (IPCC, 2001). In fact, stabilizing atmospheric carbon dioxide will not only depend on substantial cuts of GHG emissions during this century but will ultimately also require reducing emissions virtually to zero.

Enhancing energy efficiency is an important step towards reducing emissions of GHGs. Many industrial processes are highly energy intensive and so increasing efficiency is not only environmentally beneficial but also financially feasible. On a global average basis, the cement industry emits nearly 900 kg of carbon dioxide for every 1000 kg of cement produced. Industrial productions of iron and other metals account for around 10 per cent of the annual anthropogenic emissions of carbon dioxide to the atmosphere. The oil and gas industry also present opportunities for re-examination of the processes involved which reveal definite areas where emissions reductions can be achieved, for example, by fuel switching or through installation of additional equipments.

Biomass has an important role in reducing GHGs. Substituting biomass for fossil fuels, and combining biomass technologies with carbon dioxide capture and sequestration in generating electricity, are promising options. The synthesis of clean fuels, for example, carbon dioxide-rich biosyngas is also being attempted.

## Carbon Dioxide Biofixation by Microalgae

Practical microalgae-based processes that would lead to the demonstration of technical feasibility of carbon dioxide biofixation within a decade are being developed (Pedroni, 2005). Microalgae growth in mass cultures maybe used to capture and utilize carbon dioxide from power plants or other sources. GHG emissions can be mitigated by converting the harvested biomass to renewable biofuels (methane, ethanol, biodiesel) and fossil fuel-sparing products (such as fertilizers, biopolymers and lubricants). Microalgae are grown in large open, raceway-type paddle wheel-mixed ponds, similar to those already used in commercial production of speciality food and feeds (see Kumar, 1998). The algae grow in suspension culture and water containing all needed nutrients (nitrogen, phosphorus, etc.) and, most importantly, carbon dioxide. They can be harvested by settling and converted to renewable biofuels and other co-products that reduce fossil fuel use.

Microalgae cultures can potentially achieve high biomass production rates of over 100 metric tons of organic matter per hectare per year, much higher than trees or annual crops. One important application of micro-algae-based systems is in the treatment of wastewaters and in the removal and recovery of nitrogen and phosphates. Such a wastewater treatment function provides additional economic benefits and contributes to GHG abatement through energy conservation, as compared to the current energy-intensive technologies (Pedroni, 2005).

Microalgae cultures are being grown outdoors in scale-down-designed open ponds and photobioreactors using simulated natural gas-fired power plant flue gases (EniTecnologie, Monterotondo, Italy). Their cultures are being grown indoor in scale-down designed photobioreactors and also outdoors and in open ponds, using simulated coal-fired power plant flue gases (ENEL,Brindisi, Italy). The SeaTech, Inc., California, and the Clemson University, South Carolina, are growing microalgae cultures outdoors in scale-down small and large (5-3000 m$^2$) open ponds for removal of nutrients from agricultural drainage waters. Likewise, the SeaAg, Inc, florida, and the Brooklyn College, New York, are developing genetically improved algal strains and outdoor pond culture techniques for increasing productivity and solar energy conversion efficiencies and

Also, for fertilizer production. Some projects started in Brazil and India involve flue gas carbon dioxide utilization and waste treatment with microalgae (Pedroni, 2005).

Microalgae mass cultures are better than other biological options for carbon dioxide capture and utilization, in view of their potential to achieve very high productivities, their ability to capture nutrients from wastewaters, their use of waste, brackish and saline water sources unsuitable for other uses, and their high water use efficiency. The use of microalgae processes in GHG abatement will likely be part of multipurpose processes providing (additional to biofuels and GHG abatement), products and services, including wastewater treatment, animal feeds and biopolymers (Pedroni, 2005).

## Other Greenhouse Gases

Much less is known about non-carbon dioxide GHG emissions that might influence preferred solutions to direct carbon dioxide abatement. In addition, the Dutch considered indirect carbon dioxide emissions, those due to domestic energy needs but emitted outside the country (Rees and Wackernagel, 1994). By the year 2000, energy related indirect carbon dioxide and methane emissions each probably made up to 4 to 5 per cent of the total carbon dioxide-equivalent emissions attributable to the Dutch energy system.

Energy-related halocarbon, nitrous oxide, and carbon monoxide will constitute less than one per cent of the total. These other emissions are expected to increase disproportionately after 2000 due to differences in the primary energy mix and the start of natural gas imports from Russia.

### Biotechnology and Utilisation

Several new advances in the potential for biotechnology to help reduce costs of stabilising concentrations of GHG emissions have been reported for example, opportunities for increased energy-related biomass crops. Microalgae can be used for utilizing carbon dioxide, as well as removing the carbon dioxide or immobilizing its effects through biotechnology, carbon dioxide fixation by afforestation schemes is also important.

## Land Use and Sinks

Issues related to land use are causing concern. Rice cultivation, for example, requires large amounts of fertilizers and emits large amounts of methane from decomposing organic plant matter. The use of nitrogen-fixing microbes not only suppresses methane emissions but also fertilizes the land thus reducing the amount of fertilizer required.

## Utilisation

Carbon dioxide reforming and selective synthesis of valuable chemicals from carbon dioxide through catalytic hydrogenation processes are some promising options.

## Non-carbon Dioxide Greenhouse Gases

Mitigation of gases with a high global warming potential (GWP) can address the problem. Another approach is to concentrate on methane and nitrous oxide abatement. Some progress has been made in abating high GWP gases in the flat panel display and semiconductor markets in Japan and the US.

Similar successes have been reported in mitigating methane emissions from natural gas pipelines in Canada, as also developments in reducing emissions from coal mine ventilation air in Australia.

## Zero Emission Power Plants

Zero or near-zero emission power plants are the ultimate aim for future power generation. Some progress has occurred in implementing practical and affordable zero emission power plants. Noteworthy are the Canadian clean Power coalition plants for demonstrations of zero emission technology, a retrofit and a rebuild, to be operational by the year 2010. The Mitsubishi Heavy Industries has improved its flue gas carbon dioxide recovery process which includes a new absorber packing with 60 per cent lower cross sectional area.

# Ecological and Climatic Thresholds

According to Maslin (2004, 2004a), one of the most pressing questions concerning future climate change is: how is it going to affect the terrestrial vegetation, especially in the tropics, where the future of the Amazon rainforest and the continued viability of current agricultural practices are at stake? Studies of past climates may suggest how quickly vegetation is able to respond to climate change. Whereas Hughes *et al.* (2004) stated that tropical vegetation in Venezuela has in the past responded to climate change within less than half a century. Jennerjahn *et al.* (2004) reported vegetation response time of 1000 to 2000 years in north-east Brazil. It appears than such a wide difference between the estimated ecological response times in the above two cases points to how different parts of the tropics respond to rapid climate changes.

Jennerjahn *et al.* (2004) found that climate thresholds and ecological or vegetation thresholds are not always the same.

According to Maslin (2004a) and Jennerjahn *et al.* (2004), different dry season lengths probably produce a more sensitive and direct relationship between climate and vegetation in Venezuela than in northeast brazil. Ecological thresholds are by no means unique to the tropics. Tzedakis *et al.* (2004) studied pollen records in a marine core from southwest Europe and showed that the duration of tree cover can be shorter than the length of interglacial periods. Such an observed shortened period of tree cover appears to be caused by millennial-scale climate deteriorations during each interglacial. These short, cold dry events trigger an ecological threshold, causing tree populations to crash.

The vegetation does not always recover afterwards when the climate has returned to normal interglacial conditions. Since pollen records are paralleled by variations in the Antarctic atmospheric methane record, there seems to be a more global vegetation response to these events and subsequent lack of recovery (Maslin, 2004a).

In some areas, the relationship between climate change and vegetation may not be reversible, suggesting that once an ecological threshold has been crossed, a return to the previous climatic condition does not guarantee a similar reversal in vegetation. The same kind of bifurcation was proposed for the relationship between surface ocean salinity and the rate of deep-ocean circulation (Rahmstorf, 1995); it seems to be more prevalent in the climate system than believed hitherto (Maslin *et al.*, 2001). The distribution of various vegetation types (biomes) is regulated by several different climatic factors, for example, annual and seasonal temperature, annual and seasonal precipitation, and the atmospheric carbon dioxide concentration. Jennerjahn *et al.* (2004) provided a good example of a tropical ecological threshold which is chiefly controlled by the duration of the dry seasonal and not the total annual rainfall. Besides this, it is also important how these climatic factors interact. It has emerged from modeling work that the colder glacial temperatures counterbalanced the worst effects of the drier conditions and lower atmospheric carbon dioxide concentrations by reducint water and carbon loss. For Amazon, the combination of two different climatic thresholds (aridity and cooling)did not produce a significant ecological threshold (Maslin, 2004a). In the case of vegetation, while distributions can vary on timescales of less than five decades, it appears that unless we understand ecological thresholds and their relationship to climate change, it is not possible to predict how or when vegetation will change as a result of global warming; also whether or not these changes will be reversible (Maslin, 2004a).

Simple measures such as fixing of leaky gas pipes or capping of landfill sites can cut emissions of methane, a powerful and fast-acting greenhouse gas. But, under the Kyoto Protocol's rules, nations will get little rewards for such work, as compared to the much more expensive efforts to cut carbon dioxide. Doubtless, cutting carbon dioxide emissions is essential, but most nations have neglected methane and the near-term benefits it could bring.

The Kyoto Protocol was finalized in Marrakesh in November, 2001. It covers six GHGs released by human activity, including carbon dioxide, methane, nitrous oxide and hydrofluorocarbons. These gases differ from each other in respect of their warming potential and atmospheric lifetime. Under the Protocol, a 'hundred-year rule' is used to calculate their effect on greenhouse warming; but this introduces an anomaly-it increases the emphasis on carbon dioxide (atmospheric lifetime about a century) and downgrades the significance of methane. Methane is the second most important GHG, but its lifetime is only about a decade. Releasing 10 kg of methane into the air today can warm the world about as much over the next ten years as one tonne of carbon dioxide. However, since so much of the gas disappears, its warming potential over a hundred years is only a fifth of that of carbon dioxide. It is this latter figure that was used to draw up Kyoto emissions targets.

The effect of the hundred-year rule is to give a low priority to cutting methane emissions, even though these efforts can be much cheaper and can have a stronger short-term impact. It makes much sense to attempt to reduce non-carbon dioxide gases such as methane because, in some ways, it is easier. Also, perhaps the Kyoto rules should be amended. Adoption of a 20 years time horizon may significantly increase incentives for reducing methane.

## Global System Reorganisation

The ocean-atmosphere system can undergo rapid and global reorganization. Some large-scale atmospheric and oceanic processes often lead to such reorganization over the entire tropical Atlantic Ocean (Chiang and Koutavos, 2004) and cause changes in the seasonal southward migration of the Intertropical Convergence Zone (ITCZ), the rainfall band that spans the Atlantic (Hastenrath and Heller, 1977). If the southward migration of the Itcz in February-May stops short of northeastern brazil, the rainy season usually fails. While the present-day disturbances last a few years at most, I the past, wet episodes persisted for several centuries (Wang et al., 2004). Wang et al., identified an apparent synchrony of wet periods in northeastern Brazil with climate changes near and far. The timing of the wet periods can be accurately determined using uraniumthorium dating which relies on decay products of radioactive uranium-238 (U-238). Cold episodes over the North Atlantic during the last glacial period are assumed to have resulted from abrupt variations in the Atlantic's thermohaline circulation (Rahmstorf, 2002) whereby a density-driven circulation carries warm and salty surface waters to the northern North Atlantic. These waters release their heat to other atmosphere, so becoming cold and dense, then sink, and return southward as a deep current. This circulation can be shut down or weakened by freshening the ocean surface waters from glacial melt water or by some other means. The weakening reduces the heat transported to the North Atlantic and cools the climate there. These changes in the thermohaline circulation are essentially high-latitude processes apparently not connected to the tropics; but recent findings point to the presence of abrupt events in the northern tropical Atlantic. Pacific and Indian Ocean regions (Burns et al., 2003).

Wang et al. (2004) showed their existence in the southern tropics. This recent development changes our understanding of abrupt climate change. Changes in the intensity and distribution of atmospheric convection in the tropics greatly influence the global climate by altering the global atmospheric circulation-analaogous to how the ENSO system in the tropical Pacific affects global climate today (Cane, 1998). One consequence of this may be that the transport of water vapour will change and possibly affect the salinity-and hence the density-of the North Atlantic surface waters that determines the strength of the thermohaline circulation (Chiang and Koutaveas, 2004).

## Effects on Biota

It is well known that most organisms only have three types of responses that promote their survival in changing environments. According to Peck (2005), they can:

1. Cope within existing physiological flexibility;
2. Adapt to changing conditions; or
3. Migrate to sites that allow survival.

Species inhabiting coastal seabed sites around Antarctica have poorer physiological capacities to cope with change than species found elsewhere. They die when temperatures rise by only 5 to 10°C above the annual average, and many species lose the ability to carry on essential functions, for example, swimming in scallops or burying in faunal bivalve mollusks when temperatures are raised only 2 to 3°C. The ability to adapt, or evolve new characters to changing condition depends at least partly on generation time. Antarctic benthic species grow slowly and develop at rages often x 5 to x10 slower than otherwise similar temperate species. They also live long, and show deferred maturity. Longer generation times reduce the opportunities to produce novel mutations, and weaken adaptation to change. Intrinsic capacities to colonies new sites and go away from worsening conditions depend on adult abilities to locomote over long distances, or for reproductive stages to drift for extended periods. The slow development of Antarctic benthic species implies that their larvae do spend extended periods in the water column. However, whereas most continents have coastlines extending over a wide range of latitudes. Antarctica is almost circular in outline, remains isolated from other oceans by the circumpolar current, and its coastline covers only a few degrees of latitude.

Thus, in a warming environment, there are fewer places to migrate to. On all three major criteria, Antarctic benthic species appear less capable than species elsewhere to change in ways that can enhance survival (Peck, 2005).

Although species have responded to climatic changes throughout their evolutionary history, a primary concern for wild species and their ecosystems is the rapid rate of change that has occurred in the last century. Root *et al.* (2003) assembled information on species and global warming from numerous studies for meta-analyses. These analyses revealed a consistent temperature related shift, or 'fingerprint' in species ranging form mollusks to mammals and from grasses to trees. Indeed, more than 80 per cent of the species that show changes are shifting in the direction expected on the basis of their known physiological constraints. A significant impact of global warming is already discernible in animal and plant populations. The synergism of rapid temperature rise and other stresses, in particular habitat destruction, could easily disrupt the connectedness among species and lead to a reformulation of species communities, reflecting differential changes in species, and to numerous threats and possible extinctions (Penuela and filella,2001; Mccarthy *et al.*, 2001; Schneider and Root, 2002).

# Global Dimming

## Masking of the Greenhouse Effect by Air Pollution

Emissions resulting from human activities have substantially increased the numbers of aerosol particles in the atmosphere since pre-industrial times. These particles consist of a variety of chemical species, the major anthropogenic species being sulphates and particles containing carbon, such as soot and organic matter (Lohmann and Wild, 2005).

Recent measurements of the sunlight passing through the atmosphere and reaching the Earth suggest that our air is getting cleaner, largely because of reduced industrial emissions and the use of particulate filters (see Liepert, 2002; Wild *et al.*, 2004). Unfortunately, there is one concern that aerosols and dust in the air may have been protecting us from the worst of global warming. We are not sure how extra solar radiation will affect future temperatures (Schiermeier, 2005).

A reduction in the amount of sunlight reaching the Earth's surface, known as 'global dimming' has been record since the 1950s, but it s global nature dawned only recently. It aroused the apprehension that the global dimming might be used as an excuse to ignore the consequences of global warming. The fear, however, is unfounded-since 1990 the dimming has been replaced by brightening (wild *et al.*, 2005; Pinker *et al.*, 2005). The amount of radiation reaching earth's surface has fallen by about 5 per cent between 1960 and 1990,but the trend has since reversed nearly everywhere- although the total amount of radiation has not yet reached 1960 levels (Schiermeier, 2005).

Aerosol particles can influence the climate system through three different physical mechanisms. Firstly, they scatter-and-absorb-radiation from the Sun. Secondly, they can scatter, absorb and emit long-wave radiation. These two mechanisms are referred to a aerosol direct effects and cause a net negative forcing at the top of the atmosphere. In the third mechanism, aerosol particles serve as nuclei on to which cloud droplets and ice crystals form. For a given amount of cloud condensate, an increase in cloud condensation nuclei (CCN) leads to a higher surface area, and hence greater reflection of solar radiation-the so-called 'cloud albedo effect'.

At the same time, these more numerous, smaller cloud droplets retard precipitation because smaller droplets are less likely to collide and form precipitation size drops (Jones *et al.*, 1999).since this increases the lifetime of polluted water clouds, it is termed the 'cloud lifetime effect'. On the other hand, aerosols that absorb solar radiation, such as back carbon (*i.e.*, soot) or to a lesser degree mineral dust, warm up the surrounding air which prevents cloud formation because the atmosphere becomes more stable, or even leads to an evaporation of cloud droplets. This 'semi-direct effect' counteracts some of the negative aerosol forcing at the top of the atmosphere (Lohmann and Feichter, 2005).

The decline of solar radiation from 1961 to 1990 was 1.3 per cent per decade over the global land surface. While the greatest declines (3.35 per decade) occurred over the United States, surface solar radiation also declined over Europe and the former Soviet Union. During the dry winter monsoon season over the Indian Ocean region, anthropogenic aerosols-especially the highly absorbing aerosols, can decrease the average solar radiation absorbed by the surface by 15 to 35 $Wm^{-2}$ (Raman than *et al.*, 2001). The atmosphere has become cleaner and more transparent. The decrease in industrial pollutants released in many parts of Europe since the 1980s was probably a major factor.

Despite the above decrease in surface solar radiation, land surface temperatures actually increased by 0.4 K between 1961 and 1990 (Jones *et al.*, 1999); the increase in the downward long-wave radiation was not high enough to outweight decreased insolation. Surface evaporation must, therefore, have decreased, suggesting that the observed intensification of the hydrological cycle over extra tropical land is more, probably, due to increased moisture advection from the oceans than due to increased local moisture release through evaporation.

There has been continued dimming in some highly polluted areas (*e.g.* India) where extensive clouds of among from burning fossil fuels and wildfires darken the sky for long periods each year.

But there has been a brightening trend in China, despite the country's booming, fossil-fuel-intensive industry; probably, because the use of clean-air technologies in China is actually more widespread and efficient than believed hitherto.

The issue now is how the trend towards cleaner air will affect global temperatures. It is already clear that the greenhouse effect has been partly masked in the past by air pollution. Until 1990, air pollution. Until 1990, air pollution probably protected us from at least 50 per cent of the warming that would have otherwise occurred.

Various simulations using a global climate model coupled to a mixed-layer ocean model with increasing aerosols and GHGs due to human activity from pre-industrial times to the present, suggest that solar radiation decreases at the surface resulting from increases in optical depth due to direct and indirect anthropogenic aerosol effects, could be more important for controlling the surface energy budget than the GHG induced increase in surface temperature. Such components of the surface energy budget as thermal radioactive flux and sensible and latent heat fluxes decreased in response to the reduced input of solar radiation- a mechanism that could explain the observations of decreased pan evaporation over the last half a century and the historic surface temperature. Because evaporation equals precipitation on the global scale in equilibrium climate simulations, a reduced latent heat flux led to reduced precipitation.

Climate model simulations have prompted the suggestion that the decrease in global mean precipitation from the preindustrial times (1861-1880) to the present day, may reverse into an increase in global mean precipitation of about 1 per cent for the period 2031-2050 compared to 1981-2000. This is because increased warming due to GHGs will outweigh sulphate colling in the mid-21st century (see Lohmann and Wild, 2005).

In South Asia, absorbing aerosols in atmospheric brown clouds probably had an important role in the observed South Asian climate and hydrological cycle changes, and could have masked as much as 50 per cent of the surface warming due to the global increase in GHGs (Ramanathan *et al.*, 2005). Simulations suggest that should current trends in emissions continue, the South Asian subcontinent could suffer from a doubling of the drought frequency in future decades (Lohmann and wild, 2005). It appears likely that the trend of solar dimming will end in major industrialized regions, but surface solar radiation decline will continue over India and Southern Africa.

## Global Warming and Disease Incidence

Human-induced damage to the biosphere carries important implications for health. Being global, the underlying processes and the natural systems affected are part of earth's life supporting infrastructure. Unlimited exploitation of fossil fuels, cultivation, and industrialization have increased the GHGs to high levels since the advent of civilization; ruthless destruction of tropical rain forest has reduced the natural method of reduction of these gases in the atmosphere. The population growth much above the holding capacity of the planet has contributed to increasing global mean temperature over the last two centuries. This type of anthropogenic health risk due to global warming differs considerably from the more local environmental health hazards that are usually addressed on a toxicological or microbiological plane. Global warming strongly affects the distribution and incidence of many infectious diseases both directly and indirectly. Some direct effects of a rise in temperature (particularly increases in the frequency and intensity of heat waves) include deaths from cardiovascular and cerebrovascular disease among the elderly. Indirect secondary effects are exemplified by changes in vector-borne diseases or crop production, and tertiary effects by social and economic impacts of environmental refugees and conflict over freshwater supplied (Haines and Fuchs, 1991).

In general, those infections which are carried by the human host and are transmitted from person to person [*e.g.*, influenza, human immunodeficiency virus (HIV), measles, and tuberculosis (TB)] can usually appear anywhere in the world, unless a population is geographically isolated or immune (because of past infection or immunization).

Other infections require some environmental stress, climate conditions, or the presence of a specific arthropod vector, animal, or other intermediate host in order to persist in an area. These infections are geographically localized even though their area of distribution expands or shrinks in response to various factors, including weather and environmental changes. The physiochemical environment influences the pathogen, the vector, intermediate and reservoir hostes, vegetation, and human behaviour and activities (see Sur *et al.*, 2003). The physiochemical environment influences the distribution, abundance, and survival of living organisms. Temperature (mean, minimum, maximum and amplitude), humidity, rainfall (pattern as well as amount), types of vegetation, animal life, and distribution of water (such as ponds, rapidly flowing streams, etc.) often determine whether a region can support some specific vector.

The epidemiology of many vector- borne infections is characterized by seasonality through influences of weather patterns on host, vector, and microbe. the epidemiology of infections may also be periodic or cyclic, over many years, possibly tied into cyclical weather patterns (*e.g.* periodic climate perturbations such as the ENSO which causes fluctuations in temperature and rainfall) or the accumulation of susceptible people through birth or migration into an area (Sur *et al.,* 2003). Cholera is a bacterial infection occurring in response to changing environmental factors, such as warming of water temperature, increase in pH and nutrients, and decrease in salianity. The causal bacteria proliferate in areas where sanitation is poor and drinking water is untreated.

Extreme weather events also precipitate cholera epidemics by destroying land and displacing populations. Crowded living conditions enhance the spread of tuberculosis (TB). A good freshwater indicator of warming could be nuisance algal blooms, particularly those caused by cyanobacteria such as Microcystis (see Kumar, 1999). Bloomforming blue-green algae form large floating masses or flocs in village ponds. Certain species can produce toxins and cause rashes, eye irritation, vomiting and diarrhea in people who bathe in waters infested with algal blooms. The blooms are considered to be caused by a combination of calm sunny periods and sufficient nutrients, notably phosphorus (Elder *et al.,* 1993).

## Possible Impacts of Warming Climate on Water Availability and Lake Productivity

Currently available climate models predict a near-surface warming trend under the influence of increasing levels of GHGs in the atmosphere. Besides the direct effects on climate; for example, on the frequency of heatwaves, this increase in surface temperatures has substantial implications for the hydrological cycle, especially in those regions where water supply chiefly comes form melting snow or ice. Water is, of course, essential to human sustenance. More than a half of the world's potable water supply is extracted from rivers, either directly or from reservoirs. The discharge of these rivers is quite sensitive to long-term changes in both precipitation and temperature, particularly in the snowmelt-dominated parts of the world. White changes in the amount of precipitation affect the volume of runoff, temperature changes mostly affect the timing of runoff. Increasing temperatures promote earlier runoff in the spring or winter, and reduced flows in summer and autumn at least in the absence of changes in precipitation (Barnett *et al.,* 2005).

In a warmer world, thee is less winter precipitation because snow and its melting occurs earlier in spring. Even in the absence of any changes in precipitation intensity, both of these effects lead to a shift in peak river runoff to winter and early spring, away from summer and autumn when demand is highest (Barnett *et al.,* 2005). Where storage capacities are not sufficient, much of the winter runoff is immediately lost to the oceans. As over one-sixth of the Earth's population depends on glaciers and seasonal snow packs for their water supply, the consequences of these hydrological changes for future water availability could be quite severe.

A common general perception is that increasing GHGs will cause the global hydrological cycle to intensify (IPCC, 2001) with benefits for water availability (Barnett and Pennel, 2004); however, a possible exacerbation of hydrological extremes may lower the benefits to some extent. On a global scale, the strongest changes in the hydrological cycle due to warming are predicted for the snow-dominated basins of mid-to higher latitudes, because adding or removing snow cover changes the ability of snow packs to act a reservoir for water storage (Nijssen *et al.*, 2001). The stream-flow regime in snowmelt-dominated river basins is highly sensitive to wintertime increases in temperature.

## Impacts on Regional Water Supplies

Barnett *et al.* (2005) examined some case studies form the showmelt-dominated parts of the world. One of these cases was the Rhine river in Europe. Climate-change simulations have projected a warming in this river basin of 1.0 to 2.4°C over present values by the middle of the century (IPCC, 2001). As per hydrological simulations, this warming will shift the Rhine basin from a combined rainfall and snowmelt regime to a more rainfall-dominated one, resulting in an increase in winter discharge, a decrease in summer discharge, increases in the frequency and height of peak flows, and longer and more frequent periods of weak flow during summer (Middelkoop *et al.*, 2001). Some socioeconomic implications are a reduction in water availability for industry, agriculture and domestic use during the season of peak demand (which is also stressed by an increase in summertime demand due to higher temperatures): an increase in the number of weak flow days during which ships cannot be fully loaded on major transport routes (causing an increase in transportation costs); a decrease in annual hydropower generation in parts of the basin; and a loss in revenue due to a shortened ski season (Barnett *et al.*, 2005). This and other studies on western Us and Canadian prairies show that current demands for water in many parts of the world will not be satisfied under plausible future climate conditions, much less the demands of a larger population and a larger economy. The science behind this statement is temperature-driven, not precipitation-driven-this strengthens the conclusions because all current models predict a warmer future world. The other key factor affecting water availability is the lack of enough reservoir storage to manage a shift in the seasonal cycle of runoff. Current information about the climate-related water challenges facing most of the world is to strong that major future problem areas can be identified.

It appears that the most critical region in which vanishing glaciers will adversely affect water supply in the coming few decades will be China and parts of Asia, including India [together forming the Himalaya-Hindu Kush (HHK) region], because of the region's huge population. The ice mass voer this HHK region is the third-largest on Earth, after the Arctic/Greenland and Antarctic regions. Its hydrological cycle is complicated by the Asian monsoon, but there is no doubt that melting glaciers provide an important source of water for the region I the summer months: as much as 70 per cent of the summer flow in the ganga and 50 to 60 per cent of the flow in other major rivers (Singh and bengtsson, 2004). In China, 23 per cent of the population lives in the western regions, where glacial melt constitutes the major dry season water source (Gao and Shi, 1992).

The glaciers of the HHK region are melting and the melting is accompanied by a long-term increase of near-surface air temperature (see Barnett *et al.*, 2005). The entire

HHK ice mass has decreased the last two decades, and the rate of melting is accelerating (Meier and Dyurgerov, 2002). Barnett *et al.* (2005) underscored the urgency of the problem for nations in the sensitive areas, particularly those whose water supplied rely on mid-latitude glaciers. An immediate issue in need of urgent action is: how much the sensitive nations can do in the face of the uncertainty surrounding the several decades of warming that will elapse as a result of past actions, even if greenhouse emissions were halted at today's level (Hansen *et al.*, 2005)? One thing that seems certain is that strategic planning will be initiated by the looming prospect of diminished water supplies.

## Lake Productivity

McKnight *et al.* (1996) documented the effects of climate warming on the chemical and physical properties of lakes in North America. However, biotic and ecosystem scale responses to climate change have been only estimated or predicted by manipulations and models. O'Reilly *et al.* (2003) showed that climate warming is affecting productivity in Lake Tanganyika, East Africa- a lake that has in the past supported a highly productive pelagic fishery that now provides 25 to 40 per cent of the animal protein supply for the populations of the surrounding countries (Coulter, 1991; Molsa *et al.*, 1999).

Warming of surface waters and declining fish catches in Lake Tanganyika have been clearly linked to global climate change (see Verschuren, 2003). It appears that the impact of global warming on natural ecosystems is starting to affect local economies, Although the effects of global climate change on ecosystems and the geographical distribution of species can be seen clearly (Parmesan and Yohe, 2003), convincing examples of their impact on the livelihood of sizeable human populations are still very few.

O'Reilly *et al.* (2003) combined documentary and field data on long-term ecosystem dynamics in Lake Tanganyka, to show how increased heat accumulation by this deep tropical lake, linked to climate warming, has markedly reduced fish yields. In keeping with regional warming trends since the beginning of the twentieth century, a rise in surface-water temperature has enhanced the stability of the water column. A regional decrease in wind velocity has led to reduced mixing, decreasing deep-water nutrient upwelling and entrainment into surface waters. Primary productivity appears to have decreased by about 20 per cent, implying a roughly 30 per cent decrease in fish yields.

The work of O'Reilly *et al.* (2003) proves that the impact of regional effects of global climate change on aquatic ecosystem functions and services can be stronger than that of local anthropogenic activity or overfishing.

Lake Tanganyika is the second deepest lake in the world and the second largest by volume (after Lake Baikal) and the second largest tropical lake by surface area (after Lake Victoria). It holds the second greatest biological diversity (after Lake Baikal). This high species richness is accounted for by endemic fishes, snails and crustaceans found in nearshore environments. But its high ecosystem productivity, with a carbon transfer from algae to fish is comparable to that of the most efficient marine fisheries, and comes mostly from the offshore, open-water food web, which is fairly poor in species (Verschuren, 2003).

At 60 per cent by weight of overall catch, sardines are the dominant commercial fish in Lake Tanganyika. Since the late 1970s, the annual sardine catch has been falling by 30 to 50 per cent, to less than 2,00,000 tons today. This decline is being attributed to global climate change (Verburg *et al.*, 2003; O'Reilly *et al.*, 2003). According to these authors, rising air and surface–water temperatures over the past century increased the thermal density contrast between surface and deep water in this lake. Assisted by rather slack winds in recent decades, this stronger density contrast reduced the effectiveness of wind-driven mixing in bringing nutrients essential for primary production from the deep-water nutrient reservoir to the surface, which led to a decrease in algal biomass and consequent starvation of the upper levels of the aquatic food web, including the economically important zooplankton-eating fishes (O'Reilly *et al.*, 2003).

# Chapter 5

# Ozone Layer and Utraviolet-B Radiation

## General Description

Although there are several linkages between ozone depletion and climate change, ozone depletion is not a major cause of climate change. Atmospheric ozone has two effects on the temperature balance of the Earth:

1. It absorbs solar ultraviolet radiation, which heats the stratosphere, and

2. It absorbs infrared radiation emitted by the Earth's surface, effectively trapping heat in the troposphere.

Therefore, the climate impact of changes in ozone concentration varies with the altitude at which these ozone changes occur. The major ozone losses that have been observed in the lower stratosphere due to the human-produced chlorine-and bromine containing gases produce a cooling effect on the Earth's surface. On the other hand, the ozone increases that are estimated to have occurred in the troposphere because of surface-pollution gases have a warming effect on the Earth's surface, thereby contributing to the greenhouse effect (UNEP, 1999).

In the year 1974, a landmark scientific paper first pointed out that chlorine-containing substances called chlorofluorocarbons (more commonly known as CFCs) pose a threat to the ozone layer. Ozone is found in a diffuse atmospheric layer at stratospheric altitudes, between 20 and 40 km, it acts as a shield against harmful solar ultraviolet radiation. Since then, it has been proved that ozone decreases are occurring. In the Antarctic, major and unanticipated ozone holes have been detected during the spring; ozone depletions exceeding 50 per cent have also been measured. The same processes are occurring in the Arctic.

# Ozone and Climate Change

For millennia, the basic atmospheric composition probably did not change, although some rare compounds such as carbon dioxide

have varied. The ozone molecules, concentrated largely between altitudes of 15 to 35 km have determined the temperature structure of the stratosphere. By absorbing the harmful ultraviolet radiation, they safeguarded life on this planet. But over the past half a century, we humans have placed the ozone layer in jeopardy. Unwittingly, humans were releasing into the atmosphere chemicals that destroy part of the life-protecting ozone layer and upset the nature's delicate balance. It was in the early 1970s that scientific findings highlighted the potential of chlorofluorocarbons (CFCs) and halons to destroy ozone with strong environmental implications. However, it was not until the mid- 1980s that convincing evidence of ozone destruction, as demonstrated by the dramatic ozone decline in the Antarctic spring, became available. Compounds released into the atmosphere by humans are now known to deplete ozone. In fact, the various global environmental risk-management treaties concluded under theUnited Nations (UN) umbrella that took action to prevent an emerging problem, were both extraordinary and unprecedented and provided a model for further international action against global threats to the environment. The acting to defend the ozone layer will surely rank as one of the great international achievements of the 20th century.

**Table 5.1: Some 1 and marks in the history of Ozone (after Bojkov, 1995)**

| Year | Landmark |
|------|----------|
| 1839 | C.F. Schonbein discovers ozone |
| 1860 | Surface ozone measured at many allocations |
| 1880 | Strong absorption band of solar radiation between 200 and 320nm attributed to upper atmosphere ozone |
| 1913 | UV measurements prove that most ozone is located in the stratosphere |
| 1920 | First quantitative measurements of the total ozone content |
| 1926 | Six Dobson ozone spectrophotometers are distributed around the world for regular total ozone column |
| 1929 | Discovery of the Umkehr method for vertical ozone distribution |
| 1934 | Ozone sonde on balloon shows maximum concentration at about 20 km. |
| 1957 | WMO adopts standard procedures for uniform ozone observations, the Global Ozone Observing System ($GO_3OS$) established. |
| 1966 | First ozone measurements from satellites |
| 1971 | Ozone can be destroyed by NOx |
| 1974 | ClOx chemistry as an ozone-destroying mechanism |
| 1975 | First international assessment of the state of global ozone by WMO |
| 1977 | Plan of action on Ozone Layer established by UNEP and WMO |
| 1984 | Unusually low (approx. 200m atm cm) total ozone at syowa, Antarctica, in October 1982, first reported at the Ozone commission Symposium in Halkidiki, (Greece) but its significance was recognized only in 1985. |

*Contd...*

**Table 5.1–Contd...**

| Year | Landmark |
|---|---|
| 1985 | Vienna Convention for the Protection of the Ozone Layer. Data from Halley station on the existence of an ozone hole during Antarctiic springs since the early 1980s published by the British Antarctic Survey. |
| 1987 | Montreal Protocol on Substances that Deplete the Ozone Layer. Basic assessment of the state of the ozone initiated by the International Ozone Trends Panel. |
| 1988 | Decrease of ozone concentrations by ~10 per cent per decade in the lower stratosphere documented; proof from NASA Antarctic Campaign that active chlorine and bromine byproducts of human activities are the cause of the Antarctic-spring ozone hole. |
| 1990 | London amendment to strengthen the Montreal Protocol by phasing out all CFC production and consumption by 2000. |
| 1991 | The 1991 WMO/UNEP Ozone assessment: ozone is declining all year round and everywhere except over the tropics; very large concentrations of ClO measured in the Arctic confirms concerns for potential stronger ozone decline. |
| 1992 | Copenhagen amendment further reinforces the Montreal Protocol by phasing out CFCs by the end of 1995, adding controls on other compounds. |
| 1992-94 | Extremely low ozone values (approx. 100m atm cm) during Antarctic spring and largest area approx. 24m. km² covered; also the lowest ever ozone values recorded during the northern winter spring seasons all indicating increasing destructive capability by increasing chlorine and bromine concentrations in the stratosphere. |
| 1995 | Record low ozone values (exceeding 255 below long-term average) observed from January to March over Siberia and a large part of Europe. |

Ozone, as an atmospheric trace gas, has become an issue of global prominence since the 1970s because its normal atmospheric concentration is under attack from human activities. Its decline was detected from information collected by the World Metereological Organisation (WMO) Global Ozone Observing System (GOOS) since the mid- 1950s from more than 150 stations and, more recently, from specialized satellites. A connection has been established between human-made compounds and these ozone losses.

Ninety-nine per cent of the air we breathe is composed of the two gases nitrogen (78 per cent) and oxygen (21 per cent) whose ratio has not changed for millennia. Rare components such as water vapour, carbon dioxide, methane, nitrous oxide, ozone and inert gases (*e.g.*, argon, helium, neon) make up less than 1 per cent of the volume of air. In every ten million air molecules, on average, there are only three molecules of ozone.

The total ozone in the atmospheric column at any particular place is variable and is primarily determined by large-scale atmospheric dynamics. Although so rare, ozone molecules have a vital role in the life of our planet. They absorb harmful solar ultraviolet radiation (below about 320 nm) and shield us (also all other animals and plants) from damage. Ozone also determines the thermal structure of the stratosphere (10-50 km) where temperature increases with height. Most of the atmospheric ozone is located in the stratosphere, where it reaches its highest concentration between about 19 and 23 km above the surface of the Earth. The air temperature, after a rapid decrease with height in the troposphere, increase I the stratosphere because ozone absorbs radiation.

Even as the Sun's energy produces new ozone, these molecules suffer continuous destruction by natural compounds containing oxygen, nitrogen, hydrogen and chlorine or bromine. Such chemicals were all present in the stratosphere long before humans began polluting the air. Nitrogen compounds come from soils and the oceans, hydrogen comes mainly from atmospheric water vapour and chlorine comes from the oceans in the form of methylchloride and methylbromide. In more recent times, humans have greatly upset the balance between production and destruction. By releasing additional chlorine-and bromine-containing chemicals (*e.g.* CFCs) into the atmosphere, they have enhanced the destruction of ozone leading to lower ozone concentrations in the stratosphere. The opposite process is occurring in the lower part of the atmosphere (up to 10-12km) *i.e.* the troposphere, where, largely as a result of combustion, the local ozone concentrations in the northern middle latitudes have more than doubled in the past century. Although this tropospheric ozone increase cannot compensate for the stratospheric decline, the changes could influence the radiatived balance of the Earth-atmosphere system.

## Ozone Measurement and Units

Total ozone is equal to the amount of ozone contained in a vertical column of base 1 $cm^2$ at standard pressure and temperature. It is expressed in units of pressure, typically about 0.3 atmosphere centimeters. The milli-atmosphere centimeter (m.atm.cm), commonly called the Dobson unit, corresponding to an average atmospheric concentration of approximately one part per billion by volume (1 ppbv) of ozone, is concentration of approximately one part per billion by volume (1 ppbv) of ozone, is commonly used. The distribution of ozone is not uniform through the vertical column.

The world average is about 300 Dobson units. It varies geographically from about 230 to 500 dobson units. Total column ozone has, on average, its lowest values over the equatorial belt and increases with latitude.

The tropopause is the isothermal region that separates the troposphere from the stratosphere. It lies at an altitude of 8 to 10 km at polar latitudes and nearly 18 km over the equatorial belt.

## The Threat

In the mid- 1970s, F.S. Rowland and M. Molina discovered that a widely used class of very inert chemicals known as chlorofluorocarbons (CFCs) were carried to the stratosphere by convective air movements. There, they absorb high-energy photons from sunlight and release free chlorine. Once released, this free, reactive chlorine destroys stratospheric ozone through a series of catalytic reactions.

It subsequently emerged that bromine from the halaons used in some fire extinguishers is also released in the stratosphere, and has a stronger ozone-destructive effect. Some CFCs and halons can survive in the atmosphere for longer than a century. The halocarbons released over the past several decades threaten the ozone layer not only today but also long into the future. They carry thousands of tons of chlorine and bromine atoms into the stratosphere-many times greater than the chlorine reaching the stratosphere naturally from the ocean in the form of methyl chloride and bromide (Bojkov, 1995). The release of the two moss widely used CFCs, CFC-11 and CFC-12, grew from insignificant amount is the 1950s to more than 700 thousand tons a year in the early 1970s. It has declined in recent years a s a result of measures under the Montreal Protocol. But the atmospheric concentration is continuing to rise demonstrating the long lifetime of these compounds.

## Halocarbons

The generic term 'halocarbon' covers several human- produced gases that have carbon and halogen (fluorine, chlorine or bromine) atoms. Halocarbons include CFCs and halons. The CFCs were synthesized in 1928 since when they have been widely used as propellants in aerosol cans, in the manufacture of soft and hard foams, in refrigeration and air conditioning, and as cleaning solvents. So their atmospheric concentration increased rapidly. In the troposphere, halocarbons are quite inert, non-toxic, non-flammable, odorless and colourless. But, when they reach the stratosphere, particularly at and above the layer of maximum ozone (19-23 km), high- energy ultraviolet photons from the Sun release some chlorine or bromine atoms from the inert molecules. these atoms catalytically detach one oxygen atom from an ozone molecule, so converting ozone to molecular oxygen ($O_2$). In this process, ozone is depleted.

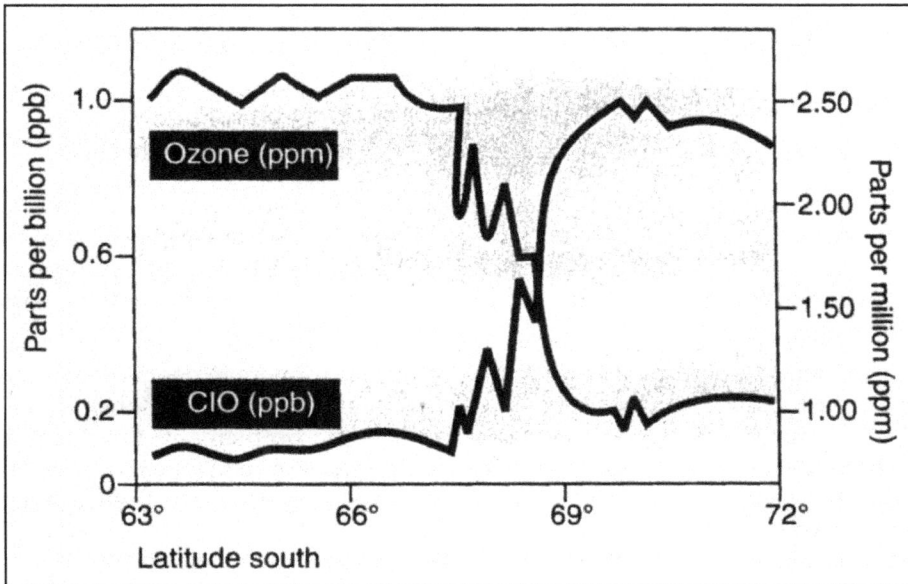

Figure 5.1: Measurements of CIO concentration and ozone from aircraft, CIO increases rapidly as the aircraft enters the polar vortex and the Antarctic ozone 'hole' (about 67°5). There is an inverse correlation of CIO with ozone decline in mid-September when the chemically unabalanced area in the vortex was sunlit.

In 1976, the WMO started its Global Ozone Research and Monitoring Project to provide advice to Members, the UN and other international organizations concerning:

☆ The extent to which anthropogenic pollutants might reduce the quantity of ozone in the stratosphere;

☆ The possible impact of changes in stratospheric ozone on climate and on solar ultraviolet radiation at eh Earth's surface.

## Observed Ozone Change

Ozone loss is very high over the frozone Antarctic continent because the winter circumpolar stratospheric vortex prevents extensive air exchange with mid-latitudes, producing low temperatures (below–80°C) which generate polar stratospheric clouds (PSC) of ice particles. Normally, chlorine and bromine are 'locked' into stable reservoir compounds (such as $ClNO_2$, $BrONO_2$, and HCl). The ice particles attract water vapour and absorb nitrogen compounds, when they drop with them to lower levels of the atmosphere, dehydrating and denitrifying the air in the stratosphere. When the sunlight returns in the early spring, these reservoir compounds change to active chlorine and bromine on the surface of the PSCs. These substances can break apart ozone molecules efficiently.

In October 1987, ozone concentrations over Antarctica fell to half their normal (1957-1978) levels, and the hole spread across an area the size of Europe. Since then, the decline has accelerated and extremes reached during the early and mid-1990s include:

☆ A record low ozone value of less than 100 m atm cm (705 deficiency) during several days;

☆ The greatest ozone hole ever of close to 24 million sq km; and

☆ An overall spring seasonal ozone deficiency of more than 40 per cent.

The ozone destruction is strongest in the lower stratosphere. During late September and October of 1992-1995, the ozone in Antarctica practically disappeared between 13 and 20 km. the ozone decline in spring, when the Antarctic stratosphere is isolated and extremely cold, is much greater than the decline in the summer season.

Extensive measurements have also been made in the northern polar region, and revealed that during winter-spring, the Arctic stratosphere experiences the same type of disturbed chemical composition, with high concentrations of destructive chlorine and bromine compounds, that causes the problems in the Antarctic. However, the ozone destruction over the Arctic is not so strong for two reasons: the stratospheric temperatures very rarely fall below –80°C due to the frequent extensive exchange of air masses with the middle latitudes; and the Arctic vortex normally dissipates in the late winter before sunlight can destroy much ozone.

In keeping with the evidence of polar ozone decline, the search for global erosion of the ozone layer was also intensified. It emerged that:

☆ Global ozone levels had declined by several per cent over the past several years-mainly during winter-spring over the middle and polar latitudes;

Uv light splits oxygen molecules (O₂) into two single oxygen atoms (O)

Free oxygen atoms combine with further O₂ molecules to form ozone(O₃)

UV Light

M

M

F Cl Cl C Cl Cl Cl

Ozone atoms also disintegrate naturally under the action of uv. but in the absence of anthropogenic compounds this process is well balanced with ozone formatin

Free chlorine atoms released from CFC Molecules (through reservoir molecules ClONO₂ and HIC) react wtih ozone, forming ClO₂ and O₂

ClO is short lived, it reacts with a free O atoms To form a further O₂ molecules, releasing the free Cl atoms ready to decompose another ozone molecule

Cl

Cl

Cl

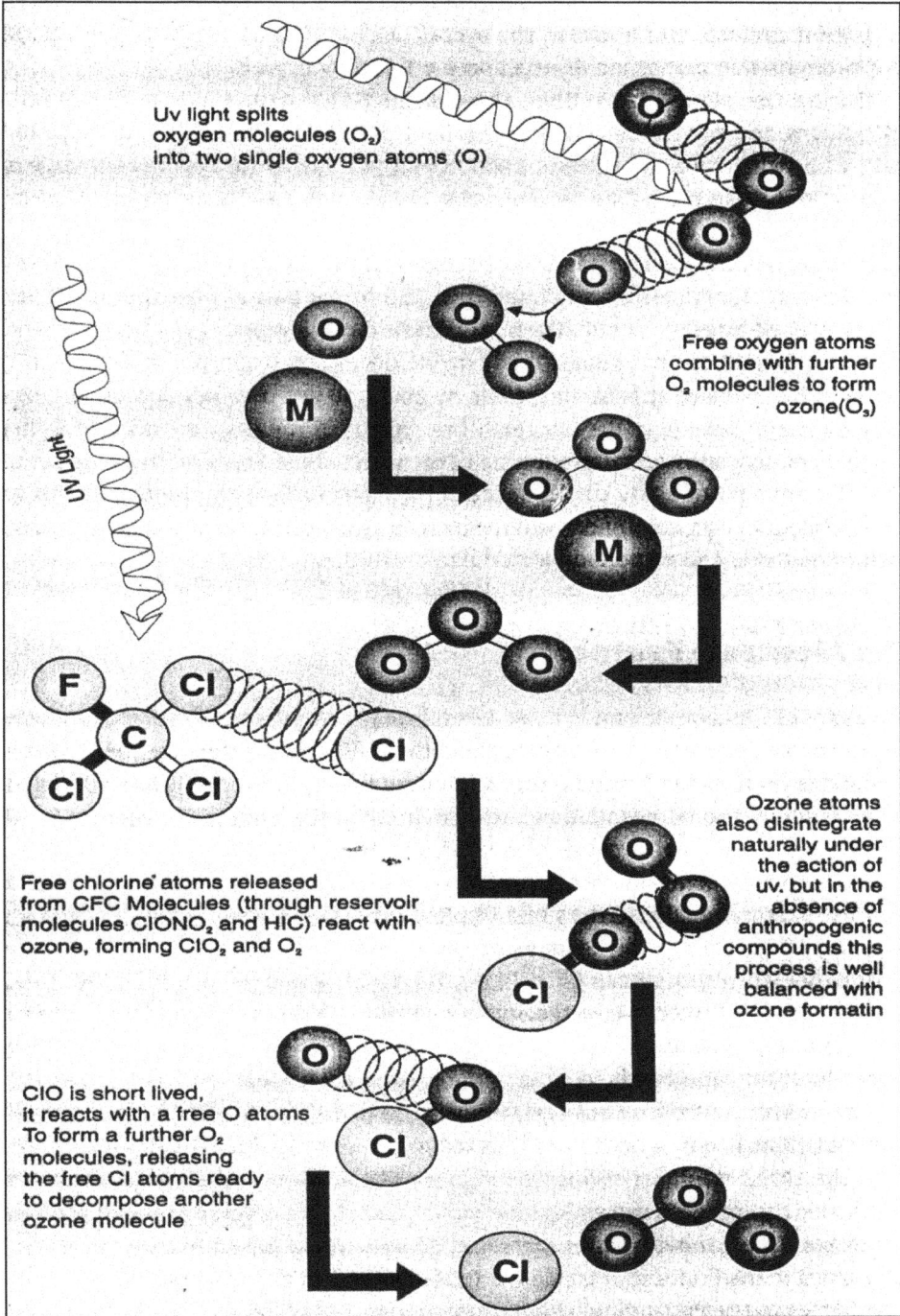

**Figure 5.2: Sequence of the destruction of ozone by active chlorine <Ci) released from a CFC-12 molecule.**

☆ Batural processes could not explain the entire ozone loss-evidence pointed towards anthropogenic halocarbons as the cause.

By the year 1991, the ozone values had dropped significantly not only in winter-spring but laos in summer. Since people spend much more time outdoors and ultraviolet-B radiation is the highest during summer, ozone loss at that time of the year poses a much greater threat to human health. An ozone decline of nearly 10 per cent has been recorded on the plot of long-term ozone values over Europe and North America. The principal interannual fluctuations are related mainly to stratospheric air transport variations, related to the phase changes of quasi-biennial oscillations (QBO) in the equatorial stratosphere; however, the overall decline is consistent with the chlorine-and bromine-iniliated ozone destruction predicted by models.

During the period 1984-1993, the overall global ozone average level fell to 297m. atm.cm. from 306 in 1964-1980 (about 3 per cent). But if we exclude the equatorial belt, where there are no significant ozone changes, the decline over the middle and polar latitudes was more than twice as large. Some continental-scale regions have even greater cumulative deficiencies.

Vertically, ozone decline has been strongest in the lower stratosphere. The ozone soundings at Hohenpeissenberg show that the ozone partial pressure in the 19 to 21 km layer declined by about 20 per cent in the past 25 years.

At the same time that stratospheric ozone decreases, tropospheric ozone-in the northern hemisphere at least-increases by about 10 per cent per decade. An ozone increase is also seen over the savannah-fire regions I the tropics. This tropospheric ozone increase in mostly a consequence of the effect of the Sun's radiation on specific air pollutants, particularly oxides of nitrogen from surface emissions, aircraft and automobile exhausts, combined with the increasing concentrations of other precursors such as methane and carbon monoxide (Bojkov, 1995).

## The Montreal Protocol

The use of CFC as a propellant in most aerosol spray cans was restricted in scandinavia and North America in the 1970s. throughout the 1980s, public pressure grew to extend limitations on use. The United Nations Environment Programme (UNEP) initiated a series of international negotiations. In March 1985, the Vienna convention for the Protection of the Ozone Layer was signed, laying the groundwork for control measures. Negotiations culminated in the signing by 24 countries, on 16th September 1987, of the Montreal Protocol on substances that Deplete the Ozone Layer. It came into force on 1st January 1989. As originally formulated, the control measures applied to two groups of ozone depleting substances: CFCs (11, 12, 113, 114, 115) and Halons (1211, 1301, 2402). Each party to the Protocol agreed to reduce consumption of controlled substances on the basis of the 1986 level.

The Protocol aimed to hold consumption at or below that level. After mid-1993, consumption of controlled CFCs was to be limited to 80 per cent of the 1986 level; and after mid-1998, it was to be further reduced to 50 per cent of the 1986 level.

## Box 1: Phase-out schedule

1. Chemicals covered by 1987 Montreal Protocol

| | | | |
|---|---|---|---|
| CFCs (11,12,113, 114,115) | Phase down 1986 levels by | 20 per cent | By the end of 1994 |

2. The Montreal Protocol (London Amendment-1990)

| | | | |
|---|---|---|---|
| CFCs (13,111,112, 211, 212,213,214, 215,216,217) | Phase down 1989 levels by | 20 per cent 85 per cent 100 per cent | 1993 1997 2000 |
| Halons (1211, 1301, 2402) | Freeze in 1992 at 1986 levels, then Phase down by | 50 per cent 100 per cent | 1995 2000 |
| Carbon tetrachloride | Phase down 1989 levels by | 85 per cent 100 per cent | 1995 2000 |
| Methyl chloroform | Freeze in 1993 Phase down 1989 level by | 30 per cent 70 per cent 100 per cent | 1995 2000 2005 |

3. Further strengthening of the Montreal Protocol (Copenhagen amendment 1992)

| | | | |
|---|---|---|---|
| CFCs | Phase out | 100 per cent | By the end of 1995 |
| Halons | Phase out | 100 per cent | By the end of 1995 |
| Carbon tetrachloride | Phase out | 100 per cent | By the end of 1995 |
| Methyl chloroform | Phase out | 100 per cent | By the end of 1995 |
| Methyl bromide | Freeze at 1991 levels | | By the end of 1995 |
| HCFCs | Phase down 1989 levels by | 35 per cent 90 per cent 99.5 per cent 100 per cent | By the end of 2004 By the end of 2014 By the end of 2019 By the end of 2029 |

In the three years following the signing of the Montreal Protocol. Scientists established the link between emissions of CFSs (and other ozone-depleting substances) and ozone depletion with much-increased confidence. In June 1990, at a meeting of the parties to the Protocol in London, attended by members of almost 100 countries, agreement was reached to eliminate consumption of the five CFCs in the 1987 agreement completely by the year 2000. This was to be achieved through intermediate step reductions of 20 per recent, 50 per cent in January 1992, 1995 and 1997 respectively. The same schedule was also agreed to concerning elimination of other CFCs- substances that have been used only in very low volume, if at all. It was also agreed to phase out the use of halons by the year 2000, starting with a 50 per cent reduction in 1995. The agreement left some scope for exemptions for essential uses of halons in high-hazard situations.

Controls on other ozone-depleting chemicals were also introduced. Carbon tetrachloride, with the exception for its use as a feedstock in the production of other substances, was to be phased out in the year 2000, with an intermediate cut of 85 per cent in 1995. Methyl chloroform, again with exemptions for feedstock use, was to be cut by 30 per cent in 1995 and 70 per cent in 2000, and eliminated altogether in 2005.

The Protocol called for 50 per cent reduction of CFCs by the year 2000, as follow:

In order to make the earlier phase-out fair for developing countries, the revised Protocol established an environmental fund paid for by developed nations. This fund gives technical assistance to developing countries as well as transfer of technology to enable them to switch over to more 'ozone friendly' replacement compounds.

Even with total international compliance with the existing agreements, chlorine and bromine concentrations will continue to increase in the stratosphere for some years. Peak global ozone losses may occur during the next ten years, when another 3 to 4 percent of the ozone will be destroyed (in some years more), before the rates of ozone decline level off and start falling (Figure 5.3).

It is hard to predict confidently when the ozone layer might recover. Figure 5.3 shows the calculated stratospheric chlorine-equivalent concentrations resulting from the predicted manufacture and consumption levels of halocarbons with full compliance to the original Protocol, its London amendment and the more stringent measures for phase out by January 1996 as accepted at the Copenhagen Conference (Nov. 1992). The stratospheric chlorine-equivalent concentration of 2 parts per billion by volume (ppbv), reached at the end of the 1970s, is a dangerous threshold for ozone depletion.

Current chlorine-equivalent concentrations exceed 3.7 ppbv,and are expected to increase for another few years. Without the Montreal Protocol, continuing use of CFCs and other ozone destructive compounds would have tripled the present stratospheric abundance of chlorine and bromine by the middle of the 21$^{st}$ century. such increases would have led to ozone depletion much larger than are being observed today.

Only full compliance with the Montreal Protocol, as reinforced by its subsequent amendments, will prevent chlorine levels from exceeding 4 ppbv and ensure that they begin to decline to reach the 2 ppbv level some time in the middle of the 21$^{st}$ century. Other things being equal, only then would the ozone layer return to normal pre-ozone-hole levels. The expected recovery of the ozone layer could not have been possible without the Montreal Protocol and its amendments.

The beneficial effect of the Montreal Protocol in checking the chlorine loading to the atmosphere can be seed from Figure 5.4.

Figure 5.5 compares the upper ranges of possible effects of various atmospheric gases and ozone changes indicated by the open bars, and the lower ranges are indicated by the solid bars.

An additional factor that indirectly links ozone depletion to climate change is that many of the same gases that are causing ozone depletion also contribute to climate change. CFCs for example, GHGs that absorb some of the infrared radiation emitted by the Earth's surface, thereby effectively heating the Earth's surface (see Zepp *et al.*, 2003; Solomon *et al.*, 2003).

Conversely, changes in the climate of the earth could affect the behaviour of the ozone layer, because ozone is influenced by changes in the meteorological conditions

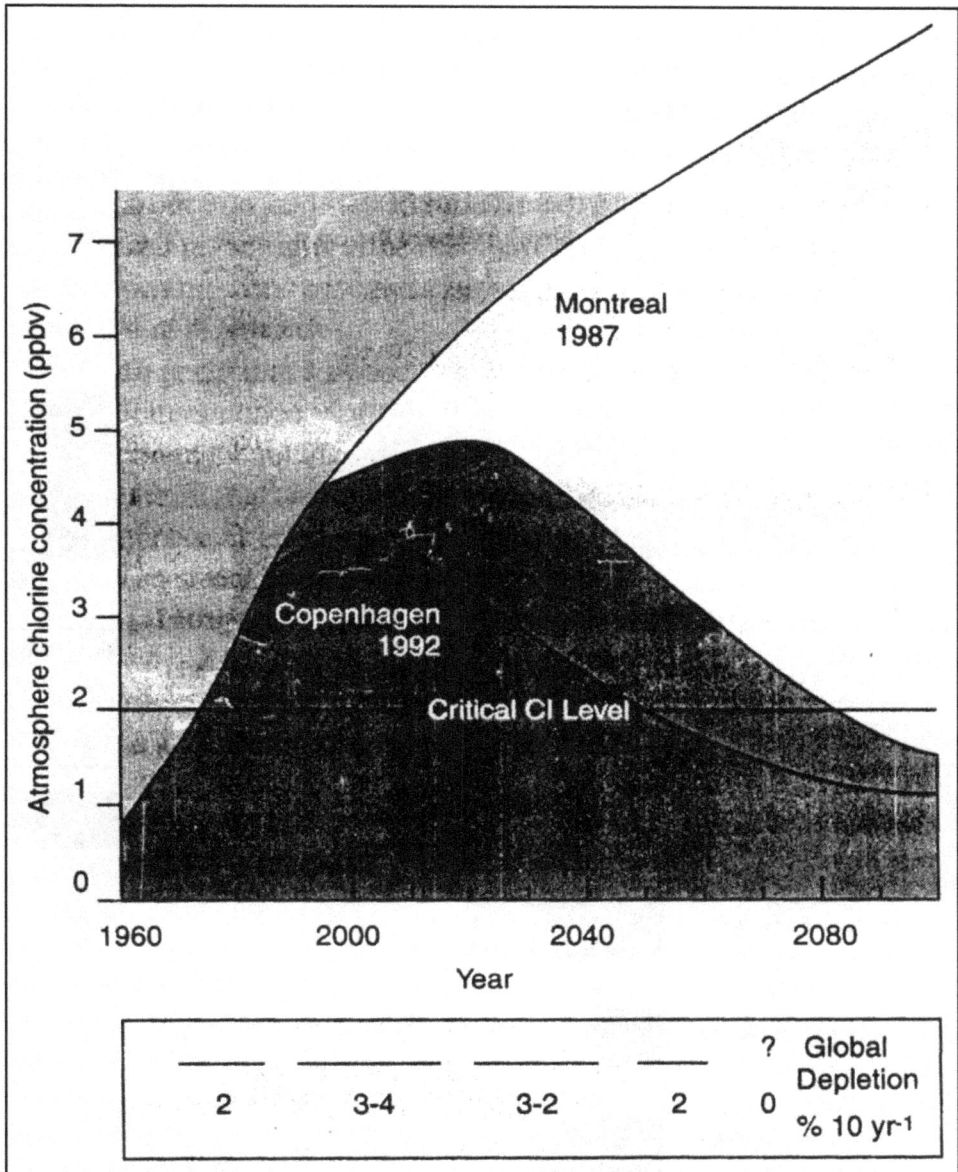

**Figure 5.3: The measured chlorine-equivalent concentrations in the atmosphere (since 1960) and projections based on various measures to phase out CFCs and other ozone-depleting substances. The numbers on the bottom line indicate the predicted global ozone depletion per decade as per strengthened Montreal Protocol (after Bojkov, 1995).**

and by climatic changes in the atmospheric composition. The major issue is that the stratosphere will most probably cool in response to climate change, therefore,

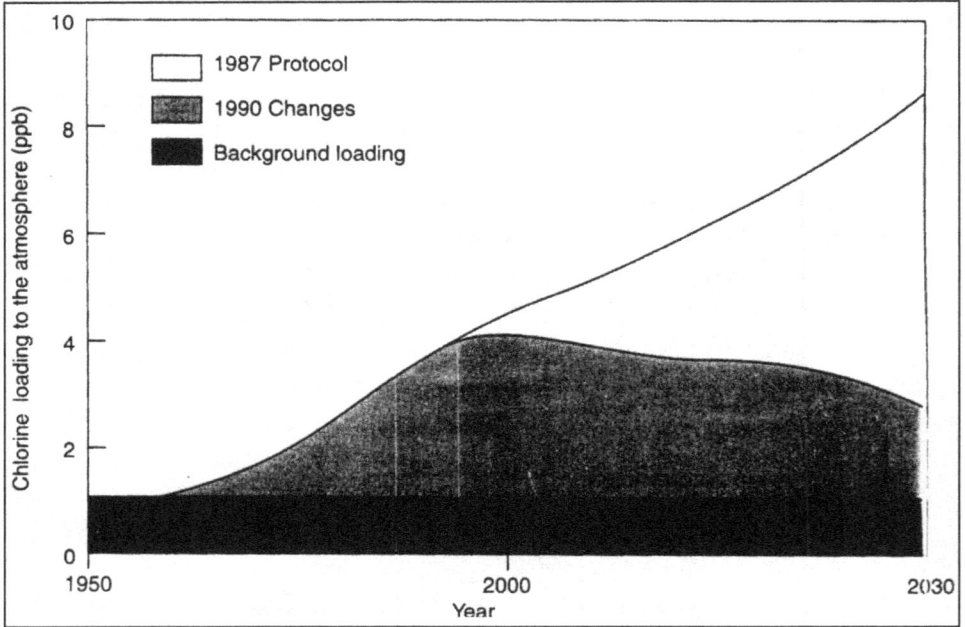

**Figure 5.4: the positive effect of the 1990 changes suggested as per the Montreal Protocol on preventing chlorine loading to the atmosphere.**

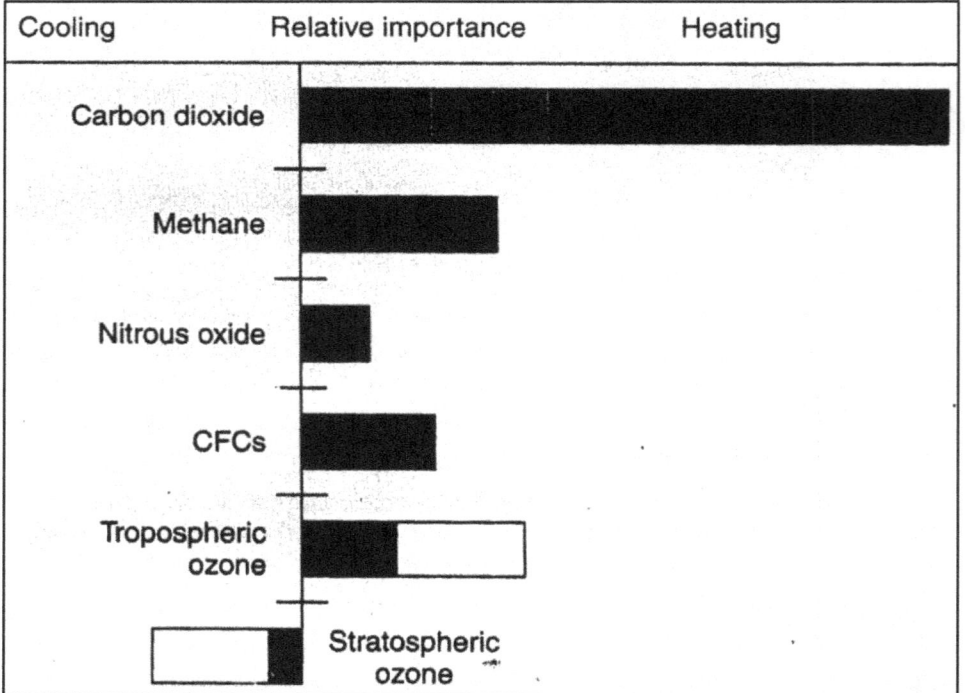

**Figure 5.5: Relative importance of the changes in the abundance of various gases in the atmosphere (after UNEP,1999).**

preserving over a longer time period the conditions that promote chlorine-caused ozone depletion in the lower stratosphere, particularly in polar regions (Kumar and Kumar, 1997; UNEP, 1999). Modifications/amendments to the Montreal Protocol were made in London (1990), Copenhagen (1992), Vienna (1995); and once again, in Montreal in 1997 (Figure 5.6). This figure illustrates:

1. The past observed abundance of atmospheric equivalent chlorine (which includes bromine, appropriately weighted and

2. The future abundances that would have been associated with each of the major decision steps of the Montreal Protocol. In any given year, the amount of human-caused ozone depletion is related to the effective chlorine abundance.

Since many of the environmental effects of ozone depletion (*e.g.* elevated incidence of skin cancer in humans) arise from long-term exposure, such impacts are related to the are under each of the curves-the larger the area, the larger the environmental effect and vice versa. Because of these associations, the summary below is linked to the figure and its messages.

Figure 5.6 also brings out the rationale for the time intervals used in the section below. The move beyond the 1987 Protocol 'freeze' provision to introduce some 'phase-

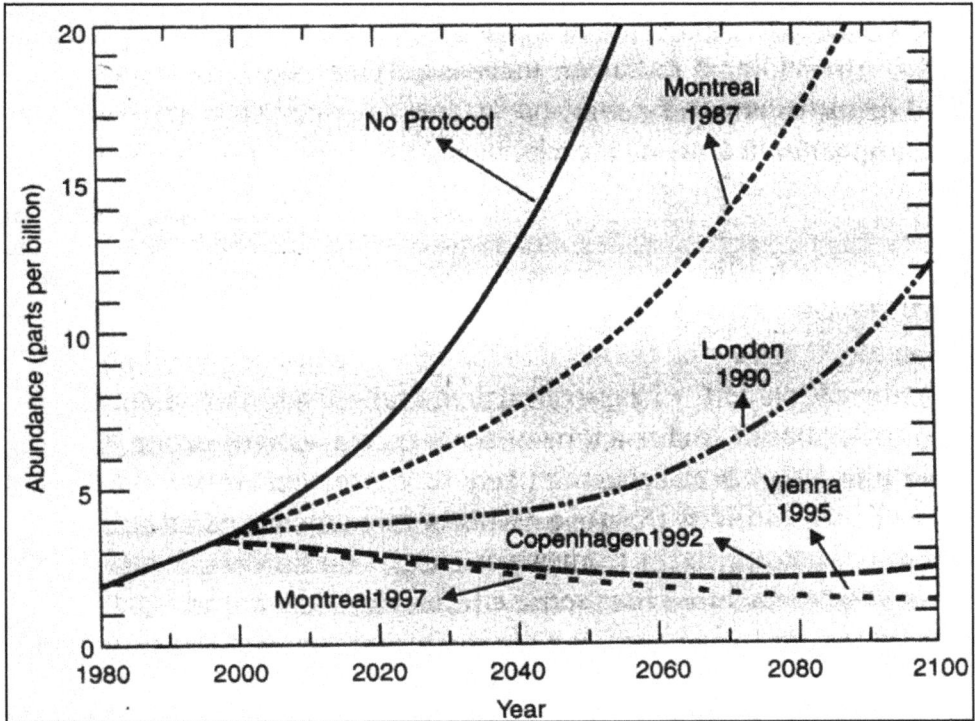

Figure 5.6: Effect of the international agreements on ozone-depleting stratospheric chlorine/bromine.

outs' dramatically lowered projected future effective chlorine growth rates over the period 2000-2040 (but still permitted a possible later return to significant growth rates). The advancement of phase-out dates and the addition of new controlled substances (Copenhagen, 1992) implied, for the first time, that a peak in effective chlorine burden would lie ahead (but still permitted a possible much later return to a positive growth). The setting of production and/or consumption caps on all controlled ozone depletory eventually for all countries implied (assuming full compliance) no return to positive growth rates. The advance of information and the corresponding policy decisions relating the effective chlorine milestones, from the Copenhagen amendment (1992) to the Vienna and the Montreal amendments are outlined below.

## Ozone Science

1. The slowing of the atmospheric growth rate of ozone-depleting gases in the lower atmosphere is detected.

2. The increasing growth rates of the CFC substitutes are documented.

3. Downward global ozone trends continue.

4. The role of particles is explained in the halogen-caused, temporarily enhanced ozone losses that followed the 1991 eruption of Mt. Pinatubo volcano.

5. Antarctic ozone losses continue unabated.

6. Non-polar ultraviolet-B radiation increases (clear sky) are recorded for low overhead ozone in non-polar regions.

7. Negative impact on the ozone layer of illegal CFC production is quantified.

## Environmental Effects

1. Deoxyribonucleic acid- the chemical in the cells of animals and plants that carried genetic information and is a type of nucleic acid–(DNA) damaging ultraviolet-B radiation in Antarctica is characterized during the period of ozone depletion.

2. Increased ultraviolet–B radiation is likely to cause substantial increases in the incidence of and morbidity from eye diseases, skin cancer, and infectious diseases, with risks now quantified for some effects (*e.g.* skin cancer).

3. Researchers have measured the increase in and penetration of UV-B radiation in Antarctic waters and have provided conclusive evidence of direct ozone-related effects on phytoplankton.

4. In terrestrial ecosystems, increased UV-B could modify production/ decomposition of plant matter, with concomitant changes in atmospheric trace gases.

## Technology and Economics

1. Many HFC-blends are identified and tested as replacements for HCFC-22 in refrigeration.

2. Hydrocarbon-based refrigerators enter the market.

3. Commercial refrigeration units are designed so that substitutes like ammonia and hydrocarbons can be used.

4. Cyclopoentane-based and hydrocarbon-blend-based foams are developed and/or commercialized.

5. Identification of halo essential-use exemptions opens consideration of earlier phase-outs.

6. Considerable increases in halo production and consumption in the article 5(1) countries are identified.

7. Except for quarantine and reshipment uses, alternatives to the uses of methyl bromide are identified.

8. Metered-dose inhalers based on HFC-134a are introduced on the market.

## Policy-Vienna/Montreal Adjustments

1. The reduction/phase-out of methyl bromide undeveloped countries is accelerated, and a phase-out schedule in developing countries is established.

2. For CFCs, halons, methyl chloroform, and carbon tetrachloride, the developing country phase-out dates are fixed to the developed country schedule that was adopted in 1990, plus ten years.

3. Establishment of lower caps on and limited uses of HCFs by developed countries and a freeze and distant-future phase-out by developing countries.

4. A licensing system is established for the control of methyl bromide trade. Therefore as of 1997, phase-outs and/or caps exist for all listed ozone depletors and for all Parties) (UNEP, 1999).

Ozone content in our atmosphere averages about 3 molecules for every 10 million air molecules. Despite this small amount, ozone plays vital roles in the atmosphere. Ozone is mainly found in two regions of the earth's atmosphere (Figure 5.7). Most ozone (about 90 per cent) occurs in a layer that begins between 8 and 18 km (5 and 11 miles) above the Earth's surface and extends up to about 50 km (30 miles). This region of the atmosphere is called the stratosphere. The ozone in this region is commonly known as the ozone layer. The remaining ozone is in the lower region of the atmosphere, called the troposphere. Figure 5.8 shows an example of how ozone is distributed in the atmosphere. Even though the ozone molecules in the above two regions are chemically identical ($O_3$), they have very different effects on humans and other living beings. Stratospheric ozone plays a beneficial role by absorbing most of the biologically damaging ultraviolet sunlight (called UV-B), allowing only a small amount to reach the Earth's surface. The absorption of ultraviolet radiation by ozone crates a source of heat, which actually forms the stratosphere itself (a region in which the temperature rises as one goes to higher altitudes). In this way, ozone plays a key role in the temperature structure of the Earth's atmosphere.

Without the filtering action of the ozone layer, more of the Sun's UV-B radiation would penetrate the atmosphere and would reach the Earth's surface. Many experimental studies on plants and animals and clinical studies of human have

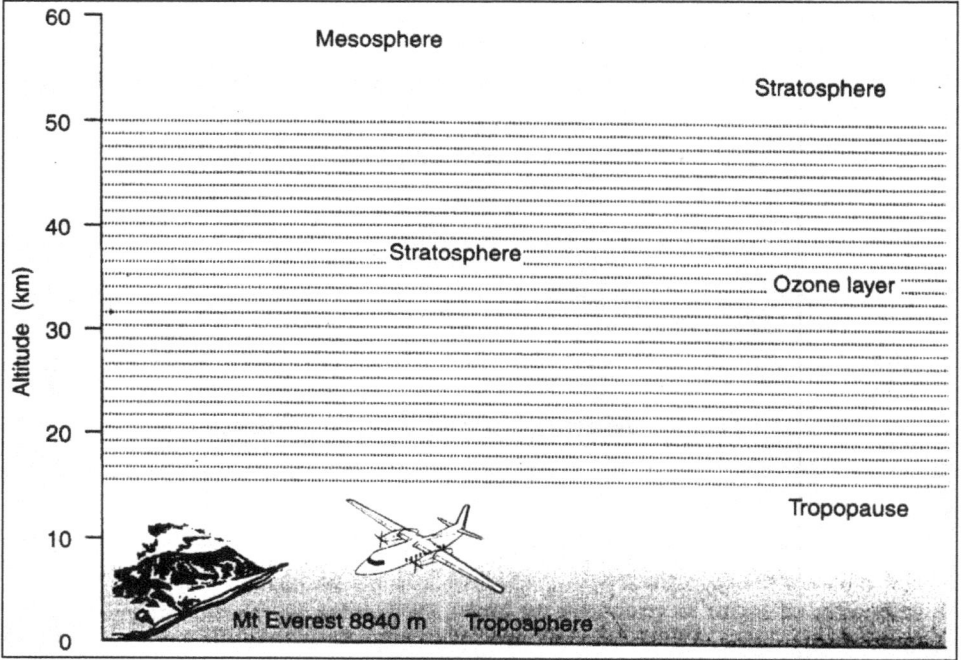

**Figure 5.7: Schematic sketch of the layers of the atmosphere. the weather we experience is influenced by the overlying stratospheric layer. The Earth's protective ozone layer is concentrated in the stratosphere (approx. 12-50 km).**

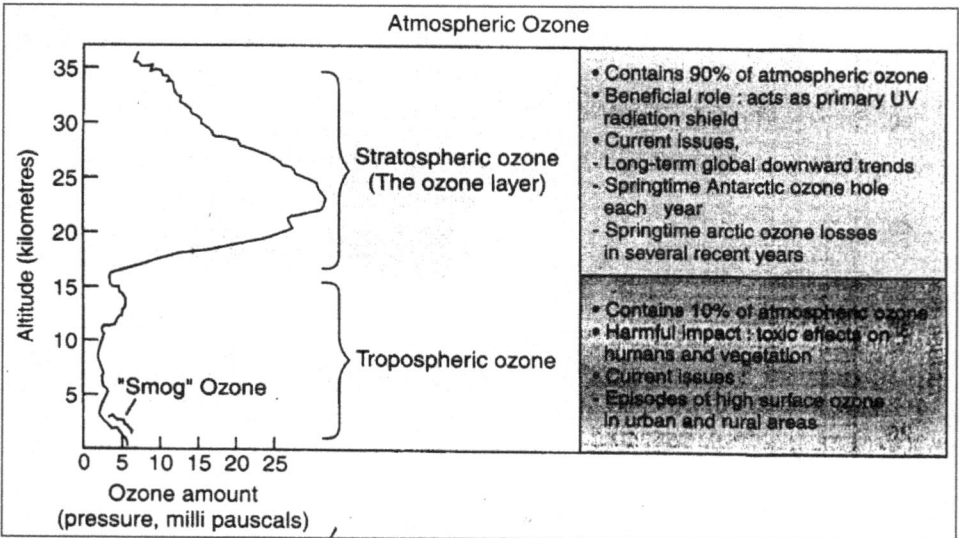

**Figure 5.8: Distribution of ozone in the atmosphere.**

shown the harmful effects of excessive exposure to ultraviolet-B radiation (see de Gruijl *et al.*, 2003; Caldwell *et al.*, 2003; Hader *et al.*, 2003).

At the Earth's surface, ozone comes into direct contact with life-forms and expresses its destructive side. Because ozone reacts strongly with other molecules, high levels of ozone are toxic to living systems, for example, on crop production, forest growth, and human health. The substantial negative effects of surface-level troposphere ozone from this direct toxicity contrast with the benefits of the additional filtering of UV-B radiation that it provides.

The dual role of ozone leads to two separate environmental issues. There is concern about increase in ozone in the troposphere. Low-lying ozone is a key component of photochemical smog which is a familiar problem in the atmosphere of many large cities. Higher amounts of surface-level ozone are increasingly being observed in rural areas. On the other hand, there is increasing concern about losses of ozone in the stratosphere. Ground-based and satellite instruments have measured decreases in the amount of stratospheric ozone incur atmosphere. Over some parts of Antarctica, up to 60 per cent of the total overhead amount of ozone (known as the column ozone) disappears during Antarctic spring (September-November). This phenomenon is known as the Antarctic ozone hole (*vide infra*).

In the Arctic polar regions, similar processes occur that have often led to significant chemical depletion of the column ozone during late winter and spring. The ozone loss from January through late March has been typically 20 to 25 per cent, sometimes even higher, depending on the meteorological conditions encountered in the Arctic stratosphere. Smaller, but still significant, stratospheric decreases have been seen at other, more populated regions of the Earth. Increases in surface ultraviolet-B radiation have been observed in association with local decreases in stratospheric ozone, from both ground-based and satellite-borne instruments.

Ever since the Antarctic ozone hole was discovered two decades ago, there has been intense global atmospheric monitoring as well as much basic research in atmospheric chemistry and physics. During the past few years, ozone level has been a little higher than expected from earlier projections based on sensitivity of ozone to influences of aerosols, halogen compounds and the solar cycle. The data seem to point to the beginning of a recovery; but this issue is complicated by several factors among which is the prominent role played by changes in meteorology, greenhouse gases and in the radiation balance, not excluding the observed recovery of the ozone layer from its perturbation by the volcanic eruption of Pinatubo in the early 1990s (see Zerofos, 2005) In fact, the extent of the future ozone recovery in a changing climate and the effect of ozone on that climate has brought out the importance of feedback mechanisms operating between water vapour content and a warmer planet.

For evaluating models and ozone loss, and its expected recovery, there is a need for well-calibrated instruments and measurements, particularly the use of satellite and ground-based data. Several recent chemistry/climate models are being used to address the issue of how changes in meterology or climate interact with changes in the chemistry of ozone.

One question is how changes in meteorology over the last 25 years might have contributed to observed ozone changes and feedback mechanisms. Models can then be used to extrapolate the knowledge gained to what may happen in the future with the expected increase in methane, nitrous oxide, and carbon dioxide (Zerefos, 2005).

Some important new work has combined satellite and in situ observations with model calculations and has provided good insight into the budget of nitrogen oxides and several halogens species, which are essential to our understanding of the global carbon and hydrological cycles. Water vapour presents a tricky issue. Recent satellite data are not consistent with trends from previous ground-based data. Understanding the feedback mechanism between water vapour content, ozone, and polar stratospheric clouds is crucial to the problem of evaluation of predictions of ozone in a future warmer global atmosphere.

Human-produced chemical have been responsible for the observed depletion of the ozone layer. The ozone-depleting compounds contain various combinations of the chemical elements chlorine, fluorine, bromine, carbon, and hydrogen and are often described by the general term halocarbons. The compounds that contain only chlorine, fluorine, and carbon are called chlorofluorocarbons (CFCs). CFCs, carbon tetrachloride, and methyl chloroform are important human-produced ozone-depleting gases that have been used in many applications including refrigeration, air conditioning, foam blowing, cleaning of electronic components, and as solvents. Another important group of human-produced halocarbons is the halons, which mainly contain carbon, bromine and fluorine and have been frequency employed as fire extinguishants. Governments have decided to eventually discontinue production of CFCs, halons, carbon tetrachloride, and methyl chloroform (except for a few special uses), and industry has developed more ozone-friendly substitutes (UNEP, 1999).

Most of the global warming that occurred over the last five decades appears to have been due to the increase in GHG concentrations. We know less about the possible impact of human-induced ozone depletion I the stratosphere on climate in the lower atmosphere (Karoly, 2003).

The climate has warmed over much of the Southern Hemisphere during the past four decades. Circumpolar westerly winds have become stronger, as a result of increasing atmospheric pressure at mid-latitudes and decreasing pressure and temperatures at high latitudes. The changes observed in Southern Hemisphere climate at high latitudes show distinct seasonal structure, with largest amplitude I the late spring and summer. Thompson and Solomon (2002) argued that they were possibly caused by stratospheric ozone depletion over Antarctica in spring. Gillett and Thompson (2003) showed that the observed spring and summertime changes can be explained as response to Antarctic ozone depletion. They noted quantitative agreement between observed climate change in the lower atmosphere and the climate response to ozone depletion in the stratosphere. Their work helps to quantity the possible influence of the stratosphere on weather and climate.

Greenhouse gases appear to have contributed to the observed Southern Hemisphere warming at mid-and lower latitudes and to the observed circulation changes [strengthening of the southern annular mode (SAM)] in winter. But the extent of the circulation response in these climatic models is not as strong as that found in the observations or in the ozone-forced model response in summer.

According to Karoly (2003), recent climate changes in the Southern Hemisphere may conceivably result from a complex combination of natural climate processes (associated with interactions between the atmosphere, oceans, and sea ice) and human influences (including decreases in stratospheric ozone and increases in atmospheric GHGs and aerosols. Understanding the separate contribution is crucial for understanding recent regional climate variations, such as the rainfall trends in Western Australia, and for predicting how climate may change in the future. Gillett and Thompson (2003) have shown that the recent summer circulation changes in the Southern Hemisphere high latitudes are likely to be caused by stratospheric ozone depletion.

## Chlorofluorocarbons

Chlorofluorocarbons (CFCs) reach the stratosphere because the Earth's atmosphere is always in motion and mixes the chemicals added into it. CFC molecules are much heavier than air. Nevertheless, thousands of measurements from balloons, aircraft, and satellites have shown that the CFCs are actually present in the stratosphere. This is because winds and other air motions mix the atmosphere to altitudes far above the top of the stratosphere much faster than molecules can settle according to their weight. Gases such as CFCs that do not dissolve in water and that are relatively uncreative in the lower atmosphere are mixed relatively quickly and, therefore, reach the stratosphere regardless of their weight (UNEP, 1999).

The two gases carbon tetra fluoride ($CF_4$), produced mainly as a by- product of the manufacture of aluminium, and CFC-11 ($CCl_3F$), used in a variety of human activities, are both heavier than air. Carbon tetrafluoride is completely unreactive at altitudes up to about 50 km in the atmosphere. It is nearly uniformly distributed throughout he atmosphere (Figure 5.9). CFC-11 is unreactive in the lower atmosphere and is uniformly mixed there. However, its abundance decreases as it reaches higher altitudes, because it is broken down by high-energy solar ultraviolet radiation. Chlorine released from this breakdown of CFC-11 and other CFCs remains in the stratosphere for several years, where every chlorine atom catalytically destroys many thousands of ozone molecules.

CFCs and halons do not occur naturally. CFCs were first developed in 1928 following the search for safe and effective refrigerants. They have also been used as blowing agents in the manufacture of plastic foam products, as propellants in aerosol sprays, and as solvents in various cleansers.

## Ozone-depleting Potential

Different formulations of CFCs and halons have different potentials to deplete the ozone layer. This is determined by the potential life of the chemical in the atmosphere. CFC-11 and CFC-12 have been used as references and assigned an ozone-depleting potential (ODP) value of 1.0. The ODP of various CFCs and halons included in the Monteral Protocol, together with their average lifetime in the atmosphere, are given in Table 5.2.

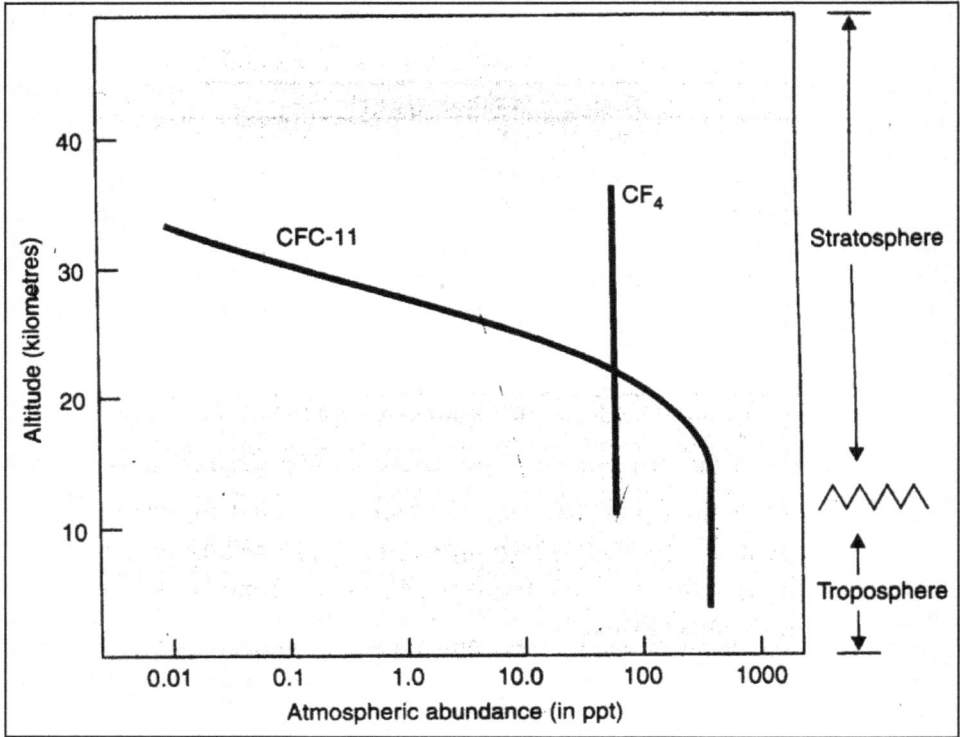

Figure 9: Atmospheric measurements of CFC–11 and CF$_4$

Table 5.2: The ozone-depleting potential (ODP) of various chlorofluorocarbons (CFCs) and halons.

|  | ODP | Lifetime (Year) |
|---|---|---|
| CFC–11 (P,R,C,B) | 1.0 | 75 |
| CFC–12 (P,R,B) | 1.0 | 111 |
| CFC–113 (R,C,B) | 0.8 | 90 |
| CFC–114 (P,R,B) | 1.0 | 185 |
| CFC–115 (R, B) | 0.6 | 380 |
| Halon 1211 (F) | 3 |  |
| Halon 1301 (F) | 10 |  |
| Halon 2402 (F) | 5 |  |

(P) = Propellant, (R) = Refrigerant (C) = Cleaner, (B) =Blowing agent (F) Fire extinguishant.

Scientific consensus is that an 85 per cent reduction in consumption of CFCs and Balons would eventually allow the ozone layer to stabilize in its present (depleted) state. Greater reductions are required to start restoring higher ozone level.

# Links with the Greenhouse Effect

Chlorofluorocarbons (CFCs) and halons in the lower atmosphere (before they start depleting ozone in the stratosphere) are highly efficient GHGs. Each molecule of CFC traps over ten thousand times more energy than a molecule of carbon dioxide. CFCs and halons are estimated to contribute over 10 per cent of overall global warming.

Ozone depletion is a global environmental problem because all users of CFCs contribute to the cause, and all countries suffer the consequences. Action by individual nations will not achieve very much: international agreements are essential. Developing countries do not use significant quantities of CFCs by world standards.

In 1985, the Vienna Convention for the Protection of the Ozone Layer was signed. It set out the parameters for action on ozone layer depletion and established a framework for further work.

**Table 5.3: Agreed Phase Outs**

| Agreed Phase Outs | |
|---|---|
| All known CFCs (including 5 previously listed) | 100 per cent by 2000 |
| 3 listed halons | 100 per cent by 2000 |
| Methyl chloroform is also known as | |
| 1,1,1-Trichloroethane | |
| Methyl chloroform | 100 per cent by 2000 |
| Carbon tetrachloride | 100 per cent by 2000 |

One crucial issue is the link between CFC production, industrialization, and economic development. Developing countries argue that industrialized nations have depleted the ozone layer. They want to link the phasing out of CFCs with access to alternative technologies and funding assistance ot use the more expensive alternatives. If developing countries as a whole do not sign the protocol, the ozone layer would continue to be depleted, even if the present parties completely banned the use of CFCs and halons. The requirements of developing countries need to be taken into account.

The Montreal Protocol established a three-step reduction in consumption of CFCs to 50 per cent of 1986 levels by 1998. Consumptions of halons were frozone at 1986 levels in 1992. The Protocol includes progressive restrictions on trade in CFCs and halons and in CFC- related technology, between parties and non-parties.

At second meeting of the parties to the Montreal Protocol, London, June 1990, other chemicals that deplete ozone were added to the Protocol. Timetables for the phase out of CFCs and halons were also tightened. The meeting also agreed to a $ 150 million fund over three years to assist developing countries to adopt ozone friendly technology.

# Development of Alternatives

The development of substitutes for CFCs is progressing. In most industries, there is a considerable optimism that CFC use can be substantially reduced, well beyond the current terms of the Montreal Protocol. There are, in many cases, existing alternatives to CFCs. For example, hydrocarbons are being used as propellants for aerosol sprays and ammonia is a refrigerant that has been used in large refrigeration systems for many years. Not all plastic foams are manufactured using CFCs, although CFCs have advantages where flexibility or insulation qualities are required.

Some new refrigerants being developed involve the addition of hydrogen atoms to the CFC. This reduces the stability of the compound making it short lived, so it breaks up before reaching the stratosphere. In the case of HCFC-22, this cuts down the ozone depleting potential by 95 per cent, but it also affects some of the desirable qualities of CFCs. Low ozone- depleting substances such as this have an important role to play in the transition to ozone benign alternatives.

Several substitutes have already been developed and are currently undergoing toxicity testing. In most cases, plant and machinery will have to be converted to enable use of the new product. Clearly where costs of conversion are high, there must be certainty that the alternative will be satisfactory.

The development of 'drop-in' substitutes, that can be readily used in place of cfcs in an existing plant, is happening. A blend with an ODP of 0.03 is expected to be available in 1992. HFC-134a with an ODP of 0 is the likely replacement for CFC-12 and is expected to be available shortly. It is likely to cost more than five times as much as CFC-12. The drop-in products will be particularly important in reducing demand for CFCs to keep existing equipment going.

## Methyl Chloroform

The use of the industrial solvent methyl chloroform (MCF) was banned under the Montreal Protocol (see Midgley and Mcculloch, 1995) because of its ozone-depleting potential. Amendments to the Montreal Protocol proposed a final phase-out of MCF in 1996 in developed countries, and 2015 in developing countries. During the 1990s, its global emissions decreased substantially and, since 1999, near zero emissions have been estimated for Europe and the United States. From 1995 onwards, the emission estimates have been based on the non-audited consumption and production data supplied by the parties under the Montreal Protocol (McCulloch and Midgley, 2001).

But Krol *et al.* (2003) presented measurements of methyl chloroform that are inconsistent with the assumption of small emissions. They estimated that European emissions were greater than 20 Gg in 2000. While these emissions are not significant for stratospheric ozone depletion, they do have important implications for estimates for estimates of global tropospheric hydroxyl radical (OH) concentrations, deduced from measurements of methyl chloroform. Ongoing emissions, therefore, cast doubt upon recent reports of a strong and unexpected negative trend in OH during the 1990s and a previously calculated higher OH abundance in the Southern Hemisphere compared to the Northern Hemisphere (Krol *et al.*, 2003).

## Flourinated Gases

With a view to protecting the ozone layer around the Earth, many ozone-depleting substances have been replaced with other gases; but, unfortunately, these other gases contribute to global warming. A new report by the IPCC, in collaboration with an expert panel under under the Montréal Protocol, has identified the danger to climate posed by the use of these ozone-friendly fluorinated gases (F-gases). It also suggests ways to minimize their warming effect.

In the business-as-usual scenario, annual CFC and HCFC emissions are together increasing from 2.1 to 1.2 billions tons carbon dioxide-equivalents from 2002 to 2014, and annual emissions of HFCs are increasing from 0.4 to 1.2 billion tons carbon dioxide equivalents in the same period. the contribution to global warming of f-gases can be halved by adopting measures that reduce emissions, such as better containment to prevent leaks and evaporation, reducing the amounts of the chemicals needed in equipment, and improving end-of-life recovery, recycling and destruction of HFCs and PFCs. The use of alternative substances with a lower or zero global warming potential needs to be increased.

The report also gives specific emission reduction and cost estimates for applications that use fluorinated gases, which include not only refrigeration and air conditioning but also foams, medical aerosols, solvents and fire protection (IPCC/ TEAP, 2005).

## Tropospheric Ozone and Crop Suffocation

The first open-air simulations of the Earth's climate in the year 2050 have revealed that adverse effects of increasing surface ozone could outweigh any benefits triggered by global warming. Rising carbon dioxide levels had previously made researchers optimistic about crop yields in coming decades-carbon dioxide promotes growth by increasing rates of photosynthesis, for instance. But much of the work was based on greenhouse trials and ignored the impact of ozone concentrations near the ground. When this shortcoming was addressed in field trials of soybean plants, yields fell sharply. This is cause for concern because soybean is an important crop and farmers growing it could face significant ozone increases.

While urban air pollutants all over the world are expected to push up levels of near-surface ozone by at least 25 per cent by the middle of this century, rises in China and in the US Midwest, where almost 3,00,000 sq km of soybeans are harvested annually could be two to three times as great (Giles, 2005).

Ozone creates reactive molecules that destroy rubisco, an enzyme essential for photosynthesis. Ozone also makes leaves age faster. To investigate ozone effects, an Illinois research team conducted open-air experiments in which $CO_2$ and ozone were released from pipes that surrounded 16 plots of 200 m$^2$. By using wind sensors to control gas release, concentrations I the air over the plot were held close to the predicted 2050 levels over a three-year study.

The preliminary data suggest that yields will be down by up to 10 per cent. Previous forecasts of crop yields made by the IPCC should therefore now be revised. The open-air studies also unraveled certain other threats that greenhouse work had missed, including delayed maturation, which puts crops at risk of forest late in the season. An pests also thrived. According to the IPCC, chewing insects will cause more damage in higher carbon dioxide. For instance, Japanese beetles like the environment-they live longer and produce more eggs.

Because all of the 22 soya varieties tested showed a similar response, it might be very difficult to breed strains with better ozone tolerance. Within decades, farmers may have to switch to different crops, although we do not know how these other types of plant may be affected (Giles, 2005).

Ozone is likely to be a particular problem in countries that are rapidly industrializing, such as India and China.

At first, dropping yields would not be serious for Europe, where countries can import food and farmers are not pushed for the highest possible yields. But, if other nations reduce production owing to ozone pollution, Europe may suffer.

Climate researchers are attempting to help farmers plan ahead. One approach combines climate simulations with software that models crop yields, to predict agricultural boom or bust (Giles, 2005).

## Long-range Tropospheric Pollution

Long-range transport of tropospheric pollution and its coupling to climate has been studied by using climate/chemistry models. Long-range transport of pollutants seems to maintain regionally high background levels of tropospheric ozone. Seasonal episodes of high ozone over the south Atlantic begin with pollution sources originating thousands of miles away (Zerefos, 2005).

Although future ultraviolet-B levels for the period 2000-2019 have been predicted to decrease in all seasons, the trends are not statistically significant, except during spring over both hemispheres. Ultraviolet-B radiation trends are determined largely by total ozone trends because in the future cloud changes are expected to be small in the coupled chemistry climate models used in these various studies. Still there is a region over western Europe which is expected to suffer from an increase in ultraviolet-B radiation mainly due to lesser cloudiness.

However, the influence of the process of interference of cloud and other physical parameters on ultraviolet-B levels at ground level is a highly compelx issue (see McKenzie *et al.*, 2003). The detection of ozone recovery requires much patience. There is need to continue quality observations of global coverage both from the ground and space. The ultraviolet-B levels in the coming decade are predicted to decrease for all seasons except during spring over hash latitudes of both hemispheres (Zerofos, 2005).

Human emissions of CFCs and halon have occurred mainly in the Northern Hemisphere. About 90 per cent have been released in the latitudes corresponding to Europe, Russia, Japan, and Northern America. Being insoluble in water and relatively uncreative, CFCs and halons are mixed within a year or two throughout the lower atmospohere.

The CFCs and halons in this well-mixed air rise from the lower atmosphere into the stratosphere mainly in tropical latitudes. Winds then move this air poleward-both north and south-from the tropics, so that air throughout the global stratosphere contains nearly equal amounts of chlorine and bromine.

In the Southern Hemisphere, the South Pole is part of a very large land mass (Antarctica) that is completely surrounded by the ocean. This symmetry is reflected in the meteorological conditions that allow the formation in winter of a very cold region in the stratosphere over the Antarctic continent, isolated by storm winds circulating around the edge of that region. The very low stratospheric temperatures lead to the formation of polar stratosphere clouds which are responsible for chemical changes that promote production of reactive chlorine and bromine which destroy ozone rapidly when sunlight returns to Antarctica in September and October of each year, resulting in the formation of the Antarctic ozone hole. The magnitude of the ozone loss has generally grown through the 1980s as the amount of human-produced ozone-depleting compounds has grown in the atmosphere.

Similar conditions do not prevail over the Arctic. The wintertime temperatures in the Arctic stratosphere are not persistently low for as many weeks as over Antarctica, so there is correspondingly less ozone depletion in the Arctic.

## Atmospheric Chemistry and Ozone Production

Hydrogen oxides and nitrogen oxides are tow basic component soft atmospheric chemistry. Hydrogen oxides ($HO_x$) consist of the hydroxyl radical (OH) and the hydroperoxyl radical ($HO_2$). Nitrogen oxides ($NO_x$) consist of nitric oxide (NO) and nitrogen dioxide (NO2). Together they determine the atmosphere's oxidizing, or cleansing, power and the troposphere's ozone production (Brune, 2000).

Concentrations of OH and $HO_2$ typically increase during the day and are higher in polluted environments than in clean ones. Two factors responsible for most observed $HO_x$ variance are: the $HO_x$ production rate, denoted as $P(HO_x)$ and the $NO_x$ concentration. The dominant HOx destroying reactions are basically determined by the $HO_x$ production rate and the $NO_x$ concentration (Figures 5.10 and 5.11).

The abundance of tropospheric hydroxyl radicals (OH) largely defines the oxidizing capacity of the atmosphere. A healthy level of tropospheric OH keeps in check atmospheric concentrations of many potent GHGs such as methane and HCFCs and prevents large amount of ozone-depleting halogenated hydrocarbons from reaching the stratosphere. Monitoring global tropospheric OH concentrations and understanding the controlling factors, both anthropogenic and natural, are necessary to protect our living environment.

Globally, the most important $HO_x$ source is the photo destruction of ozone to produce the excited state oxygen atom, $O(^1D)$, which reacts with water vapour to produce two OH molecules. It is the largest $HO_x$ source in the lower troposphere. But in the upper troposphere and in some continental environments, other sources can dominate. some of these other sources are made elsewhere by oxidation processes initiated by the reaction of OH with a hydrocarbon. They are then transported to a new location where they become important or dominant (Figure 5.11).

Another $HO_x$ source is the reaction of OH with methane. This reaction initiates an oxidation process that eventually produces water vapour and carbon dioxide.

For larger non-methane hydrocarbons, such as those found in urban areas and forests, the oxidation pathways can be much more complex than for those of methane. After the initial reaction between OH and these VOCs, $RO_2$ soon forms, where R is a hydrocarbon radical. Formation of $RO_2$ often leads to formation of $HO_2$. Whether the

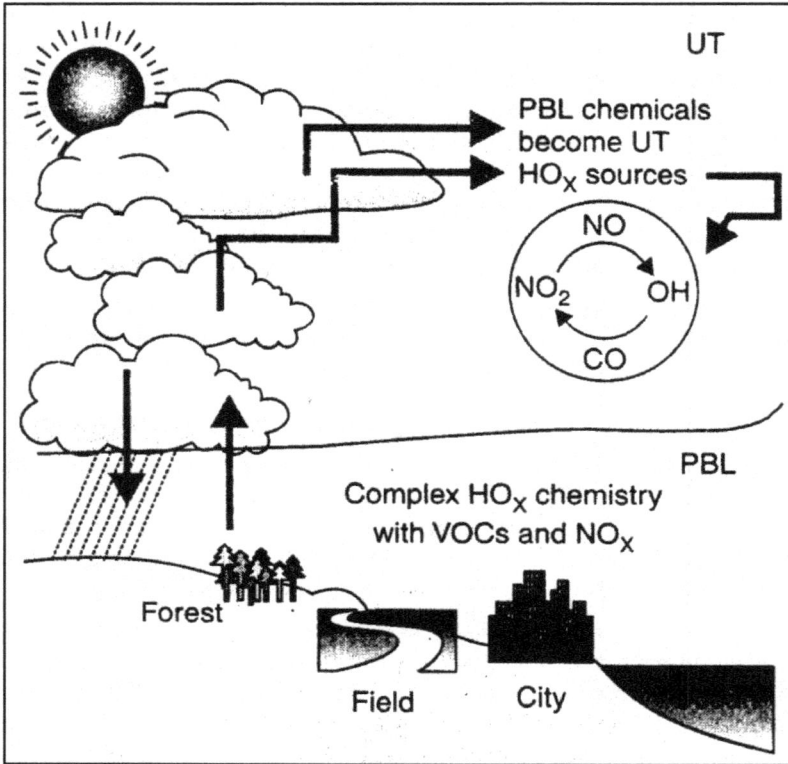

**Figure 5.10: Tropospheric HO$_x$ photochemistry. Complex chemistry in Earth's planetary boundary layer (PBL) produces oxygenated species. These can be convectively lifted into the upper troposphere (UT), where they become HO$_x$ sources. In the UT, HO$_x$ is exchanged between HO$_2$ and OH by reaction of HO$_2$ with NO and OH with CO. The reaction of HO$_2$ with NO leads to O$_3$ production (after Brune, 2000).**

oxidation of a particular hydrocarbon is a net HOx source or sink depends on several factors, for example, the amount of NO present. While VOCs are certainly important for the chemistry in the Earth's planetary boundary layer, where people live, they may have a more global role if they are carried to the upper troposphere.

Some hydroperoxyl radical sources create OH; others create HO$_x$ (Figure 5.7). Once created, HO$_x$ is partitioned rapidly into OH and HO$_2$. the reaction of OH with carbon monoxide, volatile organic compounds (VOCs), ozone, and other chemicals produce HO$_2$. Similarly, reactions of HO$_2$ with NO and O$_3$ produce OH. Throughout much of the atmosphere, the production of OH and HO$_2$ through these reactions is much faster than the OH and HO$_2$ production from other sources.

The dominant reactant with OH is usually carbon dioxide, although other species can be important. In forested regions, the dominant reactant with OH is often isoprene, a 5-carbon molecule that is emitted mostly by deciduous trees. In urban environments,

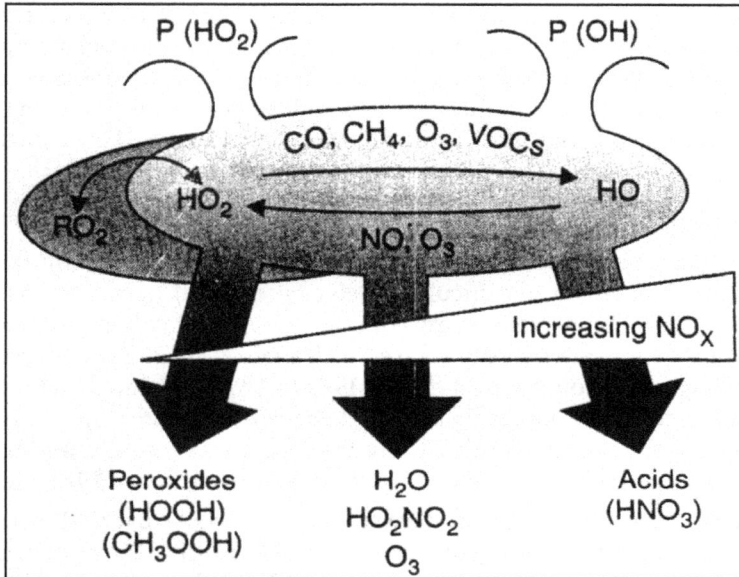

Figure 5.11: The influence of $NO_x$ and P ($HQ_x$) on $HO_x$ photochemistry, $HO_x$ is produced as either OH or $HO_2$, is rapidly exchanged between $HO_2$ and OH, creating $O_3$ in the process, and is removed by reaction that eventually form $H_2O$. The dominant removal reactions are determined by the $NO_x$ abundance and the $HO_x$ production rate (after Brune, 2000).

the dominant reactant can be oxygenated species, for example, formaldehyde and other aldehydes.

Hydroperoxyl radical ($HO_x$) is permanently lost when its hydrogen atom is recombined into water vapour; it is destroyed by several reaction, but the relative importance of the different reactions depends on the amount of $NO_x$ present (Figure 5.11).

The concentration of OH and $HO_2$, and ozone production, P ($O_3$), show a strong dependence on $NO_x$. Also, the production of ozone, P ($O_3$), is intimately tied to $HO_x$.

If P ($O_3$) increase as NO is increased, then ozone production becomes $NO_x$ limited. If P ($O_3$) is constant or decreases as $NO_x$ is increased, then ozone production is $NO_x$ saturated. Whether a region of the atmosphere is $NO_x$ limited or $NO_x$ saturated is an important question for determining the impact of future human pollution, particularly $NO_x$ on that region (Brune, 2000).

## Stratospheric Ozone Depletion

### Background

It emerged in the 1970s that human activities modify the total column amount and vertical distribution of atmospheric ozone. Ozone is the only gas in the atmosphere that prevents the most harmful solar ultraviolet radiating from reaching the surface of the Earth. An increase in the amount of UV radiation reaching the Earth's surface could exert harmful effects on human health (melanoma and non-melanoma skin cancer, eye damage, and suppression of the immune response system) and on the productivity of aquatic and terrestrial ecosystems. Changes in the vertical distribution of ozone could alter the distribution of temperature I the atmosphere and change regional and global climates.

## Global Ozone Trends

Since the concentration of ozone in the atmosphere has marked natural, temporal and spatial variations, it is difficult to determine global ozone trends, although observations of both the total column content and the vertical distribution of ozone have been made for several decades (see McKenzie *et al.*, 2003). A decrease of the annual average total column ozone from 1969 to 1988 between 3 and 5 per cent at latitudes between 30°N and 64°N became known two decades ago. The decreases were largest in the winter months. There was decrease of the column ozone total, averaged between 53°S and 53°N, of about 2.5 per cent between 1978 and 1985.

Analysis of satellite and ground based data insect 1979 revealed that the largest decrease in ozone concentration occurred near 40 km altitude in both hemispheres. Temperatures I the stratosphere between 45 and 55 km have decreased globally by about 1.7°C since 1979, consistent with the observed decreases in upper stratospheric ozone of less than 10 per cent. In addition to global trends I the total amount of ozone in the atmosphere, there are observations of its changing vertical distribution also.

Over the Antarctic, the decreases in total ozone were progressively larger in the 1980s than during the 1970s, recording 50 per cent or more loss of the total column ozone in September 1987 over an area roughly the size of the Antarctic continent. The ozone depletion in the Southern Hemisphere spring in 1989 was as large as the record-setting hole in 1987. From, early August 1989 the total ozone content decreased by about 1.5 per cent per day reaching a minimum value of ozone of 45 per cent by October. The ozone values declined still more in the 1990s, widening as well deepening the 'ozone hole'. The size of the ozone hole and the extent of depletion vary from year to year depending on the temperature and polar wind conditions. When the polar vortex (the strong atmospheric circulation centered over the polar area) is undisturbed and temperatures are especially cold, a larger decrease occurs.

The ozone losses are most marked in the lower stratosphere between about 12 to 25 km. The decrease in total ozone stops in the early October, but the residual low values persist until the final breakup of the polar vortex in November, at which point air that has more ozone is introduced from more temperate latitudes.

In 1986 and 1987, it emerged that the unique meteorology during winter and spring over the Antarctic sets up the special conditions of an isolated air mass (the polar vortex) with cold temperatures required for the observed chemical perturbation. Chlorine compounds resulting from human activities are primarily responsible for the observed decrease in ozone within the polar vortex. The chlorine comes primarily from the breakdown in the stratosphere of CFCs which are released at the Earth's surface and are transported upwards into the stratosphere without undergoing any changes. Polar stratospheric clouds (PSCs) play a crucial role in forming the Antarctic ozone hole (see Kumar and Hader, 1999). Ozone hole is now also known to be formed in the arctic. Ozone has decreased not only over the Antarctic but also by more than 5 per cent at all latitudes south of 60°S throughout the year since 1979.

## The Appearance of the Ozone Hole

The observed average amounts of ozone during September, October and November over the British Antarctic Survey stations at Halley, Antarctica, first revealed notable decreases in the early 1980s, compared with the preceding data obtained starting in 1957.

The ozone hole is formed each year when there is a sharp decline (currently up to 60 per cent) in the total ozone over most of Antarctica for a period of about three months (September-November) during spring in the Southern Hemisphere. There are no such sharp declines in the late-summer (January-March) ozone amounts. Observations from a few other stations in Antarctica and from satellite-based instruments revealed similar decreases in springtime amount of ozone overhead. Balloon-borne ozone instruments show strong changes in the way ozone is distributed with altitude. Almost all of the ozone is now depleted (Figure 5.12) at some altitudes as the ozone hole forms each springtime, compared to the normal ozone profile that existed before 1980. The ozone hole results from destruction of stratospheric ozone by gases containing chlorine and bromine, whose sources are mainly human produced halocarbon gases and CFCs (see Kumar and Hader, 1999).

Even before the stratosphere suffered from human-produced chlorine and bromine, the naturally occurring springtime ozone levels over Antarctica were about 30 to 40 per cent lower than springtime ozone levels over the Arctic. This natural difference between Antarctic and Arctic conditions was first observed in the late 1950s by Dr. Dobson of Oxford. It stems from the exceptionally cold temperatures and different winter wind patterns within the Antarctic stratosphere as compared with the Arctic. This is not at all the same phenomenon as the marked downward trend in total ozone in recent years (UNEP, 1999).

Changes in stratospheric meteorology cannot explain the ozone hole. Measurements show that wintertime Antarctic stratospheric temperatures of past decades has not changed prior to the development of the ozone hole each September. In contrast, ground, aircraft, and satellite measurements have given definite evidence of the importance of the chemistry of chlorine and bromine originating from human-made compounds in depleting the Antarctic ozone in recent times (UNEP, 1999).

An interesting question is: why has an ozone hole appeared over Antarctica when CFCs and halons are released mainly at the Northern Hemisphere?

This is because the Earth's atmosphere is continuously stirred over the globe by winds. Consequently, ozone-depleting gases are mixed up throughout the atmosphere, including Antarctica, regardless of where they are emitted. The special meteorological conditions in Antarctica cause these gases to be more effective there in depleting ozone compared to elsewhere.

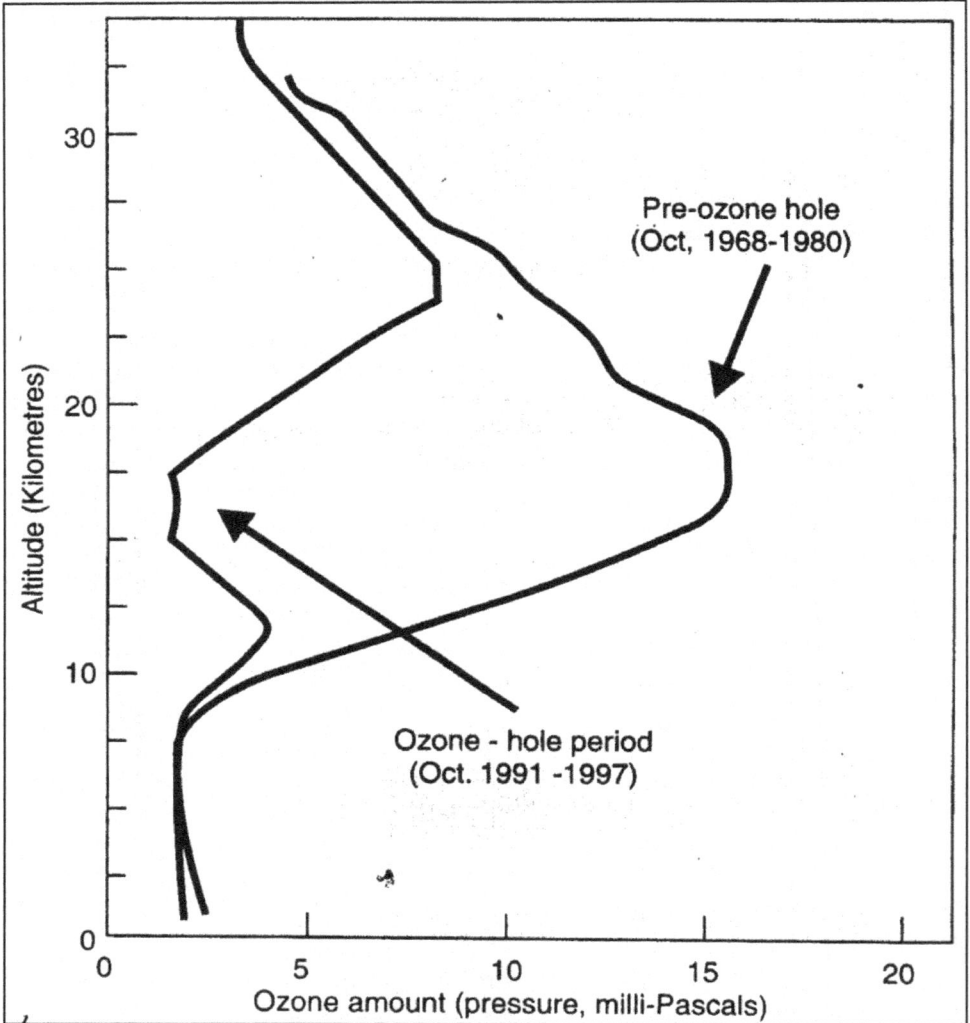

**Figure 5.12: Springtime depletion of the ozone layer over Syowa, Antarctica (after UNEPO, 1999).**

## Ozone and Radiative Balance in the Earth's Atmosphere

The decrease in stratospheric ozone in the recent past has partially offset the surface warming caused by the increase in other GHGs such as $CO_2$. This exemplifies how ozone depletion and climate change are interconnected (Figures 5.13 and 5.14).

The biggest ozone losses ever recorded over the Arctic in the year 2005 have prompted debate on whether global warming is to blame. Should there by a link, the spectra of an Arctic ozone hole looms large (Schiermeier, 2005). The unusually low Arctic temperatures in the last winter resulted by greater ozone losses than ever recorded before (Figure 5.15). An ozone hole over the Antarctic was first reported in

Stratospheric temperature

UV

Aerosols

Polar stratospheric clouds

Stratospheric ozone

Stratospheric water vapour

Stratospheric (10-5 km)

Troposphere (0-10 km)

Tropospheric temperature

Troposphere ozone

Volcanoes

Greenhouse gases

Halogens

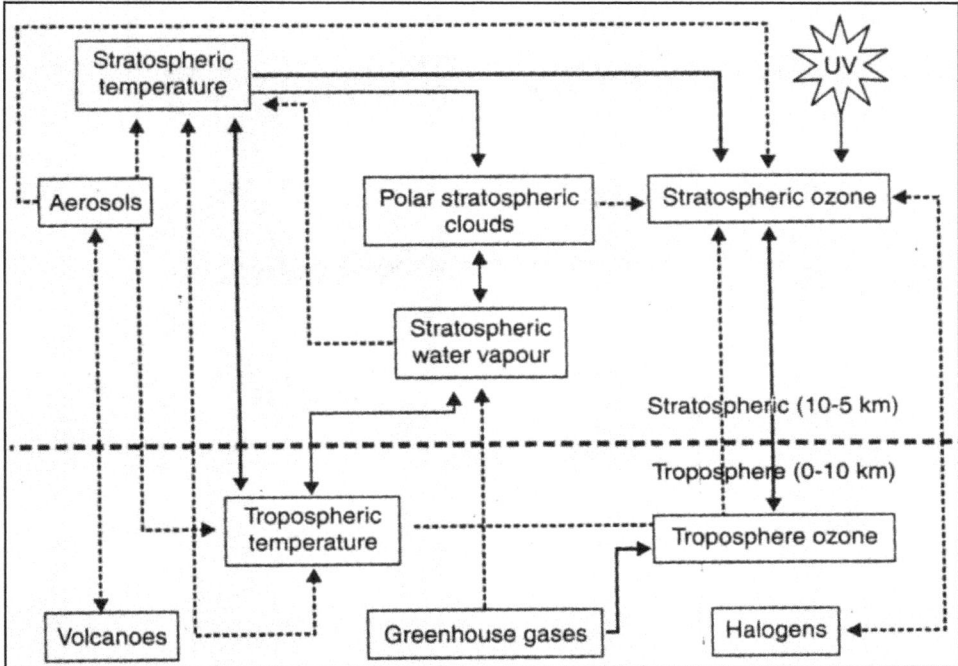

**Figure 5.13: Sketch of the interactions between stratospheric ozone and other atmospheric constituents and processes-anthropogenic emissions (*e.g.* greenhouse gases, halogens), while other factors affecting the climate system (*e.g.* volcanoes). Broken arrows indicate where one species or process affects another. In many cases process A affects B and B affecters A (a feedback, shown with bold lines). For example, decreasing polar stratospheric temperatures increase ozone depletion. Reduced ozone then causes stratospheric cooling, creating a positive feedback (after WMO/UNEP, 2002; and Bodeker, 2003).**

1985 (Farman *et al.*, 1985). The production of CFCs, which destroy ozone, has been phased under the Montreal Protocol, but remnants of the long-lived CFCs will persist in the atmosphere for at least another half a century. Ozone loss in the Arctic is less severe than in the colder Antarctic, and is more dependent on temperature variation. Stratospheric clouds provide surfaces on which CFC decay products are converted into forms that destroy ozone-but the clouds only form at temperatures below 80°C. The 2004 Arctic winter was particularly cold and in some places over a half of the ozone molecules were destroyed. By the early spring, ozone-depleted air had drifted southwards through many areas of northern and central Europe (Schiermeier, 2005).

Increasing concentrations of GHGs have a cooling effect on the stratosphere. Over the past decades, there has been four-fold increase in the area cold enough for polar stratospheric clouds. The 2005 losses may be a sign that colder winters could led to an Arctic ozone hole in the coming two decades. However, some scientists feel that any attempt to link ozone loss to GHGs is speculative-the temperature drop this

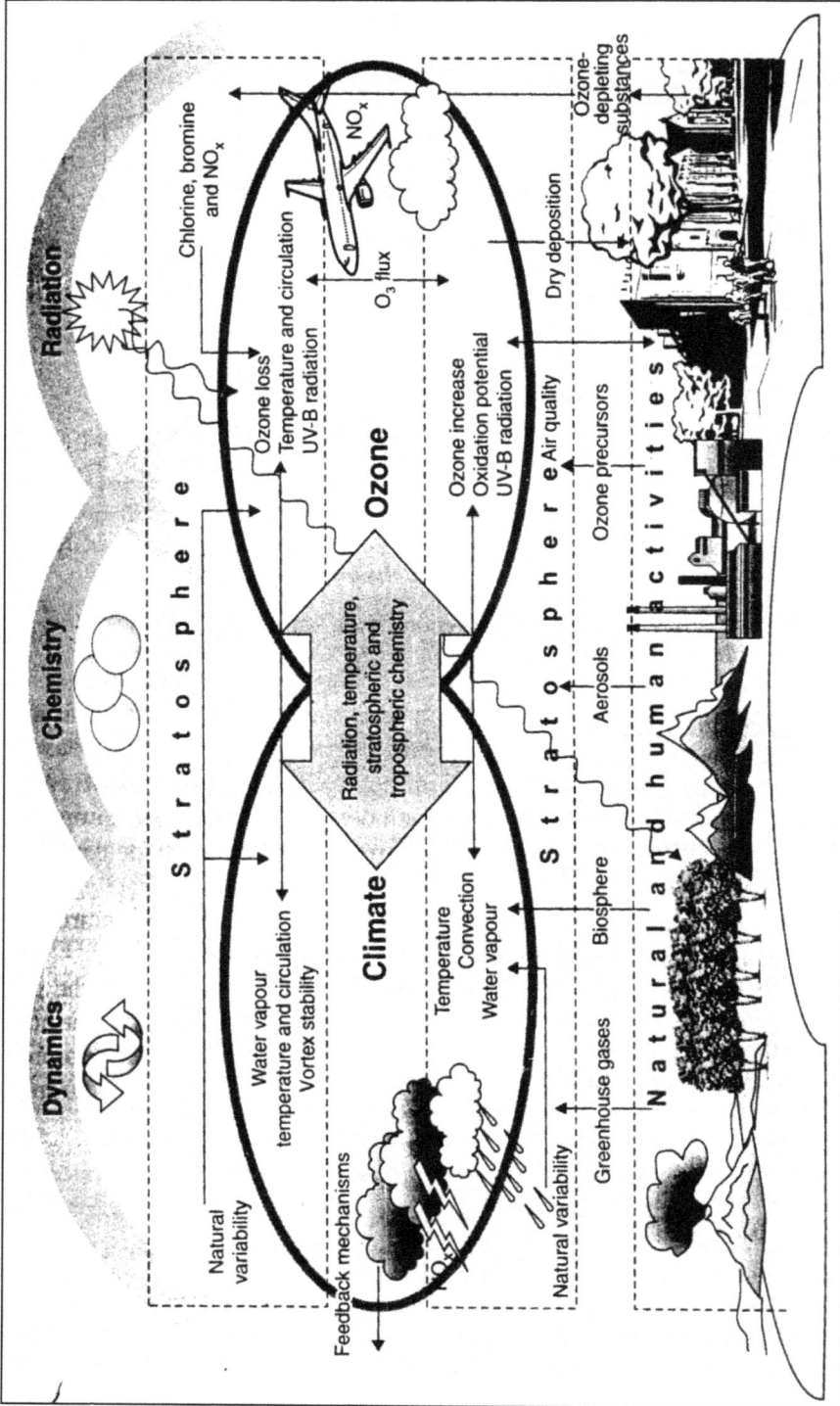

**Figure 5.14: Interactions between climate, atmospheric composition, chemical and physical processes and human activities (after Isaksen 2003).**

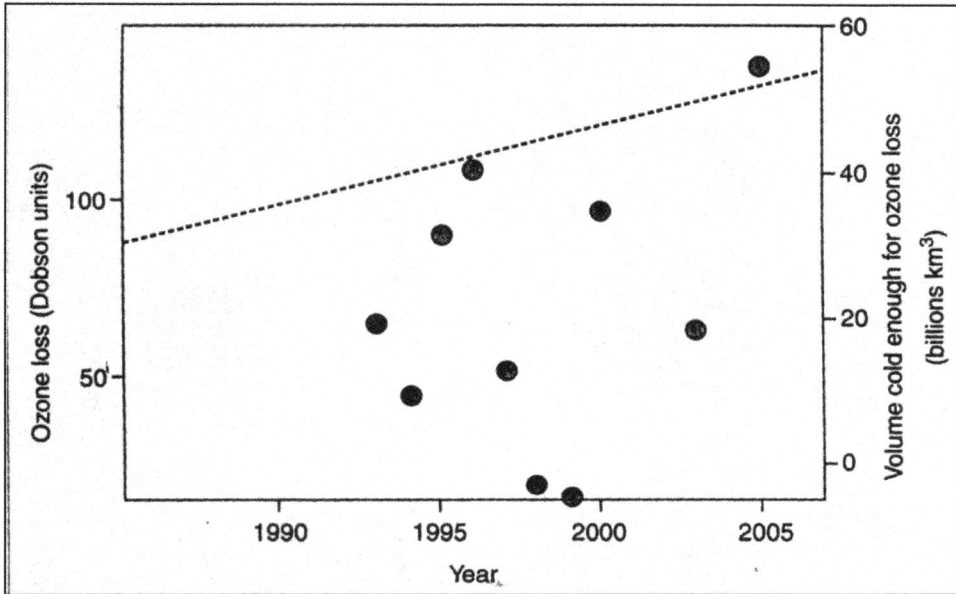

**Figure 5.15: Arctic ozone depletion (normal ozone level: 300 Dobson units).**

winter may have been much too large to be explained by greenhouse cooling and natural fluctuations in meteorological conditions, probably, far outweigh greenhouse effects.

## Ozone Depletion and Stratospheric Cooling

Measurements of air temperature is the lower stratosphere from 1979 to 1990 have shown a cooling trend that varies both spatially and seasonally (Ramaswamy *et al.*, 1996), possibly from changes in concentrations of ozone or of other greenhouse gases, as well as natural variability. The latitudinal pattern of lower stratospheric cooling for a given month through the decade matched with the pattern of the observed decadal temperature changes, confirming expectation than the observed ozone depletion leaves a spatially and seasonally varying finger print in the decadal cooling of the lower stratosphere, with the influence of increases in concentrations of other GHGs being quite small. As anthropogenic halocarbon chemicals are important causes of stratospheric ozone depletion, it appears that there is a human influence on the patterns of temperature change in the lower stratosphere over this 11 year period (Ramaswamy *et al.*, 1996).

Since 1992, springtime Antarctic ozone depletion has been particularly severe showing loss of about two-thirds of the ozone column at the South Pole during September and near total destruction of ozone in the 14 to 19 km region (Hofmann, 1996). The 1995 ozone hole lasted quite long but total column ozone measurement at the south Pole for the September-October lowest ozone period gave average values of 129 Dobson units (DU) as compared with 109 in 1993 and 119 in 1994. The interannual differences of 10 DU is significant. The rate of ozone destruction in September 1995

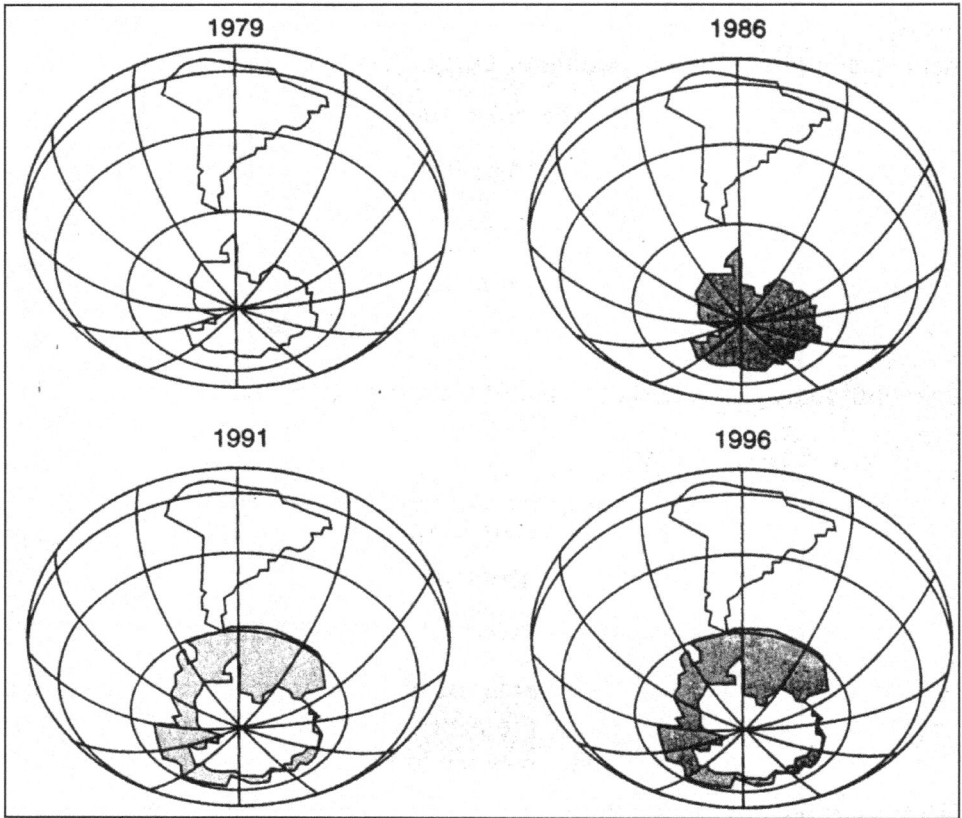

**Figure 5.16: Growth of the Antarctic ozone hole (after UNEP, 1999).**

was much less than in 1993 or 1994. The rate of halogen-catalysed ozone destruction depends both on how much halogen has been transported to the polar stratosphere and on the winter and spring temperatures which affect the availability of stratospheric cloud particle surface area and the heterogeneous reaction rates.

Integrated over the 12 to 20 km region, the September ozone loss rate has increased from about 2 to 3.5 DU per day over the past decade. This 75 per cent increase in loss rate would require about a 30 per cent increase in halogen levels if the chlorine dimer cycle were dominant, in agreement with estimates of the stratospheric halogen increase (about 2.4 to 3.2 parts per billion from 1986 to 1995). The interannual variability I the 12 to 20 km ozone loss rate is about 20 per cent, pointing to a variability in halogen levels of about 10 per cent. The depletion rates are higher during the spring following the transition of tropical winds from westerly to easterly (Hofmann, 1996).

## The Lucky Escape

The development of the ozone hole was an unexpected, unintentional result of widespread use of CFCs, as aerosols in spray cans, solvents, refrigerants and as foaming agents. If by some unfortunate chance, the world had inadvertently used bromofluorocarbons in place of CFCs, the result could well have been catastrophic. In terms of function as a refrigerant or insulator, bromofluorocarbons are as effective as chlorofluorocarbons.

But on an atom-for-atom basis, bromine is about hundred times more effective in destroying ozone as compared to chlorine. If the chemical industry had developed organobromine compounds instead of the CFCs, then without and preparedness, we would have been faced with a catastrophic ozone hole everywhere and at all seasons during the 1970s, long before the atmospheric chemists could develop the necessary knowledge to identify the problem and the appropriate techniques for the required critical measurements. It is indeed fortunate, especially in view of the fact that nobody had given any thought to the atmospheric consequence of the release of Cl or Br before 1974, that mankind was saved from annihilation !

## What Can be Done to Save the Ozone Layer ?

1. Do not buy Aerosols containing CFC, fluorocarbon, or CFC propellants.
2. Use 'ozone friendly' aerosols such as those with hydrocarbon propellants or choose brands which use roll-on or pump spray containers.
3. Do not buy those propelled by HCFC-22 (also known as chlorodifluoromethane or propellant-22).
4. Do not buy halon-filled fire extinguishers. Halons are released into the atmosphere when the fire extinguisher is used.

Be prepared to pay more for products, such as soft plastic foams, manufactures without CFCs.

## Assistance for Developing Countries

At the London meeting, mechanism for enabling developing countries to meet their obligations under the Protocol was created. This included provision for transfer of technologies needed by developing countries and arrangements to meet incremental costs incurred in moving away from ozone-depleting technologies and substances.

## Greenhouse Cooling and Ozone Hole

In theory, the ozone hole over Antarctica should gradually heal as international regulations slow down the flow of ozone-destroying chlorine compounds into the stratosphere. But the hole continues to be almost as severe as any seen before, and it now stretches over an area larger than North America. Unprecedented stratospheric cold has been driving the extreme ozone destruction. Some of the high-altitude chill may be a counterintuitive effect of the accumulating GHGs that are warming the lower atmosphere. The 1998 Antarctic ozone hole extended over about 26 million km$^2$, the largest observed since annual aholes first appeared in the late 1970s. Measured by balloon-borne instruments ascending form the South Pole, the layer of total ozone destruction extended from an altitude of 15 to 21 km. that was higher than ever seen before. By October 1998, total amount of ozone over the South Pole had dropped to 92 Dobson units; only in 1993 was the ozone hole deeper, when the catalytic effect of debris from the 1991 eruption of Mount Pinatubo pushed ozone down to 88 Dobson units (normally, there are about 280 Dobson units of ozone over the pole).

The deep chill that gripped the Antarctic stratosphere in the past austral winter may be blamed for the record ozone hole seen in 1998. Every winter, the temperature drops to below minus 78°C–to form the icy stratospheric clouds that catalytically speed up the destruction of ozone by the chlorine from CFCs. Conceivably, an underlying cooling trend in the stratosphere-induced by of all things, GHGs has probably aggravated the situation. Although GHGs warm up the lower atmosphere, they cool the stratosphere by radiating heat to space, creating an 'ice-house effect'.

Greenhouse cooling might greatly worsen the nascent ozone hole over the Arctic (see Kerr, 1998). New modeling studies have suggested that by the year 2015, ozone at lower latitudes will begin recovering as CFC controls take effect, but the chilling effect of GHGs will keep the Antarctic ozone hole as severe as ever.

## The WMO/UNEP Ozone Assessment

Once in every three or four years, the World Meteorological Organisation (WMO) and the United Nations Environment Programme (UNEP) jointly publish a report on the current understanding of the stratospheric ozone layer and its relation to humankind. This report provides useful information to the Parties of the United Nations Montreal Protocol to guide the development of policies for future protection of the ozone layer (Bodeker, 2003).

Given below is a policy-relevant summary of the issues raised in the Scientific Assessment of Ozone depletion–2002. These are:

1. *Observed ozone changes*: How has ozone been changing and how has this differed between Northern Hemisphere and Southern Hemisphere mid-latitude (35°-60°)?

2. *Understanding ozone changes*: To what extent can we explain current ozone trends?

3. *Antarctic ozone:* How has the ozone hole changed and when can we expect its recovery?

4. *Ultraviolet radiation*: How has summertime UV been changing over New Zealand, for instance?

5. *Future ozone*: How will ozone over New Zealand change in the future and will this be influenced by climate change ?

## Observed Ozone Change

The changes over the past 25 years have been substantially different in the Southern Hemisphere compared to the Northern Hemisphere.with respect to a 1964-1980 baseline, ozone averaged over 1997-2001 was 3 per cent below the baseline in northern mid-latitudes and 6 per cent below the baseline in southern mid-latitudes. The sharp decrease in southern mid-latitude ozone I the mid-1980s was not matched in the north. The decrease in the Northern Hemisphere mid-latitude ozone following the eruption of the Mt. Pinatubo volcano in June 1991 is not reproduced in the Southern Hemisphere mid-latitude.

There are also distinct differences between the north and south in the way in which seasonal ozone trends vary. Over the Northern Hemisphere mid-latitudes, larger ozone decreases are observed during winter/spring (about 4 per cent below 1964-1980 mean), with summer/autumn decreases of around 2 per cent. Over the Southern Hemisphere mid- latitudes, ozone decreases are of similar magnitude (around 6 per cent) during all season. These hemispheric differences in the seasonality of ozone changes drive different responses in surface ultraviolet radiation in the north and south. Summertime ozone decreases are far more important than wintertime decreases in relation to ozone effects on ultraviolet radiation.

## Understanding Ozone Changes

While linear trends calculated from models agree well with observed trends, the models are much less able to track interannual variability over the Southern Hemispher mid-latitudes compared to northern mid-latitudes. The two features in southern mid-latitude ozone change highlighted in earlier remain unexplained: the sudden decrease in ozone in the mid 1980s (which is not tracked by the models) and the absence of a response to the Mt. Pinatubo eruption(which the models indicate should be present). These nodel shortcomings bring out the gaps in our understanding of the key processes affecting Southern the Hemisphere mid-latitude ozone. Until our understanding has been improved, model predictions of future ozone will not be reliable.

## Antarctic Ozone

Springtime Antarctic ozone depletion has remained quite large during the 1990s and is not showing any sign of recovery. Over the last decade, the Antarctic polar vortex has persisted longer than in the 1980s and now breaks up one month later, in late November or early December. This lengthens the period during which Antarctic ozone depletion occurs with two important consequences:

1. It triggers higher peak UV levels over the Antarctic. The highest UV dose under the ozone hole is not typically observed in September-October, when maximum ozone depletion occurs, but in November-December when solar elevations are higher while ozone is still low.

2. The later the vortex break up, the closer it is to the summer solstice when ozone depleted air is mixed out over New Zealand. The combination of suppressed ozone and high solar elevations drives an increasing trend in summertime UV over New Zealand.

It should, however, be noted that the ozone hole itself has never come close to New Zealand. Emissions of most ozone-depleting gases have virtually stopped following implementation of the Montreal Protocol.

Stratospheric chlorine peaked towards the end of 20th century and is now declining. The ozone hole is expected to recover as chlorine levels decrease. However, natural variability complicates the detection of recovery, and increasing greenhouse-gas concentrations could delay the recovery. It may be at least a decade after stratospheric chlorine peaks before unambiguous recovery can be actually verified.

## Ultraviolet Radiation

Measurement soft summertime UV level over New Zealand confirmed that UV has increased in keeping with expectations following ozone decrease. Summertime UV levels over New Zealand are approximately 40 per cent higher than equivalent northern latitudes in Europe, as a result of less polluted air and lower ozone over New Zealand and because the Sun is slightly closer to the Earth in December than in June. The UV levels over New Zealand have always been high and will continue to be so even after the ozone layer has recovered.

Long-term UV changes over New Zealand are not driven by ozone along. Other factors such as changes in cloudiness, aerosols and land cover all contribute. The relative importance of each of these factors depends on local conditions.

## Future Ozone

Although future decreases in stratospheric chlorine are likely to lead to a recovery in global ozone, the rate of recovery is likely to be influenced by changes in atmospheric composition caused by increasing emissions of GHGs.

While increase in $CO_2$ will warm the troposphere, they will cool the stratosphere. Stratospheric cooling can have two effects:

1. In the upper stratosphere, the cooling is likely to enhance future $O_3$ over tropical and mid-latitudes due to a slowdown in the chemical reactions the cause $O_3$ loss.

2. A cooling in the lower stratosphere, however, is predicted to extend the period over which polar stratospheric clouds are present and will increase springtime ozone depletion.

Consequently, although total column ozone is expected to eventually recover to pre-ozone-hole levels, the distribution of ozone in the atmosphere in the future is likely to be different. Ozone will increase in the troposphere due to atmospheric pollution and remain depressed in the lower stratosphere while chlorine and bromine levels remain high. This will affect its attributes as a GHG and as an absorber of UV radiation. The element of uncertainty in the future outlook for ozone recovery is illustrated in Figure 5.17. Increase in $CH_4$ and nitrous oxide emission (primarily from agricultural activities)will cause chemical changes to the stratosphere, such as increasing water vapor and $NO_x$ levels. There changes in stratospheric composition will also influence ozone depletion. The coupling between ozone change and climate change is fairly complex (Figure 5.18).

Coupled chemistry-climate models capture this complexity by including all known feedback between climate change and changes in atmospheric chemistry. Certain models have been employed to forecast how global ozone levels over the coming decades are expected to change with decreasing chlorine levels and increasing emissions of GHGs. Coupled chemistry-climate models rely on long-term, high-

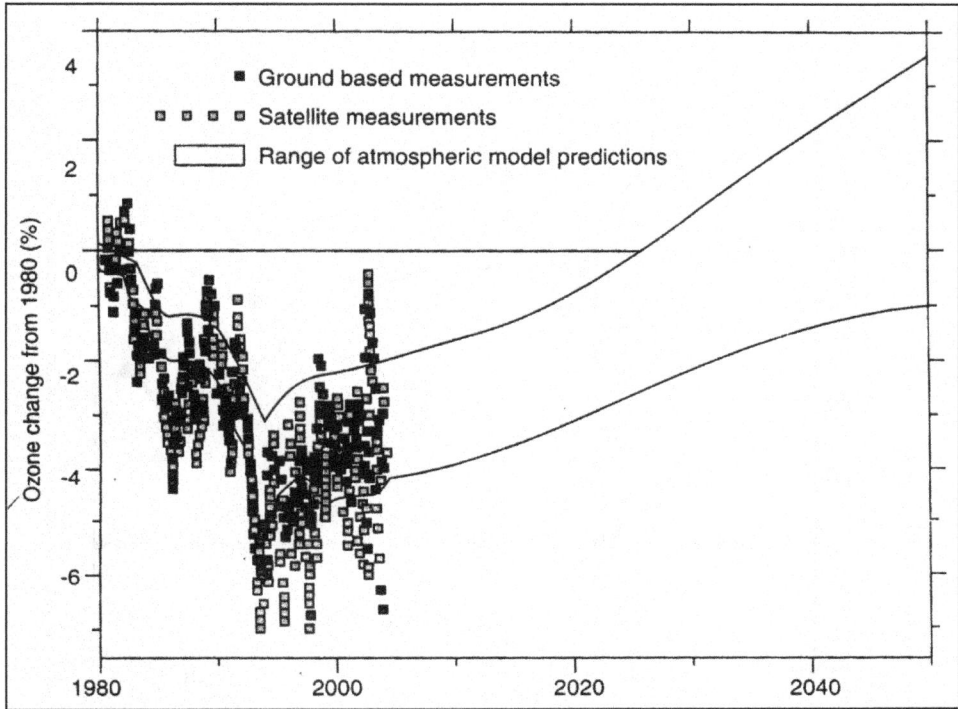

**Figure 5.17: The future outlook for ozone recovery is uncertain. Observed and modeled column ozone (60°S-60°N) amounts as percent deviations from the 1980 values, (from IPCC/TEAP, 2005).**

precision measurements for their validation, and the 22 years time series of $NO_2$ measurements at NIWa's atmospheric station at Lauder has been used for such validation.

It is gratifying to note that the Montreal Protocol appears to be working well and eventually ozone recovery is expected if all parties continue to adhere to the Protocol. Approaches to accelerate the recovery of the ozone layer are now limited; the most effective action has already been taken. However, failure to continue or comply with the Montreal Protocol would delay or could even prevent recovery of ozone. Verification that the protocol is working, through continuing global observations, is vital.

## Ultraviolet- B Effects of Ozone Changes

### Greenhouse Effect

Both ozone and halocarbons act as greenhouse gases. Like carbon dioxide, they intercept and re-radiate the Earth's outgoing infrared radiation, thereby warming the lower atmosphere. However, as ozone and its changes are not uniformly distributed, its radioactive forcing is more complicated than those of the other principal GHGs

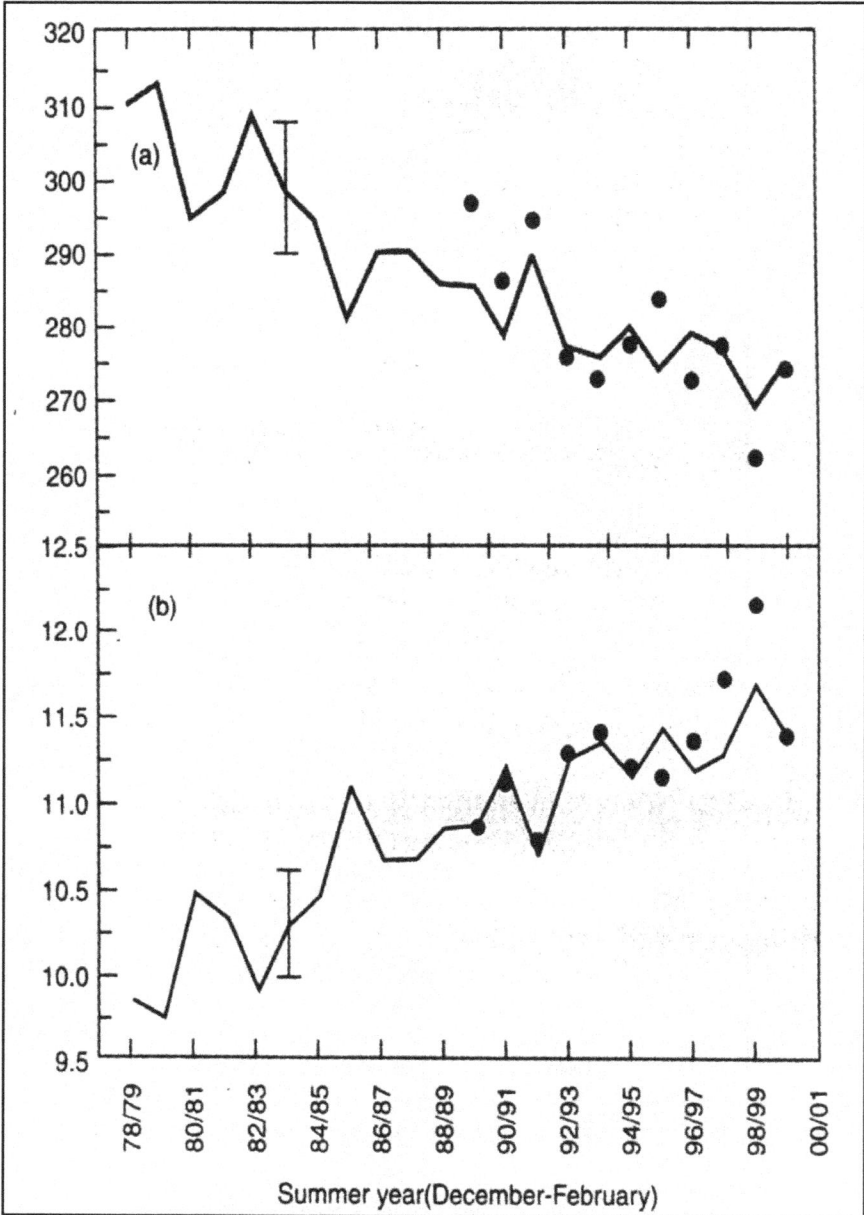

Figure 5.18: (a) Mean total column ozone and (b) noontime UV index at Lauder, New Zealand for summers (December to February) from 1978/79 to 1999/2000. the solid line in (a) shows observed summertime ozone values that have occurred since the 1970s,and the solid line in (b) shows the expected values of clear-sky UV. The symbols (.) from 1989/90 on show measured summertime value of ozone and peak UV index, both derived from measurements of the UV spectrum. Error bars (2 standard errors) are shown for reference. (Adapted from Figure 5-4 o the WMO/ UNEP 2002 assessment; see Bodeker, 2003).

which have a long life span permitting even mixing. In general, more ozone in the troposphere (especially near the tropopause) causes warming, and ozone loss in the stratosphere causes cooling. Some small temperature decline (0.6°C-0.8°C) in the 12 to 20 km layer during the past two decades is consistent with expected radiative impacts of stratospheric ozone decrease.

It appears that lower stratospheric ozone depletion in recent decades has resulted in a cooling effect on the climate and has offset, by about 15 to 20 per cent, the positive greenhouse forcing due to increases in other gases. The increase of tropospheric ozone since preindustrial times may have enhanced the total greenhouse forcing by as much as 20 per cent. Such changes could have an impact on the radiative balance of the Earth-atmosphere system and the thermal structure of the atmosphere and so cause some unpredictable changes to atmospheric circulation patterns.

## Ozone and Radioactive Forcing

Radiative forcing is stated in units of $Wm^{-2}$. Its estimations are based on net solar and thermal infrared radiance at the tropopause and do not greatly depend on uncertainties, such as the combined interactive role of clouds, aerosols or oceans, for predicting surface climate change. While a positive value signifies increased energy retention, producing warming, a negative value implies greater energy loss, *i.e.* cooling.

☆ Tropospheric ozone has increased in the Northern Hemisphere since reindustrialize times. Models and deductions from observations suggest a positive radiative forcing of about 0.5 $WM^{-2}$.

☆ Stratospheric ozone has decreased since the 1970s. Between 1980 and 1990, this has caused negative radiative forcing of about 0.1 $Wm^{-2}$ compared with positive 0.45 $Wm^{-2}$ forcing from an increase of other GHGs in the same period.

The net global mean radiative forcing arising from ozone changes is likely to have been positive since preindustrial times, contributing about 20 per cent of all the GHG caused increase of radiative forcing in that period.

### Ultraviolet Radiation

The life-protecting role of atmospheric ozone emanates from its ability to absorb harmful UV radiation of wavelengths shorter than 320 nm. The small amount of UV-B that does penetrate through the ozone shield can cause considerable harm to human health, including eye cataracts, an increase of non-melanoma skin cancers, damage to genetic DNA, and suppression of the efficiency of the immune system.

Under cloudless conditions, each 1 per cent reduction in ozone results in an increase of about 1.3 per cent I the UV-B reaching the surface of the Earth. The total ozone decline so far has resulted in smaller increase in UV-B (280-320 nm) reaching the ground except over the tropical belt.

Further ozone decline could have considerable harmful consequences, not only to humans but also to other life forms and tropospheric chemistry. Crops and the aquatic ecosystem, including marine and freshwater plankton, could be damaged (see Kumar and Hader, 1999; Hader *et al.*, 2003). Since marine phytoplanktons act as a major sink for $CO_2$, any UV-B effect on them could have a role in future $CO_2$ trends and consequently climate. The inverse correlation between UV-B intensity and the total ozone amount is fairly well established.

Although global warming and stratospheric ozone depletion increase temperature and ultraviolet (UV) in mid- to high-latitude ecosystems, very little is known about the interactive effects of temperature and UV on organisms. We known little about how temperature can alter organism-level responses to elevated UV. Williamson *et al.* (2002) exposed *Daphania catawba, Leptodiaptomus minutus,* and *Asplanchna girodi* to UV-B at four different temperatures, namely 10, 15, 20 and 25°C.

Elevated temperatures increased UV tolerance in *D. Catawba* and *L. minutus,* species that depend heavily on photoenzymatic repair (PER), but decreased it in *A. girodi,* a species that has less PER, PER is a light- dependent DNA repair process that is essential to UV tolerance in some species of zooplankton but is weak or even absent in others. Also, body size in *D. Catawba* decreased with increasing UV dose. These results show that climate change can alter responses to UV through temperature mediated effects in aquatic ecosystems, and that these effects can be species-specific and dependent on PER ability.

## Climate Change and UV-B Impacts on Arctic Tundra and Polar Ecosystems

It is generally agreed that (a) the Arctic is likely to amplify global climate warming; (b) UV-B radiation may continue to increase there because of delays I the repair of stratospheric ozone; and (c) the Arctic environment and its inhabitants would be susceptible to such environmental changes. These concerns have prompted an international assessment of climate change impacts through a four-year study [(The Arctic climate Impacts Assessment, ACIA)] (ACIA, 2004). From this Report, Callaghan *et al.* (2004) produced a major assessment by reviewing the findings on terrestrial ecosystems of the Arctic, from the treeline ecotone to the polar deserts.

The Arctic is a treeless wilderness with cold winters and cool summers. Definitions of its southern boundary vary according to environmental, geographical or political biases. The southern boundary of the circumpolar Arctic is the northern extent of the closed boreal forests. There is no clear boundary but rather a transition from South to North consisting of the sequence: closed forest? Forest with patches of tundra? Tundra with patches of forest? (Callaghan *et al.*, 2004d). The transition zone is quite narrow (30-150km). An altitudinal zonation is superimposed on the latitudinal zonation from forest to treeless areas to barren ground in some mountainous regions of the northern taiga. The transition zone from taiga to tundra stretches for more than 13,400 km around the lands of the Northern Hemisphere.

Indeed, it is one of the most important environmental transition zones on Earth (Callaghan *et al.*, 2004e)- it represents a strong temperature threshold close to an area of low temperatures. This zone is also termed forest tundra, sub-Arctic and the tundra-taiga boundary or ecotone. Vegetationally, it is characterized as an open landscape with patches of low trees with dense thickets of shrubs that together with the trees totally cover the ground surface.

Climatically, the Arctic is usually defined as that area where the average temperature for the warmest month is less than 10°C but mean annual air temperatures vary according to location, even at the same latitude. The summer decreases from about 3.5 to 1.5 months from the southern boundary of the arctic to the North. The mean July temperature decreases from 10-12°C to 1.5°C. Precipitation in the Arctic is generally low, and decreases from about 250mm in the South to around 45mm per year in the polar deserts of the north. Yet, the Arctic is not considered arid in view of low evaporation even in the polar deserts, air humidity is quite high and the soils are moist during the short growth period (Bovis and Barry, 1973). With a few exceptions, the Arctic characteristically has continuous permafrost. The depth of the soil's active layer during the growing season depends on summer temperatures, varying from 80cm close to the treeline to 40cm in polar deserts, but active layer depth varies locally within landscapes according to topography–it can reach 120cm on south-facing slopes while being as little as 30cm inbox. Topography defines habitats in terms of moisture and temperature as well as active layer dynamics (Webber *et al.*, 1980); this makes Arctic landscapes a mosaic of microenvironments. Topographical differences become more important with increasing latitude.

Ecosystem disturbances are marked in the Arctic. Mechanical disturbances include thermokarst through permafrost thaw, freeze-thaw processes, wind, sand and ice-blasts, slope processes, snow load, flooding during thaw, change in river volume and coastal erosion and flooding. Biological disturbances are exemplified by insect pest outbreaks, peaks of grazing animals having cyclic populations, and fire. These disturbances occur at various geographical entities and timescales (Figure 5.19). They affect the colonization and survival of organisms and hence ecosystem development.

The vegetation of the Arctic varies from forest tundra in the south, where plant communities contain all the known plant groups for the Arctic, and have continuous several layered canopies that extend to more than 3 m high, to polar deserts in the north where vegetation colonises 5 per cent or less of the ground surface, is less than 10cm high and is dominated by herbs, mosses and lichens (Figure 5.20).

Biodiversity in the Arctic is low, decreasing towards the north: there are bout 1,800 species of vascular plants, 4000 species of cryptogams, 75 of terrestrial mammals, 240 of terrestrial birds, 3,000 of fungi, 3,300 of insects (Walker, 2000) and thousands of bacteria and archaea. The Arctic constitutes a particularly important global pool of mossesm, lichens, and springtails whose abundance in the Arctic is higher than in other biomes. Net primary production, net ecosystem production and decomposition rates are quite low. Food chains are short with typically only a few representatives at each level of the chain. The Arctic soils are generally shallow and underdeveloped with low productivity and immature humus.

The Arctic is exposed to dramatic environmental changes which can have strong impacts on its ecosystems. In the Arctic, the dominance of climate change is a major factor affecting biodiversity (Sala and Chapin, 2000). The Arctic plants appear to be particularly vulnerable to increases in ultraviolet-B radiation because its damage is not dependent on temperature, whereas enzyme-mediated repaired of DNA damage is often constrained by low temperatures (Paulsson, 2003).

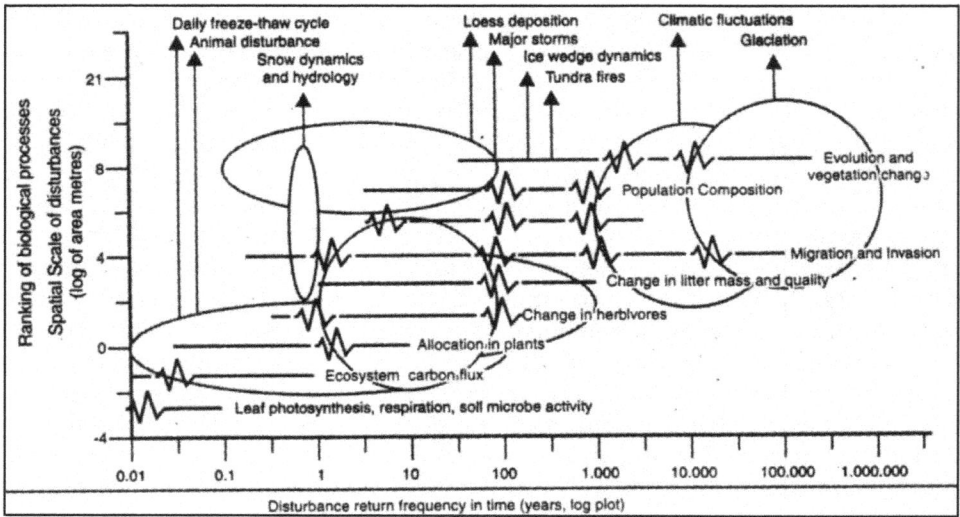

Figure 5.19: Schemewise timescale of ecological processes in relation to disturbances in the Arctic. Not shown are responses expected due to anthropogenic climate change (after Callaghan *et al.,* 2004).

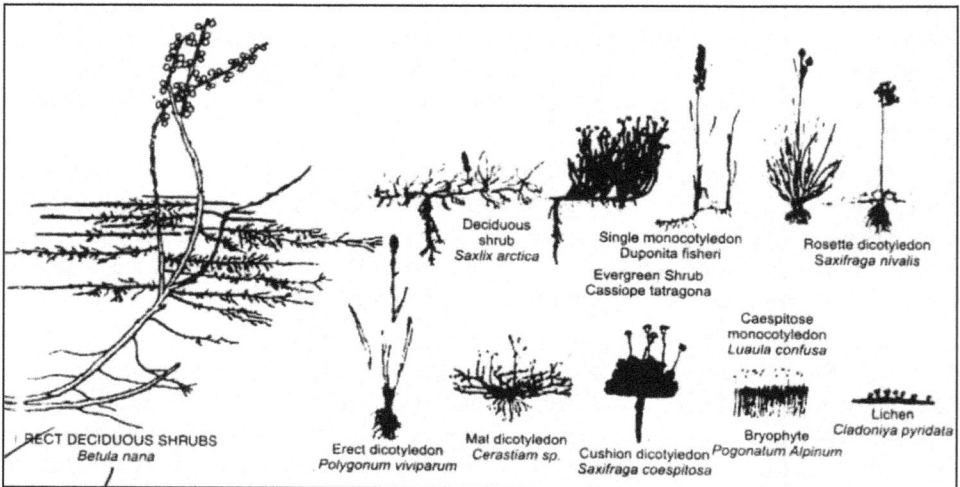

Figure 5.20: Growth forms of Arctic plants (after Webber *et al.,* 1980; and Callaghan *et al.,* 2004d).

# Climate Change and UV-B Impacts on Arctic Tundra and Polar Desert Ecosystems

The Arctic represents an important region in which to assess the impacts of current climate variability and projected global warming. This importance stems from the following reasons.

There has been considerable warming in:

1. The Arctic in recent decades (an average of about 3°C and between 4° and 5°C ove much of the landmass);

2. Climate projections point to a continuation of the warming trend with an increase in mean annual temperatures of 4 to 5°C by the year 2080;

3. Recent warming has started impacting the environment and economy of the Arctic impacts that may increase and affect also life style, culture and ecosystems, and

4. Changes occurring in the Arctic may influence other regions of the Earth; changes in snow, vegetation and sea ice are likely to influence the energy balance and ocean circulationat regional and even global scales (Callaghan *et al.*, 2004).

Prompted by the urgent need to understand and project impacts of changes in climate and UV-b radiation on the arctic, the Arctic Climate Impact Assessment (ACIA, 2004) has assessed the impacts of changes in climate and UV-B radiation on Arctic terrestrial ecosystems, including changes already occurring and those likely to occur I the future. Callaghan *et al.* (2004) have outlined these key findings as follows:

☆ The dominant response of current Arctic species to climate change (as also in the past) is very likely to be relocation rather than adaptation. Some changes are occurring even now.

☆ Some organisms, such as mosses and lichens, some herbivores and their predators face risk in some areas, but productivity and number of species is very likely to increase. Biodiversity is more at risk in some subregions than in others: Beringia has a higher number of threatened plant and animal species than any other ACIA subregions.

☆ Changes in populations are triggered by trends and extreme events, particularly winter processes.

☆ Forest is likely to replace a significant part of the turndra: this can affect the composition of species, but certain environmental and sociological processes may prevent forest from advancing in some locations.

☆ Displacement of tundra by forest will lead to a decrease in albedo which increases the positive feedback to the climate system. Such a positive feedback will generally dominate over the negative feedback of increased carbon sequestration. Forest development will also ameliorate local climate.

☆ Warming and drying of tundra soils in parts of Alaska have already changed the carbon status of this area from sink to source. Although other areas still maintain their sink status, the number of source areas currently exceeds the sink areas. Future warming of tundra soils would probably lead to a pulse of trace gases into the atmosphere, particularly in disturbed and drying areas. Current models suggest that the tundra will become a weak sink for carbon because of the northward movement of vegetation zones that are more productive than those they displace.

☆ Rapid climate change that exceeds the ability of species to relocate may lead to increased incidence of fires, diseases and pest outbreaks.

☆ Enhanced $CO_2$ and UV-B affect plant tissue chemistry and thereby exert small but long-term impacts on ecosystem processes that reduce nutrient cycling with the potential to decrease productivity and either increase or decrease herbivores (Callaghan *et al.*, 2004).

## Biodiversity, Distribution and Adaptations of Arctic Species in the Context of Environmental Change

The fairly low species diversity in the Arctic decreases from the boreal forests to the polar deserts of the extreme north. Only about 3 per cent (about 5,900 species) of the world's flora (excluding algae) occur in the Arctic north of the treeline. Primitive plant species of mosses and lichens are particularly abundant. Although the number of plant species in the Arctic is quite low, individual communities of small Arctic plants have a diversity similar to or even greater than those of boreal and temperate zones: there can be 25 species $dm^{-2}$.

Latitudinal gradients suggest that Arctic plant diversity is sensitive to climate, and species number is least sensitive to temperature near the southern margin of the tundra. The temperature gradient that has such a strong influence on species diversity occurs over much shorter distances in the arctic than in other biomes (Callaghan *et al.*, 2004).

Microorganisms are more difficult to enumerate. Arctic soils harbour large reserves of microbial biomass, although diversity of all groups of soil microorganisms is lower in the Arctic than further south. Many common bacteria and fungi are rare or absent in tundra areas. As with plants and animals, there are large reductions in number of microbial species with increasing latitude, and increasing dominance of the species that occur (Callaghan *et al.*, 2004a). An early response of diversity to warming is likely be an increase in diversity of plants, animals, and microbes together with reduced dominance of the currently widespread species. Tax most likely to expand into tundra are boreal tax now found in river valleys and which could spread into the uplands, or animal groups such as wood-boring beetles that are presently excluded due to lack of food resources.

Current extreme environmental conditions restrain the metabolic activity of Arctic microbesues but they retain enormous potential to manifest the same activity as boreal analogs immediately after climate warming. Warming could lead to extinction of some few arctic plants that today occur in narrow latitudinal strips of tundra adjacent to the sea. Some animals are Arctic specialists and could possibly face extinction. Plant and animal species with their centres of distribution in the high-or mid-arctic are very likely to suffer decline in abundance n their current locations in the face of projected warming (Callaghan *et al.*, 2004a).

## Responses of Arctic Plants to Climate Change

As many species are pre-adaptations to the Arctic climate tend to be either rare or absent. Freezing tolerance excludes approximately 75 per cent of the world's vascular plants from the Arctic. Short growing seasons and low solar angles select for long life cycles in which slow growth usually makes use of stored reserves, and development cycles extend over several growing seasons.

Some plant species occupy microhabitats, or show behaviour or growth forms that maximize plant temperatures compared with ambient. Low soil temperatures bring down microbial activity as well as nutrient availability to higher plant roots.

Low nutrient availability may be compensated by the conservation of nutrients in nutrient poor tissues, resumption of nutrients from senescing tissues, enhanced rates of nutrient uptake at low temperatures, increased biomass of roots relative to shoots, mycorrhizal fungal association, uptake of nutrients in organic forms, and uptake of nitrogen by rhizomes. Frost-heave phenomena (temperature fluctuations around 0°C) can uproot ill-adapted plants (Callaghan *et al.*, 2004, 2004a).

Many Arctic plants show pre-adaptation to fairly high levels of UV-B radiation and also have mechanisms to protect DNA and sensitive tissues from UV-B. They can repair some UV-B damage to DNA. Thick cell walls and cuticles, waxes and hairs on leaves and the presence or induction of UV-B absorbing chemical compounds in foliage, protect sensitive tissues. The Arctic plant species show no specific adaptations. Further, these species do not manifest the usually complex interactions with other organisms found in southern latitudes.

Arctic plants are adapted to grazing/browsing mainly by chemical defenses rather than by spines and thorns. Facilitation becomes more and more important relative to competition at high latitudes and altitudes (Callaghan *et al.*, 2004, 2004b).

Many of the traits of Arctic species in relation to their current environments limit their responses to climate warming and other environmental change. Several characteristics can potentially cope up with biotic selective pressures (*e.g.* climate) more than biotic (*e.g.* inter-specific competition). This renders Arctic organisms more susceptible to biological invasions so that they are very likely to change their distribution rather than evolve significantly in response to warming (Callaghan *et al.*, 2004b, c).

## Responses of Arctic Animals to Climate Change

Terrestrial Arctic animals exhibit several adaptations which enable them to live in the Arctic climate. Physiological and morphological traits in warm-blooded vertebrates (mammals and birds) include thick fur and feather plumages, short extremities, extensive fat storage before winter and metabolic seasonal adjustment (Callaghan *et al.*, 2004).

Cold-blooded invertebrates show cold hardiness, high body growth rates, and pigmented and hairy bodies, The Arctic animals often survive under surprisingly wide range of temperatures. In view of the short growing season, animal life history strategies enable individuals to complete their life cycles under time constraints and high environmental unpredictability.

The biotic environment of Arctic species is quite simple with few enemies, competitors, and available food resources. This is why, the Arctic animals have evolved fewer traits related to competition for resources, predator avoidance and resistance to diseases and parasites than their southern counterparts.

Many Arctic animals have adaptations for escaping unfavourable weather, resource scarcity or other adverse conditions either by winter dormancy or by selection of spatial refuges at a wide range of spatial scales from microhabitat selection at any given site, through seasonal habitat shifts within landscapes, to long-distance seasonal migrations within or across geographic regions (Callaghan *et al.*, 2004).

According to Callaghan *et al.* (2004), if climate changes, terrestrial Arctic animals seem likely to be most sensitive to the following conditions:

1. Warmer climate in summer that induces desiccation in invertebrates;
2. Climatic changes that interfere with migration routes and staging sites en route for long-distance migrators:
3. Climatic events that alter snow conditions and freeze-thaw cycles in winter resulting in unfavourable conditions of temperature, $O_2$ and $CO_2$ for animals below the snow, and limited resource availability (*e.g.* vegetation or animal prey) for animals above the snow;
4. Climate changes that disrupt behaviour and life history adjustment to the timing of reproduction and development that are currently linked to seasonal and multiannual peaks in food resource availability; and
5. Influx of new competitors, predators, parasites and diseases.

## Responses of Arctic Microorganisms to Climate Change

The Arctic microorganisms show resistance to freezing. Some can metabolise at temperatures down to minus 39°C–a process possibly accounting for up to 50 per cent of annual $CO_2$ emissions during winter from tundra soils. Cold-tolerant microbes are usually also drought-tolerant. Pigmentation protects lichens from high irradiance including UV radiation. Cyanobacteria and eukaryotic algae show many adaptive strategies to avoid, or at least minimize UV injury.

However, in contrast to higher plants, falvonoids do not act as screening compounds in algae, fungi, and lichens (Callaghan *et al.*, 2004). In general, microorganisms are highly adaptive, can tolerate most environmental conditions and have short generation times which facilitate rapid adaptation to novel environments associated with changes in climate and UV-B radiation (Callaghan *et al.*, 2004).

## Ultraviolet Index

Ultraviolet intensities in countries near the Soth Pole (*e.g.*, New Zealand) are relatively high as compared with Europe and other parts of the Northern Hemisphere. Some contributing factors are clear skies, relatively low levels of ozone, high solar elevations and the close Earth-Sun separation during the southern summer

These factors, combined with the temperate climate conducive to an outdoor lifestyle, are largely responsible for the quite high rates of skin cancer prevalent in New Zealand. This has generated much public interest in Ultraviolet and its effects.

As some southern latitudes, ozone levels today are 5 to 10 per cent lower than they were 20 years ago. From this reduction in ozone, sunburning ultraviolet may have increased to about 10 per cent over the same period. Ultraviolet forecasts are generally made in terms of a 'Ultraviolet index', which is proportional to the intensity of sunburning ultraviolet radiation. Ultraviolet index values greater than 10 are 'extreme'. Sometimes, values greater than 12 can occur. A ultraviolet index of 12 corresponds to an approximate burn time of 12 minutes for a fair-skinned person, the ultraviolet index of 6 corresponds to 24 minutes, etc.

Ultraviolet Index forecasting system is usually based on a sophisticated radiative transfer code (TUV) which calculates the clear-sky Ultraviolet index at any time and pale, given various geographic and atmospheric parameters (notably surface albedo, altitude, date and time, surface pressure and overhead profiles of temperature, ozone and other radiatively active species). These parameters can be obtained in various ways (Marks and McKenzie, 1997).

The forecasts are being made for clear skies only and do not include the effects of clouds or variations in aerosol optical depth. These factors are sometimes important modulators of UV at the surface. The presence of cloud covering the direct solar beam generally leads to decreases in UV, but under broken cloud conditions; reflections from cloud edges can at times increase UV. Direct measurements of UV using calibrated instruments at the surface can detect these variations. A real-time colour-coded UV display has been developed in New Zealand and installed at the Molyneux Aquatic centre in Alexandra. It incorporates a detector that measures the sun burning UV from the Sun and sky that strikes a horizontal surface. The instrument no only senses the current UV intensity but also amplifies the signals and displays the result through a pointer on a scale (using the same servo technology as is used in model aircraft) (McKenzie, 2003). The current intensity of UV is depicted in terms of the internationally agreed UV index (UVI) scale. The UVI scale was originally used in Canada, where its maximum value was designed to reach 10. In New Zealand, the peak UVI sometimes exceed 13 near midday in summer. The scale is subdivided into five different regions corresponding to five different behavioural responses, as in Table 5.4.

**Table 5.4: The Ultraviolet Index (UVI) Scale**

| UVI Values | Risk | Colour Code |
|---|---|---|
| Less than 3 | Low | Green |
| 3-5 | Moderate | Yellow |
| 6-7 | High | Pink |
| 8-10 | Very high | Red |
| 11-13 | Extremely high | purple |

# The Network for the Detection of Stratospheric Change

Although the atmosphere has undergone many changes due to human influences over the last three decades, the most spectacular has been the formation of the Antarctic ozone hole. Ozone depletion in the stratosphere, with adverse consequences for life, alerted the world community to the fragility of the global atmospheric environment. This led to the signing of the Montreal Protocol and its amendments and Adjustments with a view to restricting the release of ozone-destroying industrial chemicals into the atmosphere. Evidence of ongoing atmospheric change had led to such important issues as: How will stratospheric ozone respond as the abundance of ozone-destroying chemicals decreases? How will atmospheric composition respond to and influence climate? An international Network for the Detection of stratospheric Change (NDSC) was created to provide a consistent, standardized set of long-term measurements of atmospheric trace gases, particles, and physical parameters via a suite of globally distributed sites. These measurements constitute the scientific foundation upon which sound policy decisions can be based. Since its creation over ten year ago, the NDSC has contributed to the understanding of stratospheric ozone depletion at the poles and mid-latitudes, and documented the increase and leveling off of ozone depleting chemicals in the atmosphere and the continued growth of GHGs. Figure 5.21 show the NDSC sites. Currently, the NDSC operates at more than 60 locations worldwide. In is supported by ozone soundings, satellite measurements, and other existing ground based monitoring networks. Figure 5.22 shows the changes from the 1964-1980 levels for global total ozone as estimated from ground based measurements.

Figure 5.23 is a simplified sketch of chemical reaction involved in the activation of stratospheric chlorine with and without the presence of polar stratospheric clouds. Table 5.5 lists the instruments use in the various measurements, along with their rationale. Stratospheric nitrogen in the $NO_2$ is not only directly responsible for the catalytic control of ozone but also has important role in coupling the $NO_x$ (NO and $NO_2$), $ClO_x$ (Cl and ClO), and $HO_x$ (OH and $HO_2$) families, leading to key stratospheric reservoir species such as $ClONO_2$ and $HNO_3$. Figure 5.24 shows monthly mean slant column measurements of $NO_2$ monitored by UV/visible spectrometry at the primary NDSC station in Lauder. New Zealand, from 1981 to 1999. Diurnal (AM versus PM solar zenith angles), seasonal (minimum in winter, maximum in summer), and long-term changes are gleaned from such observational databases. The effect of major volcanic eruptions such as Mt. Pinatubo on stratospheric $NO_2$ burdens can also be determined, with substantially reduced $NO_2$ columns measured in the year following these eruptions.$N_2O$ happens to be the chief source of stratospheric $NO_x$. Its steady increase during the recent industrial era (by about 0.2 per cent per year) should lead to a similar increase in $NO_x$. This has been offset to some extent during the past few decades by the increase of $ClONO_2$. Certain spectrometers can even measure the stratospheric loading of the most important nitrogen species simultaneously.

## Vertical Profiles of Methane and Nitrous Oxide

Both Methoane and Nitrous oxide are gases with long stratospheric lifetimes and simple removal mechanisms (reaction with the OH radical for $CH_4$ and photolysis for $N_2O$). They are perfect tracers of atmospheric motions. Short-term changes in both their vertical distributions and total column abundance are pointers to atmospheric dynamic processes (meriodional circulation as well as subsidence in polar regions) that also influence the ozone field. Long-term changes in the $N_2O$ source strengths at

**Table 5.5: Measurement priorities (after Kurylo and Zander, 2001).**

| Measurement | Instrument Type | | Rationale |
|---|---|---|---|
| O₁ total column | Dobson; Blewer; other UV/visible spectrometers | | Controls penetration of UV radiation to troposphere and ground |
| O₃ profile | DIAL [1] (1<300nm) | 0-18 km | Determines stratospheric temperature, structure; influence circulation and climate; important greenhouse gas in the troposphere |
| | DIAL (1>300 nm) | 15-60 km | |
| | Mwave[2] radiometers | 12-65 km | |
| | ozonesondes | 0-32 km | |
| Temperature | Raman lidars | 15-35 km | Determines rates of chemical reaction arid dynamics of stratosphere, controls vertical transport and H₂O content near tropopause, influenced by greenhouse gases. |
| | Rayleigh lidars | 30-95 km | |
| ClO profile | Mwave radiometers | 25-45 km | Catalyzes O₃ destruction |
| H₂O profile | Raman lidars | 0-15 km | Controls radiative and chemical balance 0 the stratosphere; tracer of troposphere/stratosphere exchange; main source of OH in the stratosphere; affects climate. |
| | Hygrometer sondes | 0-30 km | |
| | mwave radiometers | 40-80 km | |
| Aerosol distribution | Backscatter lidars | 0-30 km | Influence climate; initiates heterogenous processes in particular at high latitudes affects optical senson data reduction |
| | Backscatter sonders | | |
| NO₂ stratospheric column HCl and ClONO₂ column | UV/Visible and FTIR[3] Spectrometers FTIR spectrometers | | Provides catalytic control of O₃ coupling of NO, HO, and Cl cycles. Main inorganic chlorine (Cl) reservoirs of key relevance in stratospheric heterogeneous activation processes. |
| N₂O, CH₄, and CFCs columns and stratospheric profiles HNO₃ and NO columns HF and COF₂ column | FTIR spectrometers mwave radiometers FTIR spectrometers FRIT spectrometers | | Tracers of atmospheric transport; provide reference systems for interpretation of ozone change; important greenhouse gases Important compounds of the NOₓ family Main inorganic fluorine (F) species in the stratosphere |
| Other species | UV fluorescence lidars; Mwave radiometers, FTIR spectrometers | | Regulate stratospheric oxidation capacity, sulphate aerosol loading |
| UV radiation at the ground | UV spectroradiometers | | Increase as O₃ decreases; affects oxidation capacity of troposphere and thus lifetimes of greenhouse gases, adverse effects on humans and biosphere |

(1) differential absorption lidar; (2) microwave; (3) fourier transform infrared.

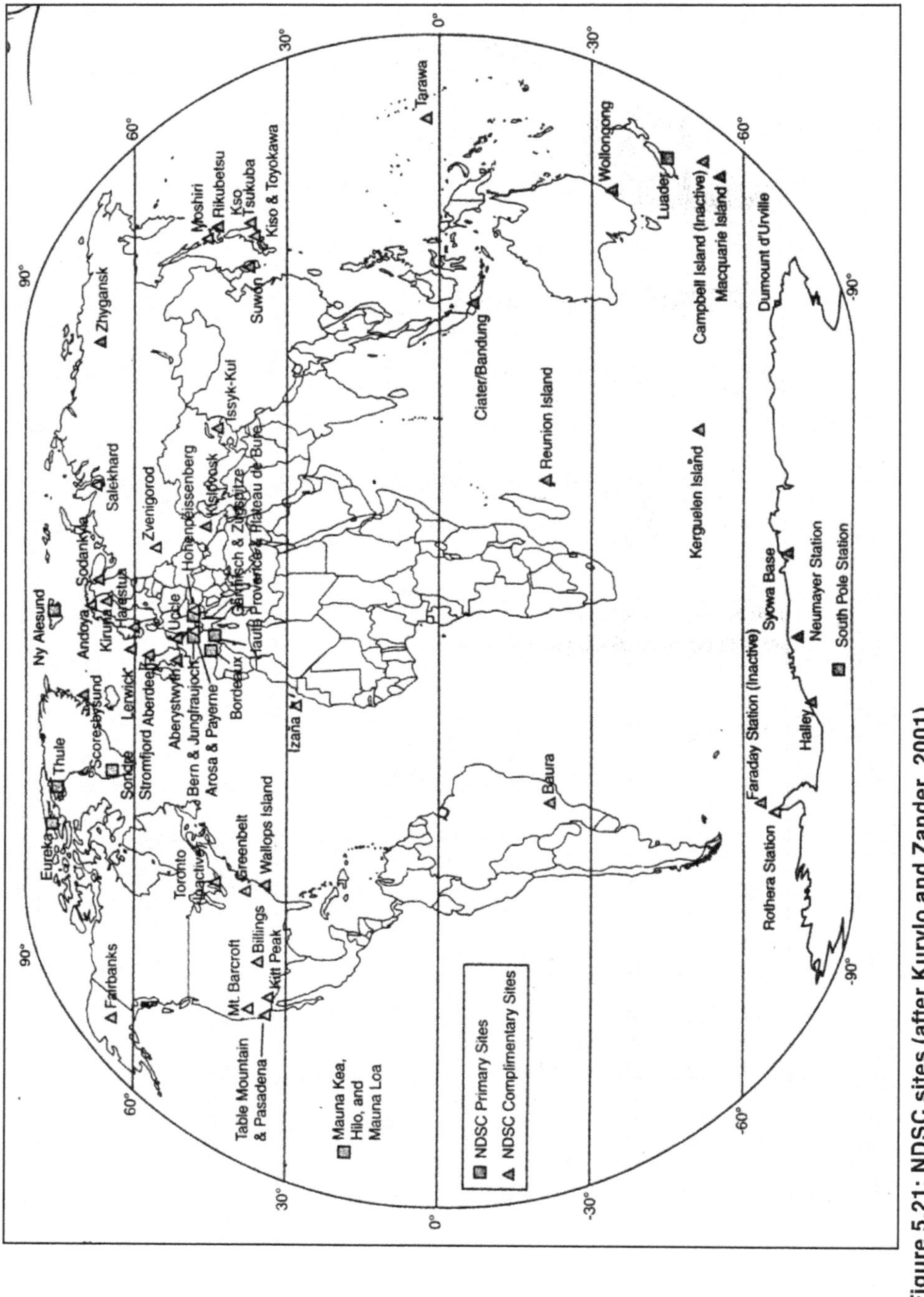

Figure 5.21: NDSC sites (after Kurylo and Zander, 2001).

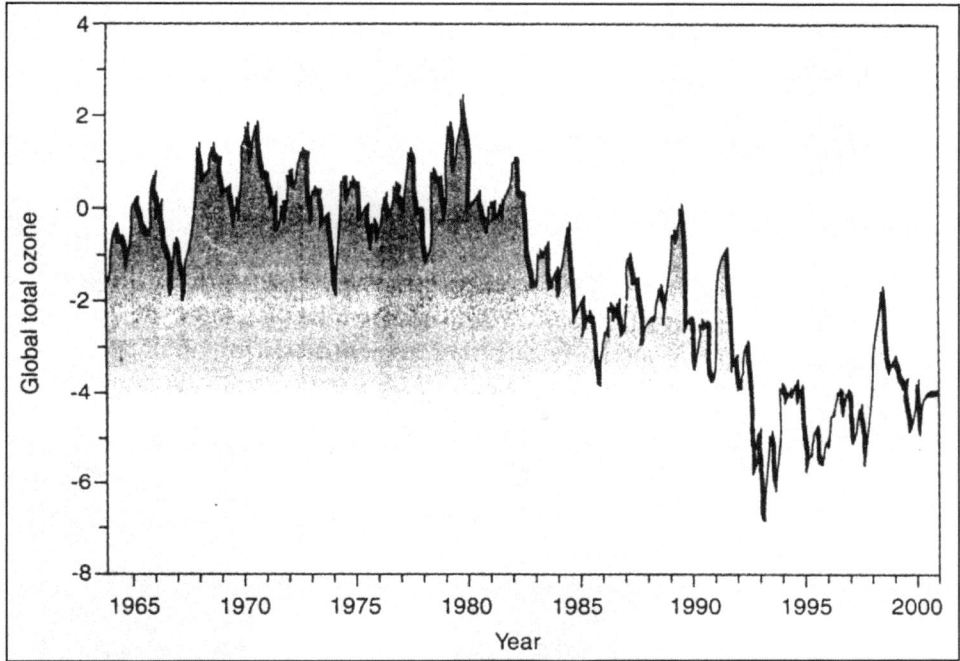

**Figure 5.22: Alterations from the 1964-1980 level for global total ozone estimated from ground-based measurements (after Kurylo and Zander, 2001).**

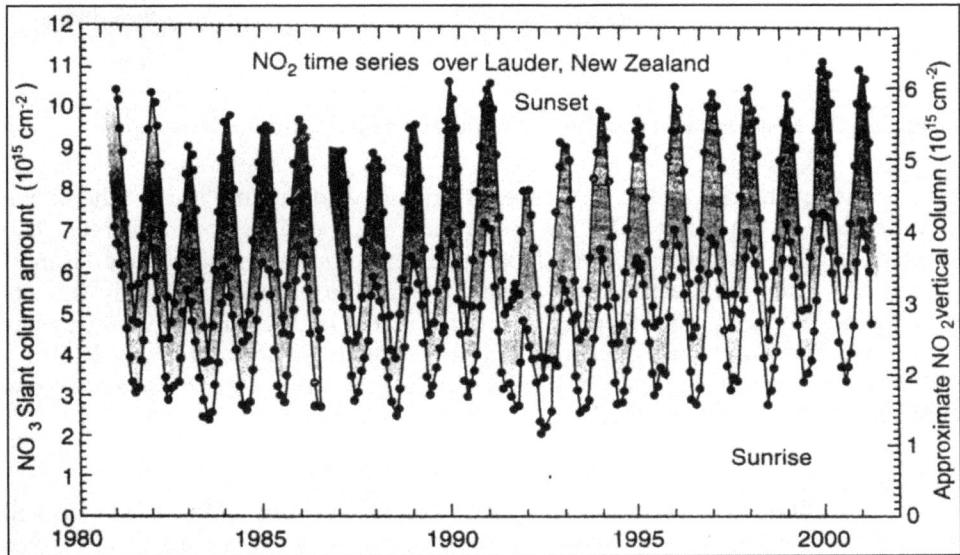

**Figure 5.23: Monthly means of sunset and sunrise slant column NO$_2$ measurements at Lauder (45°S, 170°E) by UV/visible spectrometry (after Kurylo and Zander, 2001).**

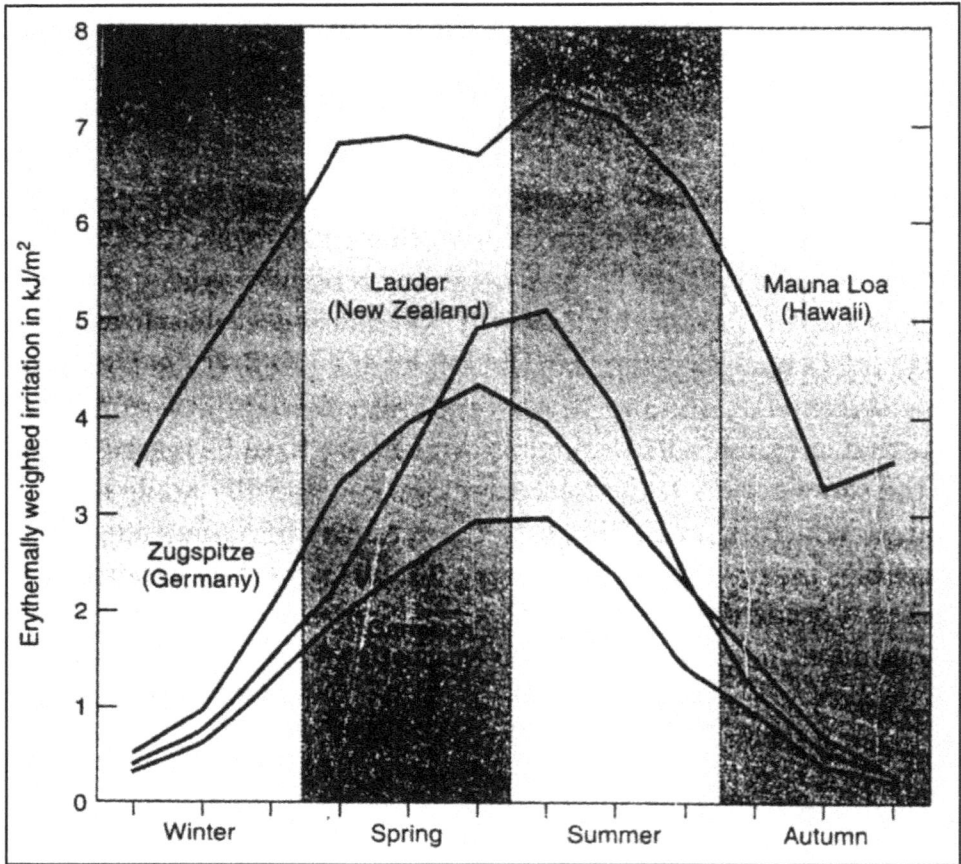

**Figure 5.24: Erythenmally weighted ultraviolet irradiation at four locations.**

the ground (primarily anaerobic processes in soils and oceans) impact the ozone layer in view of the fact that $N_2O$ is the primary source of stratospheric $NO_x$. The reaction of $CH_4$ (whose main sources include wetlands, rice paddies, animal husbandry, and fossil fuel production and use) with the OH radical directly impacts the oxidizing capacity of the troposphere; it has also an indirect effect on the lifetimes of various chlorine-bearing hydrogenated compounds. Both $CH_4$ and $N_2O$ are radioactively important gases, and increases in their atmospheric concentrations enhance the greenhouse effect (Kurylo and Zander, 2001).

## Spectral UV Radiation

As a result of the decline in stratospheric ozone, it is expected that UV levels will increase at high and mid-latitudes, producing adverse effects on the biosphere as well as on human health. The anti-correlation between ozone and UV is partly masked by the variability in cloud cover, UV-scattering and UV-absorbing aerosols in the troposphere, tropospheric pollution, and the aerosol content of the stratosphere

(Kurylo and Zander, 2001). The NDSC UV instruments can establish a UV climatology through long-term monitoring by a network of UV spectroradiometers, to detect spectrally resolved trends in global spectral UV irradiance, to provide data sets for specific process studies and for the validation of radioactive transfer models as well as satellite-derived UV irradiance at the Earth's surface, and to understand geographic differences in global spectral UV imadiance. Monthly average daily erythemal doses for the year 1996 were measured at the primary NDSC stations a Garmisch-Partenkirchen, Zugspitze, Lauder, and Mauna Loa. At all four stations, UV doses generally increase with solar elevation. Mauna Loa has the highest doses due to the highest solar elevation, lowest total ozone, and highest altitude. The UV levels in Lauder (45°S) are generally higher than in Garmisch (47°N) because of lower total zone overhead, less cloudiness, and fewer aerosols. The yearly dose at the Zugspitze is equal to Lauder's, being about 50 per cent higher than that at Garmisch (same latitude as Zugspitze, but lower altitude) (Kurlyo and Zander, 2001).

# Chapter 6
# Oceans and the Climate

## General Description

Oceans over more than 70 per cent of the Earth's surface and contain some of the Earth's most complex and diverse ecosystems. Although the coastal zone constitutes only about 10 per cent of the total oceanic area, it accounts for more than half of the ocean's biological productivity and supplies nearly all the world's catch of fish. In addition, coastal areas contain many kinds of ecosystems vital to marine life and humankind. In fact, about 60 per cent of the world population, or nearly 3 billion people, live on or within some 100 km of a sea coast. Coastal areas receive discharges from rivers, surface run-off and drainage from the hinterland, domestic and industrial effluents through outfalls, and various contaminants from ships. All these discharges have affected the marine environment in many ways. Several kinds of international and national legislation have been enacted to protect the marine environment. Although this legislation is essential, the most important prerequisite for the protection of the marine environment is the appropriate and environmentally sound planning of human activities, especially in densely populated coastal areas.

Changes of global temperature influence components of the global hydrological cycle in different ways and with different response times. Higher temperatures increase the amount of water vapour in the atmosphere. Precipitation patterns alter and affect runoff from rivers and glaciers into the sea. Ocean waters expand. Collapse of ice sheets occurs as a consequence of rising temperatures that might raise a rapid rise of sea level.

There are serious concerns that rising levels of carbon dioxide could dump even more water into the oceans. The green vegetation of the planet not only takes in carbon dioxide and release oxygen but also adds water vapour to the atmosphere. Like people, plants also lose water. An important question is, what effect will increasing levels of carbon dioxide have on this? The answer appears to be, less water in the atmosphere and more in the oceans! Measurements of the volume of water brought in by rivers to the oceans reveal that, around the world, rivers have become fuller over the past century. Research has established that the reason is not, overall, raining–or snowing, hailing or sleeting-any more than it used to. Among other possibilities, one may be changes in land use, such as deforestation and urbanization. While soil in and around villages soaks up the rain and plants transpire it back into the atmosphere, the concrete in urban areas moves rainwater into drains and from there into rivers. Another possibility could be 'solar dimming': aerosol particles make the atmosphere hazy so it holds less water. And finally, there is the direct effect of carbon dioxide on plant transpiration.

Dr. N. Gedney of the British Meteorological Office recently used the statistical technique of 'optimal fingerprinting' or detection and attribution' to identify which of the above factors matter. She made five simulations of river flow in the 20th century. In the first of these, all four explanations were allowed to very: rainfall, haze atmospheric carbon dioxide and land use. Then one of them was kept constant in each of the next four simulations. By comparing the outcome of each of these with the first simulation, she could assess the importance of its part in the overall picture. In this way, for example, the role of land use could be understood by deducting the simulation in which it was fixed from the simulation in which it varied.

Dr. Gedney admits that her model does not fully consider the use of water to irrigate crops- especially important in Asia, for instance-nor the issue of urban growth. Yet, even these aspects, taken together, would remove water from rivers and so further support the above conclusion. Indeed, it appears that fuller rivers cannot be explained by more rainfall or haze or changes in land use, but they can be explained by higher concentrations of atmospheric carbon dioxide. The underlying mechanism is straight forward. Plants breathe through stomata found I the leaves, taking in carbon dioxide. When the atmosphere is relatively rich in this gas, less effort is needed. The stomata remain closed longer, and less water is lost to the atmosphere. The implication: the plant need not draw as much moisture from the soil- the unused water flows into rivers. Dr. Gedney; work appears to be the first to identify a direct effect of carbon dioxide on ecosystems (see Gedney *et al.*, Nature, Feb. 2006).

The finding seems to have two implications. On the one hand, fuller rivers threaten more flooding. More alarmingly, if rivers dump more water into oceans, then rising sea levels could rise more rapidly still. Such changes would be felt especially in low-lying, populous and poor countries like Bangladesh, for instance. On the other hand, increased access to fresh water can be a boon, provided that rivers can be surely controlled. Should using concentrations of carbon dioxide mean that plants consume rather less water, heaving more for humans, it might not be such a bad thing after all. Most studies related to rising sea level have been directed at positive components that together could explain the observed rise of sea level.

Negative components that could extract water from the ocean for long periods, each as increases in the mass of the Antarctic ice sheet and of the level of groundwater, have received much less attention. Over the 21st century, major physical factors affecting sea level due to a postulated global increase of atmospheric temperature of 3.5°C are likely to be:

1.  Thermal expansion of ocean waters could expand the top 100m of tropical water by 10cm and the next 900m by at least 20cm, and eventually by 50cm or more. Below that level, cold deep water would continue to flow from polar regions, perhaps, at a slightly higher temperature involving a slow expansion of 10 to 20cm with little effect during the 21st century. Changes in the depth of the thermocline and hence of the vertical distribution of temperature could, however, produce significantly larger changes.

2.  Melting of smaller glaciers and ice caps could produce a rise of around 20±12cm of sea level.

3.  Changes of melting and accumulation of the ice sheets of Greenland and Antarctica may counterbalance each other. Due to uncertainty of opposing trends, the effect on sea level should be within ± 10cm/century and could well be negative.

4.  Changes of water storage on land in lakes, rivers, reservoirs and groundwater are very difficult to predict and, while unlikely to exceed ± 10cm/century, will probably be a fraction of this value.

5.  A catastrophic collapse of the West Antarctic ice sheet is not imminent, but better oceanographic knowledge is required before it can be assessed whether a global temperature rise of 3.5°C might start such a collapse by the end of the 21st century. Even then, it is likely to take at least 200 years to raise sea level by another 5 meters (Robin, 1991).

## Ocean Changes

Barnola *et al.* (1987) reported that temperatures at the Earth's surface and atmospheric carbon dioxide levels fell and rose intandem during glacial-interglacial cycles. Since then, geologists have tried to explain the connection with recent thinking centered on oceanic controls on carbon dioxide. A new hypothesis supported by palaeoceonogrphic evidence and biogeochemical modeling has emerged that could account for the reduction in atmospheric carbon dioxide during glacial periods.

Milankovitch cycles are small variations in solar radiation reaching the Earth due to changes in the Earth's orbit; these cycles explain the pacing of the oscillations between glacial and interglacial intervals. But understanding of the large temperature shifts involved has required further amplification within the Earth's climate system. Once analyses of air trapped in ice cores revealed that carbon dioxide had fallen and risen in keeping with the glacial-interglacial rhythm, the attention shifted to the greenhouse effect by changes in carbon dioxide as the amplifier.

Certain parts of the world's oceans have an especially strong influence on climate; one such part is the Southern Ocean- the waters around Antarctica south of the polar front. Reduced leakage of carbon dioxide from ocean to atmosphere in such areas could contribute significantly to the lower levels of atmospheric carbon dioxide in glacial periods (Ganeshram, 2002). In the Southern Ocean, carbon dioxide-charged, nutrients-rich deep waters are reaching the surface and releasing carbon dioxide to the atmosphere. This leak could effectively be plugged by an increase in biological production I the surface waters through fuller house of the available nutrients by phytoplankton. Increased production would move more carbon dioxide back to the deep ocean in the form of sinking organic particles. Nutrients such as nitrate that reach the sunlit surface ocean around Antarctica are underused by phytoplankton because of a deficiency of the iron needed by their photosynthesis. But, during glacial periods, the larger amounts of dust in the atmosphere, probably, provided enough iron, increasing nutrient use and yielding a net transfer of carbon dioxide back to deep waters. Isolating the deep waters from the atmosphere by capping them with less-dense water at the surface (stratification) has the same effect.

Stable isotope ratios of nitrogen and silicon in organic and diatom remains, respectively, point to the nutrient status during the geological past. Sediment cores from the southern Ocean have higher $^{15}N$ values in the glacial intervals, indicating increased $NO_3$-uptake as compared to that in interglacial. But the $^{30}Si$ trends are the opposite: Si $(OH)_4$ use seems to have been lower during the glacial. Other arguments that invoke, for example, increased ocean stratification, could account for the higher $^{15}N$, but they fail to explain lower $^{30}Si$ values. According to Brzezinski *et al.* (2002) the additional of iron greatly changes the uptake ratio of $NO_3$ and Si $(OH)_4$ by diatoms from as much as 4:1 to about 1:1. Given that ratio of these nutrients in the Southern Ocean today are 2:1, uptake with a ratio of 1:1 under iron-replete conditions, as might have occurred during glacial periods, would result in higher use of $NO_3$- but relative underutilization of Si $(OH)_4$- a view that is quite consistent with the isotope data.

According to Brzezinski *et al.* (2002), the Southrn Ocean contributed to the drop in atmospheric carbon dioxide during glacial periods but not as a direct effect of carbon dioxide uptake in Antarctic surface waters. Rathr, as shown by the modeling work of Matsumoto *et al.* (2002), when ocean circulation patterns are considered, the effects of this peculiar nutrient biogeochemistry appear further field. The ocean changes on all scales, both of time and distance. However, some scales are more important than others, because they have large variations or 'signals' associated with them. For instance, tides have a very clear cycle with a dominant 12.4 hour period. The annual cycle is also obvious as the ocean warms and cools once a year with the seasons. Some length scales dominate-there are characrteristic length scales associated with ocean 'weather'.

The ocean is stratified, with lighter water overlaying denser water. Lighter water is typically warmer and/or fresher, while denser water is colder and/or saltier. The uppermost layer of the ocean, or mixed layer, has uniform temperature and salinity and is in direct contact with the atmosphere. The mixed layer can be shallow as 20 to 30m in summer; while in winter, it extends down to over 100m. There are two reasons for this. Winter cooling makes surface waters denser so that they sink; and stronger winds in winter result in deeper mechanical stirring.

Below the mixed layer is the thermo cline in which the water cools markedly and its density increases with depth. Because light water floats on top of denser water, the water in this intermediate layer is locked in this depth range. Finally, the deepest waters (>2,500m) have nearly constant properties. These waters last contacted the atmosphere in the Greenland sea or around Antarctica thousands of years ago. These deep waters are critical in climate stability and change because they store heat and carbon dioxide for such long periods.

*Is there climate change in the ocean?* The main problem in measuring climate change is discriminating the signal from the noise. Predictions of warming in the ocean due to global warming are 0.5°C per decade. Should this be correct, then in a 15 year ocean temperature time series we would need to discriminate a signal of 0.75°C from the noise. The noise included: 5°C from the annual cycle; ± 15°C variability on 4-8 year timescales apparently resulting from ENSO; and other unknown responses from longer period changes. The sheer variability of the ocean makes the subtle signals of global warming extremely hard to discern (Sutton, 2001).

## Ocean Circulation and Climate Change

Although the ocean is in perpetual motion, the tides and waves that we see are just a tiny part of the movement in the ocean as a whole. Most movement comprises large-scale, unseen shifts to equal out water density differences caused by variation in temperature and salinity from the poles to the tropics. Dense (cold and salty) water tends to sink and then flow sideways as its appropriate density level, generating currents that flow from basin to basin, redistributing heat and salt as they go. This is called thermohaline circulation. Cold, salty deep water is produced in the seas around Antarctica and the high North Atlantic.

Freezing winds cool the surface water and increase its density. The formation of salt-free sea ice causes the surrounding water to become saltier and denser. The dense water falls to abyssal depths and then flows along the seabed, hugging the western margins of ocean basins due to the 'steering' effect of the Earth's rotation (the Coriolis effect). The currents created by this process are called Deep Western Boundary Currents.

Some 40 per cent of the densest deep water enters the world's ocean via the southwest Pacific, and the gateway of this huge flow is New Zealand.

Giant currents in the deep ocean transport so much heat around the globe that they play a critical role in shaping the Earth's climate. From the moment of sinking, deep water can take up to 1,500 years to complete its journey around the globe and re-emerge at the surface. During this time, the deep water collects nutrients from material that rains down from the upper ocean. When the deep water nears the surface again in upwelling regions, it supplies nutrients for massive blooms of plankton (Manighetti, 2001).

Oceanographers view of global ocean circulation as a gigantic conveyor belt. Deep water from both the arctic and Antarctic joins the huge Ocean current the encircles the Southern Ocean. This Antarctic circumpolar current links the Atlantic, Pacific and Indian Oceans. Little or no deep water forms in the Pacific and Indian Oceans, but differences in water temperature, rainfall and evaporation still lead to large flows in and out of their basins via the Southern Ocean.

Thermohaline circulation is finely balanced, especially in relation to saltiness. A three per cent boost in salinity increases the density of seawater as much as a 5°C cooling! The vigour and extent of the deep and shallow flows depend upon a balance between evaporation and freshwater supply, temperature distribution through the ocean, and wind patterns. Any or all of these factors may change as global warming continues (Manighetti, 2001).

Conceivably, the efficiency of the ocean at redistributing heat may be delaying and regulating global warming. Events of thousands of years ago make it possible that, as global temperature increases, the accompanying changes in rainfall, evaporation and sea-ice cover will slow down deep water formation in the North Atlantic. The insulating effect of the Southern Ocean means that Antarctic deep-water production is less likely to be effected unless or until warming is further advanced. However, all the oceans communicate via the Antarctic circumpolar current.

First effect may be a greater contrast between shallow and deep-water temperatures, leading to changes in the sinking and upwelling behaviour critical for chemical and nutrient cycling in the ocean. Through feedback effects via the atmosphere, however, change sin the ocean can generate abrupt climate events as the system moves into a new mode of operation. In future, switching off the 'global heat pump' through greenhouse warming could, ironically, help initiate a new Ice Age (Manighetti, 2001). Table 6.1 shows various sea-level drivers and responses with timescales.

**Table 6.1: Various sea-level 'drivers' and responses with timescales.**

| Driver | Sea-level response | Timescale |
|---|---|---|
| Global warming | Long-term sea-level change | Centuries |
| Interdecadal Pacific Oscillation (IPO) | Interdecadal oscillations | decades |
| *El Nino*/Southern Oscaillation (ENSO) | Interannual oscillations | Years |
| Annual temperature cycle | Annual cycle | 1 year |
| Changing atmospheric pressure | Inverfted barometer | 1 to 7 days |
| Wind | Set-up | 1 to 7 days |
| Gravitational attraction of astronomical bodies | Tides | 24, 12, 8, 6 3 hour periods |
| Chaotic interactions | Seiche | 2 to 4 hours |
| Submarine earthquakes, avalanches and volcanoes | Tsunami | Minutes to 1 hour |

The world ocean is a major constituent of the climate system. It affects climate in many ways (Clement and Cane, 1999; Seidov *et al.*, 2001). In view of its enormous size, most of the solar radiation received at the Earth's surface goes into the ocean and warms the surface waters. The ocean is able both to store and redistribute this heat before it is released to the atmosphere (much of it in the form of water of water vapour) or radiated back into space (Rahmstorf, 2002).

The heat storage effect is best seen on the seasonal timescale. While the mid-latitude temperature range between summer and winter is typically around 8°C over the ocean and at the coast, this range goes up to several tens of degrees in the continental interiors. The extent of the temperature deviation from the zonal mean brings out the effect of ocean heat transport on surface temperatures, with warm anomalies over the three main regions of deep-water formation of the world ocean: the northern North Atlantic, the Ross Sea and the Weddell Sea. These are crucial for the thermohaline circulation of the world ocean where surface waters after releasing heat to the atmosphere reach a critical density and sink. Besides its heat storage and transport effects, the ocean influences the Earth's heat budget by its sea-ice cover, which changes the planetary albedo and so can affect the steady-state global-mean temperature. Sea ice is an effective thermal blanket that insulates the ocean from the overlying atmosphere.

The large-scale ocean circulation is a combination of currents driven directly by winds (mostly confined to the upper several hundred metres of the sea), currents driven by fluxes of heat and freshwater across the sea surface and subsequent interior mixing of heat and salt (the so-called thermohaline circulation) (see Clark *et al.*, 2002) and tides (driven by the gravitational pull of the Moon and Sun).

Wind-driven currents are thought to lead to climatic changes through their effect on upwelling (ekman divergence) near coasts and the equator, changing sea surface temperatures. It is this mechanism that plays a part in the *El Nino*/Southern Oscillation (ENSO) cycle. The ENSO is the strongest mode of natural climate variability today. It is a coupled ocean- atmosphere mode centered on the tropical Pacific, with a variable period of 3 to 7 years and worldwide ecological and societal impacts due to its effect on the global atmospheric circulation. Tides constitute a primary source of turbulent energy (in addition to that provided by the wind) to mix the ocean (Munk and Wunsch, 1998). The ocean affects the climate systems not only by being part of the planetary energy cycle but also by affecting the biogeochemical cycles and exchanging gases with the atmosphere, so influencing its GHG content. The ocean contains about fifty times more carbon than the atmosphere (Rahmstorf, 2002). Ocean circulation seems to be involved in abrupt and dramatic climate shifts that occurred repeatedly during the last glacial cycle (Paillard, 2001; Rahmstorf, 2002). The climate system is sensitive to forcing and responds with large, often abrupt, changes in surface conditions. The ocean circulation acts as a highly nonlinear amplifier of climatic changes.

## Thermohaline Circulation

As the thermohaline circulation (THC) helps in driving the ocean currents around the globe, its importance to the world's climate cannot be overemphasized. It is possible that the North Atlantic THC might weaken considerably during this century. In the future, global change via atmospheric pathways is likely to increase the freshwater supply to the Arctic, which will reduce the salinity and hence the density of surface waters, and thereby may reduce ventilation.

Such a weakening of the THC could extract bad effects on our climate-perhaps some cooling over northern Europe (Hansen *et al.*, 2004). The THC is involved in ocean currents. Both cooling and ice formation at high latitudes increase the density of surface water sufficiently to make them sink. Various different processes termed 'ventilation' are involved collectively. Active ventilation maintains a persistent supply of dense waters to the deep high-latitude oceans. At low latitudes, in contrast, vertical mixing heats the deep water whose density decreases. Thus, both high-latitude ventilation and low latitude mixing generate horizontal density differences in the deep ocean, which generate forces. In the North Atlantic, these forces drive the North Atlantic Deep Water (NADW) that supplied a large part of the deep waters of the world oceans (Hansen *et al.*, 2004). Whether or not the THC is an important driving mechanism for the NADW flow, is debatable.

Conceivably, the north- south density differences seen at depth might be generated by the flow rather than driving it (Wunsch, 2002). A potential weakening of the North Atlantic THC might affect the deep oceanic waters of the world over the long term, but would have more immediate effects on the climate in some areas. The dense overflow waters feeding the deep Atlantic are being replenished by a compensating northward flow in the upper layers. These currents take warm saline water northward to those areas where ventilation and entrainment occur. It is this oceanic heat transport that enables large Arctic areas to be ice-free and also for some parts of the North Atlantic to be several degrees warmer than they would otherwise have been (Seager *et al.*, 2002). If the THC is significantly weakened, it reduces this heat transport and regionally counter balances global warming. In some regions, it might even cause some cooling (Vellinga and Wood, 2002).

According to Siegenthaler and sarmiento (1993), the oceans are a significant sink for atmospheric carbon dioxide. The strength of this sink varies on interannual timescales as a result of regional and basin-scale changes in the physical and biological parameters that control the flux of this GHG into and out of the surface mixed layer (Gruber *et al.*, 2002).

Karl and Lukas (1996) and Dore *et al.* (2003) analysed a 13-years time series of oceanic carbon dioxide measurements form stations ALOHA in the substropical North Pacific Ocean near Hawaii, and found a significant decrease in the strength of the carbon dioxide sink over the period 1989-2001. Much of this reduction in sink strength could be attributed to an increase in the partial pressure of surface ocean carbon dioxide caused by excess evaporation and the accompanying concentrating of solutes in the water mass. Their results suggest that carbon dioxide uptake by ocean waters can be greatly influenced by changes in regional precipitation and evaporation patterns brought on by climate variability.

## The Ocean in a Warming World

There is an intimate link between the ocean and atmosphere. Powerful weather systems, outlined by swirls of white cloud, sweep eastward across the ocean. Winds within these systems create turbulent seas which help transfer GHGs into one of the world's major carbon stores, the ocean. Those wind-affected seas, aided by the general oceanic circulation, shift solar heated and polar cooled waters around the Earth. The world's climate is warming rapidly. The IPCC predicts increases in surface air temperatures of 1.4 to 5.8°C over the 21st century. But these are globally averaged increases. In reality, the Earth is far from average, with the hot, arid brown triangle of

Africa contrasting starkly with cold, white Antarctica. Continued research on diverse aspects of our oceans is essential to verify future computer models and to provide an even better understanding of the complexities of the global climate system.

At the start of the 21st century, the world is a warmer place than at any time since the dark Ages. For the Northern Hemisphere, the 20th century was the warmest of any century over the pas thousand years. For the world as a whole, 1998 was the warmest year since the start of reliable records, and the 1990s the warmest decade. Over the past hundred years, average sea level has risen by 0.1 to 0.2m.

Extremes of weather are now becoming more common. In mid-and high latitudes of the Northern Hemisphere, cloud cover has increased and heavy rain occur more often. There have been decreases in snow cover, in annual duration of lake and river ice cover, in the frequency of very low temperature events and in the daily temperature range. There is uncertainty about how much the climate will change in response to present and predicted GHG levels- the Earth's ocean and atmosphere influence the climate in complicated ways.

The ocean and atmosphere make most of the Earth livable by distributing the heat form the tropics to the poles. Although the atmosphere has a much bigger volume and moves much faster than the ocean, water holds much more heat than air. Consequently, the ocean is as important as the atmosphere in transporting heat, each moving the equivalent of the output of several million power stations. Together, the ocean and atmosphere form a 'coupled' system: changes in one force changes the other. Much of the ocean's circulation is driven by the wind.

When the winds vary, the velocities of the ocean currents also vary. Changes in the currents alter the ocean's temperature, which in turn adjusts the winds by increasing or decreasing the transfer of heat and moisture to the atmosphere. Transfers of heat and moisture between the ocean and atmosphere can drive oscillations in the climate. The ENSO is the best known.

Besides transporting heat, the ocean affecters the climate by absorbing or releasing gases to the atmosphere. In response to evaporation and cooling form the atmosphere, ocean waters in polar regions become cooler and saltier (*i.e.* denser) and sink very deep. This sinking removes carbon for long periods from contact with the atmosphere and biosphere. Marine plants and animals take up or release dissolved carbon through photosynthesis or respiration. Some of the carbon is removed from the water as biologically formed particles sink. These and other processes interact in determining conditions in the world's ocean and atmosphere, now and in the future.

Climate variations occur over a range of timescales. The changes causes by humans are likely to be long-term trends. But, such trends can be affected in the shorter term by natural fluctuations such as ENSO.

## *El Nino-*Southern Oscillation

The *El Nino-*Southern Oscillation (ENSO) has a 'normal' state with 'warm' and 'cold' departures known as *El Nino* and *La Nina*, respectively. During the normal state, the trade winds, which below on either side of the Equator, force the South Equatorial Current to flow from east to west along the equator. It is heated by the Sun, making surface waters in the western Pacific the warmest on the planet. During *El Ninos*, the trade winds weaken. The South Equatorial Current collapses, leading to warming in the eastern and cooling in the western equatorial Pacific. These sea-surface temperature changes lead to increases in the ocean-atmosphere heat flux in the east, and decreases in the west. This alters the atmospheric heat balance and changes wind patterns over the Pacaific Ocean. These wind-field changes can often be felt at great distances from the Equator. During *La Nina*, the opposite happens: the trade winds increase, with cooling of sea-surface temperature in the eastern Pacific and w4arming in the west.

## Dimethylsulpohide in Ocean-Atmosphere Interactions

Dimethylsulphide (DMS) is a trace sulphur gas that is present in many atmosphere and surface water samples. It is derived from dimethyl-sulphonioproprionate (DMSP), but its natural production, consumption and cycling are not well understood. Until recently, DMS was thought to originate mainly from marine waters, but it is now known that estuaries and lakes also release DMS. DMS also originates form terrestrial plants, such as maize, wheat and lichens, but it is not known why. DMS impacts the global environment by influencing factors such as the acidity of the atmosphere, cloud condensation nuclei (CNN) and solar insulation. DMS is affected by temporal and geographical factors, as well as physical factors, such as salinity and wind speed; yet, when studied under *El Nino* conditions, which modify these physical factors in vivid, no fluctuation was found in the concentration of DMS in the water column (Buckley and Mudge, 2004).

### Oceans and Carbon Dioxide Exchange

In 1750, the concentration of carbon dioxide in the Earth's atmosphere was about 280 ppm. Today, it is about 378 ppm and it is continuing to increase every year by 1.5 ppm. Being a GHG, it absorbs infrared radiation emitted by the Earth, leading to increased surface temperature, changing the climate. While the increase in atmospheric carbon dioxide is mainly due to the burning of fossil fuesl, not all such carbondioxie stays in the atmosphere. Global carbon budget is measured in PGC (petagrammes of carbon); where 1 Pg equals 1 billion tones. According to current estimates, fossil-fuel burning replaces 6 PGC every year. Land-use chine, mainly tropical deforestation, adds another 2PgC. Of the total of about 8 PGC, 3 PGC, 3 PGC accumulates in the atmosphere, 2 PGC is taken up by the terrestrial biosphere, and 2 PGC is absorbed by the ocean. (The numbers do not balance exactly because they are estimates).

Carbon dioxide is much more soluble in water than is oxygen or nitrogen. This high solubility means that the total amount of carbon dioxide held in the ocean is huge-about 60 times that held in the atmosphere and about 20 times greater than the Earth's biosphere. The transfer of carbon dioxide through the ocean surface is driven by its 'partial pressure'- that is the pressure exerted by each element in a gas mixture. If the partial pressure of carbon dioxide ($CO_2$) just above the ocean surface is greater than that just below, then carbon dioxide is transferred from the atmosphere into the ocean (and vice versa if the partial pressure is reversed).

## Modeling of Carbon Dioxide Exchange

The layer immediately below the ocean surface is called the mixed layer because it is continually stirred by winds, waves and convection. The mixed layer is at its deepest when the surface temperature is lowest. When the surface water warms, a new shallow mixed layer forms. When it dies, some of its fixed carbon is converted back to carbon dioxide, while some sinks into deeper water. In winter, mixing brings up nutrients into the mixed layer; but, since this layer is very deep, most of the phytoplankton do not get enough light. In summer, there is plenty of light, but because they are out of contact with the deeper water, the phytoplankton tends to run out of nutrients. In this model, there is normally a burst of phytoplankton growth- the 'spring bloom'–when the mixed layer becomes shallower in spring (Hadfield *et al.*, 2001).

Coastal and marginal seas are bristling with biological activity which is partly triggered by terrestrial and human impacts. These seas have an important role in the global carbon cycle- they link the terrestrial, oceanic and atmospheric carbon reservoirs (Gattuso *et al.*, 1998). The high biological activity causes high carbon dioxide fluxes between the coastal and marginal seas, the atmosphere and the adjacent open oceans, respectively. The North Sea (of the northwest European shelf), is amongst the best-studied coastal areas in the world in terms of its physical, chemical and biological conditions. Much of the North Sea acts as a year-round carbon dioxide sink (Thomas *et al.*, 2004). In terms of the surface area, coastal sea seem to have a disproportionately high contribution to the open ocean storage of carbon dioxide via a mechanism called the 'continental shelf pump' (Tsunogai *et al.*, 1999). Only limited information is available on these carbon dioxide fluxes at the global scale (Anderson and Mackenzie, 2004; Thomas *et al.*, 2004). Extrapolating the North Sea's carbon dioxide uptake across coastal areas of the global ocean (75), suggests that coastal seas absorb about 20 per cent of the global oceanic uptake of anthropogenic carbon dioxide. They appear to significantly enhance the open ocean sequestration of carbon dioxide.

The lion's share of global carbon dioxide is found in the ocean, which holds 50 times as much carbon dioxide as the Earth's atmosphere. Oceanic absorption varies widely depending on the water temperature and seasons. Colder water soaks up more carbon dioxide than warmer water, and deep water holds more carbon dioxide than surface water-except in winter, when surface water becomes colder and heavier, and forces deep water to the surface where it releases carbon dioxide into the atmosphere. Some experts are concerned that global warming may raise ocean temperatures and reduce the ocean sink. a close watch is needed on industrial emissions. New technologies such as hybrid cars and high efficiency electrical generation should be developed. Energy conservation should be actively promoted.

Wave conditions can vary immensely along a rocky shore-large ocean swell can pound the exposed parts of the reef, allowing only the most robust marine plants to survive. Within a short distance in more sheltered areas, a quite different ecosystem may flourish. In both the exposed and sheltered habitats, too much sediment may prevent seaweeds from settling and growing on rocks. Since the seaweeds provide biodiversity and functioning of the whole rocky-reef ecosystem.

Coastal currents and waves are the dominant physical control for many coastal habitats. With applications from the aquaculture industry for ever-larger farms, there is a need to quantify their effects on the currents and waves. As usual, in environmental science, the greatest challenge in analyzing the data is separating the influence of the farm structures from the natural variability in the data. The effects of the farm structures on waves also need to be examined.

The world's oceans are in peril. Pollution has crippled coral reefs and created 'dead zones' where a few living things can exist. Over fishing has depleted the stocks of cod, tuna and many other species. There is urgent need for a new commitment to ocean research and a wholesale reform of fisheries management, including greater reliance on scientific data when determining the levels of allowable catches (Kumar, 2003).

Loureno and Jorge (2004) analysed land-use change and sustainability in the coastal regions of India with regard to societal pressures and the nature of coastal ecosystems, identifying regions where socio-economic and biophysical problems led to coastal vulnerability. They defined indicators of both the pressure and state of vulnerability for all coastal regions. These indicators were then used to classify 66 districts of the west and east coasts to identify the most vulnerable districts. The three most vulnerable regions also turned out to represent the primary forces responsible for change in coastal areas, *viz.*, tourism in north Goa, industry and urbanization in Thane (Mumbai) and intensive agriculture and aquaculture in east Godavari. In north Goa, degradation of coastal vegetation, land form changes, and deterioration of surface water quality have arisen from land-use and land-cover change and ground-water use. In east Goad, rib, deterioration of the quality of groundwater, coastal and fresh surface waters has accompanied mangrove swamp conversion and rising groundwater level. In Thane, polluted coastal and groundwaters have been found to be associated with changes in land use and land cover (Lourenco and Jorge, 2004).

According to TERI (2003), while well-managed tourism could improve vegetation cover and biodiversity (despite the disappearance of some beach vegetation near Goa), it is mainly industrial expansion that degrades vegetation. In Thane, where plant biomass is declining and vegetation cover is fragmenting, although agricultural expansion has had significant negative impacts. It has enhanced plant diversity. But plant diversity has declined near villages in east Godavari, and mangrove forest area has decreased due to expansion of shrimp breeding ponds and paddy fields. Sustained development imposes increasing demands on water, increases sewage loads and levels of fertilizer and pesticide use that often exceed the assimilative capacity of the environment. Because these thresholds greatly depend on the nature and health of the ecosystem, it is important to understand ecosystem health before assessing the environmental impacts of future development. It has been commonly observed that the relationships between development pressures and environmental responses tend to be nonlinear and are mediated by intermediate pressure driver relationships.

The TERI (2003) also suggested suitable tools and approaches to improve sustainable ecosystem management. Examples: a novel method for delineating groundwater well head protection areas, optimization models to help meet the growing water demand, and a coastal groundwater management policy based on optimization and protection zoning. Three types of tools were developed: (a) visualization, (b) spatial analysis, and (c) advanced modeling; of these, the type (a) tools deal with the study area from different perspectives, define, the main biophysical properties, highlight direct cause-effect relationships, and established the basis for condition assessments. Type (b) tools allow exploration of the spatial and attribute relationships between datasets; and the type (c) tools are employed for multi-criteria analyses (see Noronha *et al.,* 2003).

Attempts have been made to differentiate the patterns of natural resource consumption arising from differences in tourist accommodation infrastructure (TERI, 2003). Analyses of 1989/1990 and 1999/2000 land cover showed that the diversity of species has decreased due to loss of the original vegetation. Coastal vegetation has increased in tourism locations. Tourism-related activities have impacted forest-related activities and traditional activities including salt extraction, agriculture and aquaculture. Land reclamation is becoming popular and is broadening the coastal tourism belt eastward towards the hinterland.

## Anthropogenic Carbon and Ocean pH

Houghton *et al.* (2001) showed that most carbon dioxide released into the atmosphere from the burning of fossil fuels is eventually absorbed by the ocean, with potentially adverse consequences for marine biota (Seibel and Walsh, 2001). Caldeira and Wickett (2003) found that oceanic absorption of carbon dioxide from fossil fuels may result in larger pH changes over the next several centuries than any change deduced from the geological record of the past 300 million years, with the possible exception of those resulting from some rare or extreme events.

Dissolving carbon dioxide in the ocean lower the pH, making the ocean amore acidic. Coral reefs, calcareous plankton and other marine organisms whose skeletons contain calcium carbonate are affected. Although most biota live near the surface where the greatest pH change would be expected to occur, it is the deep-ocean biota which are probably more sensitive to pH changes (Seibel and Walsh, 2001).

Based on the record of atmospheric carbon dioxide levels over the past 300 million years (Myr) and some modeling and simulation. Caldeira and Wickett (2003) found no evidence that ocean pH was more than 0.6 units lower than today. Their modeling results indicate that continued release of fossil-fuel carbon dioxide into the atmosphere could lead to a pH reduction of 0.7 units. They concluded that unabated carbon dioxide emissions over the coming centuries could produce changes in ocean pH that are greater than any experienced in the past 300 Myr, with the possible exception of those resulting from certain catastrophic events in the Earth's history. Indeed, it appears that the coming centuries may see more ocean acidification than the past 300 Myr.

## Interdecadal Pacific Oscillation

Climate researchers have discovered that there may be longer timescale oscillations in the ocean-atmosphere system. These are manifest in a variety of climate indicators (such as 10 year variations in the frequency of cyclones) in both the Pacific and Atlantic. One example is the Interdecadal Pacific Oscillation (IPO), which shifts climate every one to three decades. The IPO is strongest in the north Pacific centered near the dateline at 40°N. The IPO has positive (warm) and negative (cool) phases. Positive phases tend to be associated with an increase in *El Ninos*, and negative phases with an increase in La Linas. Three phases of the IPO have been identified during the 20th century: a positive phase (1922-44), a negative phase (1946-77) and another positive phase (1978-98). It is not known what causes the IPO.

## Anthropogenic Ocean Acidification

Although the surface ocean is saturated in respect of calcium carbonate, increasing atmospheric carbon dioxide concentrations, by lowering ocean pH and carbonate ion concentrations, reduce calcium carbonate saturation. No doubt, ocean uptake of carbon dioxide can moderate future climate change, but the hydrolysis of carbon dioxide in seawater increases the hydrogen ion concentration $[H^+]$. Surface ocean pH is already 0.1 unit lower than preindustrial values. By the end of the century, it will fall by another 0.3 to 0.4 unit (Haugan and Drange, 1996) under the IS92a scenario, which translates to a 100 to 150 per cent increase in hydrogen ion concentrations $[H^+]$.

At the same time, aqueous carbon dioxide concentrations will increase and carbonate ion concentrations will decrease; this will make it more difficult for marine calcifying organisms to form biogenic calcium. Should these trends continue, such key marine organisms as corals and some plankton will find it difficult to maintain their external calcium carbonate skeletons. On the basis of modeling work. Orr *et al.* (2005) projected that Southern Ocean surface waters will begin to become under saturated with respect to aragonite, a metastable form of calcium carbonate, by the year 2050. while recent predictions of future changes in surface ocean pH and carbonate chemistry have largely focused on global average conditions (Caldeira and Wickett, 2003) or on low-latitude regions, where reef-building corals are abundant, Orr *et al.* (2005) have focused on future surface and subsurface changes in high latitude regions where plank tonic shelled periods are prominent components of the upper-ocean biota in the Southern Ocean, Arctic Ocean and sub-Arctic Pacific Ocean.

By the year 2100, this undersaturation could even extend throughout the entire Southern Ocean and into the sub-Arctic Pacific Ocean. Orr *et al.* (2005) also conducted experiments on live periods which were exposed to the predicted level of undrsaturation for two days in a shipboard experiment. Their aragonite shells showed significant dissolution. These findings indicate that conditions detrimental to high latitude ecosystems could develop within decades, not centuries as suggested previously (Orr *et al.*, 2005).

## The Ocean Surface: The Greenhouse Bottleneck

A crucial piece in the global warming puzzle is the way GHGs are transferred between air and sea. The oceans act as a remarkable songe holding 50 times more of carbon dioxide than the atmosphere. Somewhere between 30 and 40 per cent of the carbon dioxide produced by human activity is absorbed by the world's oceans. For predicting the future climate, it is important to determine where and how GHGs are exchanged between the atmosphere and ocean.

Gas exchange at the sea surface is a vital link between the ocean and atmosphere, especially in relation to climate change. But measuring what is actually going on at the surface can be very difficult (Smith *et al.*, 2001). The sea surface acts as a gas-transfer bottleneck governed by two key properties:

1. The differences in gas concentration between air and sea. This controls the direction and rate of the gas movement. A larger difference pushes gas at a faster rate. The gas concentration in the water depends both on temperature and on biological processes (such as photosynthesis and reparation by the plankaton).

2. The strength of ocean stirring processes. Stirring, or mixing, relieves the buildup of gas at the thin surface layer by transporting dissolved gas down to deeper water. Stirring is affected by waves and wave-breaking, which in turn are controlled by wind. The mixing and bubbles created by waves and wave-breaking near the sea surface are crucial in determining the rate that gases and heat are exchanged between the atmosphere and ocean. Strong turbulence thins the diffusive skin layer and aids transport, allowing a more rapid exchange.

## Southern Ocean sponge

Some regions of the ocean are more important than other in the transfer of gases between the air and the sea. In the Southern Ocean, for example, the cool cool waters have many of the right properties for gas absorption. In particular, localized plankton blooms there can use much of carbon dioxide in the water, thus increasing the air-sea concentration difference. Also, the stormy, wind-ravaged surface of the Southern Ocean makes it easier for gases to be absorbed by the ocean. It seems that the transfer processes will increase markedly when measured in storm conditions. So, despite bad weather, the Southern Ocean often becomes a good laboratory.

These transfer processes apply not only to carbon dioxide, but also to heat and to other climatically important gases (Smith *et al.*, . 2001). Dimethylsulphide (DMS) is the most intriguing of these since it is produced abundantly by plankton in the ocean. Some belived that when DMS escapes into the atmosphere, it aids in the formation of clouds and alters their reflective properties by providing just the type of particles on which water will condense. This could counteract some of the warming effect caused by the carbon dioxide increase. There is the possibility that microscopic life in the ocean could act as a climate regulator.

## Future Changes

As carbon dioxide dissolved more easily in cold water than in warm water, a warming ocean will store less carbon dioxide than if it remained cool. Changing wind and wave patterns will also change the balance between concentration gradients and mixing. Models predict higher rainfall over the Southern Ocean and reduced sea-ice coverage. These combine to produce fresher, lighter water that gets less easily mixed by wind and wave action. This could mean a reduction in gas exchange at the ocean surface. Globally, the oceans will not be able to mitigate the effect of human production of carbon dioxide as well as they do now. As the ocean temperature changes, phytoplankton types and numbers may also change, along with the amount of carbon dioxide produced and used (Smith *et al.*, 2001).

## Potential Impact on Plankton and Carbon Export

The ocean covers 70 per cent of the Earth's surface and stores a huge quantity of atmospheric carbon dioxide ($CO_2$), especially at southern latitudes. The sea surface acts as a bottleneck for the transfer of carbon dioxide and other gases into the ocean. But once there, how does carbon dioxide leave the surface waters so that it stays out of contact with the atmosphere for a long time ?

## Physical and Biological Carbon Dioxide Pumps

Dissolved carbon dioxide is removed from the surface ocean in two ways:

1. *The physical pump.* Because cold water can absorb more carbon dioxide than warm water, and is also denser, so cold water (relatively rich in carbon dioxide) sinks more easily to the deep ocean.

2. *The biological pump.* In sunlight, surface waters, marine phytoplankton absorb dissolved carbon dioxide during photosynthesis, converting it to the organic carbon material used to build up living organisms. About 10 to 30 per cent of the carbon sinks out of surface waters, either directly as organic particles or indirectly after this materials has been eaten by marine animals. The carbon eventually reaches the deep ocean, and is effectively removed from further contact with the atmosphere for centuries to millennia.

The organic particles in the surface ocean consist of a wide range of plant parts and animal wastes. Many particles stick together to make large flakes. These composite particles can sink rapidly through the water, falling hundreds of meters each day. Setting particles-called 'marine snow'- take 20-30 days to reach the ocean floor 2 to 3 km down. As they sink, they are broken down by animals and bacteria in the cold, dark waters of the deep ocean. Also they provide food for such benthic animals as worms, crabs and starfish. Less than 1 per cent of the carbon fixed during photosynthesis in surface waters is buried in deep ocean sediments.

The strength of the biological pump has varied during past fluctuations in climate. But how might the pump, and hence atmospheric carbon dioxide levels, alter in response to future global climate change ? We can assess this by using what we know about production and carbon loss in subtropical and sub Antarctic waters in computer models that allow predicting how the ocean and marine ecosystems might change.

Future global climate change is likely to lead to change in basic biological processes, such as water and nutrient cycling, plant productivity and species interactions. It is extremely difficult to predict about particular regional effects: the interactions within marine food webs are too complex, and levels of uncertainty about how marine biological communities might respond are too high (Nodder and Boyd, 2001).

# Global Mean Temperature and Sea-level Rise

As pointed out by Hansen *et al.* (1985) and Wigley and Schlesinger (1985), oceanic thermal inertia makes climate change lag behind and changes in external forcing. It also causes the response to be damped relative to the asymptotic equilibrium response. The effect of this lag, or damping effect, and of the changes in atmospheric composition (and radioactive forcing) that have already occurred, is that the climate system will continue to change for many decades (centuries for sea level) even if there were no future changes in atmospheric composition. For global-mean temperature, this is referred to as the unrealized warming, residual warming, or committed warming (Wigley, 2005). Wigley prefers the term 'warming commitment' or, to include sea level rise 'climate change commitment' (see Warrick *et al.*, 1993).

The warming commitment idea is based on unrealistic assumption of constant atmospheric composition. An alternative indicator of the commitment to climate change is to assume that emissions (rather than concentrations) of radioactively important species will continue to be constant. Wigley (2005) discussed the constant-composition (CC) warming and sea level commitments, the constant emissions (CE) commitments, and the uncertainties which arise from uncertainties in the climate sensitivity, the rate of ocean heat uptake, the magnitude of past forcing, and the ice melt contribution to sea-level change. He quantified CC commitments as well as their uncertainaties. The CE commitments were also considered.

According to Wigley (2005), the CC warming commitment could exceed 1°C. The CE warming commitment is 2 to 6°C by the year 2400. For sea level rise, it is 10 centimeters per century (extreme range approximately 1 to 30 cm per century) and the CE commitment is 25cm per century (7-50 cm per century). Avoiding these changes will require, eventually, a reduction in emissions to substantially below present levels. For sea-level rise, a substantial long-term commitment may be impossible to avoid. The CE results reinforce what is already well-known that in order to stabilize global mean temperatures, we will have to reduce emissions of GHGs to well below present levels (Wigley et. al. 1996; Wigley, 2005).

The CC results are potentially more disturbing- they are based on an unachievable future scenario and so represent an extreme lower limit to climate change over the next few centuries. For temperature, they show that the inertia of the climate system along will guarantee continued warming which may eventually exceed 1°C. Sea level will continue to rise by about 10 cm/century for several centuries. Such a slow rate will probably allow many coastal communities to adapt; but profound long-term impacts on low-lying island communities and on vulnerable ecosystems (such as coral reefs) may be inevitable (Wigley, 2005).

Houghton *et al.* (2001) and Wigley *et al.* (2005) showed that increases of GHGs in the atmosphere produce a positive radiative forcing of the climate system leading to warming of surface temperatures and rising sea level caused by thermal expansion of the warmer seawater, in addition to the contribution from melting glaciers and ice sheets.

Even if concentrations of GHGs could be stabilized at some level, the thermal inertia of the climate system would still lead to further increase in temperatures as also in sea level rise. Meehl *et al.* (2005) have performed multimember ensemble simulations with two global coupled three-dimensional climate models (Parellel Climate Model, PCM, and Community Climate Model version 3, CCM3; see Meehl and Tebaldi, 2004 and www.ccsm.ucar.edu) to quantify how much more global warming and sea-level rise (from thermal expansion) could be expected under several different scenarios. These climate models showed that even if the concentrations of GHGs in the atmosphere had been stabilized in the year 2000, we are already committed to further global warming of about another half degree and an additional 320 per cent sea level rise caused by thermal expansion by the end of the 21st century. Projected weakening of the meridional overturning circulation in the North Atlantic Ocean does not lead to a net cooling in Europe. At any given point in time, even if concentrations become stabilized, there is a commitment to future climate changes that will be greater than those already observed. The chief unknown factor in making predictions of rising sea levels in response to global warming relates to the role played by the massive ice sheets that cover Antarctica and Greenland. If some of these melted, the sea level would rise more rapidly than in the past (when much of the rise occurred because water expands as it warms). Till recently, it was assumed that any melting taking place in the ice caps is rather gentle. But, four years ago a small Antarctic ice shelf suddenly disintegrated, forcing many experts to rethink. More recently, another very alarming news has come from the other end of the world.

In the past, researchers used data from aeroplanes that fly criss-crossing paths over Greenland to assess the extent of its ice sheets- a challenging task. The Greenland ice sheet cover 1.7 m km$^2$ and the surface of the ice rises to an altitude of 3 km. Not surprisingly, the flights can leave some areas unmonitored, and so computer models are used to fill the gaps and to estimate the role played by these patches. It was concluded that the Greenland ice sheet is relatively stable in the centre, but thinning slowly at the edges. That conclusion has recently been challenged by E. Rig not of the California Institute of Technology and P. Kanagaratnam of the University of Kansas, who used satellite data concentrating on Greenland's coastline to examine how fast the thinning is taking place. They found that the flow-speed of 12 glaciers which are collectively responsible for about a half of the discharge of water from the ice sheet, is increasing fast. In fact, the speed at which the glaciers flow has doubled to 12 km a year. Consequently, the volume of ice falling into the sea from Greenland has also doubled over the past decade!

Furthermore, Rig not and Kanagaratnam also observed that the Greenland ice sheet experienced a greater area of surface melting in 2002 and 2005 than at any time during the past two decades. Most of this happened in the south of the island. Where the accelerating glaciers lie. Water flowing from the surface could make it easier for the glaciers to pas into the sea. When both factors are taken into account, it seems the contribution made by the Greenland ice sheet to the rise is global sea levels has increased from 0.23mm a year in 1996 to 0.57mm in 2005. Beyond this, because glacial ice lacks salt, the melt water is fresh. Such an increased flow of freshwater from Greenland could, according to the best available models of ocean circulation, change the way that currents flow in the North Atlantic, and prove detrimental to the Gulf Stream, the current that keeps north-west Europe warmer than its latitude suggests it should be. Indeed, if the Gulf stream is weakening, the news from Greenland is doubly disturbing.

## Impacts of Sea-level Rise

Recent work on the impacts of sea-level rise (SLR) (Yohe and Schlesinger, 1998; West and Dowlatabadi, 1999) has resulted in a better and improved recognition of:

☆ Regional and local variations in sea level which, in some locations, can dominate the global signal. So, projections for SLR need to account for local processes affecting sea level such as subsidence, uplift or continental arebound.

☆ Spatial in homogeneity in SLR associated with global warming. The thermal expansion component of sea level rise is a dynamic quantity and in a transient scenario varies by location (West and Dowlatabadi, 1999; Patwardhan, 2006).

☆ The interaction between sea level and storm surge. Realistic, more accurate assessments of SLR impacts should consider the joint effect of secular trends in sea level and storm regimes (see Patwardhan, 2006).

In fact, similar considerations of regional heterogeneity and the combined effect of extreme events are also important for most climate variables and impact sectors. The importance of the interaction between storms and sea level comes not only in terms of the physical impacts but also in terms of human responses which modulate the socio-economic impacts.

In many locations, sea-level change could possibly exert a greater impact by modifying storm damage than by directly causing inundation or erosion. Patwardhan (2006) illustrated this idea by means of the following two physical processes:

☆ Erosion is usually modeled by the Bruun rule and treated as a continuous process. In reality, however, it is episodic and often triggered by wave action during storms. Increased sea level enhances the effect of wave attack, leading to stronger erosion- a straightforward application of the Bruun rule could underestimate the actual erosion that might take place for a particular sea level rise scenario.

☆ Increase in local sea level produces stronger waves and flooding, so amplifying the impact of a particular storm. Viewed another way, a storm of a particular severity may happen more often.

Consequently, in the absence of any response, the damage associated with sea-level rise may be higher when considered together with a particular storm regime than in isolation.

## Coral Reefs

Coral reefs are shallow subtidal ecosystems of the tropical oceans formed at the edge of the land and sea. They provide many valuable resources to millions of people. They constitute a unique marine ecosystem that is characterized by a geologic component, the deposition of calcium carbonate by corals, mollusks, foraminifera and algae. These geological structures leave good fossils that enable paleontologists to track evolutionary change over millennia. Evolutionary change on coral reefs has been well documented and ancient reefs have attracted many investigations of past global change. As an ecosystem and geologic structure, reefs have persisted over time although their species composition has changed over time (McClanahan, 2002).

Although coral reefs have persisted through changes in water temperature and as level that accompanied glacial and other cycles, some species suffered extinction over these cycles. Such extinction was associated with synergistic losses in habitat and climate change. The Earth's atmosphere goday happens to be one of the most carbon dioxide rich in recent geologic history the current level is greater than at any time in the last 4,20,000 years. The level projected for the year 2100 is greater than that seen in the last 20 million years (McClanahan, 2002). Carbon dioxide and other greenhouse gases (GHGs) are reversing the slow cooling that has been occurring over the past 50 million years. The current rate of carbon dioxide emissions is expected to produce an atmospheric concentration in one century not experienced during the past 20 million years, as also water temperatures above those of the past interglacial 1,30,000 years before present. Human influences on water temperatures, seawater chemistry (toxic substances, nutrients and aragonite saturation), the spread of diseases, extinction of species, and food web alterations are changing reef ecology.

A significant ecological reorganization is taking place and changes include a reduction in calcifying and zooxanthellae-hosting organisms, their obligate symbionts, and species at higher tropic levels, with an increase in generalist species of low tropic level that are adapted to variable environments (McClanahan, 2002). Late-successional brown algae of low net productivity or non-commercial invertebrates, such as sea urchins, starfish and coral eating snails, will dominate many reefs. These changes will be associated with a loss of both net benthic and fisheries production and inorganic carbonate deposition and this will simplify reef complexity, lower species richness, decrease reef growth and increase shoreline erosion. Critical management is needed to avert these changes globally and locally; at both these levels, there is need to reduce GHGs and other waste emissions. Renewal of efforts is required to improve resource management including restrictions on the use of resources and globalization of resource trade, run-off and waste production, and balancing potential reef production and resource consumption(McClanahan, 2002).

According to Moberg and Folke (1999), coral reefs provide valuable ecosystem goods and services to maritime tropical and subtropical countries, but these reefs have been declining; an estimated 30 per cent are already severely damaged, and close to 60 per cent may be lost by the year 2030 (Wilkinson, 2002). According to Hughes *et al.* (2003), the diversity, frequency, and scale of human impacts on coral reefs have increased to such an extent that reefs are threatened globally. Projected increased in carbon dioxide and temperature over the next 50 years exceed the conditions under which coral reefs have flourished over the past half-million years. However, reefs will change rather than disappear entirely, with some species already showing much greater tolerance to climate change and coral bleaching than others.

International integration of management strategies that support reef resilience need to be vigorously implemented, and complemented by strong policy decision to reduce the rate of global warming.

There is an incontrovertible link between increase GHGs, climate change, and regional scale bleaching of corals (Watson *et al.*, 2001). Future changes in ocean chemistry caused by higher atmospheric $CO_2$ may further weaken coral skeletons and reduce the accretion of reefs, particularly at higher latitudes (Kleypas *et al.*, 1999). The strongest impact of climate change, however, is episodes of coral bleaching and disease that have already increased greatly in frequency and magnitude over the past 30 years (Knowlton, 2001; Harvell *et al.*, 2002).

Regional-scale coral bleaching is intimately associated with elevated temperatures, particularly during recurrent ENSO. Stressed, overheated corals expel most of their pigmented micro-algal endosymbionts (zooxanthellae), becoming pale or white. Under severe thermal stress, most of the corals on a reef bleach and may die. One strong concern is that the increasing rate of environmental change may exceed the evolutionary capacity of corals and their zooxanthellae to adapt. Most corals are too long-lived to evolve quickly, geographic differences in temperatures tolerances have evolved over much longer time frames than the decadal scale of current changes in climate. High levels of asexual reproduction are common traints that can retard rapid evolution in many coral species.

Climate change is a global issue, yet local conservation efforts can be helpful in maintaining and increasing resilience and in limiting the longer-term damage from bleaching and related human impacts. Managing coral reef resilience through a network of no take areas, integrated with management of surrounding areas, is essential to any workable solution. The focus should be on reducing pollution, protecting food webs, and managing key functional groups (such as reef constructors, herbivores, and boarders) as insurance of sustainability (see Gunderson and Pritchard, 2002).

Coral reefs constitute highly productive hotspots of biodiversity that support social and economic development. Their protection is not only a socioeconomic imperative but also an environmental one. Global warming, coupled with pre-existing human impacts, is a grave threat that has already caused much damage. Much available evidence indicates that, at a global scale, reefs will undergo major changes in response to climate change rather than disappear entirely (Hughes *et al.*, 2003).

## The Seaward Icy Antarctica

When the ocean eats at one end of a glacier, it tends to attract far- distance ice towars the sea, with potentially dangerous consequences. Polar ice is a major worry for glaciologists concerned at global warming. The ice sheet of West Antarctica which projects like a muscular arm from the huge mound of ice in East Antarctica is particularly worrisome. Warmer air per se is not a troublesome because even the thinner West Antarctic Ice Sheet (WAIS) is likely to hold out against its effects for millennia. But researchers to not know whether warming could somhow affect the WAIS indirectly, destrbilise it, and send its ice into the sea to melt, raising sea level up to a disastrous 5 metres in a few centuries! Most glaciologists agree that warming may somehow accelerate the movement of at least some of the WAIS ice toward the sea (Kerr, 2004). According to Robert Thomas of NASA contractor EG&G at Virginia, the half-dozen glaciers flowering into the Amundsen Sea have been getting thinner and thinner in recent years and that one of them, termed the Pine Island Glacier, has been flowing faster and faster for more than 100 kilometres inland. It could possibly lead to a collapse of the WAIS.

Recent observations of Thomas using modern motion science and ice penetrating radars and laster and rader altimers mounted on satellites and aircraft have indicated that some of the glaciers (*e.g.* the Pine Island and other) are hauling away about 253 km$^3$ of ice per year- 90 km$^3$ more than accumulates each year from snowfall (see Kerr, 2004).Also that ice withdrawals have been accelerating at least through the Pine Island Glacier, the largest of the group; it sped up by 3.5 per cent between April 2001 and the early 2003 making for a 25 per cent increase since the mid- 1970s. And the draw-down is not limited to areas near the coast. There has been a thinning, presumably induced by the faster flow, that extends along the main trunk of the Pine Island glacier and averages about 1.2 metres per year between 100 and 300 km inland. Some West Antarctic glaciers are flowing faster to the sea, breaking into more icebergs, and raising sea level faster.

## Corals and Sea-level Change

Changes in the amount of seawater that ocean basins contain form an integral part of the global climate system. Like the rest of that system, such changes are influenced by regular cycles in the Earth's orbit. By altering the seasonal distribution of incoming solar radiation, these orbital changes produce changes in sea level with a frequency of 21,000 years or longer (see Thompson and Goldstein, 2005; Henderson, 2005). Being a sensitive index of global climate, sea level has been connected to the Earth's orbital variations, with a minimum periodicity of about 21,000 years.

Although there are ample evidences for climate oscillations that are too frequent to be explained by orbital forcing, suborbital frequency sea-level change has been difficult to resolve, primarily because of problems with uranium/thorium coral dating. What Thompson and Goldstein (2005) did was to correct coral ages for the frequently observed open-system behaviour of uranium series nuclides, substantially improving the resolution of sea-level reconstruction.

Many coral species survive only in shallow after. Fossil corals found above or below present reefs reflect variations in the past sea level. It is possible to date corals by radiocarbon for the past 40,000 years and by the formation of 230$^{Th}$ through radioactive decay of uranium for the past 500,000 years. Coral alteration can sometimes be detected by changes in mineralogy, but a more subtle geochemical alteration commonly affects the uranium-thorium ages of corals that are mineralogically pristine. Such alteration changes the uranium isotope ratio of the coral and so can be screened for; however, for periods older than about 70,000 years, a few samples pass this screening (Henderson, 2005). For the period from 55,000 to 30,000 years ago, several corals pass the screening to indicate millennial sea-level changes of 10 to 15 metres.

Thompson and Goldstein (2005) demonstrated the presence of millennial sea-level changes at times of high sea level and relative climate stability. They developed a geochemical model to date corals that have not behaved as chemically closed systems. Instead of using uranium isotopes simply to reject samples, they used them to correct ages, thereby enabling dating of many altered corals for the interval from 250,000 to 70,000 years ago.

## Ocean Flow and Climate Amplification

Paleoceanographers have been using elements or isotopes preserved in deep-sea sediments as markers of the workings of past climate. This kind of 'proxy' approach has worked only up to a small extent because both the climate system and paleoclimate proxies are sometimes highly complex.

Recently, paleoceanographers have advanced another proxy-isotopes of the rate-earth element neodymium, which may faithfully trace the ups and downs of the heat-carrying Gulf stream flow. Their study with neodymium suggests that changes in the speed of the Gulf Stream- a popular mechanism for altering climate- came too late in major climate transitions to have set the climate change sin motion. This importance of neodymium as a circulation tracer manifested when it offered a prized trait *viz.* immutability. The ratio of neodymium-143 to neodymium-144 in North Atlantic and Pacific waters differs so much (thanks to the range of ratios of surrounding continental rocks) that it is used to follow the mixing of waters as current flow from basin to basin (Kerr, 2005). The idea soon emerged that ocean circulation changes should dominate the changes in the neodymium ratio. For example, plankton cannot change the ratio-as it does the isotopic composition of carbon-because neodymium is too massive an element for biology to separate its isotopes. Indeed, the isotope ratio preserved in the microscopic bits of iron-manganese in a classic sediment core from the south-eastern South Atlantic is quite consistent with the inferences drawn from work on the previous tracers. During each of four temporary warmings during the last Ice age, the ratio declined and then rose just as it should have done if the warm, north flowing Gulf Stream had temporarily sped up, as more North Atlantic water flowed south in the deep arm of the 'conveyor belt' flow (Kerr, 2005).

In the run-up to the Ice Age, by contrast, the core pointed to a more complicated story. Starting about 70,000 years ago, bottom waters cooled as glacial ice grew on the polar continents, as indicated by oxygen isotopes of microscopic remains of bottom dwelling organisms. A few thousand years later, carbon isotopes shifted as the growing ice and climatic deterioration shrank the mass of plants of land, sending their isotopically light carbon into the sea. Only after another couple of thousand years did thee conveyor belt flow slow down, according to neodymium (Kerr, 2005).

Given that millennia long lag behind the growing cold and ice, ocean circulation did respond to climate change. At least, at glacial transitions, the slowing of warm current would have finally chilled the Ice Age, but it could not be the trigger of climate change. Presumably, the initial cooling was an indirect response to the decline of solar heating over high northern latitudes generated by the so-called Milankovitch orbital variations, the ever-changing orientation of the Earth's orbit and rotation axis. It is, however, quite possible that changes in ocean circulation could have triggered abrupt climate shifts once the Ice Age was under way. Today, many paleoceanographers like the idea of ocean circulations a follower rather than a leader; yet, a single core may not be the explanation. Neodymium appears to be working remarkably well, but the history of climate proxies and a few hints in the South Atlantic record suggest that neodymium may not be the perfect ocean circulation proxy.

## Marine Biota and Geochemical Cycles

Watson and Liss (1998) discussed the influence of the marine biota on climate, particularly their role in mediating surface temperatures via their influence on atmospheric carbon dioxide and dimethyl-sulphate (DMS) concentrations. Variations in natural carbon dioxide concentrations occurring over $10^3$ to $10^5$ years are set by oceanic processes, and in particular by conditions in the Southern Ocean, so it is this region that is relevant for understanding the glacial-interglacial changes in carbon dioxide concentration. Marine productivity in the Southern Ocean appears to be limited by a combination of restricted iron supply to the region and insufficient light.

Plankton-produced DMS probably influences climate powerfully by changing the numbers of cloud condensation nuclei available in remote regions. It has a much shorter timescale than the carbon dioxide effect, and consequently may well be a player on the global change timescale. The direction of both the carbon dioxide and the dimethyl-sulpohide (DMS) mechanisms is such that more marine productivity would lead to lower global temperatures, and it is plausible that the overall effect of the marine biota today is to cool the planet by about 6°C as a result of these two mechanisms, with one-third of this figures being due to carbon dioxide effects and two thirds due to DMS (Watson and Liss, 1998).

While the marine biota influence climate, climate also influences the marine biota, mainly by changing atmospheric circulation. This is turn alters ocean circulation patterns, responsible for mixing up subsurface nutrients, and also influences the transport of nutrients (*e.g.* iron) in atmospheric dust. A stronger atmospheric circulation would be expected to increase the productivity of the marine biota on both counts. More production leads to greater cooling by reduction in carbon dioxide and increase in DMS.

The phytoplankton of the world's oceans are mostly small, unicellular organisms and they live only a few days and may be individually insignificant. Their total mass is also very small as compared to the mass of vegetation on the land surfaces. But globally, the productivity of the plankton is comparable to that of the land vegetation (Longhurst *et al.*, 1995) and collectively their effects on the climate, chemistry and geology of the planet have been profound.

The marine biota affect and interact with the climate on many different timescales ranging from the seasonal, for example, the annual 'blooming' of the North Atlantic or the response to the monsoons in the Indian Ocean, up to the periods of more than $10^7$ years over which carbon is removed from the atmosphere-ocean system to form carbonate rock and organic deposits–part of the rock cycle which over long time spans, probably, controls atmospheric carbon dioxide concentrations.

Marine algae produce DMS from the precursor compound dimethyl sulphoniopropionate (DMSP), which is metabolized by many species, but in different amounts. It is an intracellular osmoregulant. The ocean surfaces are always supersaturated with DMS relative to its atmospheric concentration, so there is a perennial net flux of the gas from the sea to the air. Once in the atmosphere, DMS in oxidized by free radicals such as OH and $NO_3$ to form various products including methane sulphonic acid (MSA) and sulphur dioxide ($SO_2$), which itself is oxidized, mainly in water droplets, to form sulpahte. The cycle of production of DMS in seawater and its oxidation in the atmosphere is illustrated in Figure 6.1. The major oxidation products are acidic, and in the pristine marine atmospohere, aerosols and rain obtain most of their acidity by this route. The formation of $SO_2$ by the man's burning of fossil fuel has considerably increased this acidity over major parts of the Northern Hemisphere, although for much of the Southern Hemisphere, DMS oxidation is still the major source (Bates *et al.*, 1992).

The DMS affects climate through the sub-micron sulphate particles resulting from its atmospheric oxidation. These particles interact both directly with sunlight

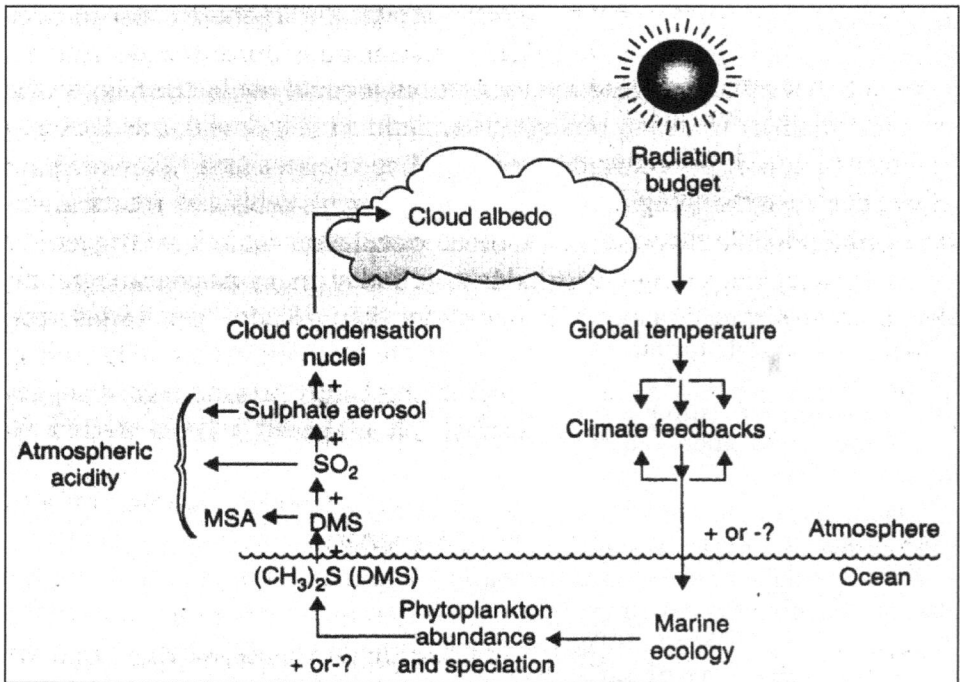

Figure 6.1: The mechanisms by which marine DMS emissions from plankton influence climate via cloud properties. DMS oxidizes in the atmosphere to form acidic aerosols-the major source of CCN in remote areas. Higher amounts of CCN lead to more reflective clouds, and thereby a decrease in the absorption of solar radiation, lower surface temperatures and possibly other climatic effects too. This may in turn have an influence on the rate of DMS emission, but the sign of this effect is unknown (after Charlson *et al.*, 1987 and Andreae, 1990).

and indirectly as the nuclei on which cloud droplets form cloud condensation nuclei (CCN). The number and density of CCN determine the cloud formation and hence the planet's albedo (Charlson *et al.*, 1987).

Even when they are emitted in the same areas, marine DMS emissions, probably have greater climatic impact than the equivalent quantity of sulpohur gas emitted from anthropogenic sources (Watson and Liss, 1998).

The processes involved in production and transformation/removal of DMS in the ocean surfaces are shown in Figure 6.2. As DMS is formed biologically, it is affected by availability of light and nutrients. But not all phytoplanktons have the same ability to form DMSP and hence DMS. Diatoms are poor producers, whereas prymnesiophytes are prolific; probably, because DMSP is made by organisms as an osmolyte and alternative compounds can serve a similar function. One of these is the nitrogen analogue of DMSP, glycine betaine (GBT); it appears that plankton can switch between GBT and DMSP depending on the availability of nitrogen in the water and sulphate, from which DMSP is synthesized, is never in short supply. Iron also affects the production of DMS.

The main release of DMSP and DMS formed by phytoplankton into the water occurs when the cells senesce and die, or are grazed by zooplankton. Many of the above processes can undergo modification as a result of climate change. But while for some processes the sign of the change is predictable, its magnitude is not and in some instances even the sign is not clear. As a general rule of thumb, any change which

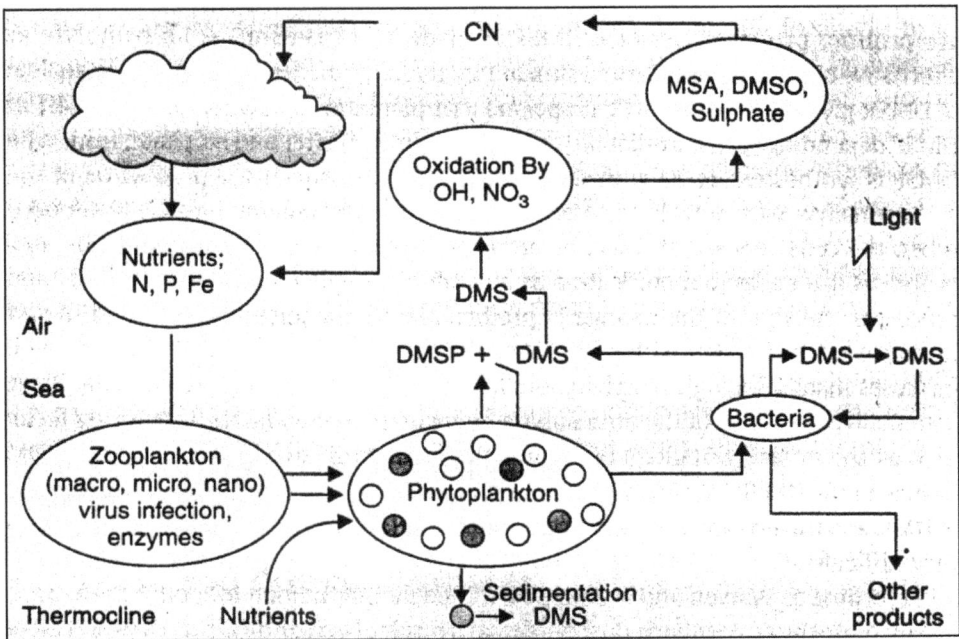

Figure 6.2: Major processes in the cycling of DMS by the marine ecosystem (after Liss *et al.*, 1997).

increases marine biological activity will lead to enhanced DMS production and hence, potentially, to cooling of the atmosphere by increase in cloudiness. Interfering factors, such as the differing abilities of various phytoplankton species to form DMSP/DMS, change in the DMSP: GBT ratio under different N-nutrient regimes, bacterial oxidation of DMS, and the nonlinear nature of atmospheric effects, make quantitative predictions very difficult.

According to Watson and Liss (1998), it is likely that human influence in the future will increase the atmospheric dust burden as a result of agricultural practices and indeed that may already be taking place. This might slow down global warming. Modelling studies carried out by Sarmiento and Lequere (1996) of the uptake of carbon dioxide by the oceans forced by increasing carbon dioxide, have also indicated that the chief influence of the marine biota may be a negative feedback. The issue concerning the marine biota's influence on climate will continue to be important for the foreseeable future in climate research (Watson and Liss, 1998).

## The Ocean and Climate Regulation

Between 1,30,000 and 90,000 years ago, the record shows sea-level change at higher frequency than can be explained by orbital changes. A 10-m lowering of sea level in the middle of the last interglacial period illustrates this. It appears that most of the approximately 0.5m of sea-level rise expected by the year 2100 will be due to heating of the oceans. On millennial timescales, temperature change extends to the deep ocean and may well cause several metres of sea-level change. The ocean is a major component of the Earth system and regulates not only the Earth's climate but also the biogeochemical cycling of key elements. It is a huge storehouse of heat and gases which have important impacts on the climate. It contains the most extensive and least known biosphere. In general, the ocean reacts more slowly than the atmosphere to anthropogenic forcing.

In fact, it is often considered as that part of the Earth system which buffers and modulates physical and geochemical phenomena. But the ocean can also amplify atmospheric effects and trigger additional global change (Hall *et al.*, 2002). The atmosphere and ocean are coupled and exchange certain physical and chemical signals that are not accurately quantified, particularly with respect to their feedbacks to the atmospohere. The ocean and its continental shelf are intensely exploited for food and resources. Fisheries leave distinct footprints in marine ecosystems and influence the dynamics of geochemical cycles and food webs whose capacity to respond and adapt to global change is thereby compromised. Some direct human perturbations of the ocean are exemplified by over fishing, increasing nutrient and sediment loading in river runoff, coastal pollution caused by waste disposal and lowering of pH due to increasing atmospheric carbon dioxide. Indirect examples come from the expected climate change due to human-induced changes in atmospheric chemistry, such as increasing emissions of greenhouse gases (GHGS) and aerosols. Changes to ocean ecosystems, physical processes and biogeochemistry, whether induced by human activities or natural causes, may have grave consequences. Not only do ocean ecosystems provide a great variety of goods and services necessary to sustain humans but also play a pivotal role in the functioning of the Earth System (Hall *et al.*, 2002).

## Hot Spots and Critical Domains

Some areas of the ocean appear to be particularly sensitive to gradual long-term changes in climate and deserve intensive studies. These 'hot spots' often occur in critical domains such s continental margins, the mesopelagic zone, intermediate waters, regions of upwelling and deep mixing, high latitude areas and the sediment water interface (Hall *et al.*, 2002). The world's coastal zone is exposed to heavy and increasing pressure from both human use and habitation and from global climate changes. The resources and amenities of the coastal zone are critical to the societal and economic needs of the world population.

Although it represents less than 20 per cent of the land surface, the coastal zone presently is: the location of more than half the global population; a major food source (most croplands and much agriculture, most of the global fisheries); a focus of transport and industrial development; a source of minerals and geological products including oil and gas; a location for growing tourism; and an important repository of biodiversity and ecosystems that support the functioning of the Earth system (Lindeboom *et al.*, 2002). Box 1 brings out the heavy impact of humankind on the coastal zone. Society faces a strong challenge in maintaining the flow of benefit from the goods and services of the coastal domain within the context of increasing uncertainty over the scale and nature of climatic and consequent human impacts on coastal systems.

Besides anthropogenic influences, natural processes and resource, and structure and transition all contribute to the dynamics of the coastal zone at regional and global scales. Resources, products and amenities are as heterogeneously dispersed at local and regional scales as the natural setting sand processes on which they depend. They are subject to changing patterns of availability, quality limitations and pressures. Human dimensions and natural systems closely interact. They are bound together by the various pressure and resultant sate of the coastal domain. We need to go in for an analysis of existing information, on key issues concerning human uses of the coastal zone.

## The Coastal Zone

Even a pristine coastal area suffers from the impact of human activities that originate in the 'hinterland'. Activities in river basins modify the flow of materials to the coast with impacts on coastal morphology (*e.g.*, sediments) or ecosystems (*e.g.* nutrients and contaminants). There is need to study the river basin and coastal zone as one system because while the major impacts are no doubt at the coast, part of the solution lies within the river basin. Likewise, although most of the benefits of scientific management are at the cost, many of the costs are incurred within the river basin (Lindeboom *et al.*, 2002). Another major challenge is how to disentangle the regional scale cause-effect relationships from those emanating from much wider external pressures on the river coast system, such as climate change, population pressure and the global economy. Thus, climate-induced changes in hydrology affect the flow of sediment and water, whereas the changes in land cover and use modify the flow of carbon and nutrients (Lindeboom *et al.*, 2002).

Managing human influences (Box 1) and targeting of sustainable uses required reliable information on actual and predicted options and the valuation of change and options. The limits of marine sustainability, plans for combining goods and services in the coastal zone without irreversible damage to marine nature,and resolution of cross-boundary spatial conflicts associated with coastal zone management are some key issues to be addressed.

Being most directly relevant to society, continental margins are a critical boundary. For example, inputs of nutrients, sediments and pollutants are large in these areas and fishing pressure can be intense. These regions are also responsible for significant draw-down of atmospheric carbon dioxide, release of other climatically active gases and cross-shelf export of carbon with large unquantified deposits of terrestrial carbon along the shelf slopes. Little, if any attention, has been given to the role of marine food

---

**Box 1: Human influences in the coastal zone (after Lindeboom *et al.*, 2002)**

The coastal domain of land and coastal seas is influenced by human activities in the maritime seaboard and in river basins where alterations to upland water resources are causing marked changes in the timing, the flux, and the dispersal of water, sediments, nutrients and contaminants. Human influences include:

☆ Changes in the timing of when water is transported to the coastal zone, through flood/ wave mitigation, via reservoir storage, or entire water diversion schemes;

☆ Changes in the amount of water transported to the coastal zone due to water use for urban development, industry and agriculture;

☆ Regional decreases in the delivery of sediment to the coastal zone through sediment trapping within reservoirs;

☆ Regional increases in the delivery of sediment to the coastal zone through increased soil erosion driven by agriculture, construction (urban development, roadways), mining and forestry operations;

☆ Changing the flux of nutrients to the coastal zone (*e.g.* storing carbon within reservoirs; elevated nitrogen flux through agriculture activities);

☆ Building of shoreline engineering structures, ports and urban developments;

☆ Harvesting and often over harvesting of marine resources;

☆ Increased competition for marine space;

☆ Increased pollutants, contaminants and atmospheric emissions from industries and urbanization;

☆ Modification of the type and quantity of coastal discharges from surface and groundwater flows;

☆ Alienation of coastal wetlands and ecosystems through land use change; and

☆ Modification of habitat structure and functioning through introduction of non-indigenous species.

The consequences of the human impacts in the coastal zone are far-reaching and vital to societal and global functions. These include changes in: ecosystem health and diversity; vitality of coastal wetlands, mangroves, and reefs; coastal stability, biostability and shoreline modification; dispersal area of riverine particulate and dissolved loads; the fate and distribution of materials in coastal and shelf waters; yield of resources and products than sustain society and economics: and uncertainty or diminished options for sustainable development.

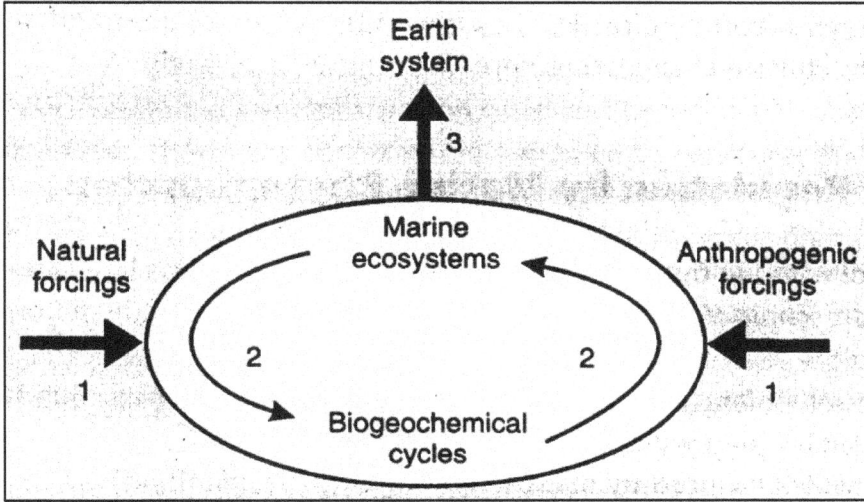

**Figure 6.3: The impacts of natural climatic and anthropogenic forcing on biogeochemical cycles and marine ecosystems (arrows 1), particularly on how these forcing alter the relationships between elemental cycles and ecosystems (arrows 2) and the feedbacks to the Earth System (arrow 3) (after Hall *et al.*, 2002).**

webs in continental margins that are heavily impacted by fishing activities, and how closely these may relate to the biogeochemical cycles in this region. Continental margins also characteristically have a more intense nutrient cycling between the water column and shelf sediments, and can suffer changes in oxygen concentrations with marked effects on chemical and biological processes.

The mesopelagic zone of the ocean is important for the recycling of nutrients. This region has been largely ignored in previous research. It is also important for pelagic food webs. Intermediate waster masses of the world's oceans are major storage reservoirs for anthropogenic carbon dioxide. The climate sensitivity of intermediate water mass circulation to climate change, therefore, warrants critical study.

## Climate Regulation by Marine Phytoplankton

Every drop of water in the top 100 metres of the ocean contains large numbers of free-floating, microscopic algae. These single-celled organisms include diatoms, green and blue-green algae and other algae and inhabit three-quarters of the Earth's surface, and yet they account for less than 1 per cent of the 600 billion metric tons of carbon contained within the Earth's photosynthetic biomass (Falkowski, 2002).

Chlorophyll is produced by microscopic algae (phytoplanakton). Its concentration is related to the abundance of phytoplankton in the water. These organisms lie at the base of the marine food china which makes it important to estimate how much phytoplankton is there. Away from the coast, phytoplanktons are the main determinants of water colour- the greener the ocean, the more the chlorophyll- so one can use data from an ocean colour satellite to estimate chlorophyll concentrations. But near the coast, other factors become important: while sediment in the water scatters visible light, dissolved organic material absorbs it. This can make a difference in the coastal Oceana. It is now possible to distinguish and separately estimate the water constituents affecting coastal ocean colour.

One of the most notable activities of marine phytoplankton is their influence on climate. These organisms can very efficiently draw carbon dioxide out of the atmosphere and store it in the deep sea. The fact that marine phytoplankton are so highly sensitive to changes in global temperatures, ocean circulation and nutrient availability, has prompted researchers to manipulate phytoplanakton populations- by adding nutrients to the oceans-with a view to mitigating global warming. A two- month experiment conducted in the year 2002 in the Southern Ocean has confirmed that injecting surface waters with trace amounts of iron stimulates phytoplankton growth; however, the efficacy and prudence of widespread, commercial ocean- fertilisation schemes are still hotly debated. Understanding how human activities can alter phytoplankton's impact on the planet's carbon cycle is crucial for predicting any long-term ecological side effects of such actions (Falkowski, 2002). Green plants play a major role in drawing carbon dioxide out of the atmosphere. In 1997, National Aeronautics Space Administration (NASA) launched the Sea Wide Field Sensor (Sea WiFS), the first satellite capable of observing the entire Earth's phytoplankton population every single week.

Phytoplankton and al land-dwelling plants split water molecules into hydrogen and oxygen. The oxygen is liberated as a waste product and makes possible all animal (and human) life on earth. The planet's cycle of carbon (and, largely, its climate) depends on photosynthetic organisms using the hydrogen to help convert the inorganic carbon in carbon dioxide into the organic constituents that make up their cells. This conversion of carbon dioxide into organic matter is termed primary production. Until about five years ago, most biologists greatly underestimated the contribution of phytoplankton relative to that of land-dwelling plants.

Today, modern satellites can quantity chlorophyll and by association, phytoplanakton abundance. By 1998, there was considerable evidence to conclude that every year phytoplankton incorporate approximately 45 billion to 50 billion metric tons of inorganic carbon into their cells-nearly double the amount cited in the most liberal of previous estimates (Falakaowski, 2002).

Earlier researchers suggested that land plants assimilate some 100 billion metric tons of inorganic carbon a year. Recent satellite analyses show that they assimilate only about 52 billion metric tons. This means that phytoplankton draw nearly as much carbon dioxide out of the atmosphere and oceans through photosynthesis as do trees, grasses and all other land plants combined (Falkowski, 2002).

Dead phytoplankton strongly modifies the planet's cycle of carbon and carbon dioxide gas. Because phytoplankton direct virtually all the energy they harvest from the Sun toward photosynthesis and reproduction, the entire marine population can replace itself every week. In contrast, land plants invest copious energy to build wood, leaves and roots, taking some two decades to replace themselves. As phytoplankton cells divide-every six days on average- half the daughter cells die or are eaten by zooplankton that in turn provide food for shrimp, fish and larger carnivores.

The rapid life cycle of phytoplankton enables them to influence climate. Researchers have been quantifying the oceanic carbon cycle, in which the organic matter in the dead phytoplankton cells and animals' faecal matter sinks and is consumed by microbes that convert it back into inorganic nutrients, including carbon dioxide (see Hanson *et al.*, 2000). Much of this recycling happens in the sunlit layer of the ocean, where the carbon dioxide is instantly available to be photosynthesized or absorbed back into the atmosphere. The total volume of gases in the atmosphere is exchanged with those dissolved in the upper ocean approximately every six years.

Most influential to climate is the organic matter that sinks into the deep ocean before it decays. When released below about 200 meters, carbon dioxide remains there for much longer because the colder temperature and higher density of this water prevents it from mixing with the warmer waters above. Through this process, termed the biological pump, phytoplankton remove carbon dioxide from the surface water and atmosphere and store it in the deep ocean. The material pumped into the deep sea amounts to between seven billion and eight billion metric tons, or 15 per cent, of the carbon that phytoplankton assimilate every year (see Hanson *et al.*, 2000).

Within a few centuries, almost all the nutrients released in the deep sea find their way via upwelling and other ocean currents back to sunlit surface waters, where they stimulate more phytoplankton growth. This cycle keeps the biological pump at a natural equilibrium in which the concentration of carbon dioxide in the atmosphere is about 200 parts per million, lower than it would be otherwise- a significant factor, considering that today's carbon dioxide concentration is about 365 (PPM) (Falkowski, 2002).

Over millennia, however, the biological pump leaks slowly. About one half of 1 per cent of the dead phytoplankton cells and faecal pellets settle into seafloor sediments before it can be recycled in the upper ocean. A fraction of carbon becomes incorporated into sedimentary rocks such as black shales, the largest reservoir of organic matter on earth. A smaller fraction is deposited as petroleum and natural gas.

When we burn fossil fuels, we bring buried carbon back into circulation about a million times faster than volcanoes do. As neither forests nor phytoplankton can absorb carbon dioxide fast enough to keep pace with these increases, its atmospheric concentrations have risen rapidly, thereby contributing significantly to the global warming trend of the past half a century.

Policymakers, seeking ways to make up for this shortfall turned to the oceans, which can potentially hold all the carbon dioxide emitted by the burning of fossil fuels. It was proposed that artificially accelerating the biological pump could exploit this extra storage capacity. Hypothetically, this enhancement could be achieved in two ways: by exogenously adding nutrients to the upper ocean or ensuring that nutrients not fully consumed are used more efficiently. It was supposed that in this way, more phytoplankton would grow and more dead cells would be available to take carbon into the deep ocean.

The vast majority of phytoplankton can use atmospheric nitrogen to build proteins only after it is fixed *i.e.* combined with hydrogen or oxygen atoms to form ammonium ($NH_4$), nitrite ($NO_2$) or nitrate ($NO_3$). The vast majority of nitrogen is fixed by small subsets of bacteria and cyanobacteria that convert $N_2$ to ammonium, which is released into seawater as the organisms die and decay.

Phytoplankton growth is usually limited by the availability of fixed nitrogen. To catalyse the reaction, cyanobacteria use the enzyme nitrogenase that relies on iron to transfer electrons. This is why, iron controls how much nitrogen these special organisms can fix.

In the mid- 1980s, the late John Martin, an American chemist, hypothesized that the availability of iron is low enough in many ocean realms to limit phytoplankton production. He found that its concentration in the equatorial Pacific, the northeastern Pacific and the Southern Ocean is so low that phosphorus and nitrogen in these surface waters are never used up. Martin and his associates pointed out that practically the only way iron reaches the surface waters of the open ocean is via windblown dust. In vast areas of the open ocean, far removed from land, the iron concentration seldom exceeds 0.2 part per billion- a fiftieth to a hundredth the concentrations of phosphate or fixed inorganic nitrogen.

Martin and other researchers also noted an inverse correlation between dust and carbon dioxide. In 1993, Martin and colleagues conducted the world's first preliminary open-ocean manaipulatin experiment by adding iron directly to the equatorial Pacific. Their research ship carried tanks containing a few hundred kilograms of iron dissolved in dilute sulpohuric acid and slowly released the solution as it traversed a 50 square-kilometer patch of ocean like a lawn mower. The outcome of this first experiment was promising but inconclusive.

When the experiment was repeated for four weeks in 1995, the additional iron dramatically increased phytoplankton photosynthesis, leading to a bloom of organisms that coloured the waters green. Since then, three independent groups, from New Zealand, Germany and the USA have proved that adding small amounts of iron to the Southern Ocean greatly stimulats phytoplankton productivity. The most extensive fertilization experiment to date took place in January-February 2002. The project, called the Southern Ocean Iron Experiment (SOFEX) involved three ships and 76 scientists. Preliminary results indicate that one tone of iron solution released over about 300 km$^2$ resulted in a 10 fold increase in primary productivity insight weeks' time. This shows that iron indeed stimulates phytoplankton growth at high latitudes, but no one has so far proved whether this increased productivity enhances the biological pump or increased carbon dioxide storage in the deep sea.

## Fertilising the Ocean

Oceanic iron fertilization is one possible option to sequester carbon and so help to mitigate climate change. By increasing phytoplankton primary productivity in iron-poor oceanic regions, the carbon flux to the deep sea may be enhanced and excess carbon dioxide drawn out of the atmosphere. However, concerns have been voiced that manipulations of the oceans at large scale can alter marine ecosystems dramatically: and therefore, ocean fertilization should not be made eligible for carbon credits on the global carbon-trading market. This arguments is based on marine ecology. The carbon cycle is intimately coupled with those of other elements, some of which have critical roles in climate regulation. The likelihood of unintended climatic and atmospheric change is a forceful argument against ocean fertilization that needs extensive discussion and critical debate.

Dimethylsulphide ($CH_3SCH_3$ or DMS) is an important precursor for maritime sulphate aerosols and cloud condensation nuclei (CCN). It is produced by phytoplankton and influences cloud properties and climate (Charlson *et al.*, 1987). Any increase in phytoplankton primary productivity (and hence in DMS levels) may lead to some cooling of the sea surface waters (Lawrence, 1993).some other chemicals that may also be affected are volatile organohalaogens such as methyl halides ($CH_3Cl, CH_3Br$, and $CH_3I$) which photolyse to produce reactive halogens, which contribute not only to lower stratospheric $O_3$ depletion (Solomon *et al.*, 1994) but also to marine boundary layer $O_3$ destruction (Vogt *et al.*, 1996).

Another chemical carbonyl sulphide (OCS) contributes to the stratospheric aerosol layer (see Lawrence, 2002) and thus to heterogeneous $O_3$ loss. Any increase in these gases can increase stratospheric ozone depletion, leading to higher ultraviolet levels at the Earth's surface, with negative biological health consequences.

Ocean fertilization may also directly influence the atmosphere-ocean system radiative budget. The extreme scenario of removing 600 mmol/mol of atmospheric carbon dioxide over a century by fertilizing 30 per cent of the world's ocean would require a sustained increase in photosynthetic energy equivalent to approximately $1.5 W/m^2$ over the fertilized region. Most of this would transfer as heat to the ocean's surface waters through respiration and so increase regional sea surface temperatures. But this scenario for complete removal of anthropogenic $CO_2$ may be realized only if limitations on the availability of nitrogen and phosphorus were also overcome (Lawrence, 2002).

Inspired by the promising results, one company has proposed a scheme in which commercial ships that routinely traverse the southern Pacific would release small amounts of a fertilizer mix. Other groups have debated the possibility of piping nutrients, including iron and ammonia, directly into coastal waters to trigger phytoplankton blooms. It is not clear whether such ocean-fertilisation strategies will ever be technically feasible. To be effective, fertilization would have to be conducted year in and year out for decades. Because ocean circulation will eventually expose all deep waters to the atmosphere, all the extra carbon dioxide stored by the enhanced biological pump would return to the atmosphere within a few centuries of the last fertilizer treatment (Falkowski, 2002). Moreover, such efforts are not easily controlled. It is almost impossible to properly manage fertilizing a patch of grubulent ocean water. Hence, some ocean experts have argued that once initiated, large scale fertilization could produce long term damage that might be impossible to fix, as there could be major disruptions to the marine food web. Also enhancing primary productivity could lead to local problems of severe oxygen depletion. The microbes that consume sinking dead phytoplankton cells sometimes consume oxygen faster than ocean circulation can replenish it. Creatures that cannot escape to more oxygen rich waters may die so suffocation. Such conditions also encourage the growth of microbes that produce methane and nitrous oxide-greenhouse gases with even greater heat-trapping capacity than carbon dioxide.

For these and other reasons, the idea of designing large, commercial; ocean fertilization projects to change climate is still being seriously debated. Many scientists feel that the potential temporary human benefit of commercial fertilization projects may not be worth the unpredictable consequences of altering natural marine ecosystems (Falkowski, 2002).

# Chapter 7

# Climate, Energy and Aerosols

## General Description

The composition of the Earth's atmosphere is known to play a central role in establishing the surface temperature and climate of the earth. Direct observations over the past few decades show that the concentrations of certain atmospheric gases, the so-called greenhouse gases (GHGs), have been increasing to a degree that might alter the natural flow of radiation that controls the balance between heating and cooling in the atmosphere. Human emissions of these GHGs have contributed to the observed buildup. These emissions occur as a byproduct of fundamental human activities in energy, industry and agriculture, that are expected to increase in the future to meet the economic development needs of a growing global population. That this ongoing build-up may lead to significant future climate change has generated considerable concern. The issue is controversial because uncertainty and ignorance about critical aspects of the science and about future human behaviour limit our ability to project not only future emissions but also impacts through climate change. Much of what we project about the future has to be evaluated in large computational models. Although these models are the best available tools to make forecasts, they can be incomplete and unconfirmed, and are likely to remain so in the future because scientists do not fully understand the workings of critical climate components, such as oceans, clouds, and the response of the biosphere to climate change. Also, some forecasts of future human activity and technology over multi-decade timescales are questionable. Model projections are uncertain concerning the magnitude, timing, and regional distribution of climate change. But even if the climate change could be successfully projected, it would still be difficult to judge its impact on society and natural ecosystems-while some effect, for example, a rise in sea level, would be mostly negative, other changes in temperature and precipitation levels could, in some regions, be positive.

From analyses of potential response options, it has emerged that the effort needed to eliminate the threat of climate change by limiting the growth of future emissions would be economically costly, would require major technical innovations, and might require some basic changes in current human behaviour. Indeed, the underlying uncertainty in scientific, social and technical aspects of climate change may require decades to resolve. Consequently, the debate over the appropriate response to a potentially serious threat is likely to continue Table 7.1. Pie charts (Figures 7.1 and 7.2) show percentage contributions to carbon dioxide emissions (1980-85) from human activities, and percentage annual contributions of methane emissions from different sources.

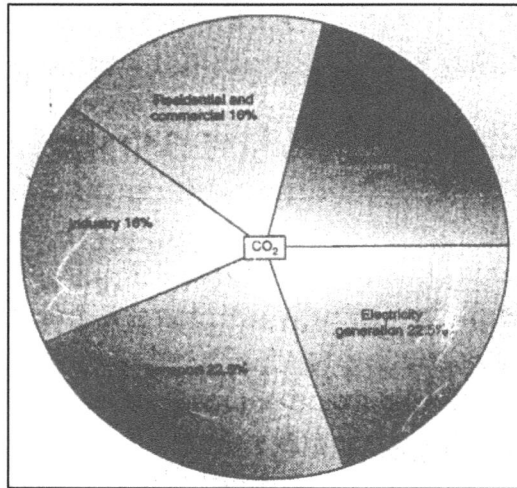

**Figure 7.1: Per cent contribution to carbon dioxide emissions from human activities (after IPIECA, 1991).**

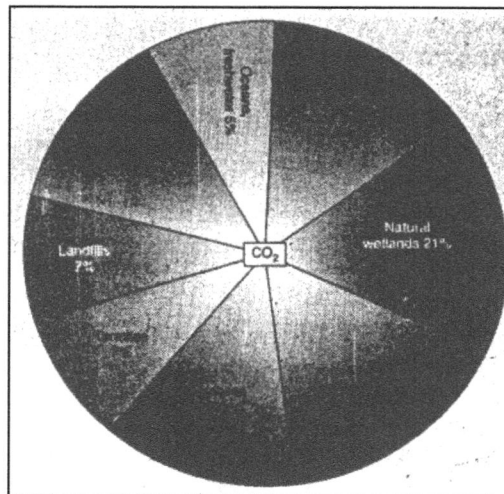

**Figure 7.2: Per cent annual contributions of methane emissions from different sources (after IPIECA, 1991).**

**Table 7.1: The Major Greenhouse Gases**

| | $CO_2$ | $CH_4$ | CFC-11 | CFC-12 | $N_2O$ |
|---|---|---|---|---|---|
| Concentration | | | | | |
| Pre-industrial | 280 | 0.79 | 0 | 0 | 280 |
| Present | 353 (ppmv) | 1.72 (ppmv) | 280 (pptv) | 484 (pptv) | 310 (pptv) |
| Radiative forcing per molecule $(CO_2^{-1})$ | 1 | 21 | 12400 | 15800 | 206 |
| Lifetime in atmosphere (years) | 50-200 | 10 | 65 | 130 | 150 |
| Global warming potential[1] relative to $CO_2$ | | | | | |
| 20 years | 1 | 63 | 4500 | 7100 | 270 |
| 100 years | 1 | 21 | 3500 | 7300 | 290 |
| 500 years | 1 | 9 | 1500 | 4500 | 190 |
| Percentage contribution to total radiative forcing 1980-1990 | 55 | 15 | 24 (all CFCs) | 6 | |
| Current increase (Per cent/year) | 0.5 | 0.9 | 4 | | 0.25 |

1: The warming effect of an emission of 1 kg of each gas relative to $CO_2$ based on the atmosphere as in the 1990s.

# Radiative Forcing

Radiative forcing of climate forms the basis for the evaluation of human influence. The forcing caused by emission of long-lived greenhouse gases is global and is already equivalent to an increase in solar radiation of about 1 per cent. The radiative forcing pattern is strongly regional because of an increased aerosol load over and leeward to industrial centres centres and areas with burning vegetation; here, the uncertainty is large and adding up both the greenhouse gas and the aerosol forcing can be misleading, since the climate system might react differently to regional forcing. Radiative forcing by tropospheric ozone increase has, on global average, been nearly as important as the methane accumulation since industrialization began. However, this does not lead to similar reaction, because of the strongly dissimilar patterns of forcing. Stratospheric ozone depletion leads on average to a slight negative radiative forcing which partly offsets the positive radiative forcing by chlorofluorocarbons (CFCs) (Grassl, 1999).

# Interactions Between Aerosols and Climate

Besides carbon dioxide, aerosols have contributed significantly to recent large-scale temperature changes (Stott *et al.*, 2000) and may have been locally important. The most important aerosols are sulphates from volcanic and anthropogenic sources in the stratosphere and troposphere, respectively. They increase the scattering of sunlight away to space, either directly or in the case of the troposphere, via some effects on clouds (Ramaanathan *et al.*, 2001). They provide a surface cooling and so a greater precipitation response compared to changes in carbon dioxide concentrations (Allen and Ingram, 2002).

Aerosols ocean also produce direct tropospheric heating, with similar effects to carbon dioxide on the hydrologic cycle. The effects of warming on the atmospheric moisture content, maximum rainfall and the zonal-mean precipitation should all apply irrespective of the cause of the warming (Allen and Ingram, 2002).

## Mount Etna

Some volcanoes release enormous quantities of invisible carbon dioxide in a surprisingly surreptitious manner. One good example is that of Mount Etna, an active but, at present relatively subdued volcano in Sicily. Mount Etna is one of the Earth's most potent natural sources of carbon dioxide. It quietly pumps some 25 million tons of carbon dioxide into the atmosphere each year, roughly 20 times than the amount unleashed by far more spectacular volcanoes, such as Kilauea in Hawaii.

Because concentrations of carbon dioxide in the volcano's plume are difficult to measure directly, researchers have looked instead at sulphur dioxide emission in the plume and then combined these with estimates of the relative abundance of the two gases. It appears that Mount Etna's copious emissions might come from carbon dioxide being cooked out of carbon-rich rocks under the volcano. Another possibility is that the volcanoa's behaviour is just an extreme example of a class of volcanoes whose lavas are relatively alkaline.

Human activities have been releasing carbon dioxide over 900 times the rate that Mount Etna does. In the past, however, the long-term flow of carbon oxide from volcanoes was crucial to sustaining the natural greenhouse effect. Slow but steady weathering processes remove carbon dioxide from the air. The huge output of Mount Etna helps pinpoint where the carbon dioxide that made up for this deficit came from.

## Uncertain Climate Sensitivity and the Need for Energy without Carbon Dioxide Emission

Climate sensitivity is defined as the global mean climatological temperature change resulting from a doubling of atmospheric carbon dioxide content. It is thought, based on models, to lie in the range of 1.5°C (Houghton *et al.*, 2001). Cloud feedbacks introduce high uncertainty in model predictions of global mean warming (Cess *et al.*, 1996). Aerosols, non-$CO_2$ GHGs, internal variability in the climate system, and land use change also influence the Earth's temperature. Uncertainty in aerosol radiative forcing makes it impossible to accurately estimate climate sensitivity to a $CO_2$ doubling (Forest at al. 2002). Caldeira *et al.* (2003) focused on $CO_2$ induced climate change because $CO_2$ is the dominant source of change in the Earth's radiative forcing in all Intergovernmental Panel on Climate Change (IPCC) scenarios of the future (Nakicenovik *et al.*, 2000), and future aerosol emissions decrease in all of the IPCC SRES scenarios. The UN Framework Convention on Climate Change has called for 'stabilisation of greenhouse gas concentrations at a level that would prevent dangerous anthropogenic interference with the climate system'. Even if a 'safe' level of interference in the climate system could be determined, the sensitivity of global mean temperature to increasing atmosphere $CO_2$ is known only roughly. Caldeira *et al.* (2003) showed how a factor of high uncertainty in climate sensitivity introduces even greater uncertainty in allowable increases in atmospheric $CO_2$.

According to Caldeira *et al.*, if climate sensitivity is 1.5°C, stabilization at 2°C of $CO_2$-induced warming might be achieved at $CO_2$ concentrations of 700 ppm; however, if climate sensitivity is 4.5°C, then $CO_2$ would need to be leveled off at only 380 ppm, a level that is only slightly higher than today's value of 370 ppm. Top-down models of global energy systems indicate that we can stabilize climate with $CO_2$ concentrations well below 500 ppm and still grow the economy by an order of magnitude over this century (Azar and Schneider, 2002). However, several physico-chemical, engineering, and environmental considerations suggest that this will be quite difficult to achieve (Hofert *et al.*, 2002). Unless climate sensitivity is low and acceptable amount soft climate change are high, climate stabilization will require a massive transition to $CO_2$ emission free energy technologies.

If climate sensitivity should apply in the upper half of the accepted range, climate stabilization at a 2°C warming would require immediate reductions in fossil fuel carbon emissions (Figure 7.3). Even if the climate sensitivity were low, allowable end-of-century $CO_2$ emissions are roughly half of the emissions implied by the IPCC IS92a reference scenario assumptions (Hoffert *et al.*, 1998).

On the basis of present understanding, Caldeira *et al.* (2003) determined that climate sensitivity uncertainty exceeds carbon cycle uncertainty in its impact on allowable emissions. For $CO_2$ stabilisation scenarios, the IPCC estimated that carbon cycle uncertainties translate into uncertainty in year 2100 allowable emissions 'approaching an upper bound' of –14 to +31 per cent. For the climate stabilization scenario described by Caldeira *et al.*, climate sensitivity uncertainty in the 1.5° to 4.5°C range introduces 100 to +429 per cent uncertainty in year 2100 allowable $CO_2$ emissions relative to results at a 3°C climate sensitivity. Hoffert *et al.* (1998) had shown that to stabilize atmospheric $CO_2$ content, massive amounts of carbon-free energy and massive improvements in the efficiency of energy use will have to be made. Long-term economic projections are unreliable because they cannot anticipate unforeseen technological or socioeconomic developments. Still, emission scenarios' frameworks have tried to limit these uncertainty problems by providing range of GHG emission [*e.g.* the IPCC IS92 and SRES future emissions of greenhouse gases and aerosols precursors (Nakicenovik *et al.*, 2000)]. These scenarios illustrated several assumptions about economics, demography, and policy on future emissions.

In summary, it appears that climate sensitivity uncertainty introduces much greater uncertainty in allowable $CO_2$ emissions than does carbon cycle uncertainty. For $CO_2$ stabilisation by the year 2150 leading to a $CO_2$-induced global mean warming of 2°C, estimated allowable carbon emissions later this century could be less than GTC or greater than 13 GTC ($1GtC=10^{12}kg$ C) per year, depending on whether climate sensitivity is 4.5° or 1.5° per $CO_2$ doubling, respectively (Caldeira *et al.*, 2003). With this climate stabilization scenario and IPCC IS92 a 'business-as-usual' economic

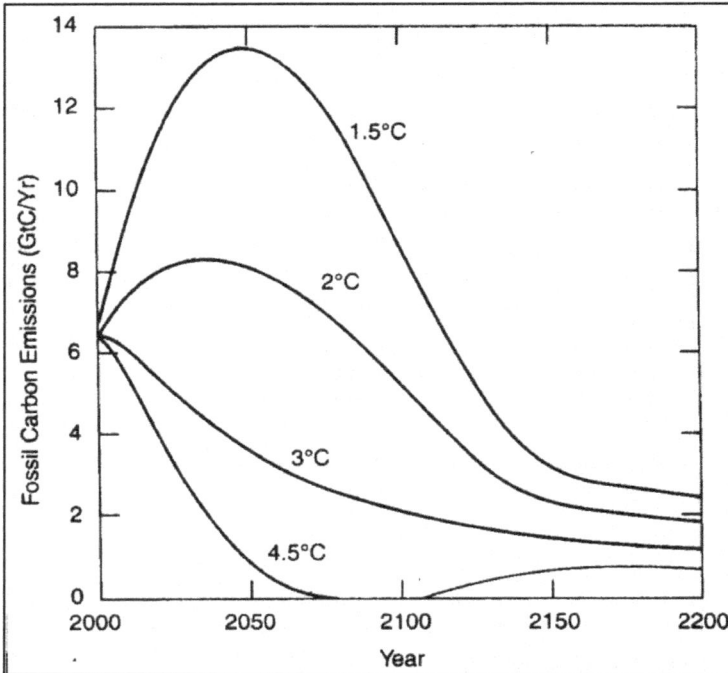

Figure 7.3: Allowable emissions of carbon dioxide to the atmosphere to produce climate stabilization at a 2°C global mean warming relative to the reindustrialize state, shown for different climate sensitivities. To achieve such climate stabilization, today's emission rate may either be allowed to double by mid-century or we have to bring down emission near zero, depending on whether climate sensitivity is 1.5° or 4.5°C per $CO_2$ doubling (after Cladeira *et al.*, 2003).

assumptions, if climate sensitivity is at the high end of the IPCC range, then by the end of this century, most or all of our primary power will have to come from non-$CO_2$ emitting sources. Surprisingly, even if climate sensitivity is at the low end of the accepted range, by the end of this century, over three-quarters of our primary poor will have to come from sources that do not released $CO_2$ into the atmosphere. We still lack cost-effective $CO_2$ emission-free energy technologies that can be applied today at the required scale (Hoffert *et al.*, 2002). As it takes quite long to implement new energy technologies, we must develop appropriate energy technologies immediately. With such technologies, the industrialized world can evolve to and the industrializing world can develop with an environmentally acceptable energy infrastructure and one 'that would prevent dangerous anthropogenic interference with the climate system' (Caldeira *et al.*, 2003).

## Linking Energy, Development and Climate

The decade since the Rio Earth Summit has witnessed a marked technical and social change. Although modern forms of energy are a necessity for development, the implications of how that energy is derived- and how it is used- still elude us. The

United Nations Environment Programme (UNEP) and the UNEP Collaborating Centre on Energy and Environment (UCCEE) promote policies that can move the world towards energy systems based on cleaner forms of energy.

The UNEP's Energy Programme (UNEP Energy) is focused on renewable energy, energy efficiency, transport, energy finance and policy issues that address the environmental consequences of energy production and use, for example, global climate change and local air pollution (Johanson and Radka, 2002). A new network to link existing institutions focused on energy, development and environmental issues has been launched during the 2002 World Summit on Sustainable Development (WSSD).

The Global Network on Energy for Sustainable Development (GNESD) will promote efforts to provide energy for sustainable development by linking existing centres working on energy, development, and environment issues. The Network is committed to help place energy for sustainable development more firmly on the global agenda and also link the provision of clean energy services to other development goals. Such networks can facilitate the exchange of knowledge at relatively low cost, provided that there is a shared purpose and vision among partners.

In many developing countries today, climate change attracts a much lower priority than more pressing issues of food security, poverty, natural resource management, energy access, and urban transport. Policies and actions are needed that can aid development and at the same time address the challenge of climate change. Of course, there do exist some good sustainable development projects that do not directly address climate change issues but still produce positive climate impacts. But a new project aims to find the development path that links to positive climate outcomes; it is inspired by the Marakesh declaration, which emphasizes than actions to limit climate change should be in the context of sustainable development. This new project is based on a strong international collaboration among many partners (Christensen, 2002).

## Achieving Energy Efficiency

There are two ways to achieve energy efficiency while pursuing energy efficiency projects. One relates to payback times for technical solutions, such as installing heat economizers, back-pressure turbines or efficient motors. The second focuses largely on technical improvements. This sometimes can leave untapped a part of the savings to be gained thought he low-risk, low tech solution of training for proper maintenance and operation. Plant managers have very often focused on technical innovations which generally lead to greatly improved energy efficiency. As a result of this, energy intensity (energy use per unit of production) in the manufacturing sector fell steadily from 1973 to 1985, where after it stabilized. Reductions in energy intensity increased again in 1993.

Even with such good track record, facility managers can rely on technical solutions to solve only some of the energy use problems. Many problems arise from lack of training related to system optimization or from ineffective training programmes. Establishing an effective, low-cost, low-tech training and maintenance programmes within a plant can generate a fast payback and lasting results.

The true value of training is often underestimated by regarding it as a cost, rather than an investment. Investing in a training programmes can go a long way in improving energy efficiency, minimizing costs, increasing profit, improving productivity and reliability, and enhancing operational safety.

## Climate Variations and the Sun

The Earth's climate system is driven by the Sun. every second, the Sun loses about four million tons of weight which are irradiated into space mainly in the form of visible light. One billionth of this power (1,017 w) reaches the top of the Earth's atmosphere; this amount corresponds to about 10,000 times humankind's present global consumption. This solar power arriving at the top of the atmosphere is known as the solar constant (1.365 $Wm^{-2}$).

Till recently, it was generally believed that the Sun had nothing to do with climate change. Some scientists tried to test whether the solar constant really is constant, but their attempt failed because they were only able to observe the Sun from the surface of the Earth- absorption of the sunlight crossing the atmosphere fluctuates. It was the advent of the satellite era that made it possible to continuously monitor the solar constant from outside the Earth's atmosphere. Now we know that the solar constant is indeed not constant but varies with solar activity as indicated by the presence of some sunspots (Frohlich and Lean, 2004).

However, over an eleven-year cycle the observed change is fairly small-only about 0.1 per cent. These findings have prompted many people to conclude that though solar forcing of the climate does occur, it is negligible (Beer,2005).

However, besides direct measurements of the solar constant [more appropriately called total solar irradiance-TSI], there is much indirect evidence to show that the Sun is a variable star and that these variations can influence climate change.

In contrast to TSI, changes in the ultraviolet part of the solar spectrum are quite large and influence the amount of ozone in the stratosphere. These changes can ultimately affect the circulation of the lower atmosphere (Haigh, 1999). These change in irradiance are caused by processes on the solar surface, and models describing the lifetime of the Sun(approximately 10 billion years) have shown that 4.5 billion years ago when the Soalr System was created, TSI was lower by about 30 per cent, where after it has steadily increased, and will continue to do so for about another 4 billion years. Here, an interesting question is: how the Earth System could avoid becoming an 'ice house'; this problem is termed the 'faint young Sun paradox' (Beer, 2005).

For much shorter-but still fairly long-timescales, the amount of solar radiation arriving at the top of the atmosphere is related both to the emission from the Sun and to the position of the Earth relative to the Sun. Because of the gravitational forces of other planets in our Solar System (mainly Jupiter and Saturn), the orbital parameters of the Earth change with periodicities ranging from 100,000 to 400,000 years (eccentricity), through approximately 40,000 years (tilt angle) to periodicities of around 20,000 years (precession of the Earth's axis) (Beer, 2005).

Orbital forcing is the only forcing that can be calculated precisely, not only for the past several million years, but also for the future (Berger *et al.*, 2003). The measured d$^{18}$O record of the GRIP ice core (Greenland) points to the temperature changes during the past 100,000 years. this record is chiefly characterized by the abrupt change soft the so-called Dansgard-Oeschger events due to changes in the deep-water formation of the North Atlantic, and the general long-term trend agrees well with insolation changes, except in the past 10,000 years (Beer, 2005).

The greatest effect of the orbital forcing is the cyclic change between glacial periods and interglacial periods over the past 700,000 years, with a periodicity of 100,000 years-surprising, because the corresponding mean annual change in forcing is very small ($\sim 0.2$w m$^{-2}$). This has prompted some doubts about the sensitivity of the climate system. One basic problem in assessing the effect of any change in forcing is that the climate system contains many components which interact in nonlinear ways on very different spatial and temporal scales.

Due to positive feedback mechanisms, even very weak but persistent forcing signals can be amplified and produce strong effects. The sensitivity of the climate system can be studies by means of climate models designed to simulate reality (Rind, 2002). But the closer the climate models approach the complexity of the climate system, the less they can simulate orbital forcing effects on timescales of 20,000 to 100,000 years (Beer, 2005).

Several high-resolution and well-constrained reconstructions of the palaeoclimate during the Holocene have uncovered considerable climate changes suggesting external forcing. Prior to the industrial era, the anthropogenic influence on the climate was, probably, very low, so we are basically left with solar and volcanic forcings. Since many palaeorecords show a relatively high correlation with the reconstructed solar activity, it appears that solar forcing dos indeed play an important role (Beer, 2005). But solar forcing is just one forcing factor-understanding climate change also required considering all the other forcing factors. A better knowledge of the natural forcing factors during the industrial era can lead to a better quantification of the anthropogenic forcing, and ultimately facilitate better predictions of the future climate change.

## Nuclear Energy and Climate Change

A major source of GHG's, particularly $CO_2$ is the fossil fuels burned by the energy sector. Energy demand is expected to increase dramatically in the 21$^{st}$ century, especially in developing countries, where population growth is fastest and where over 1.6 billion people have no access to modern energy services. In the absence of any worthwhile efforts to limit future GHG's emissions from the energy sector, therefore, the expected global increase in energy production and use could well destabilize the global climate.

With a view to reducing the risk of global climate change, industrialized countries have agreed to reduce GHG's emissions under the Kyoto Protocol (1997) as an addition to the 1992 United Nations Framework convention on Climate Change (UNFCCC). In this protocol, industrialized countries have committed to reduce their collective emissions during 2008-2012 by the least 5.2 per cent below 1990 levels.

In contrast to fossil fuels, nuclear power produces virtually no GHG's emission and is, therefore, an important part of future strategies to reduce GHG's emissions. Nuclear power is already contributing to the world's electricity needs. In 1999, it supplied more than one-sixth of global electricity and a substantial 30 per cent of electricity in Western Europe. Yet, despite this contribution, nuclear power's future role is uncertain.

In an increasingly liberalized electric power industry, return on the investment needed to build a new power plant is a critical factor in deciding which power technology to invest in. the high up-front capital costs for building new nuclear power plants, their relatively long construction time and payback period, and the lack of public and political support in some countries for new construction often make nuclear power a less attractive alternative than fossil fuelled power plants (IAEA, 2000).

The nuclear industry is making earnest efforts to reduce costs and increase political and public acceptance for nuclear power. The near absence of GHG's emissions from nuclear power further brightens its future competitiveness.

As of 1999, there were 433 nuclear power plants in operation around the world. They represented a total capacity of 350 gigawatts of electricity [GW(e)] and produced 16 per cent of the world's nuclear power capacity were in Western Europe, where they generated 30 per cent of the region's electricity supply.

In North America, 118 reactors provided 20 per cent of the electricity supply in the US and 12 per cent in Canada. In Eastern Europe and the Newly independent States, there were 68 nuclear power plants. There were 84 in the Middle East, South Asia, and the Far East, where a further expansion of nuclear power is being planned especially in China, India, and the Republic of Korea. Latin America and Africa account for less than 2 per cent of global nuclear electricity capacity (IAEA, 2000).

No new reactors are under construction in North America or in western Europe. In fact, Belgium, Germany, the Netherlands, and Sweden have plans to phase out nuclear power. In Austria, Denmark, Greece, Ireland, Italy, and Norway, national policy restrictions prevent the use of nuclear power. The main regions of the world where the use of nuclear power is expected to grow tin the short term are the Far East and South Asia. Energy demands will rise considerably, largely because of demographic and economic growth in developing countries. The latest median projection of the United Nations estimates an additional 4.4 billion people by the year 2100, an increase of almost 75 per cent relative to 1998 (U.N., 1998).

The IPCC Special Report on Emission Scenarios (SRES) provides a comprehensive picture of true energy needs (IPCC, 2000). The report presents a set of 40 scenarios developed as reference projections of GHG's emissions. The 40 scenarios reflect a broad range of different assumptions on population growth and economic development, environmental priorities, technological progress, and international cooperation on global energy use. However, none of the scenarios, intentionally includes any climate change policies. The SRES reveals that global primary energy use in these scenarios will grow from 1.7 to 3.7 fold between 2000 and 2050, with a median increase by a factor of 2.5 (Figure 7.4). Electricity demand grows almost 8 fold

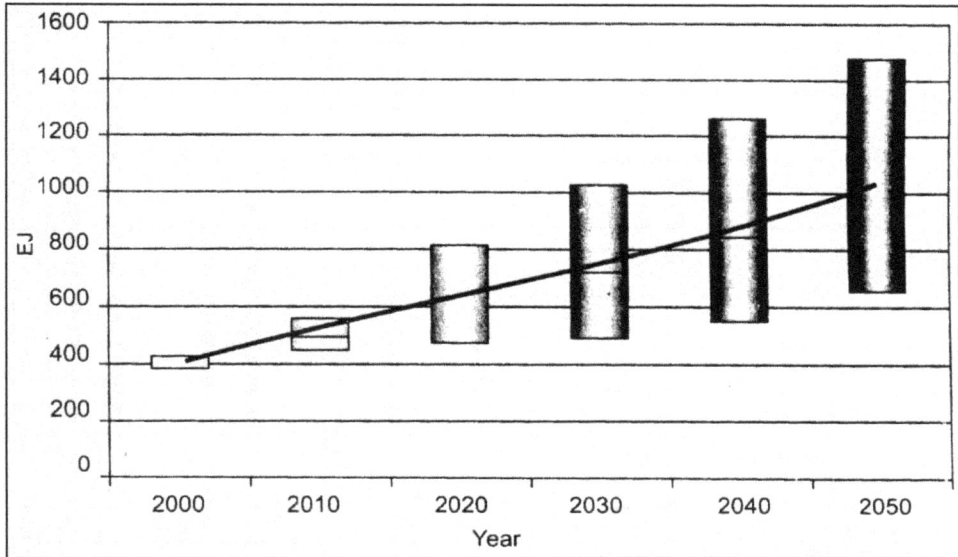

**Figure 7.4: Range of future primary energy demand in SRES scenarios, 2000-2050. Solid line represents median (after IPCC, 2000).**

in the high economic grown scenarios and more than doubles in the more conservation oriented scenarios at the low end of the range. The median increase is by a factor of 4.7.

Most of the scenarios also include substantial increases in the use of nuclear power (Figure 7.5). Thirty-five of the 40 scenarios report results clearly for nuclear power, not just 'non-carbon technology'. The projections for the year 2050 range between current capacity levels of 350 GW9e) up to more than 5,000 GW(e) (with a median of more than 1,500 GW(e). Such growth levels would require added global nuclear power capacity of 50 to 150 GW(e) per year from 2020 to 2050, even in the absence of any policies to reduce GHG emissions. Thy might well be higher if nuclear power generated more than just electricity (*i.e.* chemical fuel and desalination) (IAEA, 2000).

## The Second European Climate Change Programme

The European Commission launched the Second European Climate Change Programme (ECCP II) in Brussels in October 2005. The new programme aims to provide a new policy framework for European Union (EU) climate change policy, with a scope and perspecptive beyond the year 2012.

Two of the five groups under the programme are assigned the task of suggesting ways of reducing emissions form aviation and passenger road transport, respectively. A third group is to formulate proposals for geological carbon capture and storage, while the fourth will examine the best ways to adapt to climate change. The fifth group will analyse what has been achieved under the first European Climate Change Programme that was launched in 2000 and resulted in a list of 40 recommended,

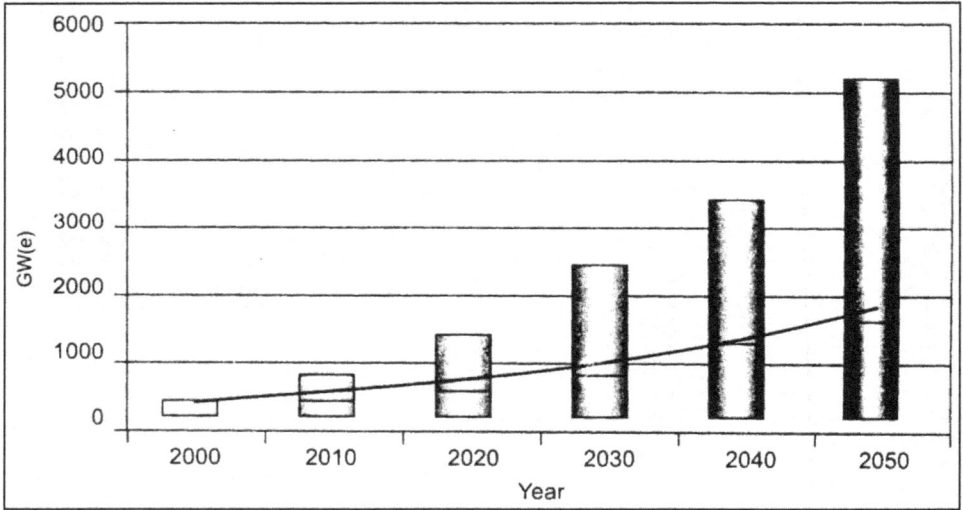

**Figure 7.5: Range of nuclear power in SRES scenarios, 2000-2050. Solid line represents median (after IPCC, 2000).**

cost-effective measures estimated to have a reduction potential of over 700 million tons of carbon dioxide, twice the reduction undertaken by the EU under the Kyoto Protocol. The EU would not be able to meet its Kyoto undertaking with existing measures, and much more remains to be done. The EU needs much greater reductions so as to limit the extent to which the European climate will change. In March 2005, the European Council suggested possible reduction pathways for developed countries in the order of 15 to 30 per cent by 2020-ECCP II can prove helpful in facilitating the task of achieving the set targets (Box 1).

Although commendable success has been recorded in some areas, the EU faces environmental challenges in terms of meeting its long-term environmental commitments, particularly in achieving its targets in respect of climate change and the use of renewable energy. The climate change issue has turned out to be the most problematic. It is intimately linked with a variety of human activities, including energy use, transport, and agriculture. It seems difficult to achieve the target of an 8 per cent reduction in GHG's emissions by the year 2010, to which the EU countries are committed under the Kyoto Protocol. The reduction that may be actually achieved could be as little as three per cent. Further measures must be implemented if the target is to be met.

Europe not only can phase out nuclear power but also simultaneously bring down its emissions by 30 per cent by the year 2020, half of Europe's energy demand could be supplied from renewable energy sources and $CO_2$ emissions could decline by nearly 75 per cent (see ITT, 2005). Should the EU fail to reform its energy sector, $CO_2$ emissions could increase by 50 per cent by 2050.

The electricity sector in the EU is dominated by large power plants use fossil and nuclear fuels. Some 75 per cent of Europe's primary energy supply comes from fossil

**Box 1: Clean air for Europe- the CAFÉ Programme
(Source http://europa.eu.int/comm/environment/air/cafe/
index.htm and Acid News No. 4 p. 13 Nov. 2005).**

The Clean Air for Europe (CAFÉ) programme was launched by the European Commission in 2001, with the aim of reviewing current air quality policies and assessing progress towards attainment of the EU's longterm air quality objectives, as laid down in the Sixth Environment Action Programme. CAFÉ has dealt with health and environmental problems related to fine particles (PM), ground-level ozone, acidification, and eutrophication.

CAFÉ has provided the analysis for the EU's thematic strategy on air pollution, which was adopted by the Commission in September 2005. the idea is that café should evolve into an ongoing five-year cyclical programme, in which the 2005 thematic strategy on air pollution simply marks the first milestone.

The activities of the programme include:

☆ Developing and collecting scientific information on the effects of air pollution, making inventories and projections of emissions and air quality, doing studies of cost-effectiveness and carrying out integrated assessment modeling-all leading to new and/or revised objectives in respect of air quality and pollutant deposition,and identifying the measures required for reducing emissions.

☆ Supporting the implementation of existing legislation and reviewing its effectiveness, especially in view of the directives on air quality and on national emission cellings, and developing new proposals for measures to abate emissions.

☆ Determining at regular intervals an integrated strategy to define appropriate air quality objectives for the future and cost effective measures for meeting those objectives.

☆ Disseminating the information emerging from the programme.

fuels. But many of these power plants are over 20 years old. So, the decisions taken now on new power production over the coming ten years will have a crucial role in Europe's energy system for a long time to come.

According to ITT, a clean energy future requires that governments fix legally binding targets for the use of renewable energy for power, heat and transport and they must assure renewable energy a high priority access to the grid, while also shifting investment away from fossil and nuclear fuels.

A very important first step is to remove all subsidies to fossil fuels and nuclear power. Only long-term solutions and investments are feasible for the power sector. While renewable energies are quite costly now, most of them will be cheaper in the coming years. These results can only be achieved in time, if we start this drastic shift in the power sector without further delay. In fact, a sincere commitment to renewable energy and efficiency improvement can lead to a whole string of benefits including climate protection and insulating national economics against the fluctuations of the global markets for fossil and nuclear fuels.

But it is not just the supply of energy that is crucial. A widespread commitment to energy efficiency is equally important. Proper exploitation of existing energy efficiency potentials such as the insulation of houses, the use of 'waste-heat' from power plants for district heating instead of discharging it via cooling towers and the efficient use of electricity can go a long way in reducing the current primary energy demand by more than one third by the year 2050.

## The Kyoto Challenge

The possibility of widespread climate change resulting from increased atmospheric concentrations if GHGs is now a major global concern. It aims to achieve stabilization of GHG concentrations in the atmosphere at a level that would prevent dangerous anthropogenic interference with the climate system. The emission limits, established in the 1997 Kyoto Protocol, are, however, only a first step towards that goal, and uneven progress has occurred towards implementing even this first step (Figure 7.6).

Among OCED countries, the greatest progress on limiting emissions has been recorded by the European Union (EU), where 1999 GHG emissions were only 0.4 per cent above their 1990 level (WCI, 2000). Yet, as per the 2008-2012 Kyoto commitment, the EU must actually reduce emissions to at least 8 per cent below 1990.

The situation is worse in other OECD countries. The US has to reduce its emissions by 7 per cent between 1990 and the commitment window, but through 1999, emissions actually rose by 12.7 per cent from their 1990 level (WCI, 2000). In Canada, emissions were up 12.4 per cent in 1999, compared to a required 6 per cent reduction. In Japan, the 1999 increase was 13.6 per cent, against a required 6 per cent reduction. Australia recorded a 15.4 per cent increase. Fortunately, countries with economics in transition are, by and large, still below their Kyoto limits. Emissions

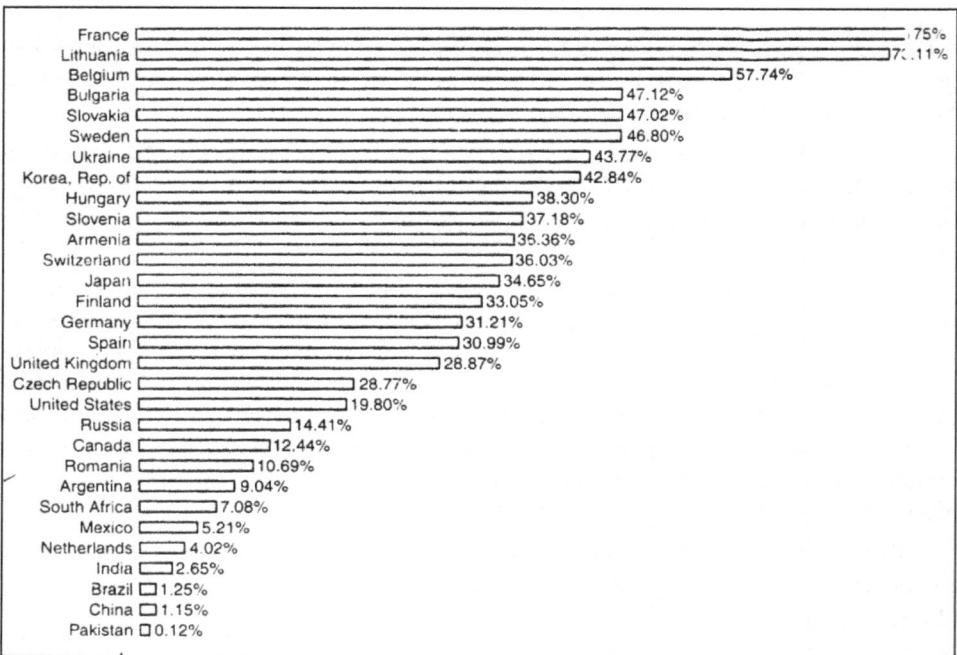

**Figure 7.6: Nuclear share of electricity generation as of April 2000.**

during 1999 from economics in transition fell collectively to 35.8 per cent below their 1990 level (WCI, 2000).

**Table 7.2: Nuclear power status around the world (after IAEA, 2000)**

| | Reactors in Operation | | Reactors Under Construction | |
|---|---|---|---|---|
| | No. of Units | Total Net MW(e) | No. of Units | Total Net MW(e) |
| Arjentina | 2 | 935 | 1 | 692 |
| Armenia | 1 | 376 | | |
| Belgium | 7 | 5712 | | |
| Brazil | 1 | 626 | 1 | 1229 |
| Bulgaria | 6 | 3538 | | |
| Canada | 14 | 9998 | | |
| China | 3 | 2167 | 7 | 5420 |
| Czech Republic | 4 | 1648 | 2 | 1824 |
| Finland | 4 | 2656 | | |
| France | 59 | 63103 | | |
| Germany | 19 | 21122 | | |
| Hungary | 4 | 1729 | | |
| India | 11 | 1897 | 3 | 606 |
| Iran | | | 2 | 2111 |
| Japan | 53 | 43691 | 4 | 4515 |
| Korea, | 16 | 12990 | 4 | 3820 |
| Lithuania | 2 | 2370 | | |
| Mexico | 2 | 1308 | | |
| Netherlands | 1 | 449 | | |
| Pakistan | 1 | 125 | 1 | 300 |
| Romania | 1 | 650 | 1 | 650 |
| Russian federation | 29 | 19843 | 3 | 3375 |
| South Africa | 2 | 1842 | | |
| Slovakia | 6 | 2408 | 2 | 776 |
| Slovenia | 1 | 632 | | |
| Spain | 9 | 7470 | | |
| Sweden | 11 | 9432 | | |
| Switzerland | 5 | 3079 | | |
| United Kingdom | 35 | 12968 | | |
| Ukraine | 14 | 12115 | 4 | 3800 |
| United States | 104 | 97145 | | |
| World Total* | 433 | 349063 | 38 | 31128 |

* This total includes Taiwan, china where six reactors totaling 4,884 MW(e) are in operation. Two units are under construction. Table reflects status as of April 2000 as reported to the IAEA.

For the Kyoto Protocol as well as for its successor agreements, the energy supply sector will bear a major part of the burden of reducing carbon emissions. This will require much ingenuity and innovation, and substantial contributions from all possible present and future mitigation options. In other words, in addition to end-use efficiency improvements, the current mix of energy supplies will have to change. Such restructuring will mean added costs.

To help active the Kyoto limits cost effectively, there are three flexible mechanisms in the Protocol: emission trading, joint implementation, and the clean development mechanisms (CDM). All are relevant to nuclear power. Annex I countries could gain credits toward their Kyoto limits by investing in nuclear power plants in other countries, if these investments meet the criteria being negotiated for joint implementation and CDM projects (IAEA, 2000).

Nuclear power generation produces almost no GHG emissions. The entire nuclear chain has among the lowest emissions per kilowatt-hour (kw.h) of any generating option including renewables. Countries with large nuclear and hydroelectric capacities have markedly lower $CO_2$ emissions per unit of energy than do those depending heavily on fossil fuels. Currently, nuclear and hydroelectric power each annually avoid GHG emissions equal to about 8 per cent of the total global emissions from fossil fuels; together, they avoid about 1.2 billion tons (Gigatonnes, GTC) of carbon each year that would otherwise have been produced through burning of fossil fuels (IAEA, 2000).

However, although nuclear power is no more able than other mitigation options to solely provide all the GHG reductions called for in the Kyoto Protocol, it can substantially contribute under a plausible scenario.

Nuclear power certainly carries the potential to bridge a major part of the gap between where emissions from Annex 1 countries are now headed, and where they should be in 2008-2012 under the Kyoto Protocol. If we take CDM into account, nuclear power's potential approximately doubles. And if the path charted by the Kyoto Protocol is to continue beyond the 2008-1012 commitment window, the potential importance of nuclear power can only grow.

## Pollution Transport by Convective Clouds

Cumules clouds greatly influence not only the vertical redistribution of tracers but also their subsequent dispersion. Rapid up draughts effectively vent the boundary layer and raise air pollutants upward by several kilometers in tens of minutes (Greenhut, 1986; Hauf *et al.*, 1995). Evaporatively cooled downdraughts can bring tracers from the middle troposphere to the surface levels. Also, associated precipitation can bring hygroscopic contaminants to the surface itself. These mechanisms lead to a more polluted boundary layer and/or surface (Stith *et al.*, 1996).

Modelling work facilitates realistic predictions of local air quality or long-range tracer transport. This applies also to the meteorological variables themselves-quantities such as potential temperature, total water content or potential vorticity often behave as conserved tracers themselves. Sometimes, the tracer is subject to sources and sinks through, for instance, precipitation scavenging, photochemistry or latent heating.

Then its advection also needs to be treated appropriately. For instance, the acidity of rainwater depends not only on the density of gases present but also on the amount of liquid water into which the gas dissolves and subsequently oxidizes (Gibson, 1997).

Gimson (1997) considered the effects of cumulus convection on the dispersion of passive tracers in the framework of a mesoscale model, where the subgrid-scale clouds were parameterized. The parametrisation used a 'mass-flux' scheme, where guidance on vertical fluxes and cloud cover is obtained from a cloud-resolving large-eddy model run, and applied to the one-dimensional time-dependent vertical transport model. He presented a set of three one-dimensional schemes, the simplest of which contains approximations appropriate to and commonly used in large-scale models. On the mesoscale, however, these approximations are no longer valid and hence a less idealized scheme is warranted. Comparison of the schemes revealed that they diverge as the mass flux or grid-box area is decreased. There was some sensitivity to mixing between the clouds and their surrounding environment, where differences between the schemes are decreased if entrainment dominates and increased if detrainment dominates at in cloud levels (Gibson, 1997). Gimson's results imply a necessity for the scheme with fewest approximations when the host code is at mesoscale resolution.

## Pollution and Global Change

In a sense, global change may be considered as nothing but the sum of all environmental pollution. Although human beings do not know how long a technological civilization can be sustained, the experiment currently in progress at the centre of global change is mostly one on global climate change. Enormous amounts of non-renewable fossil fuels are being consumed every day, mostly by industrialized countries and countries in cold northern climates, for heating in winter and cooling in summer. The number of vehicles (automobiles) powered by fossil fuels has galloped tremendously in recent years, and the human population has already crossed 7 billion. To sustain these activities and numbers, enormous amounts of pollutants are released into the biosphere every day. All these activities taken together lead to warming of the biosphere, weather extremes, loss of biodiversity and pollution of the global water, air and soil components of the biosphere. Indeed, virtually everything in our daily lives depends on some form of energy derived from non-renewable fossil fuels (Trevors, 2003).

The incidence and severity of droughts in some regions has been increasing. Droughts increase the dependence of previous highly fertile agricultural soils on intensive irrigation which requires both water (which may or may not be available) and costly fuels to run irrigation systems. Warming of previously cold waters often disperses infectious microorganisms into new environmental aquatic locations, where they previously could not grow, multiply and survive. This has ominous implications of far-reaching consequences for potable water usage for humans and for domestic and wild animals. The enhanced UV-B radiation resulting from ozone depletion in the stratosphere can limit the amount of time humans can safely spend outdoors. This has a strong impact on the length of time we may exercise outdoors. The sum of all environmental pollution may already have exposed us to global change pollution, which should attract the highest priority to be addressed and solved. All people everywhere will have to stop or minimize destruction of global resources and life.

## Aerosols in the Climate System

Anthropogenic aerosols are intricately linked both to the climate system and to the hydrological cycle. Whereas anthropogenic greenhouse gases (GHGs) reduce the emission of thermal radiation to space, thereby warming the surface, aerosols mainly reflect and absorb solar radiation (the aerosol direct effect) and modify cloud properties (the aerosol indirect effect), cooling the surface. The net effect of aerosols is to cool the climate system by reflecting sunlight. Depending on their composition, aerosols can also absorb sunlight in the atmosphere- this further cools the surface but warms the atmosphere (Kaufman *et al.*, 2002). These effects of aerosols on the temperature profile, along with the role of aerosols as cloud condensation nuclei, impact the hydrologic cycle through changes in cloud cover, cloud properties and precipitation.

Unravelling these feedbacks has proven difficult because aerosols assume many different shapes and forms, ranging from desert dust to urban pollution; also because their concentrations vary greatly over time and space. Accurate study of aerosol distribution and composition, therefore, requires continuous observations from satellites, networks of ground-based instruments and dedicated field experiments. Increases in aerosol concentration and changes in their composition, driven by industrialization and an expanding population, may adversely affect the Earth's climate and water supply (Kaufman *et al.*, 2002).

Carbonaceous aerosol particles may be considred to be the big unknowns of the global atmosphere. Even the concept of carbonaceous aerosol has just recently grown out of atmospheric pollution studies, without ever being anchored to solid nomenclature and terminology. No major breakthrough can be expected in resolving climate issues without a better understanding of the nature and atmospheric role of carbonaceous particles.

Aerosol effects on climate differ from those of GHGs in two other ways. Most aerosols being highly reflective, raise the Earth's albedo, thereby cooling the surface and effectively offsetting GHG warming by anywhere from 25 to 50 per cent (IPCC, 2001). But some aerosols containing black graphitic and tarry carbon particles (present in smoke and urban haze) are dark in colour and hence strongly absorb sunlight. The effects of this type of aerosol are two-fold, both warming the atmosphere and cooling the surface before a redistribution of the energy occurs in the column. Heating the atmosphere and cooling the surface below reduces the atmosphere's vertical temperature gradient and, therefore, may reduce evaporation and cloud formation (Hansen *et al.*, 1997).

The second difference rests on the aerosol on clouds and precipitation. In polluted regions, the numerous aerosol particles share the condensed water during cloud formation and reduce cloud droplet size, so increasing cloud reflectance of sunlight and cooling the Earth's surface. The smaller, polluted cloud droplets are inefficient in producing precipitation, so they may ultimately modify precipitation patterns in populated regions that are adapted to present precipitation rates (Kaufman *et al.*, 2002).

To assess the aerosol effect on climate one should first distinguish natural from anthropogenic aerosols., Satellite data and aerosol transport models reveal that plumes of smoke and regional pollution have distinguishably large concentrations of aerosols, in particular of fine (submicron) size. In contrast, natural aerosol layers may have concentrated coarse dust particles and only widespread fine aerosols from oceanic and continental sources (Chin et al., 2002). As satellites can observe the spatial distribution of aerosols and distinguish fine from coarse particles, this ability can be exploited to separate natural from anthropogenic aerosols.

Most aerosols are regional in nature owing to their short lifetime, the regional distribution of sources, and the variability in their properties. Seasonal meterological conditions determine how they vertically distributed through the atmosphere.

1. Urban haze is made of mainly fine hygroscopic particles that are found downwind or populated regions in air polluted, for example, by acar engines, industry cooking and fireplaces.

2. Smoke from vegetation fires is dominated by fine organic particles with varying concentrations of light absorbing black carbon emitted in the hot, flaming stage of the fire, In forest fires, the flaming stage is followed by a long, cooler smouldering state in which the thicker wood, not completely consumed, gives off smoke (composed of organic particles without black carbon) in larger quantities than during the flaming stage.

3. Dust is emitted from dry lakebeds in the Sahara, east Asia and the Saudi Arabian deserts that were flooded in the Pleistocene era (see Kauafman et al., 2002). Almost no dust is observed form Australia, when the topogrphy is flat, because the arid regions are old and highly weathered. An unknown amount of dust is emitted from disturbed soils in Africa and east Asia.

4. Oceanic aerosol is made of coarse salt particles emitted from bursting sea foam in windy conditions and fine sulphate particles from oceanic emissions.

The cooling influence of aerosols on climate, directly through the reflection of sunlight to space and indirectly through changes in cloud properties, has prompted many observations, simulations and analyses.

The effect of anthropogenic aerosols is not limited to cooling by sulphates. Instead, carbonaceous compounds that include light-absorbing black carbon cause warming and the sign of the temperature change from aerosols can vary depending on the aerosol' radiative properties and their distribution over the dark ocean and reflective land (see Kaufman et al., 2002). The cooling of the Earth's surface from absorbing aerosols (compared with the top of the atmosphere) and consequential warming of the atmosphere causes a flattened vertical temperature profile in the troposphere, which may slow the hydrological cycle, reduce evaporation from the surface and reduce cloud formation (Hansen et al., 1997).

## Aerosols and the Earth's Energy Budget

Aerosols directly influence climate by scattering or absorbing incoming solar radiation. They indirectly influence it by acting as nuclei on which clouds can form.

Together, these effects represent the largest uncertainty in our understanding of the Earth's energy balance. We badly need better understanding of both the aerosol-cloud interactions that determine the optical properties of clouds and affect precipitation, and of the processes that determine the amounts and optical properties of atmospheric aerosols themselves. It is equally important to measure the extent of human influence on aerosol and cloud properties at the global scale.

The best way to develop the necessary process of understanding is to employ both field and laboratory measurements and model investigations. Global scale measurements are best obtained using satellite-based observing systems, but satellite data, however valuable and necessary, cannot replace in situ process studies for understanding how aerosols affect climate.

## Soot

While most aerosols cool the atmosphere by increasing the Earth's reflectivity, aerosols containing black carbon (BC) or soot warm it by absorbing sunlight (Ackerman *et al.*, 200). The contribution of soot to global warming may be substantial, perhaps second only to that of $CO_2$ (Jacobson, 2001).

Although black carbon warms the atmosphere, it is not really a GHG. It is a solid which differs from GHGs by absorbing sunlight-GHGs cause warming by absorbing infrared or terrestrial radiation. Black carbon, therefore, has a slightly different climatic effect. Unlike most GHGs, black carbon, is short lived and its concentration varies greatly from urban-industrial areas to remote locations; so its effect on climate also varies spatially (Chameides and Bergin, 2002).

Black carbon is generated through incomplete combustion of biomass, coal, and diesel fuel. Unlike $CO_2$ emissions, black carbon emissions are largest in developing countries (Table 7.3) (on a per capita basis, black carbon emissions from China and the United States are roughly equivalent).

**Table 7.3: Carbon dioxide and BC annual emissions (after Chameides and Bergin, 2002).**

|            | USA   | Germany | Japan | China | India | Global |
|------------|-------|---------|-------|-------|-------|--------|
| $CO_2$(TgC)* | 5.576 | 983     | 1.285 | 3.749 | 991   | 26.939 |
| BC(TgC)    | 0.32  | 0.05    | 0.09  | 1.2   | 0.56  | 7.00   |

* $1Tg=10^{12}G$

## Primary Biological Aerosol Particles

Atmospheric aerosols have a key role in regulating the global climate. They can either enforce or suppress anthropogenic forcing. Although their influence on natural and anthropogenic climate forcing has been estimated, a better understanding of the composition and sources of atmospheric aerosols is essential to improve climate models, Jaenicke (2005) showed that particles injected directly from the biosphere make up a major portion of atmospheric aerosols. Cellular (protein) particles injected directly into the atmosphere can include dandruff, skin fragments, plant fragments, pollen spores, bacteria, algae, fungi, viruses, and protein crystals. These particles can range in size from tense of nanometers to millimeters. Our knowledge of 'dead' or fragmented biological particles in the atmosphere is very limited. Tropical forests could be a possible source and filter samples taken in Siberia at ground and aloft show-3 mg/m³ of protein: however, cellulose and protein make up only a fraction of primary biological aerosol particles (PBAPs).

The meteorological relevance of cellular particles may be high. Since pollen grains attract water at relative humidity below 100 per cent, they might act locally as cloud condensation nuclei (CCN) and a influence cloud formation. Other biological particles including, for example, decaying vegetation (and associated bacteria) are excellent ice nuclei which trigger precipitation and hence remove water from the atmosphere they might influence global cloud cover, climate forcing, feedbacks and precipitation distribution if the source and distribution of cellular atmospheric particles varies one regional to global scale (Jaenicke, 2005).

The biosphere has usually been considered to be a minor source of primary particles (Graedel and Crutzen, 1993). Bioaerosols were believed to occur only in very low concentrations, with insignificant global emissions for the year 2,000 [56 Tg/year of biogenic carbonaceous aerosols (>1 mm in size) compared with 3,300 Tg/year for sea salt and 2,000 Tg/year for mineral dust]. But, in recent years, a greater contribution to the atmosphere of particles from biological activities of the oceans has been reported (O'Dowd *et al.*, 2004). Still, some surveys have found as much as 20 to 40 per cent of the aerosol measured as compositionally unidentified (Jaenicke, 2005).

Jaenicke (2005) observed PBAP at several geographical locations and aloft, covering most seasons and many characteristic environments. He found a total absence of a pronounced annual cycle, despite the expectation that concentrations in spring or summer should be higher than in winter. The fractions of different biological compounds do vary, though: whereas in spring, pollen is more abundant, in winter, decaying cellular matter prevails. Further, resuspension from exposed surfaces acts as a source in winter and in dry periods.

Measurements (2001) in a tropical revealed that particles smaller than 1mm constitute up to 40 per cent, and particles larger than 1mm up to 80 pre cent, of the total aerosol number concentration of PBAPs (see Jaenicke, 2005). It now emerges that the biosphere is a major source for primary aerosol particles, and cellular (protein containing) particles make up a major fraction of the atmospheric aerosol.

## Measurements of Global Atmospheric Aerosol

Exploratory and monitoring observations of atmospheric aerosols over the last several decades have highlighted three features of those particulate systems: chemical composition and concentration, physical characterizations (primarily particle size distribution and number population) and cloud nucleating properties. The non-cloud aerosol is a major factor governing the amount of sunlight reaching the ground; however, most such observations are sporadic, for short periods of time and for only a few places on Earth. There are large variations in both the extensive and intensive properties of the aerosols, and these variations occur on many time and space scales, especially in the vertical direction. Despite this high degree of variability, aerosols occur globally and have regional to global effects, ranging from climate forcing (see Charlson *et al.*, 1999) to acidification of precipitation and modification of biogeochemical balances. Quantification of large-scale effects requires extension of observations from the regional to global scale.

Till recently, three disparate approaches have been adopted to study atmospheric aerosols: (a) in situ observations of micro-physical and chemical properties (b) long path/column and remote sensing of their influence on propagation of radiation in the atmosphere (including satellite observations); and (c) modeling on all spatial/ temporal scales, which depends on accurate and realistic knowledge from (a) and (b) or on certain assumptions. These different approaches have been fairly useful in the exploratory stages of research on atmospheric aerosols, but their applications to current scientific questions regarding climate forcing and other global-scale issues require an integrative strategy defined by the scientific issues themselves.

Simultaneous and coordinated use of all three approaches is essential for a complete description of these complex systems, particularly for understanding and quantifying global effects of aerosols. Accomplishing the needed integration requires consideration of the spatial/temporal variability of aerosols as well as the influences of and correlations with, the thermodynamic state (particularly the relative humidity). Relationlisation of the data from the three required approaches in the context of clearly stated scientific questions cannot be achieved without coordination (Charlson, 2001).

No single mode of observations or modeling can provide a complete or adequate integrated output. It is essential to coordinate in situ and satellite observations so that the resultant multivariate data sets strictly and demonstrably apply to the same exact air parcel, and so that the integrated data set includes all of the independent variables that control the column integral properties. This includes both the relevant extensive properties (EP)- such as scattering and absorption coefficients, lidar backscatter and species mass concentrations- and intensive properties (IP) (see Table 7.4) and their dependence on the thermodynamic state (TS), particularly RH.

Thermodynamic state (TS) has been defined as the aggregate of temperature and vapour pressure related quantities that influence phase changes, particle size, and refractive index in the multiphase aerosol system. The optical properties of any column or path within the atmosphere are determined by both the EP and IP of the aerosol particle substances and TS and both are functions of length along the path, such that measurements of one variable alone (whether at a point or over the path) cannot describe the whole path (Charlson, 2001).

The key variables to be measured (Table 7.4) are all parameters that are needed either for modeling climate forcing (see Charlson *et al.*, 1999) or for inputs into the retrieval of data from satellite borne instruments. Other scientific questions would require a different set of variables. It is usually not possible to utilize data acquired for one problem (*e.g.* health effects) for another scientific question (*e.g.* relationship to satellite data for climate forcing estimates) (Charlson, 2001).

**Table 7.4: Aerosol measurements for direct forcing of climate (after Charlson, 2001).**

| *Properties* | | |
|---|---|---|
| **Extensive Properties** | | |
| $\sigma_{sp}(1)$ $(m^{-1})$ | | Scattering component of extinction, scattering coefficient |
| $\sigma_{bsp}(1)$ $(m^{-1})$ | | Hemispheric backscatter coefficient |
| $\sigma_{ap}(1)$ $(m^{-1})$ | | Absorption coefficient* |
| m: | | Mass concentration |
| $M_1$: | | Species mass concentration (chemical composition as f(r)) |
| $\beta_{180}$ $(m^{-1}\,\sigma p^{-1})$ | | Lidar backscatter coefficient |
| **Intensive Properties** | | |
| a: | D log $\sigma_{sp}$/d long $\lambda$ | Wavelength dependence (angstrom exponent) |
| f (RH) | $\sigma_{sp}$ (RH)/$\sigma_{sp}$ (low RH) | Humidity dependence |
| B: | $\sigma_{bsp} V_{sp}$ | Backscatter ratio |
| $\omega$: | $\sigma_{ap}/(\sigma_{sp} + \sigma_{ap})$ | Single scatter albedo* |
| $\alpha_m$: | $\delta\sigma_{sp}/\delta m (m^2 g^{-1})$ | Mass scattering efficiency |
| $\alpha_1$: | $\delta\sigma_{sp}/\delta m_1$ (mg) | Species scattering efficiency |
| S (sr) | $(\sigma_{sp} + \sigma_{sp})/\beta_{180}$ | Lidar ratio |
| | | Batios of chemical components |

*: Most uncertain property.

# Aerosols as Source of Copper

One of the gravest problems of modern civilization is the degradation of air quality in urban areas of developing countries. Human activities release several gas and liquid/solid particles (aerosols) to the atmosphere which exert cumulative effects on the climate of the Earth and on the health of its life system. The north-western part of India suffers frequent dust storms in the summer months (Figure 7.7).

Yadav and Rajamani (2003) have determined the concentration of such heavy metals as Cu, Pb, Cr, V and Ni in three different size fractions, PM$_{10}$, suspended particulate mater (SPM) and dry-deposition dust. They recorded abnormally high concentration of copper in the aerosols.

Considerable dust-laden material is transported by SW-W summer winds in the that desert and adjoining regions. These storms usually deposit salty materials in the downwind direction, for example, on the quartzite ridge in the Delhi area (Yadav and Rajamanai, 2003). To characterize the aerosols of various size classes along the dominant SW-W wind trajectory, some sampling sites were chosen along Bikaner in Thar desert, Jhunjhunu and Delhi area.

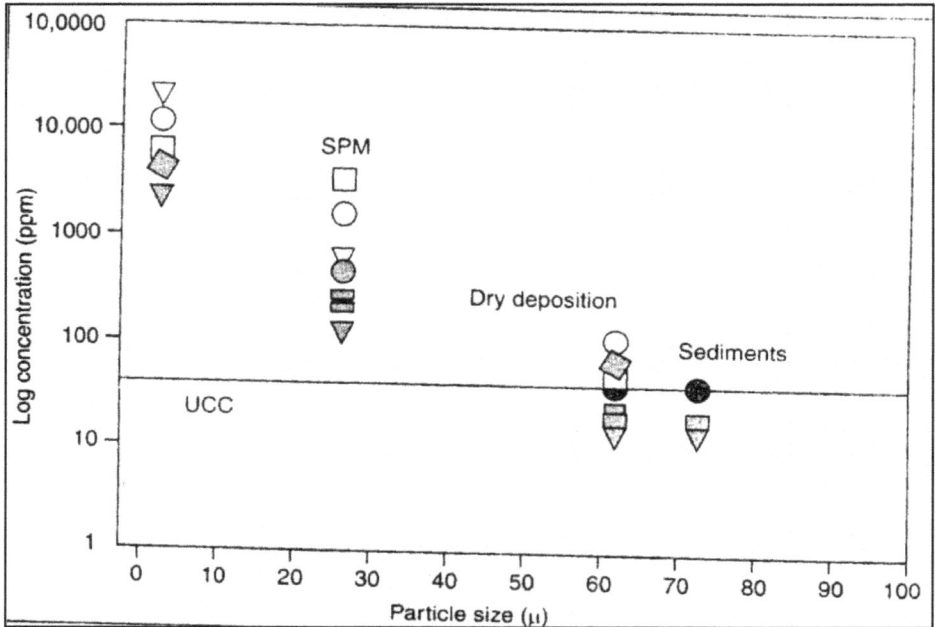

Figure 7.7: Variation in copper concentration as a function of particle size of aerosols in four sampling sites of NW India. Open and filled symbols represent summer and winter samples. Bikaner:; Jhunjhunu:; Delhi; Garhmuktesar: (after Yadav and Rajamani, 2003).

Different size fractions of aerosols were collected using standard sampling techniques. At each site and for each season, 5 to 7 samples of each size fraction were collected. The $PM_{10}$ size fraction everywhere had the highest copper concentration relative to the other size fractions. A general decrease in copper concentration with increasing particles size diameter was seen. Samples collected during winter had higher copper concentration relative to the summer samples. Site dependency of copper concentration relative to the summer samples. Site dependency of copper concentration was not very significant. Finally, the deposited surface sediments contained the lowest concentration of copper, even lower than that in the average UCC.

It appears that the anthropogenic source of copper in the aerosols of NW India is chiefly power sector-production, transmission and consumption. Worldwide copper ranks fourth, after Pb, Zn and Ni in terms of atmospheric emission from anthropogenic source. But, in their study, Yadav and rajamani, found that copper is the most abundant among the four trace metals in the aerosols.

Conceivably, such high level of copper could be related to its chemical affinity for organic phases found in large amounts in the atmosphere in this region (the so-called 'asian brown' cloud) (Srinivasan *et al.*, 1991).

The $PM_{10}$ dust samples studied by Yadav and Rajamani (2003) had a high content (6,000 and 15,000 ppm) of copper in the summer and winter seasons respectively.

These values, nearly 1.5 per cent copper in the aerosols, show their superiority event copper ores in respect of Cu content (any rock containing > 0.5 per cent copper is a potential copper ore today). The high copper levels in the reparable fraction of aerosols ($PM_{10}$) could cause gastrointestinal disturbances, including nausea, vomiting and liver or kidney damage depending on exposure time.

## Levels of Particulates in Europe

Every day we inhale about 10,000 liters of air which is supposed to be clean and, in fact, may not be so. A recent analysis for the Clean Air for Europe (CAFÉ) programme revealed that one type of air pollutant alone, namely fine particles, accounts for nearly 350,000 premature deaths each year in the EU25 (see Box 1). Besides, a large number of morbidity effects impact a much larger number of people.

Although fine particles are covered by European Union (EU) air quality legislation, and the air quality limit values for maximum allowed concentrations of this pollutant are not to be exceeded on more than a limited number of days anywhere in the Union; in reality, monitoring stations reveal frequently recurring accidences. This calls for urgent measures leading to effective integration of structural change in the sectors of energy, transport and agriculture into a coherent air pollution strategy (Agren, 2005).

Air pollution resulting from particulate matter (PM) claims an average of 8.6 months from the life of every person in the EU. The Germans lost even more: 10.2 months of life in the year 2000, according to a report presented by the WHO Regional Office for Europe at a press conference in Berlin on 14 April 2005.

In Europe, a new EU limit value for particulates in air has come into force which sets a maximum average threshold of 50 micrograms of particulate matter ($PM_{10}$) Per $M^3$ of air over a 24-hour period; this cannot be exceeded more than 35 days over the course of a year, as well as a limit value of 40 $\mu/m^3$ for the annual average concentration (Directive 1999/30/EC). But even by February 2005, several cities in Italy had reached their 35th day of excessive levels. By March and April, Budapest, Munich, Stuttgart, Berlin and dusseldorf had reached their 35th day and by May-June Leipzig. Hanover, braunschweig, Dortmund, Dresden Augsburg, Halle and Borna had reached the 35th day. In Sweden also the limit was passed in April in Stockholm and Gothenburg. Exceedances are also reported in some places in Belgium. At present, it is not clear how many cities have already exceeded the particulate limit set by the directive. In fact, the problem is quite widespread:

☆ In Thessaloniki (Greece), the limit value was exceeded on 219 days during 2003. The situation was also bad in Heraclion (184 days), Patra (173 days) and Athens (165 days).

☆ In the Czech Republic, several stations recorded more than 100 days of exceedance in 2003.

☆ In Lisbon (Portugal), the threshold was exceeded on 129 days in 2003.

☆ More than 100 days of exceedances were also recorded in 2003 in many other cities in Europe, including Paris (103 lays).

Several German cities have either already introduced, or are planning to soon introduce, restrictions on trucks and diesel cars using local roads with a view to combating rising air pollution. Dusseldorf plans to ban trucks from certain streets, and in case the problem still persists, it intends to make much of the city centres a no-drive zone for all vehicles without particle filters. The proportion of diesel cars in Germany has risen greatly in recent years. These cars now account for about half of one car sales. Many environmental organizations in Europe have evinced fairly low interest in using biofuels in the EU transport sector. The European Federation for Transport and Environment has taken the stand that if the goal is to reduce vehicle emissions of greenhouse gases (GHGs), then it is much more effective to legislate on fuel consumption standards for vehicles.

Under the Italian law, city mayors can attract liability for health threatening environmental damage. In several cities, driving is banned on Sundays in order to reduce air pollution. In Stockholm (Sweden), the measured levels of $PM_{10}$ are among the highest in Europe, but exhaust fumes from traffic are less of a problem. The chief problem is caused instead by somewhat larger particles that are abraded from the road surface by studded tyres during November-April. The city needs to improve road sweeping and to inform people about the drawbacks of studded tyres (Elvingson, 2005). London (UK) introduced a congestion charge in 2003. The charging system has successfully reduced traffic by around 15 per cent. It is possible that the calculated number of deaths caused by PM in the air has been greatly underestimated (see Forsberg *et al.*, 2005). In the case of Sweden, calculation based on commonly used risk coefficients for $PM_{10}$ and $PM_{25}$ showed that current concentration of long-range transported anthropogenic particles give rise to around 3,500 premature deaths each year, with a reduction in average life expectancy of around seven months.

The significance of local sources cannot be accurately estimated partly due to large variations in concentration and exposure. Nevertheless, recent estimates indicate that about 1,800 deaths are brought forward each year, with a life-span reduction in of about two to three months. However, it should be remembered that some parts of the population are, probably, exposed to much higher levels of the particles that several studies have shown to cause the most harm, namely those from combustion processes. This applied particularly to people who live near busy roads.

Current models indicate that long-range transported sulphate rich particles are the chief reason for the negative health effects of particles in most countries, which means that abatement strategies tend to focus primarily on these (Forsberg *et al.*, 2005). But, at the same time, some recent studies also indicate that engine exhaust particles may have a stronger health effect than anticipated when using the common risk coefficients for PM.

An Eu directive has set down limit values for $PM_{10}$ of 50 μg/m$^3$ for the annual average. Effective from 2005, the 24-hour average value should not be exceeded more than 35 times a year, a limit that had already been exceeded in many European cities.

By the year 2010, the reduction in health damage through implementation of current emission reduction legislation in the EU is expected to save 2.3 months of life for the EU population and 2.7 months of life for the population of Germany (see Acid News No. 2, June 2005). This is the equivalent of preventing 80,000 premature deaths each year and saving over one million years of life in the EU as compared to the situation in 2000. Besides, reducing long-term PM concentrations and exposure would bring important financial savings.

The World Health Organisation (WHO) has estimated the financial benefits of decreased mortality at between 58 and 161 billion euro per year, plus 29 billion euro for reduced diseases. The corresponding figures for Germany are 13 to 34 billion euro and 6 billion euro per year, respectively. There is evidence to suggest that PM in the air increases deaths from cardiovascular and respiratory diseases.

Even a short-term rise in PM concentrations enhances the risk of emergency hospital admissions for these causes. PM comprises tiny particles that vary in size, composition and origin. When inhaled, the coarse fraction ($PM_{10}$-particles with a diameter smaller than 10 μm) may rich the upper part of the airways and lungs. Fine particles ($PM_{25}$-with a diameter smaller than 2.5 μm) are more dangerous because they penetrate deeper into the lungs and sometime even reach the alveolar region.

No threshold concentration has been identified below which ambient PM has not effect on health. Although the reduction of PM levels to the EU limit values for 2005 will benefit health, it will by no means eliminate all significant health effects of PM exposure. Transport and use of fossil fuels in households are the primary contributors to PM air pollution. Diesel combustion contributes a third of total emission of $PM_{25}$ (see WHO fact sheet, www.who.dk/document/mediacentre/fs0405e.pdf)

Owing to the transboundary movement of PM, a substantial part of concentrations in a country originates in emissions from other countries. For example, while on average, about 40 per cent of $PM_{25}$ concentrations in Germany is of domestic origin, the rest is due to transboundary air pollution. In the same way, German emissions also contribute to $PM_{25}$ in other countries.

## Health Effects

Swedish researchers have demonstrated a direct link between air pollution and acute cardiac failure. A study in Stockholm revealed that a person who has lived by a busy street for 30 years has a 50 per cent higher risk of dying from an acute myocardial infarction than another person who has lived in the countryside.

This link has been further reinforced by correlating air pollution levels at every address in the city with the home addresses of a group of Stockholm residents who died as a result of myocardial infarction.

One possible theory to explain this link is that air pollution contributes to low level lung irritation, which in turn causes circulatory problems. The second is that air pollutants irritate nerve endings in the airways, which leads directly to heart arrhythmia (see an abstract entitled "Environmental factors in cardiovascular diseases", at http://diss.kib.ki.se/2005/91-7140-292-6/).

# Benefits from Emission Reductions

In the year 2000, air pollution accounted for nearly 370,000 premature deaths in the 25 member countries of the EU. Overall, the concentrations of fine particles have a much more important effect than ozone with respect to mortality-the former being responsible for 348,000 and the latter for 21,000 premature deaths (Agren, 2005).

Estimates of the total monetary damage from health impacts for the baseline scenario, *i.e.*, the benefits from current policies through the year 2020 are presented as an annual impact in million euro for the EU25 (Table 7.5).

**Table 7.5: Expected health damage due to air pollution in 2000 and 2010 in EU25 (billion euro).**

|  | 2000 Low Estimate | High Estimate | 2020 Low Estimate | High Estimate | Difference Low Estimate | High Estimate |
|---|---|---|---|---|---|---|
| O$_3$ mortality | 1.1 | 2.5 | 1.1 | 2.4 | 0 | 0.1 |
| O$_3$ morbidity | 6.3 | 6.3 | 4.2 | 4.2 | 2.1 | 2.1 |
| PM mortality | 190.2 | 702.8 | 129.5 | 548.2 | 60.7 | 154.6 |
| PM morbidity | 78.3 | 78.3 | 54.1 | 54.1 | 24.2 | 24.2 |
| Total | 275.8 | 789.9 | 188.8 | 608.9 | 87.0 | 181.0 |

*Note*: The result are based on 1997 meteorological data. For acute mortality (O$_3$), two alternative values are presented, based on a range reflecting the median and mean values. For chronic mortality (PM), two alternative values are presented, based on value of life years lost (VOLY) and numbers of premature deaths, the latter using the mean of a statistical life (VSL) value (after Agren, 2005).

Significant reductions in concentrations and impacts are expected over the period 2000 to 2020, especially for fine particles. The annual health benefits of implementing current legislation up to 2020 are valued at between 87 and 181 billion euro for the year 2020. This translates to an estimated annual average benefit across the EU of 191 to 397 euro per person.

Two additional types of air pollution impact have been quantified in economic terms: the effects of crop yield and the damage to modern buildings. For the year 2000, this damage was valued at 2.8 and 1.1 billion euro, respectively.

Emission of CO$_2$ from new cars in the EU continue to fall, but are still a long way from what is actually needed to meet the target of 140 km by 2008-09. Emission fell or average by 1.8 per cent in 2004, to a level of 160 g/km. Annual 3.3 per cent improvements are needed to meet the target of 140g/km. annual 3.3 per cent improvements are needed to meet the target of 140 g/km by 2008/09.

The European Federation for transport and Environment, T&E, reiterated its call for obligatory consumption standards to replace the voluntary agreement, and said that there is now support from the Commission for forcing car makers to limit emissions from new cars to 120 g/km, as this is justifiable on economic and social grounds. It appears that the costs are not too high and that benefits for society appear to outweigh costs.

The EU intends to establish a cap-and-trade system to limit emissions of $CO_2$ from large industrial sources. This so-called Emissions trading System became effective on 1 January 2005. Companies must not emit more $CO_2$ than their allowances. If they cut emissions more, they can sell their surplus permits on the open market. If they pollute more, they must buy them. The idea of the trading system is that emissions are cut where it is cheapest.

The trade in $CO_2$ emission rights has been modest so far but it is growing steadily. The price has risen from around seven euro per emission right at the start of the year to nearly 20 euro by the end of May 2005. The EU emissions trading system is not expected to result in any substantial investments in cleaner processes and plants over the next three years because of the relatively generous levels of free allowances being granted by governments for the first period. Stronger caps are, however, needed for the period 2008 to 2012, when the trading system runs in parallel with the Kyoto Protocol (Elvingson, 2005).

## EU 2020 Greenhouse Gas Target

The EU member countries have set a target for avoiding global warming for the period after 2012. at a meeting in Brussels in March 2005, the EU environment ministers stated that developed nations should aim for cuts in the order of 15 to 30 per cent by 2020 and 60 to 80 per cent by 2050, compared to the baseline contemplated in the Kyoto Protocol. No targets were set for the EU itself.

At their meeting on 22-23 March, the EU Heads of State (European Council) backed the environment council's calls to aim for a 15 to 30 per cent reduction in emissions from industrialized countries by 2020. But they dropped the target of a 60 to 80 per cent cut by 2050, possibly because of pressure from Germany and Austria.

The Heads of State also endorsed- for the first time- the goal of keeping global temperature rise below 2°C over pre-industrial levels. 'Despite the recent rejection of GHG reduction targets by former climate champions Germany and Austria, and the continued sabotage efforts of Italy, the EU has stood strong and sent a clear political signal that it will push for further action. Much more is urgently needed, but this is a good start". Stated Mahi Sideriodou, the EU Climate Policy Director at Greenpeace (see Elvingson, 2005). The EU now needs not only to speedily implement policies and measures to not only fully comply with Kyoto targets but also to comply with this longer term objective of 2020.

## Acidifying Pollutants–Global Emission Trends

Although emissions of sulphur and nitrogen oxides are falling in Europe, they are rising in Asia. Emissions of sulphur dioxide have fallen since 1990, not only in Europe, they are rising around the world, and a continuing reduction is expected in coming years. This conclusion of a new analysis by Cofala *et al.* (2005) contradicts many previous assumptions of a sharp rise in emissions in the future.

The critical factors in the forecasts of future emissions are expected economic activity (primarily energy use) and emission control requirements. The report give two scenarios for the period 1990-2030.

The 'current legislation' (CLE) scenario reflects the current perspectives of individual countries on economic development and duly considers the anticipated effects of already agreed emission control legislation. The 'maximum technically feasible reduction' (MFR) scenario outlines the scope for emission reduction offered by a full implementation of the best available emission control technologies (Elvingson, 2005).

When national expectations for the development of energy demand within the next 30 years are compiled, they show an increase of global energy-related annual emissions of $CO_2$ of about 4.4 billion tons of carbon in 2030 (current annual emissions are around 6 billion tons). (The study concentrated on national emissions, so the emissions from international shipping and aviationwere not included.).

## Sulphur Dioxide

Calculations made by the IIASA showed global $SO_2$ emissions in 1990 to be about 122 million tons. In 2000, the emissions were about 20 per cent lesser, mainly due to strict controls implemented in Western Europe, but also due to economic restructuring in Central and Eastern Europe and in Russia and the Newly Independent Sates (Table 7.6 and Figure 7.8).

**Table 7.6: Emission of sulphur dioxide by regions: Current legislation(CLE) and maximum technically feasible reduction (MFR) scenarios. Million tons $SO_2$. Emissions from biomass burning, international shipping and aircraft not included (after Elvingson, 2005).**

|  | Emissions | | CLE Scenario | | | MFR Scenario | | |
|---|---|---|---|---|---|---|---|---|
|  | 1990 | 2000 | 2010 | 2020 | 2030 | 2010 | 2020 | 2030 |
| Western Europe | 17.9 | 7.9 | 3.8 | 3.1 | 2.9 | 1.4 | 1.3 | 1.2 |
| Central and Eastern Europe | 11.1 | 5.9 | 4.1 | 2.6 | 2.3 | 0.8 | 0.6 | 0.6 |
| Newly Independent States | 19.5 | 11.1 | 7.8 | 6.0 | 6.3 | 1.8 | 1.7 | 1.7 |
| Centrally Planned Asia[1] | 22.0 | 28.4 | 30.9 | 31.1 | 29.4 | 6.7 | 6.6 | 6.4 |
| South Asia | 4.8 | 7.6 | 11.0 | 17.3 | 22.5 | 1.9 | 2.6 | 3.3 |
| Pacific OECD | 2.7 | 2.6 | 2.8 | 2.1 | 1.5 | 0.6 | 0.5 | 0.5 |
| Other Pacific Asia | 5.1 | 4.3 | 5.4 | 6.9 | 8.7 | 1.5 | 1.8 | 2.0 |
| North America | 24.4 | 18.5 | 16.4 | 15.9 | 17.5 | 2.9 | 3.2 | 3.3 |
| Latin America and Caribbean | 6.7 | 6.2 | 6.8 | 5.8 | 5.3 | 1.8 | 1.8 | 1.7 |
| M. East and N. Africa | 3.1 | 5.0 | 3.5 | 2.8 | 2.4 | 0.8 | 0.7 | 0.7 |
| Sub-Saharan Africa | 4.8 | 5.4 | 4.9 | 5.2 | 5.8 | 1.3 | 1.3 | 1.3 |
| World Total | 122 | 103 | 97 | 99 | 104 | 22 | 22 | 23 |

1 Including China.

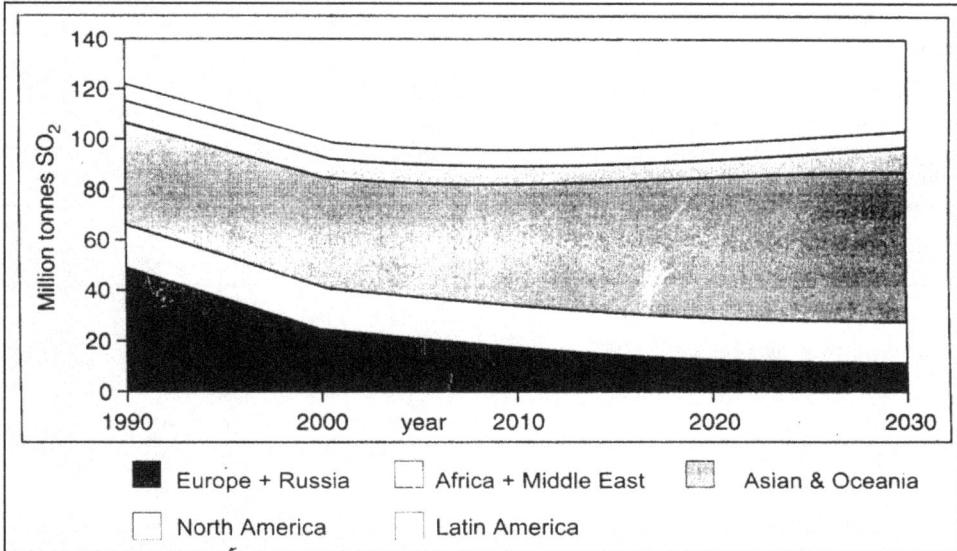

**Figure 7.8: Global emissions of sulphur dioxide (after Elvingson, 2005).**

Under the baseline assumptions (CLE), global emissions will fall to about 97 million tons in 2010 and then increase again to 105 million tons by 2030 unless stricter controls are enforced in South-East asia. If the best available control technology (MFR) is successfully implemented, the emissions might be reduced to about one-fifth of the 1990 level by 2030 (Elvingson, 2005).

Future development of emissions is strongly region-dependent. In Pacific OECD. Russia and the Newly Independent Sates, Middle East and North Africa, central and Eastern Europe, and Western Europe, emissions are likely to decrease by 40 to 60 per cent. The decrease in North and Latin America may be only about 5 to 15 per cent.

The emissions from Centrally Planned Asia and China stabilize at a level close to year 2000 emissions. However, if policy targets from the recently announced pollution control plans in China materialize, the resulting emissions might significantly decline. National projections anticipate a substantial increase in the consumption of coal in many countries in South and Pacific Asia, where the sulphur dioxide emissions are expected to increase greatly.

## Nitrogen Oxides

The calculations reveal no change in global emissions between 1990 and 2000 where after the CLE scenario predicts a slight reduction until 2010, followed by rising emissions (Table 7.7 and Figure 7.9). The rise is significantly lower than in previous estimates-many developing countries could enforce stricter control requirements for mobile sources (Elvingson, 2005).

The moderate increase expected in emissions from developing countries would be partly offset by the decrease in European emissions, so that global anthropogenic emissions would emissions would grow by no more than 13 per cent up to the year 2030.

**Table 8.7: Emissions of nitrogen oxides by regions: Current legislation (CLE) and maximum technically feasible reduction (MFR) scenarios. Million tons NO$_2$. emissions from biomass burning, international shipping and aircraft not included (after Elvingson, 2005).**

| | Emissions | | CLE Scenario | | | MFR Scenario | | |
|---|---|---|---|---|---|---|---|---|
| | 1990 | 2000 | 2010 | 2020 | 2030 | 2010 | 2020 | 2030 |
| Western Europe | 14.1 | 10.8 | 7.5 | 6.0 | 6.1 | 2.8 | 3.0 | 3.1 |
| Central and Eastern Europe | 3.5 | 2.8 | 2.0 | 1.7 | 1.8 | 0.6 | 0.6 | 0.7 |
| Newly Independent States | 11.2 | 7.2 | 6.7 | 5.9 | 6.7 | 1.5 | 1.6 | 1.8 |
| Centrally Planned Asia[1] | 7.8 | 12.3 | 13.8 | 15.0 | 16.2 | 3.7 | 4.2 | 4.5 |
| South Asia | 3.1 | 5.4 | 7.6 | 9.7 | 11.6 | 1.9 | 2.4 | 2.9 |
| Pacific OECD | 3.7 | 3.7 | 3.4 | 3.2 | 2.8 | 1.2 | 1.1 | 1.1 |
| Other Pacific Asia | 3.5 | 5.6 | 5.8 | 6.9 | 8.2 | 1.6 | 2.0 | 2.4 |
| North America | 23.4 | 19.9 | 18.3 | 20.8 | 22.2 | 5.7 | 6.3 | 6.8 |
| Latin America and Caribbean | 5.5 | 6.4 | 6.0 | 5.8 | 6.3 | 1.3 | 1.5 | 1.7 |
| M. East and N. Africa | 2.6 | 3.3 | 2.6 | 2.8 | 3.1 | 0.7 | 0.8 | 0.8 |
| Sub-Saharan Africa | 2.6 | 3.7 | 4.8 | 6.7 | 1.1 | 1.2 | 1.4 | |
| World Total | 81 | 81 | 77 | 82 | 92 | 22 | 25 | 27 |

1 Including China.

Implementation of the best available control technology (the MFR scenario) could reduce these emissions to about one tired of the 1990 level by 2030 (Elvingson, 2005).

Like sulpher dioxide, the future development of NOx emissions is strongly region-dependent; a strong decline in Europe and stabilization in North America due to present emission control legislation, despite the underlying economic growth and the corresponding increase in transport volumes.

For Asia, a growth in transport demand by a factor of four to five is anticipated.

## Carbon Monoxide

Carbon monoxide contributes to the formation of ground-level ozone at the hemispheric scale. As per the CLE scenario, global emissions will fall by around 15 per cent between 2000 and 2030, despite increased economic activity. As for regional differences, the highest decline (55 per cent) occurs in Latin America, chiefly due to a switch from fuel wood to other energy sources in the residential sector. The only region with increasing emissions is Africa, with a rise of 10 per cent. This results in a global decrease in the anthropogenic emissions of 16 per cent from 1990 to 2030 under the CLE scenario.

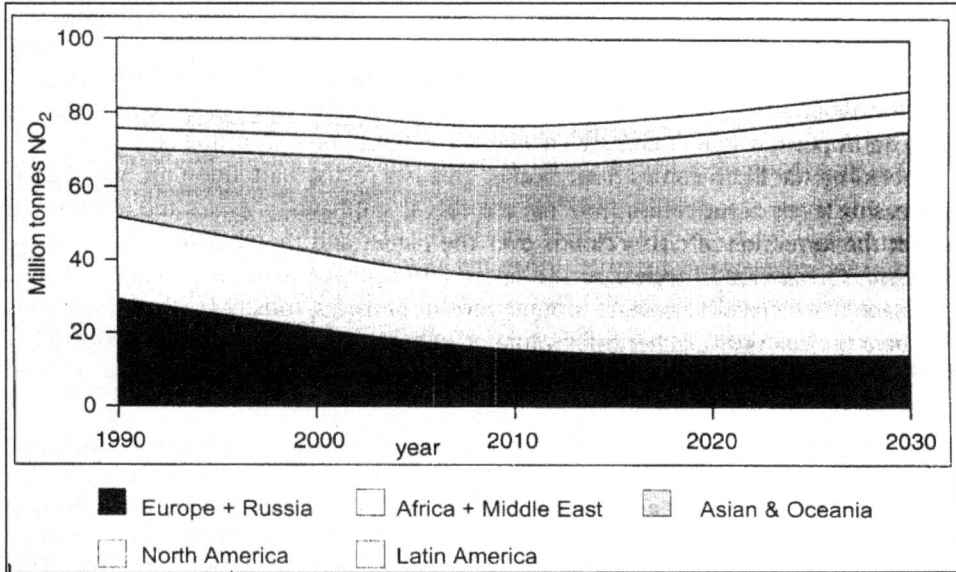

Figure 7.9: Global emissions of nitrogen oxides (after Elvingson, 2005).

## Methane

Like carbon monoxide, methane also contributes to the formation of ground level ozone at the hemispheric scale. Methane is also a greenhouse gas. 'current legislation' (CLE) will result in a continued increase in global anthropogenic emissions, leading to 35 per cent higher emissions in 2030 than in 2000. Overall, emissions from all sectors are expected to grow due to increased economic activity and absence of widespread emission control measures.

## Emission Control

Implementation of the best available control technology (MFR) would facilitate further reducing the emissions of all these pollutants. Compared with 2000 levels, the achievable reductions are 77 per cent for $SO_2$, 67 per cent for NOx, 53 per cent for CO and 6 per cent for $CH_4$.

## Cirrus Clouds and Jet Contrails

The formation of cirrus clouds which warm the Earth is not well understood, but they play an important role in climate change. More than one half of the solar radiation reflected by the Earth comes from clouds. This has focused attention on how might increasing levels of such anthropogenic aerosols as sulphates, organics and trace gases affect the formation of cirrus clouds over the planet and thereby alter its radiation balance. Cirrus clouds appear in the upper troposphere when super cooled water droplets or ice crystals nucleate around aerosol particles, mostly $H_2SO_4$ droplets. At temperature below 0°C air becomes saturated with respect to water and supersaturated with respect to ice. If an ice crystal is formed in a cloud other that has mostly super cooled water, it will grow much more rapidly tan its super cooled-water neighbours. As ice crystals grow more and more rapidly, they lower the water vapour pressure below saturation relative to supercooled water, resulting in the rapid transfer of water from droplets to crystals. In this way, the available water tends to become concentrated in a small number of large ice crystals (Seinfeld, 1998).

Human-induced emission of aerosols can raise droplet concentration in clouds by increasing the number of particles that act as cloud condensation nuclei. That promotes scattering and hence cloud-top reflection, so lowering the amount of energy absorbed by the Earth. The magnitude of this effect on the world's radiation budget is highly uncertain, and estimated to be between 0 and $-1.5$ Wm$^{-2}$ (for ready reference, the effect of anthropogenic GHGs is about $+2.4$ Wm$^{-2}$). The uncertain by is due to our poor understanding of how aerosols affect the nucleation of different types of cloud (Seinfeld, 1998).

Low-level stratus clouds cool the Earth because they reflect more of solar radiation back to space than they absorb rising infrared radiation. In contrast, cirrus clouds-ice-particle clouds that he between 10 to 20 km above our head exert a general heating effect, because their reflectivity at solar wavelengths is low, and they absorb more rising infrared radiation than they emit to space from their very cold cloud tops. This promoted the possibility that if anthropogenic gases and aerosols alter the properties and lifetimes of clouds, they might exert positive or negative radiative effect, depending on the extent to which low- and high-level clouds are affected. It cannot yet be predicated how increasing levels of trace gases and particles will influence the properties and lifetime of any type of cloud. However, a kind of laboratory to study how this happens is provided by jet aircraft.

Jet contrails are analogous to artificial cirrus. When jet planes emit oxidized sulphur, soot, metal particles and nitrogen oxide, contrails form as the hot, humid exhaust gases mix with the cold, dry air. Soot and sulphuric acid particles in the exhaust activate into water droplets that letter freeze. Sulphur and soot emissions from aircraft are quite small compared with global sources; but in the relevant altitude range of emissions (10-12km), aircraft sources are comparable to natural and anthropogenic sources (NASA, 1995). So aircraft emissions may affect cirrus formations. Although, being small individual contrails cannot appreciably affect the Earth's radiation balance, if large areas of persistent cirrus are produced from contrails, then jet exhaust can exert a significant effete on climate. Satellite observations of contrails have shown that persistent contrails can indeed evolve into extensive cirrus clouds that otherwise would not have formed. A considerable amount of ice-supersaturated air seems to exist in the upper troposphere that can produce cloud in the presence of nucleating aerosols.

The nitrogen oxides emitted by jet planes may also promote cirrus formation. Sulphate happens to be the predominant aerosol in the upper troposphere, but as much as 25 per cent by mass is other species, such as crystal compounds, metal, soot, ammonium and nitrate. These particles (less than 0.3mm in diameter, see Strom *et al.*, 1997) may act as cirrus cloud condensation nuclei (CCN).

# Fog

Fog is formed when moist air comes into contact with a cool surface. The cool air condenses into visible water droplets. Fogs form after radiative cooling, when the ground loses heat rapidly at night in still conditions, and through advection, when moist air intrudes over a cold surface. Radiation fogs usually lift after daylight when the Sun heats the moist air, but advection fogs typically clear after a change in wind direction, with the arrival of a drier air mass. Dense fogs over airports reduce visibility, thereby disrupting take-off and landing of aircraft.

## Sulphuric Acid Particles

Atmospheric aerosols have figured prominently in the past three decades to identity potential human influence on climate (see Wilson, 1971). Their mass balances are appropriate for mass-dominated sources of primary aerosol particles, such as the Earth's crust or the sea surface, as well as of their influence on processes controlled by particulate mass, such as iron fertilization of southern oceans (Watson *et al.*, 2000). However, such crucial processes as cloud formation and certain health effects are regulated by the number of aerosol particles, including secondary particles (see Pruppacher and Klett, 1978).

The source processes controlling the number of secondary particles in the atmosphere are much more complex than those of the primary particles:

1. A host of gas-phase particle precursors with their respective formation processes is involved.
2. Particle nucleation from the gas phase probably comprises several species concurrently, including ions; and
3. Experimental techniques to size, quantify, or chemically specify the newly formed particles *ab* initio, are lacking (Berndt *et al.*, 2005).

Sulphuric acid is a gas-phase particle precursor. It is an important atmospheric nucleating species. The paucity of data on binary sulphuric acid nucleation rates with water precludes explanation of atmospheric nucleation (Ball *et al.*, 1999).

Ternary nucleation involves the additional ubiquitous species ammonia and has been invoked on theoretical grounds as an explanation for observed particle nucleation in the atmosphere (Korhonen *et al.*, 1999).

Some other possible pathway for atmospheric particles formation include nucleation processes involving iodine oxide for coastal areas, ion-induced nucleation, and nucleation assisted by products from the atmospheric oxidation of aromatics (Lee *et al.*, 2003; Zhang *et al.*, 2004).

Berndt *et al.* (2005) have studied the formation of new particles in a laboratory study, starting from sulphuric acid ($H_2SO_4$) produced in situ through the reaction of OH radicals with sulphur dioxide. Newly formed particles were observed for sulphuric acid concentrations above $7 \times 10^6$ per $cm^3$. At 293 kelvin, a rough estimate yielded a nucleation rate of 0.3 to 0.4 particles per cubic centimeter per second for approximately $10^7$ particles per $cm^3$ of sulphuric acid (particle size=3 nanometres).

These results show that under laboratory conditions, similar to the atmosphere, particle formation occurs at atmospheric sulphuric acid concentration levels.

The Kyoto Protocol has already come into force. The governments and people of the world have rightly decided than a do-nothing approach to a changing climate is not acceptable. Governments are obliged, in the five year period from 2008 to 2012 to equal their 1990 GHG emissions or assume responsibility for any excess over the 1990 level. This responsibility obligates nations either to buy emission units on the international market from those countries that have exceeded in achieving their target or to offset the defaulting country's emissions with young forests that will effectively absorb carbon as they grow. With a view to complying with the requirements of the Protocol, some governments have already decided to introduce a carbon tax in 2006-2007 to help change the current pattern of rising GHG emissions. The tax may be levied on fossil fuels and on some GHGs emitted by certain industrial processes.

All governments will have to work on a system for carbon monitoring and accounting. Such an accounting system is essential if sink credits from forests have to be used to offset a country's GHGs emissions.

With a view too making international action effective in slowing the rate at which the climate changes, countries need to look at their own backyard on a high priority basis. Because changes in rainfall and more extreme weather events such as drought and floods can be caused by changing climate patterns, any neglect or delay in looking at the backyard can prove detrimental to the interests and future of the nation.

## The Scientific Consensus

It is sometimes asserted that climate science is highly uncertain and, therefore, strong measures need not be adopted. Some big companies whose revenues might be adversely affected by controls on carbon dioxide emissions, claim that there are significant uncertainties in the science (see van den Hove *et al.*, 2003). But many such claims, allegations and arguments are baseless-in fact, most scientists agree that human activities are heating the Earth's surface (se Oreskes, 2004). The scientific consensus has been explicitly stated in the reports of the intergovernmental Panel on Climate Change (IPCC). The purpose of IPCC is to evaluate the state of climate science as a basis for informed policy action.

Mc Carthy *et al.* (2001) noted that the consensus of scientific opinion is that the Earth's climate is being affected by human activities which have been modifying the concentration of certain atmospheric constituents that absorb or scatter radiant energy. Much of the observed warming over the past half a century appears to have been due to the increase in GHG concentrations. Indeed, many another scientific bodies in the United States and elsewhere have also issued similar statements (see NAP, 2001).

The NAP points to continuing accumulation of GHGs in the Earth's atmosphere as a result of human activities, causing surface air temperatures and subsurface ocean temperatures to rise. The NAP also supports the conclusions and assessment of the IPCC. The American Meteorological Society (2003) and the American Geophysical Union (2003) have also concluded that the evidence for human modification of climate is compelling. Jones and Moberg (2003) reported that global surface temperature increased in the past century by >0.5°C and the warming was at least partly a result of anthropogenic climate-forcing agents (Houghton *et al.*, 2001).

A climate forcing refers to an imposed (natural or anthropogenic) perturbation of the Earth's energy balance with space. Increasing anthropogenic GHGs, as well as aerosols cause large positive forcing. Continued growth of GHGs is likely to provide the major global anthropogenic climate forcing in the 21st century, because the levels of aerosols rise in some places but fall in others.

The thermal inertia of the climate system, mainly due to the ocean, tends to delay the global response to forcing. This inertia results in some additional global warming 'in the pipeline' due to gases already in the atmosphere. The climate system appears to be out of energy balance by 0.5 to 1 Wm$^{-2}$, *i.e.* solar energy absorbed by the Earth exceeds outgoing thermal radiation by that amount (Hansen and Sato, 2004).

According to Hanson and Sato (2004), feasible reversal of the growth of atmospheric methane and other trace gases can help in averting dangerous anthropogenic interference with global climate. Such trace gas reductions can potentially allow stabilization of atmospheric carbon dioxide at an achievable level of anthropogenic carbon dioxide emissions, even if the added global warming constituting dangerous anthropogenic interference is as small as 1°C. A decline of non-$CO_2$ forcing allows for climate forcing to be stabilized with a higher transient level of $CO_2$ emissions. Global warming is likely to enhance 'natural' emissions of $CO_2$, $N_2O$ and $CH_4$ - emissions that are an indirect effect of all climate forcings and are small as compared to human-made climate forcing; they occur on timescale of a few centuries, but they tend to aggravate the task of stabilizing atmospheric composition (Hansen and Sato, 2004).

## Planetary Energy Balance

As early as 1827, Jean-Baptiste Joseph fourier had discussed the problem of planetary temperatures. He introduced planetary temperature as a proper object of study in physics. He established the framework of energy balance that is still being used today: a planet obtains energy at a certain rate form various sources, and warms up until it loses heat at the same rate. He correctly inferred that a planet loses heat virtually exclusively by infrared radiation and can do so in a vacuum.

Concerning the Earth's heat source. Fourier correctly deducted that the internal heat remaining from the formation of the Earth no longer has a significance off surface temperature. He knew that sunlight carriers heat, that the atmosphere is essentially transparent to sunlight, that the light is converted to infrared when it gets absorbed by the surface and that the atmosphere is relatively opaque to the infrared which moves the received heat away to space (see Pierrehumbert, 2004). He also assumed correctly that the temperature must increase (compared with the no-atmosphere case) to allow enough infrared radiation to balance the heat budget.

Fourier quite correctly grasped the meaning of the greenhouse effect-the principle of energy balance and the asymmetric effect of the atmosphere on incoming light versus outgoing infrared. However, he did not comprehend some phenomena, for example, the role of convection in causing atmospheric temperature to decrease with height, the importance of this decrease in reducing the mean temperature at which the planet radiates to space, the role of minor atmospheric constituents (for example, carbon dioxide and water vapour) in determining the infrared opacity quantum theory relating to infrared absorption and emission, the dynamic nature of water vapour and its consequent radiative feedback, and optical and microphysical properties of clouds (see Pierrehumbert, 2004).

In fact, we still do not understand much about vertical temperature gradient, and water vapour and clouds, in relation to global climate.

While Fourier did correctly deduce a few things, he was quite wrong in some other ways. According to him, the Earth receives a significant amount of heat directly from interplanetary space, which he thought had a temperature comparable to that of the polar winter. In arriving at this conclusion, he rejected certain alternative explanations: thermal inertia and atmosphere-ocean heat transport, which keep the poles and the night warm without any need to invoke an influx of heat from interplanetary space (Pierrehumbert, 2004). It appears that the Earth's climate can change much more strongly and catastrophically than current models predict (Pierrehumbert, 2004).

## Responses of Grassland Biodiversity to Climate and Atmospheric Changes

Human activities have been altering the global atmosphere and climate in many ways. Anthropogenic increase in global atmospheric carbon dioxide concentrations are raising both global temperatures and precipitation over some continents (IPCC, 2001). Global anthropogenic nitrogen fixation now exceeds all natural sources of nitrogen fixation: its products includes the greenhouse gas $N_2O$ which further affects climate change. These change in the global atmosphere and climate are not only co-occurring but are also causally linked.

Anticipating ecosystem responses to combined climate and atmospheric changes, therefore, becomes critical in the context of management, policy and conservation efforts to mitigate their effects (Zaveleta *et al.*, 2003).

Biodiversity responses to various ongoing climate and atmospheric changes affect both ecosystem processes and the delivery of ecosystem good and services. Combined effects of co-occurring global changes on diversity, however, are not well understood. Three classes of interactions appear to charactrerise biodiversity and ecosystem responses to multiple, simultaneous global changes. Biodiversity effects of climate and atmospheric changes may not interact (additive response) and so may be predicted directly from single factor experiments. Alternatively, effects of multiple global changes could be synergistic (amplifying) or antagonistic (canceling or damping), producing larger or smaller biodiversity change, respectively, than those expected on the basis of single-factor results. Changes in environmental conditions can change the diversity of plant communities by changing resource availability and by affecting individual species performances and in turn also the strength and outcomes of interspecific competition (see Tilman, 1988; Reich *et al.*, 2001).

Net primary production (NPP) responses to some combination of simulated global changes, *e.g.*, N-fertilization and elevated carbon dioxide, show characteristic interactions suggestive of alleviated resource colimitation in certain systems, for example, grassland, but not in others. Diversity seems quite sensitive to NPP changes and its responses to combined global changes might be converted to NPP responses, producing similar interactions (see Zavaleta *et al.*, 2003).

Zavaleta *et al.* (2003) examined plant diversity responses in a California annual grassland to manipulations of four global environmental changes, singly and in combination: elevated carbon dioxide, warming, precipitation and nitrogen deposition. After 3 years, while elevated carbon dioxide and nitrogen deposition each reduced plant diversity, elevated precipitation increased it, an warming had no significant effect. Diversity responses to both single and combined global change treatments were driven overwhelmingly by gains and losses of for species, which make up most of the native plant diversity in California grasslands.

Diversity responses across treatments also showed no consistent relationship to net primary production responses, suggesting that the diversity effects of these environmental changes could not be explained simply by change in productivity. In two to four way combinations, simulated global changes did not interact in any of their effects on diversity. These results show that climate and atmospheric change, can rapidly alter biological diversity, with combined effects that can be simple, additive combinations of single-factors effects (Zavaleta *et al.*, 2003).

## Soot and Dust Aerosol in East Asia

Atmospheric aerosols contain elemental carbon (EC, *i.e.* soot), emitted from biomass and fossil fuel combustion. The Elemental carbon (EC) absorbs light in visible wavelengths. This absorption and the subsequent atmospheric heating lead to top of atmospheric radiative forcings in the range of 0-0.8 $Wm^{-2}$ (IPCC, 2001: Jacobson, 2002). Source of China seem to account for about a quarter of the global anthropogenic EC emissions (Cooke *et al.*, 1999). This means that understanding the effects of elemental carbon in the plume downwind of east Asia is a significant step towards understanding both the global budget of short wave absorption and the climate in the western Pacific region (Chuang and Schauer, 2005).

East Asia significantly contributes to the global budget of both EC and Aeolian dust (Chuang and Schauer, 2005). Chuang *et al.* (2003) and Chuang and Schauer (2005) have summarized observations of these important aerosol types downwind of continental east Asia, including periods when particles with dust and soot mixed together were common. The efficiency with which soot absorbs light and soot mixed together were common. The efficiency with which soot absorbs light increase as the amount of soot decreases consistent with theoretical expectation. Should this turn out to be a general relationship, then reducing elemental carbon emissions in order to mitigate future global warming may be less effective than simple scaling would suggest. Results from modeling of mixed soot and dust particles suggest that while these particles absorb significantly more light than dust alone, the overall change in absorption is a decrease of 10 to 40 per cent compared to separate dust and soot particles (Chuang and Schauer, 2005).

One interesting observation made by Chuang and Schauer (2005) was that on some of the days which are strongly influenced by long-range dust transport, a significant fraction of elemental carbon is the coarse (>1 μm diameter) particle samples. The average ratio of coarse particle elemental carbon to total elemental carbon was around 40 per cent. Because elemental carbon is emitted from combustion sources as particles typically smaller than 0.1μm, observing a large fraction of the elemental carbon associated with coarse particles was unexpected. A possible explanation may be that the same meteorological conditions that lead to the lofting of dust result in the lofting of pollution aerosols in the industrialized area of china farther east, and the coagulation of elemental carbon from the pollution plume with the previously lofted dust leads to dust particles coated with EC (and other pollutants) (Chuang and Schauer, 2005).

Since elemental carbon happens to be the most important absorber of visible light, and because East Asia appears to represent about a quarter of the global elemental carbon budget (see Cooke *et al.*, 1999), it is important to examine the efficiency with which elemental carbon absorbs light. As elemental carbon becomes more dilute, its absorption efficiency increase. This general trend was observed for total particles and seemed to be present for fine particles (<1 μm diameter) as well. Conceivably, as elemental carbon mass fraction increases, the additional elemental carbon shields the pre-existing elemental carbon from light thereby leading to a smaller fraction of the elemental carbon that can effectively absorb light.

Should this be a consistent feature of absorption by atmospheric aerosols, it would be crucial for understanding the future impact of light absorption on climate. Hansen *et al.* (2000) suggested that control of elemental carbon emissions may be an effective strategy for mitigation of climate change. According to Chuang and Schauer (2005), the possible concomitant increase in the average elemental carbon absorption efficiency would lead to a net benefit smaller than that assuming a linear response of absorption efficiency would lead to a net benefit smaller than that assuming a linear response of absorption with changing elemental carbon mass concentration. In fact, it seems plausible that a 50 per cent decrease in elemental carbon concentration (with everything else held constant) could increase the absorption efficiency by a factor of about 2, which would completely negate the intended decrease in short-wave absorption.

Besides their radiative impacts, the associations of elemental carbon on dust particles during certain periods may possibly have other important climate implications I the western Pacific regions. The lifetime of aerosols is usually determined by removal by cloud and precipitation processes. The elemental carbon that is attached to dust might well be removed at a very different rate, which would thereby significantly alter its lifetime in the atmosphere; what is not clear is in which direction?

# Chapter 8

# Climate Change, Farming and Forestry

## General Description

Everywhere climate influnces crops and livestock. Projected changes in the global climate including rise in mean temperature, unpredicted rainfall pattern, high $CO_2$ and rise in pests and diseases are anticipated to negatively influence crop productivity, but this is particularly relevant in marginal agriculture of the tropics including India. Use of global circulation models to predict climatic behaviour and crop growth models for extrapolating growth patterns of crop plants in a changed climate can help in evaluating the complex interaction between the environmental variables influencing crop growth and yield.

No doubt crop plants do have enough genetic plasticity to withstand the climatic interference, still efforts need to be undertaken to develop environment-friendly technologies for sustaining crop productivity (Adhya *et al.*, 2003). Climate change can involve alterations in temperature, precipitation and sea-level rise as well as increased incidence of ultraviolet-B (UV-b) radiation. Agriculture being a highly climate-sensitive sector, any change in climatic parameters can severely affect global agricultural output and food security, especially for the more fragile tropical agriculture that is already stretched to its limits due to adverse interference of climatic parameters, poor soil resources and wide array of pests and diseases.

Over the last two decades, tropical agriculture, particularly cereal production, has shown a downward trend as revealed by analysis of several long-term experiments conducted throughout Asia.

Such decline/stagnation in yield coupled with the recent evidences of climatic interference, has generated concerns about the impact of climate changes on crop production system and food security. Climate influences plants in many ways; it can inhibit stimulate, alter or modify crop performance (Adhya *et al.*, 2003).

Greenhouse gases (GHGs) that trap infrared radiation from the Sun leading to heating up of the earth cause a global climate forcing, *i.e.* an imposed perturbation of the earth's energy balance with space. There are many competing natural and anthropogenic climate forcing, but increased GHGs are thought to be the strongest forcing, especially during the past few decades. Evidence supporting this interpretation is provided by some positive heat storage in the ocean, which is of the magnitude of the energy imbalance estimated from climate forcing for recent decades (Hansen *et al.*, 1997).

The relationship between climatic change and agriculture is a crucial issue in view of the fact that the world's food production resources are already stressed from rapidly growing population. Paddy farming is affected by climate change and it also contributes to global warming through the release of methane into the atmosphere. In the Philippines, the International Rice Research Institute has investigated relationship between climate change and paddy production, as also explored mitigation options to help reduce methane emissions from paddy fields (Matthews and Wassmann, 2003).

One crucial component of these studies involved the quantification of these interactions between climate change and rice production into simulation models, and their subsequent use to upscale field measurements to national and regional levels. Such a model was developed to integrate existing knowledge of effects of increased levels of carbon dioxide and temperature on rice growth, and was used to predict the impact of several climate change scenarios on rice farming in South-East Asia. Routines describing the dynamics of methane production and emission from the soil could be linked to a crop simulation model to estimate the effect of different crop management scenarios on national methane emissions from various countries in the region (Matthews and Wassmann, 2003).

Salinger *et al.* (2005) reviewed the assessments of climate variability and climate change, and their impacts on agriculture and forestry, and recommended appropriate adaptation strategies for reducing the vulnerability of agriculture and forestry to climate variability and climate change. Among other solutions, the authors offered management strategies to mitigate GHG's emissions from different agro ecosystem, and proposed the use of seasonal climate forecasts to reduce climate risk.

Agricultural production is highly sensitive to weather and climate-related disasters such as drought, storm and flood. While it is not possible to prevent the occurrence of natural disasters, the resultant disastrous effects can be reduced or mitigated through good planning.

Sivakumar *et al.* (2005) have discussed ways to reduce the vulnerability of agriculture to disaster and extreme events, both the accurate and timely warning, and by impact-reducing countermeasures.

Some current flash points at the intersection of environment science, economics, and public policy are sustainable development, climate policy and biodiversity conservation. Forest resources constitute a perfect starting point for economic analysis of sustainability.

Kant and Beery (2005) have proposed that new economic theory, rather than a new public policy based on old theory, is needed to guide humanity toward sustainability.

## Agriculturally Driven Global Environmental Change

The span of next 50 years is likely to be the final period of rapid agricultural expansion. During this period, demand for food by a weather and 50 per cent larger global population will be a major driver of global environmental change. Global population, which increased 3.7 fold during the 20th century to 6 billion people is forecast to increase to 7.5 billion by the year 2020 and to about 9 billion by 2050 (UN, 1999).

If past dependences of the global environmental impacts of agriculture on human population and consumption continue, some $10^9$ hectares of natural ecosystems would be converted to agriculture by the year 2050. This would be accompanied by 2.4 to 2.7 fold increases in nitrogen and phosphorus driven eutrophication of terrestrial, freshwater, and near shore marine ecosystems, and comparable increases in pesticide use. This eutrophication and habitat destruction would cause unprecedented ecosystem simplification, loss of ecosystem services, and species extinctions. Significant scientific advances and regulatory, technological and policy changes are urgently needed to control the environmental impacts of agricultural expansion (Tilman *et al.*, 2001).

Besides its effects on GHGs (Conway, 1997), agriculture influences ecosystems by the use and release of limiting resources that influence ecosystem functioning (nitrogen, phosphorus, and water), release of pesticides, and conversion of natural ecosystems to agriculture. These sources of global change may rival climate change in environmental and social impacts (NRC, 2000).

Population size and per capita consumption may be the two strongest drivers of global environmental change. Humans currently appropriate over a tired of the production of terrestrial ecosystems and about a half of usable freshwaters; they have doubled terrestrial nitrogen supply and phosphorus liberation and have manufactured and released globally significant quantities of pesticides.

The doubling of world food production during the past 35 years was accompanied by large increases in global nitrogen (N) and phosphorus (P) fertilization and irrigation. If past trends in nitrogen and phosphorus fertilization and irrigation and their dependence on population and gross domestic product (GDP) continue, the mean forecast is for global nitrogen fertilization to be 1.6 times present amounts by 2020 and 2.7 times present values by 2050 (Tilman *et al.*, 2001). By the year 2050, nitrogen fertilization alone would annually add $236 \times 10^6$ MT of nitrogen to terrestrial ecosystems compared with $140 \times 10^6$ MT from all natural sources. Individual forecasts for nitrogen fertilization in the year 2050 range from a 1.9- fold increase to a 3.9 fold increase.

Phosphorus fertilization is forecast to be 2.4 times current amount in 2050. P estimates for 2050 range from 1.6 fold to 3.4 fold increases. Irrigated area which is a measure of agricultural demand for water, is forecast to be 1.9 times the current area in 2050 (Tilman *et al.,* 2001).

Although society benefits from pesticides, some cause environmental degradation on or affect human health (WHO, 1990). Some pesticides, depending on persistence and volatility, disperse all over the world, bioaccumulate in food chains, and have impacts on human health and the health of other species far from points of release and for years after release. If past patterns continue, global pesticide production which has increased for 40 years, would be 2.7 times the present amount by the year 2050.

Projections for 2050 range from 1.9 to 4.8 fold increases. World trade in pesticides, another estimate of trends in pesticide use, would be 1.6 times present levels by 2020 and 2.7 times present levels by 2050. Should trends continue, by 2050, humans and other organisms in natural and managed ecosystems would be exposed to markedly higher levels of pesticides (Tilman *et al.,* 2001).

Projections point to an average global agricultural land base in 2050 that would be 18 per cent larger than at present. The net loss of natural ecosystem to cropland and pasture in developing countries by 2050 is likely to be $10^9$ ha., about half of all potentially suitable remaining land being mainly in Latin America and Africa.

Just as demand for energy happens to be the chief cause of increasing atmospheric GHGS, demand for agricultural products may be the major driver of future nonclimatic global change. The projected 50 per cent increase in global population and demand for diets richer in meat by a wealthier world are projected to double global food demand by 2050 (Alexandrator, 1995, 1999). This can create a serious environmental challenge and one that may rival and significantly interact with climatic change.

Substantial advances in agricultural production are needed during the coming several decades to assure a sufficient, secure, and equitable global food supply (Conway, 1997). An environmentally sustainable greener revolution that is based on the total cost and benefits of agriculture, including agriculture dependent gains and losses in values of such ecosystem goods and services as potable water, biodiversity carbon storage, pest control, pollination, fisheries, and recreation will be needed.

According to Walsh (1991) and Woomer and Swift (1994), existing knowledge, if widely used, may significantly reduce the environmental impacts of agriculture and lead to increased productivity. Integrated pest management, application of site- and time appropriate amounts of agricultural chemicals and water, use of cover crops on fallow lands and buffer strips between cultivated fields and drainage areas, and appropriate deployment of more productive crops can increase yields while reducing water, fertilizer and pesticide use and movement to nonagricultural habitats (Tilman *et al.,* 2001).

The research needs are diverse. There is need to seek, by breeding and biotechnology, gains in the fundamental efficiency of crop N, P, and water use (Postal, 1999; Ruttan, 1999). Advances in precision agriculture that decrease N and P inputs are required, also methods that manage soil organic matter and microbial communities to reduce nutrient leaching and to optimize soil fertility (Cassmanm, 1999).

We badly need methods to efficiently close the nutrient cycle from soil to crop to livestock and back to agricultural soil. We also need better approaches to control crop pathogens and pests such as by grater use of natural enemies, crop diversity and biotechnology, if deployed so as by greater use of natural enemies, crop diversity and biotechnology, if deployed so as to reduce evolution of pest resistance.

Should global population stabilize at 8.5 to 10 billion people, the next 5 decades may well be the final episode of rapid global agricultural expansion. During this period, agriculture can potentially exert massive, irreversible environmental impacts. How to minimize these impacts, while providing sufficient and equitably distributed food is going to be a strong challenge (Tilman *et al.*, 2001).

## Impact of Climate Change on Ecosystems

It seems certain that a future change in climate of the order of magnitude obtained from climate models for a doubling of the atmospheric carbon dioxide concentration could have profound effects on global ecosystems. Prediction of the future impacts on ecosystems is precluded by the absence of reliable estimates of climatic changes at regional scales, and by the lack of knowledge concerning the interactive effects of carbon dioxide and climate variables on vegetation. Despite our inability to predict, sensitivity analyses can produce useful information for judging the possible directions and magnitude of effects for given changes in carbon dioxide levels or climate variables, and so for identifying those regions and environmental changes which may warrant policy attention in the future. From a world perspective, geographic differences in agricultural regions have significant implications for assessing the effects of increased carbon dioxide and climatic change. Whereas rainfall is the major constraint on agriculture in the tropics and subtropics, in the temperate and higher latitudes, temperature exerts a greater influence.

## General Approaches to Assessing the Impacts of Increasing Carbon dioxide and Climatic Change

In relation to agriculture, there are four general approaches to assessing the impacts of increasing carbon dioxide and climatic change (a) crop impact analysis (b) marginal spatial analysis (c) agricultural sector analysis and (d) historical case studies.

### Crop Impact Analyses

Crop impact analyses deal with the effects on plant growth and crop yields. It has been estimated from laboratory experiments on individual plants that a doubling of carbon dioxide from 340 to 680 parts per million by volume (ppmv) could result in a 0-10 per cent increase in the growth and yield of $C_4$ crops (*e.g.* maize, sorghum, sugarcane) and a 10 to 50 per cent increase for $C_3$ crops (*e.g.* wheat, soybean, rice).

These beneficial effects are obtained under most environmentally stressful as well as non-stressful conditions (Table 8.1) and would therefore, benefit both the environmental margins and the core of crop regions. Higher yield benefits would accrue to those regions of the world where $C_3$, rather than $C_4$, crop predominate.

**Table 8.1: Relative effects of increased carbon dioxide on growth and yield: a tentative compilation[1] (after Warrick *et al.*, 1991)**

|                                  | $C_3$  | $C_4$   |
|----------------------------------|--------|---------|
| Under non-stressed conditions    | ++     | 0 to +  |
| Under environmental stress:      |        |         |
|    Water (deficiency)        | ++     | +       |
|    Light intensity (low)     | +      | +       |
|    Temperature (high)        | ++     | 0 to +  |
|    Temperature (low)         | +      | ?       |
| Mineral nutrients:               | 0 to+  | 0 to +  |
|    Nitrogen (deficiency)     | +      | +       |
|    Phosphorus (deficiency)   | 0?     | 0?      |
|    Potassium (deficiency)    | ?      | ?       |
|    Sodium (excess)           | ?      | +       |

1 Sign of change relative to control carbon dioxide under similar environmental constraints.
++ = strongly positive; + = positive; 0 = no effect; ? = not known or uncertain.

Considering the sensitivity of crop yields to climatic change without including the direct $CO_2$ effect, crop impact analyses have focused largely on grain yields in temperate and higher latitudes; the tropics and subtropics have tended to be neglected. It has been estimated that, with no precipitation change, a warming of 2°C could reduce average yields of wheat and maize in the mid-latitudes of North America and Western Europe by 10.7 per cent assuming instantaneous warming and no change in cultivars, technology, or management. These yield reductions would be offset by wetter conditions and aggravated by drier conditions. These estimates pertain to core regions; in contrast, average yields at the cool margins of cereal production, for example, might well benefit from a lengthening of the growing season and a reduction of damaging frosts.

## Marginal-spatial Analyses

Marginal-spatial analyses consider the margins of production where conversion to other crops (or genotypes) or land uses is most likely to take place. A few studies in the mid and high latitudes suggest potential shifts in the boundaries of cereal regions of the order of magnitude of several hundred kilometers per °C change. Other studies of high altitude locations with steep environmental gradients suggest potential altitudinal shifts or more than a hundred metres per °C change. Although these estimates are highly uncertain, any such shifts at the margins would certainly modulate the effects of climatic change on regional crop yields and production.

## Agricultural Sector Analyses and Historical Case Studies

Agricultural sector analyses and historical case studies examine the range of environmental, agricultural and socioeconomic impacts and explore the ways in which agricultural systems adopt to climatic change and availability. There are many feedback mechanisms that can enhance or diminish the potential impacts of environmental changes on crop yields and food production.

Many past studies have considered the impacts of climatic change separately from primary responses of plants to increased carbon dioxide. But the effects are interactive and nonlinear, not simply additive. There are many uncertainties in extrapolating laboratory results to field conditions. In general, given the uncertainties in regional scale estimates of climatic change and many deficiencies in methodologies of impact assessment, there is presently no firm evidence for believing that the net effect of higher carbon dioxide and climatic change on agriculture in any specific region of the world will be adverse rather than beneficial. However, it is certain that some will gain and others will lose, although we know neither where they will be found not the magnitude of the impacts.

In general, the direct effects of enhanced carbon dioxide concentrations on crop yields may be beneficial. In the absence of climatic change, a doubling of the carbon dioxide concentration is expected to cause a 0-10 per cent increase in growth and yield of $C_4$ crops (*e.g.* maize, sorghum, sugarcane) and a 10 to 50 per cent increase for $C_3$ crops (*e.g.* maize, soybean, rice) depending on the specific crops and growing conditions.

In analyzing the sensitivity of crop yields to possible changes in climate without including the direct effects duty to higher carbon dioxide concentrations, most research has focused on average yields of cereal grains in core crop regions of the temperate latitudes. Much less attention has been given to the tropics and subtropics, to the climate sensitive margins of production and to possible changes in year to year climatic extremes. Models of agricultural production and trade suggest that several feedback mechanisms exist in many regions through which agriculture can adjust and adapt to environmental change. Over the long term, food production in such areas appears more sensitive to technology, price or policy changes than to climatic changes, and these factors are largely controllable, whereas climate is not. However, for the lands marginal for food production in the developing world, agriculture may be highly sensitive to climatic change, as evidenced by the tolls taken by year to year variations in climate. If these regions can go in for suitable measures to reduce the ill effects of current, short term climatic variability, they may be better prepared to adapt to some adverse effects of future changes in climate, should they occur.

## Climate Change and Biological Diversity

Human activities all over the world have caused and will continue to cause a loss in biodiversity through, *inter alia*, land-use and land-cover change; soil and water pollution and degradation (including desertification),and air pollution; diversion of water to intensively managed ecosystems and urban systems; habitat fragmentation; selective exploitation of species; the introduction of non-native species;

and stratospheric ozone depletion. The current rate of biodiversity loss exceeds the natural background rate of species extinction. What is not definitely knows is to what extent might natural or human induced climate change enhance or inhibit these losses in biodiversity (Gitay *et al.*, 2002).

Changes in climate exert additional pressure and affect biodiversity. The increase in atmospheric concentrations of GHGs resulting from human activities is the primary factor involved. Also, land and ocean surface temperatures have warmed, the spatial and temporal patterns of precipitation have changed, sea level has risen, and the frequency and intensity of *El Nino* events have increased. These changes, especially the warmer regional temperatures, have disturbed the timing of reproduction in animals and plants and/or migration of animals, the duration of the growing season, species distributions and population sizes, and the incidence of pest and disease outbreaks.

For the wide range of IPCC emissions scenarios, the Earth's mean surface temperature is projected to warm 1.4 to 5.8°C by the end of the 21st century, with land area warming more than the oceans, and the high latitudes warming more than the tropics. The sea level can rise by 0.09 to 0.88 m. In general, precipitation could increase in high latitude and equatorial areas and decrease in the subtropics, and heavy precipitating events may increase. Climate change is projected to affect individual organisms, populations, species distributions, and ecosystem composition and function both directly (*e.g.* through increases in temperature and changes in precipitation and in the case of marine and coastal ecosystems, also changes in sea level and storm surges) and indirectly (*e.g.* through climate induced changes in the intensity and frequency of wildfires). Loss, modification and fragmentation of habitats and the introduction and spread of non-native species can modify the impacts of climate change.

The general effect of projected human induced climate change is that the habitats of many species will move upward (*i.e.* toward higher latitudes) from their present locations. Species will be affected differently by climate change. They will migrate at different rates through fragmented landscapes, and ecosystems dominated by long lived species (*e.g.* long-lived trees) will change slowly. The composition of many current ecosystems is expected to change because different species will not necessarily shift together. The most rapid changes may occur where they are speeded up by changes in natural and human induced non climatic disturbance patterns.

Changes in the frequency, intensity, extent, and sites of disturbance will determine whether, how and at which rate the existing ecosystems will be succeeded by new plant and animal communities. Disturbances can increase the rate of species loss and make possible the establishment of new species.

By the year 2080, about 20 per cent of the world's wetlands could be lost due to sea-level rise. The impact of sea-level rise on coastal ecosystems (*e.g.* mangrove/coastal wetlands, sea grasses) will vary regionally and will depend on erosion processes from the sea and depositional processes from land.

The risk of extinction will increase for many currently vulnerable species. Species having limited climatic ranges and/or restricted habitat requirements and/or small populations (endemic mountain species and birds restricted to islands, peninsulas) tend to be the most vulnerable to extinction.

On the other hand, species with extensive, non-patchy ranges, long range dispersal mechanisms, and large populations face less risk of extinction. Climate change is likely to increase species losses.

Change in biodiversity at ecosystem and landscape scale, in response to climate change and other stresses, would further affect global and regional climates through changes in the uptake and release of GHGs and those in albedo and evapotranspiration.

Climate change mitigation activities can have substantial impact on biodiversity. Land-use, land use change, and forestry activities (afforestation, reforestation, and improved forest, cropland, and grazing land management practices) and popularization of renewable energy sources (hydro, wind-, and solar power and biofuels) could affect biodiversity depending upon site selection and management practices. For example:

1. Afforestation and reforestation projects can have positive, neutral or negative impacts depending on the level of biodiversity of the non-forest ecosystem being replaced, the scale one considers, and other design and implementation issues;

2. Avoiding and reducing forest degradation in threatened/vulnerable forests that contain communities that are unusually diverse, globally rare or unique to that region can provide substantial biodiversity benefits along with the avoidance of carbon emissions:

3. Large-scale bioenergy plantations that generate high yields would have negative impacts on biodiversity where they replace systems with higher biological diversity, whereas small scale plantations on degraded land or abandoned agricultural sites would have environmental benefits; and

4. Increased efficiency in the generation and/or use of fossil-fuel-based energy can reduce fossil-fuel use and thereby reduce the impacts on biodiversity resulting from resource extraction, transportation (*e.g.* through shipping and pipelines), and combustion of fossil fuels (Gitay *et al.*, 2002).

Climate change adaptation practices can promote conservation and sustainable use of biodiversity and decrease the impact of changes in climate and climatic extremes on biodiversity. These include the establishment of a mosaic of interconnected terrestrial, freshwater, and marine multiple use reserves designed to take into account projected change in climate, and integrated land and water management activities that reduce non-climate, and integrated land and water management activities that reduce non-climate pressures on biodiversity and hence make the systems less vulnerable to changes in climate. Some of these adaptation activities may make people less vulnerable to climatic extremes (Gitay *et al.*, 2002).

Integrating adaptation and mitigation activities with broader, more sustainable strategies can increase their effectiveness. Potential environmental and social synergies and tradeoffs take place between climate adaptation and mitigation activities (projects and policies), and the objectives of multilateral environmental agreements (*e.g.* the conservation and sustainable use objective of the convention on Biological Diversity) as well as other aspects of sustainable development. It is possible to evaluate these synergies and tradeoffs for the various potential activities such as energy and land-use, land-use change, and forestry projects and policies.

The following information needs and assessment gaps have been identified:

☆ Better understanding of the relationship between biodiversity, ecosystem structure and function, and dispersal and/or migration through fragmented landscapes, as also the response of biodiversity to changes in climatic factors and other pressures.

☆ Development of suitable climate change and ecosystem models particularly for quantifying the impacts of climate change on biodiversity at all scales, taking into account feedbacks.

☆ Improved understanding of the local to regional scale impacts of climate change adaptation and mitigation options on biodiversity.

☆ Development of improved assessment methodologies, criteria and indicator stop assess the impact of climate change mitigation and adaptation activities on biodiversity and sustainable development.

☆ Biodiversity conservation and sustainable use activities and policies that would benefit climate change adaptation and mitigation options (Gitay *et al.*, 2002).

## Amphibian Declines

Among the several different factors that have been proposed as causes for global amphibian population declines, global climate change has received fairly little attention. Recent studies have attributed declines of amphibian populations to variations in temperature and moisture caused by climate change in general or by the *El Nino* phenomenon in particular (Carey and Alexander, 2003).

Many workers have correlated amphibian declines with climatic factors over short time periods, including a single climate event. It appears that global climate changes can serve as both direct (the actual cause of death) or indirect (contributory factor that facilitates the direct cause of death) causes of amphibian declines.

## Parasitism, Climate Oscillations and the Structure of Natural Communities

The North Atlantic climate Oscillation (NAO) generates interannual and decadal fluctuations in winter temperatures, precipitation, wind conditions, and distribution and fluxes of currents on both sides of the North Atlantic Ocean. A strong positive NAO index (the deviance from the average sea level pressure difference between the Azores and Iceland) is associated with high temperatures, strong winds and high precipitation in northern Europe, and low temperatures in eastern North America. On the other hand, a low NAO index creates the opposite conditions (Mouritsen and Poulin, 2002). The NAO is analogous to the *El Nino*-Southern Oscillation (ENSO) in the Pacific Ocean which was also results in temporal and geographical fluctuations in temperature and precipitation from the Indian Ocean to the gulf of Mexico.

Such large scale climatic fluctuations must surely affect a variety of ecological processes. Indeed, besides the direct influence on the general biological performance of different organisms, some studies have demonstrated highly complex and indirect cascading effects in which NAO mediates the outcome of interspecific competition and predator-prey/herbivore plant interactions. However, despite the strong evidence that parasites can influence the composition and structure of natural communities, and that parasite transmission can be greatly affected by weather conditions, few if any have studied or considered the possible impact of NAO on host parasite interactions. However, some have shown that parasite induced host population dynamic or distribution are governed by short term changes in weather conditions of general climate change (see Mouritsen and Poulin, 2002). Obviously there is a need to direct attention to the potential influence of climate oscillations on host parasite interactions host population dynamics, and hence also on community structure.

## Consequences for Forests

The forests of the world constitute a complex system with many possible responses, both to the direct effects of an increase in atmospheric carbon dioxide concentration and to the possible changes in climate. These response originate from phenomena that operate on very different scales of time and space. Serious difficulties are faced in the 'scaling up' of the short term physiological and biochemical response of leaves and individual plants to estimate the intermediate and long term responses of forests. The difficulties arise from the large uncertainties involved in the method of extrapolation and from the complex interactions that occur at larger scales. The two uncertainties are presently large enough to preclude meaningful estimates of the effects on forests of higher carbon dioxide concentrations and climatic change except in a very general way.

With respect to the direct effects of carbon dioxide, these problems of scaling up are compounded by the lack of experimental evidence for relevant forest species, particularly for plants that have been allowed to acclimate to enhanced carbon dioxide concentrations over one or more growing cycles.

Although higher concentrations of carbon dioxide increase the growth rates of individual trees in controlled conditions over the short term, it is very uncertain whether such effects would be sustained and would lead to increased productivity in actual forest environments over the long term. In uncontrolled environments, the direct carbon dioxide effects are complicated by micro-meteorological differences in the degree of coupling between forests and atmosphere (within as well as between forest systems, and by species competition and interaction. If, indeed, elevated carbon dioxide, concentrations are conducive to long-term growth enhancement, increases in productivity would be more likely to occur in commercial forests than tin mature forests in which the capacities for increased carbon storage are more limited.

In general, the results of a small number of modeling studies suggest that climatic changes of the order of magnitude predicted by climate models for a doubling of atmospheric carbon dioxide can suffice to produce substantial intermediate and long term changes in the composition, size, and location of the world's forests. The natural forest of the high latitudes in general and the boreal forests in particular, appear sensitive to predicted temperature changes and it is at these latitudes that climate models predict the largest warming to occur as a result of increased concentration of GHGs (Figure 8.1). Warmer conditions could possibly lead to large reduction in the area extent of boreal forests and a poleward shift in their boundaries. In contrast, the forests of the tropical and subtropical zones would be more sensitive too changes in precipitation than temperature.

## Soil Warming Feedbacks to the Climate System

The acceleration of global warming due to terrestrial carbon-cycle feedbacks is likely to be an important component of future climate change (Woodwell and Mackenzie, 1995). Experiments with three dimensional carbon climate models have suggested that carbon cycle feedbacks could either accelerate or slow climate change over the 21$^{st}$ century (Melillo *et al.*, 2002).

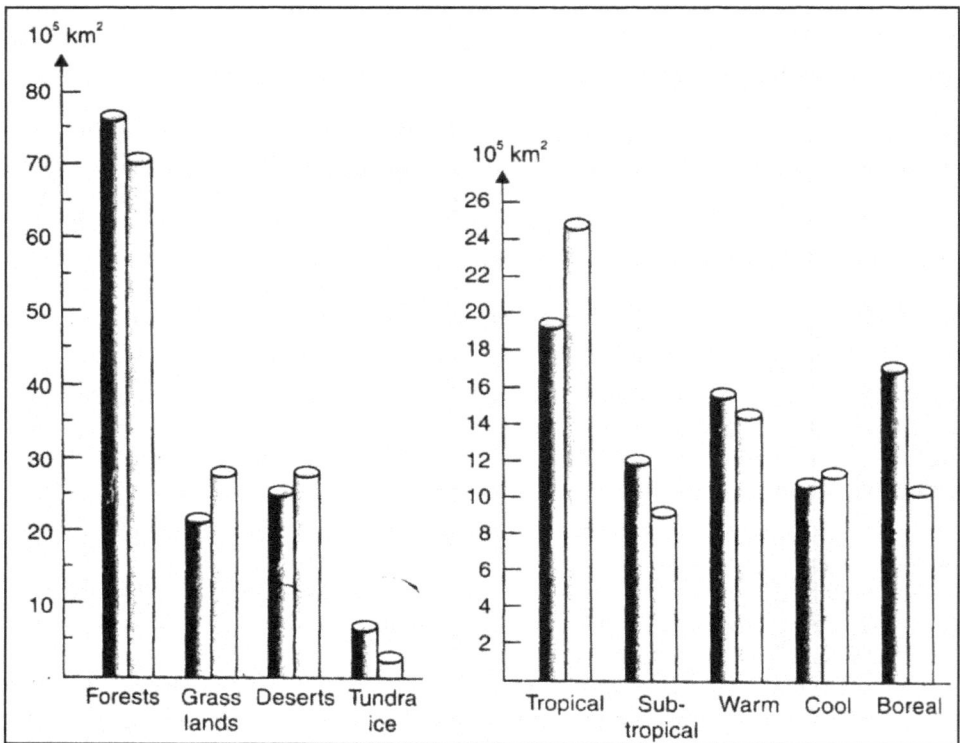

Figure 8.1: Estimation of the change in life zone extents for a carbon dioxide induced warming from present climate (hatched columns) to that for a doubling of the atmospheric concentration of carbon dioxide (open columns). In calculating these changes in potential vegetation the changes in precipitation were not taken into account. Note also that potential vegetation corresponds to extension of ecosystems unaffected by man's direct impact (after Bolin *et al.*, 1991).

The sign and magnitude of these feedbacks in the real Earth system remain uncertain because of gaps in our understanding of terrestrial ecosystem processes for instance, the potential switch of the terrestrial biosphere from its current role as a carbon sink to a carbon source will depend upon the long-term sensitivity to global warming of the respiration of soil microbes (Giardina and Ryan, 2000).

With a view to exploring this feedback issue in an ecosystem context (Melillo *et al.*, 2000) studies, in a decade long soil warming experiment in a mid latitude hardwood forest changes in soil carbon and nitrogen cycling with a view to investigating the effects of these changes on the climate system (Melillo *et al.*, 2002). They showed that whereas soil warming speeds up soil organic mater decay and carbon dioxide fluxes to the atmosphere, this response is weak and short lived for a mid-latitude forest, because of the limited size of the labile soil carbon pool. They also showed that warming increases the availability of mineral nitrogen to plants. Because plant growth in many mid-latitude forests is nitrogen limited, warming can potentially stimulate enough carbon storage in plants to at least compensate for the carbon losses from soils. These results are at variance with those in some climate models showing projections of large long term release of soil carbon in response to warming of forest ecosystems.

Of course, the carbon balance of forest ecosystems in a climate changed world would depend on more than soil warming. Carbon storage in woody tissue is affected by other factors related to climate change, including the availability of water, the effects of increased temperature on both plant photosynthesis and above ground plant respiration and the atmospheric concentration f carbon dioxide. Reductions in soil moisture and increased plant respiration associated with warming can reduce carbon storage in mid latitude forests, whereas moderate increases in soil moisture and increased concentrations of carbon dioxide may increase carbon storage in these systems, especially when nitrogen limitation is relieved (Melillo *et al.*, 2002).

Warming experiment as in mid-latitude ecosystems without a dominant woody vegetation component have shown either small carbon losses or little change in carbon storage. Warming, probably, has its largest positive feedback effects in high-latitude ecosystems that contain small stature and/or sparse woody vegetation and large pools of partially decomposed soil carbon that have accumulated under cold, wet conditions. If, these soils undergo both warming and drying, they can potentially lose large amounts of carbon as carbon dioxide to the atmosphere (Davidson *et al.*, 2000).

## Human Actions and the Earth's Climate

The heralding of the industrial era with the construction of coal burning factories and power plants started releasing carbon dioxide and other GHGs into the air. Later, automobiles contributed to such emissions. Human beings who have lived during the industrial era have been responsible both for the gas buildup in the atmosphere and for some part of the accompanying global warming.

It is not true that GHGs released by human activities have perturbed the Earth's delicate balance only during the past 200 years; in fact, even our ancient agrarian ancestors had been adding these gases to the atmosphere over several millennia, thereby altering the Earth's climate.

Concentrations of $CO_2$ started rising some 8,000 years ago. About 3,000 years later the same thing happened to methane. These surprising rises have exerted profound effects. Without them, current temperatures in northern parts of North America and Europe would be cooler by 3 to 4°C–high enough to adversely affect agriculture. Further, an incipient Ice Age would probably have begun several thousand years ago in parts of northeastern Canada. But in reality, the Earth's climate has been fairly warm and stable in recent millennia (Ruddiman, 2005).

It appears that human farming activities-primarily agricultural deforestation and crop irrigation-contributed extra carbon dioxide and methane to the atmosphere. As a result, human beings kept the planet significantly warmer than it would have been otherwise- and possibly even averted the start of a new Ice Age. Three predictable variations in the Earth's orbit around the Sun have exerted the major control over long-term global climate for millennia. These orbital cycles operate over 100,000; 41,000 and 22,000 years.

As a consequences, the amount of solar radiation reaching different parts of the Earth during a given season can differ by more than 10 per cent. Over the past three million years, these regular changes in the amount of sunlight reaching the earth have produced a long sequence of Ice Age (when great areas of Northern Hemisphere continents were covered with ice) separated by short, warm interglacial periods (Ruddiman, 2005).

Concentrations of carbon dioxide and methane rose and fell in a regular pattern during virtually all of the past 40,000 years. These increase and decreases in GHGs occurred at the same intervals as variations in the intensity of solar radiation and the size of the ice sheets. This is particularly true for methane.

Methane concentrations rose and fell over the past 2,50,000 years in near harmony with the precession induced ups and down of solar radiation in the Northern Hemisphere (precession refers to the 22,000 year tempo of an orbital cycle). The highest temperatures stimulated extreme methane production in wetlands which are the atmosphere's primary natural source of this GHG.

Carbon dioxide concentration, which fluctuated in cycles over the past 3,50,000 years, varied in response to precession as well as to shifts in the tilt of the Earth's rotational axis and in the shape of its orbit. These other cycles occur every 41,000 and 1,00,000 years, respectively (Ruddiman, 2005).

The GHG concentrations fell during the last four interglaciations, yet rose only during the current one. One plausible explanation of this is the impact of the new factor of farming that has been operating in the natural working of the climate system during the past several thousand years (interglaciation). Agriculture originated in the Fertile crescent region of the eastern Mediterranean around 11,000 years ago and shortly thereafter in China.

Several thousand years later, it originated in the Americas; and through subsequent millennia it spread to other parts of the world.

Farming activities, notably paddy fields, produce methane. As in natural wetlands, flooded vegetation decomposes in the standing water. Methane is also generated when grasslands are sometimes burnt to attract game. Domesticated animals emit methane with faces and belches. All these factors may have contributed to a gradual rise in methane as human populations grew slowly; the onset of rice irrigation in South Asia probably accounted for the abruptness of the reversal from a natural methane decline to an unexpected rise around 5,000 years ago (Ruddiman, 2005).

Another farming related process-deforestation-may explain the start of the anomalous carbon dioxide trend. As growing of crops in naturally forested areas requires cutting of trees, farmers started clearing forests for this purpose in Europe and China by 8,000 years ago. Regardless of whether the fallen trees were burned or left to rot, their carbon would soon oxidize and end up in the atmosphere as carbon dioxide. Europe and southern Asia suffered heavy deforestation long before the advent of the industrial era, and the clearance process was well under way throughout the time of the unusual carbon dioxide rise (Ruddiman, 2005).

If farmers were indeed responsible for such large GHG anomalies (250 ppb for methane and 40 ppm for $CO_2$ by the 1700s), the effect of their practices on the Earth's climate would have been considerable. Based on the average the combined effect from these anomalies could have been an average warming of almost 0.8°C just before the industrial era. That amount is larger than the 0.6°C warming actually measured during the past century. This implies that the effect of early farming on climate rivals or even exceeds the combined changes registered during the time of rapid industrialization. This strong warming effect escaped detection for so long probably because it was masked by natural climatic changes in the opposite direction-the Earth's orbital cycles were driving a simultaneous natural cooling trend, especially at high northern latitudes. The net temperature change was a gradual summer cooling trend lasting until the 1800s. Had GHGs followed their natural tendency to decline, the resulting cooling would have augmented the cooling being driven by the droop in summer radiation, and the Earth would have become much cooler than it is now. According to Ruddiman (2005), nature would have cooled the Earth's climate, but our ancestors kept it warm by discovering agriculture.

It is conceivable that the rapid warming of the past century may persist for at least 200 years, until the economically accessible fossil fuels become scarce. Once that happens, the Earth's climate may begin to cool gradually as the deep ocean slowly absorbs the pulse of excess carbon dioxide from human activities. Whether global climate will cool enough to produce the long-overdue glaciations or remain warm enough to avoid that fate cannot be predicted (Ruddiman, 2005).

## Climate Change and Russian Agriculture

It has been traditionally believed that climate change would bring clear net benefits to Russian agriculture and water resources. Recently, however, Alcamo *et al.* (2003) have challenged this view. According to them, the predicted wetter and warmer climate over much of Russia may indeed result in higher crop yields and in the expansion of crop-growing areas, but expansion may be limited by poor soils, lack of infrastructure, and/or remoteness from agricultural markets. If conditions for crops are congenial, it could also mean better conditions for pests, diseases and weeds.

Meanwhile, a dryer and warmer climate is predicted for the current crop growing and exporting areas of southeastern Russia. This is likely to threaten productivity and cause more frequent years of bad harvest. Thus, any gains in Russia's potential new crop areas may well be counterbalanced by losses in current crop production areas.

## Agriculture and Biodiversity in the European Union

The agricultural sector is a source of heavy pressure on Europe's environment. Through trade, European agriculture also impacts other parts of the world.

In Western Europe, agricultural expenditure is shifting from market support towards subsidies supporting farmers' income, and farmers are rewarded for being managers of Europe's landscape and environmental, rather than just food producers. Also, the EU funding for rural development is rising.

In Central and Eastern Europe, farming today involves lower nutrient input, lower productivity, and often land with a higher nature value than in the West (EEA/UNEP, 2004). With the EU enlargement, more stable market returns and new funding could induce better off farmers in Central Europe to expand and intensify. At the same time, an ageing population and new economic opportunities in cities may stimulate land abandonment (OECD, 2004). These developments can potentially impact Europe's biodiversity, which is already under considerable pressure.

To reduce the stress on Europe's biodiversity, more agri-environment programmes and support to farmers in less-favoured areas are essential. While doing so, funds should be better targeted towards high nature value farmland (EEA/UNEP, 2004).

## Predictions from Climate Change Models

Estimations of the agricultural impacts of long-term global climate changes depends on understanding the direction and magnitude of climate changes. Climate change projections rely on highly complex global circulation models (GCMs) that have been successful in depicting the observed large-scale climatological features. However, considerable uncertainty is associated with these projections on regional scale, since GCMs are yet to realistically reproduce the observed features at regional scale, particularly over the monsoon region. The more popular GCMs are GISS (Goddard Institute for Space Studies, National Aeronautics and Space Administration) and GFDL (Geophysical) Fluid dynamics Laboratory, National Oceanic and Atmospheric Administration), UKMO (United Kingdom Meteorological Office).

But many GCMs have certain limitations in predicting climate change. The most significant limitations include:

1. Poor spatial resolution;
2. Inadequate coupling of atmospheric and oceanic processes;
3. Poor simulation of cloud processes, and
4. Inadequate representation of the biosphere and its feedbacks.

Several modeling groups constantly revise and improve the GCMs. In general, GCMs can at best be used to suggest the likely direction and rate of change.

All GCMs predict increases in mean global precipitation. The crop water regime may be affected by changes in seasonal precipitation, within season patter of precipitation and interannual variation of precipitation. Moisture stress during he flowering, pollution and grain filling stages is especially harmful to maize, soybean, wheat, rice and sorghum (Decker *et al.*, 1986).

Global climate change exacerbates the demand for irrigation water. Higher temperatures, increased evaporation and yield decreases contribute to this projection but assured supply of other required irrigation water under climate change is highly uncertain. Where water supplied are diminishing, extra demand could require that some land be withdrawn from irrigation.

Carbon dioxide enrichment in the atmosphere tends to close plant stomata. By doing so, it reduces transpiration per unit leaf area while enhancing photosynthesis. Morison and Gifford (1984) observed marked decreases in the stomatal conductance of 18 agricultural species (by 36 per cent on average) in an atmosphere enriched by doubled carbon dioxide. However, crop transpiration per ground area may not be reduced commensurately, because decreases in individual leaf conductance are offset by increases in crop leaf area (Adhya *et al.*, 2003).

In any case, highly carbon dioxide usually improves water-use efficiency, defined as the ratio between crop biomass accumulation or yield and the amount of water used in evapoatranspiration.

Because climate variables regulate the geographic distribution of pests, climate change can alter their ranges. Insects may extend their ranges in conditions where warmer winter temperatures allow their over-wintering survival and increase the possible number of generations per season (Stinner *et al.*, 1989). Pests and diseases at low latitude regions (where they are much more prevalent) may also shift to higher latitudes. When pests increase, there may be substantial rise in the use of agricultural chemicals in both temperate and tropical regions to control them (Adhya *et al.*, 2003).

Global estimates of climate impacts on agriculture have been fairly approximate because of lack of consistent methodology and uncertainty about the physiological effects of carbon dioxide (see Kumar, 2006). Modelling studies undertaken by the US Environmental Protection agency (see Rosenzweig *et al.*, 1991) showed that the climate change scenarios without the physiological effects of carbon dioxide cause decreases in estimated national production, while the physiological effects of carbon dioxide mitigate the negative effects.

The UKMO climate change scenario (mean global warming of 5.2°C) generally causes the largest production declines, while under the GFDL and GISS (4.0 and 4.2°C mean global warming, respectively), production changes are more moderate. When embedded in a global agricultural food trade model (the Basic Linked System, see Fisher *et al.*, 1994), the production change estimates based on IBSNAT crop model results will allow for projection of potential impacts on food prices, shifts in comparative advantage, and altered patterns of global trade flows for a suite of global climate change, population, growth and policy scenarios (Adhya *et al.*, 2003).

## Plants and Atmospheric Carbon dioxide

Due to its radiative properties carbon dioxide not only affects average global temperature, but as the main plant nutrient, it also influences vegetation directly. Two outstanding features of this direct impact are the stimulation of plant biomass production ($CO_2$ fertilisation) and the improvement of water use efficiency of plants. As carbon dioxide is a plant nutrient, any increase in its supply causes plants to take up and utilize more of this gas (fertilization effect). This lead to an enhanced fixation of carbon dioxide in the biomass of terrestrial vegetation (Figure 8.2).

The practice of stimulating plant growth through artificial increase of carbon dioxide levels in greenhouses has raised hopes that globally elevated atmospheric carbon dioxide levels could stimulate photosynthesis and so increase biomass production. Conceivably, this could enhance agricultural production and boost efforts to meet the food crisis caused by the population explosion.

Most researches on the effects of an increase carbon dioxide supply for plants suggest that under favourable growth conditions, plants respond with enhanced photosynthesis and growth, so there could be a tendency towards greater biomass production. Nevertheless, there are still some unresolved questions about such enhanced biomass production and possible interactions between a rise in carbon

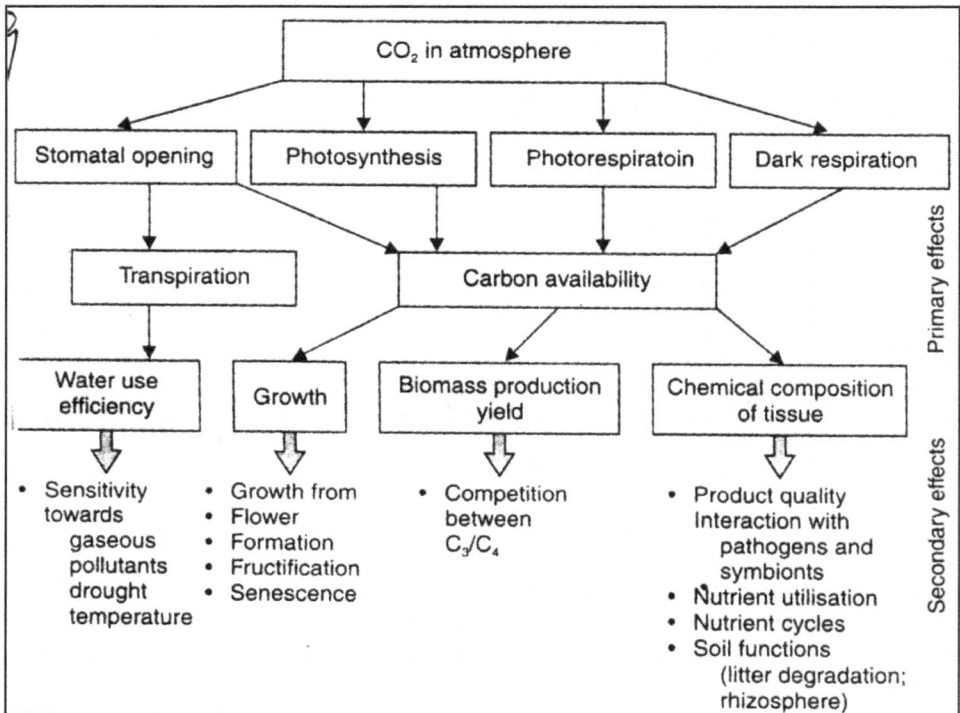

**Figure 8.2: Some effects of rising atmospheric carbon dioxide concentrations on plants (after Brunnert and Weigel, 1997).**

dioxide and other growth factors (Brunnert and Weigel, 1997). Some of these questions are:

1. *Acclimation and adaptation phenomena*: Will the stimulation of growth observed in short time experiments be maintained over the long term in a carbon dioxide enriched atmosphere, and how do plants react to a gradual rise in carbon rise in carbon dioxide concentration?

2. *Consequences for plants' water budgets*: Whether the improvement in water-use efficiency observed in the single plant also applies to whole stands? To what extent growth stimulation is modified by the water supply? Whether plants can become less sensitive to drought stress?

3. What are the possible changes in quality of the plant tissue associated with alterations of its chemical composition as a result of an increased carbon dioxide supply and the consequences of this for product quality, nutrient cycles, and litter degradation, as well as for host pathogen systems and symbiotic interrelations?

4. What are the influence of elevated carbon dioxide on growth stimulation and the consequences of this for the plant's nutrient balance?

5. The influence of elevated carbon dioxide on structural and functional soil parameters (root growth, rhizosphere, humus and total C turnover)

Table 8.2 shows the effect of enhanced carbon dioxide level on biomass and yield in important crop plants.

**Table 8.2: Effect of elevated carbon dioxide concentration on biomass and yield in the world's nine most important crop plants. Per cent increment upon doubling of the carbon dioxide supply in relation to ambient air (mean ± standard deviation) (after Brunnert and Weigel, 1997).**

| Crop | Biomass Production | Yield Increase (per cent) |
|---|---|---|
| Maize | + 9±5 | +29±64 |
| Wheat | + 31±5 | +3±514 |
| Soybean | +39±5 | +29±8 |
| Millet | +9±29 | - |
| Barley | +30±17 | +70 |
| Cotton | +84±126 | +209 |
| Rice | +27±7 | +15±3 |
| Potato | +15 | +51±111 |
| Sweet potato | +59±18 | +83±12 |

Plant biomass grown under elevated carbon dioxide often shifts its elemental composition towards carbon, *i.e.* the tissue is impoverished in certain minerals (alkali and earth alkali ions, nitrogen, sulphur, trace elements), possibly due to the diluting effect of carbohydrate enrichment, a reduction in nutrient uptake as a result of reduced transpiration or in the case of nitrogen, a reduction of photosynthesis in adaptation to the plant's sink metabolism.

The biomass produced at elevated carbon dioxide concentration show a strong shift of the C/N ratio towards carbon. The shift in the C/N ratio of plants may have far-reaching economic and ecological consequences.

An elevated C/N ratio can influence the feeding behaviour of insect herbivores: these insects tend to compensate for nitrogen deficiency by an increase consumption of biomass. In this way, they avoid reduced larval and pupil weights and an increased mortality, thus retaining their reproductive efficiency (Brunnert and Weigel, 1997).

Non-biotic regulation of the Earth's global climate on a multimillion-year timescale occurs by the long term inorganic carbon cycle through which the atmospheric concentration of carbon dioxide is controlled by its supply from volcanoes and metamorphic degassing, and its removal is similarly controlled by the chemical weathering of calcium and magnesium silicate rocks (Berner *et al.*, 1983; Berner, 2004; Beerling and Berner, 2005).

Atmospheric carbon dioxide concentration over the last half billion years have remained within finite limits as a consequence of the coupled evolution of plants, carbon dioxide and climate. This indicates the involvement of a complex network of geophysiological feedbacks. Unfortunately, our insight into this important regulatory network is very limited (Beerling and Berner, 2005). The rise of vascular land plants added a strong biotic feedback into climate regulation, with the capacity to alter the long-term atmospheric carbon dioxide concentration through the production of organic matter for burial in sediments, and acceleration of the chemical weathering of silicate rocks. But long-term changes in carbon dioxide and climate also drive terrestrial plant development and evolution(Kenrick and Crane, 1997; Osborne *et al.*, 2004).

Plant evolution not only creates global change sin environmental conditions but also feeds back on itself. This codependency generates a tight regulatory system for the long-term carbon cycle, with several feedback mechanisms that check run away changes in carbon dioxide and catastrophic planetary warming.

Beerling and Berrner (2005) characterized the network of geochemical effects of plants on atmospheric carbon dioxide and the physiological effects of carbon dioxide on plants by using a system analysis to reveal positive and negative geophysiological feedbacks involved with regulating the long-term carbon cycle (Figure 8.3).

According to Beerling and Berner (2005), positive feedbacks accelerated falling carbon dioxide concentrations during the evolution and diversification of terrestrial ecosystems in the Paleozoic and enhanced rising carbon dioxide concentrations across the Triassic Jurassic boundary during flood basalt eruptions. The existence of positive feedbacks reveals the unexpected establishing influence of the biota in climate regulation that led to environmental modifications accelerating rates of terrestrial plant and animal

evolution in the Paleozoic. Their systems analysis identified five important previously unrecognized positive feedback loops (PFLs) involving land plants and

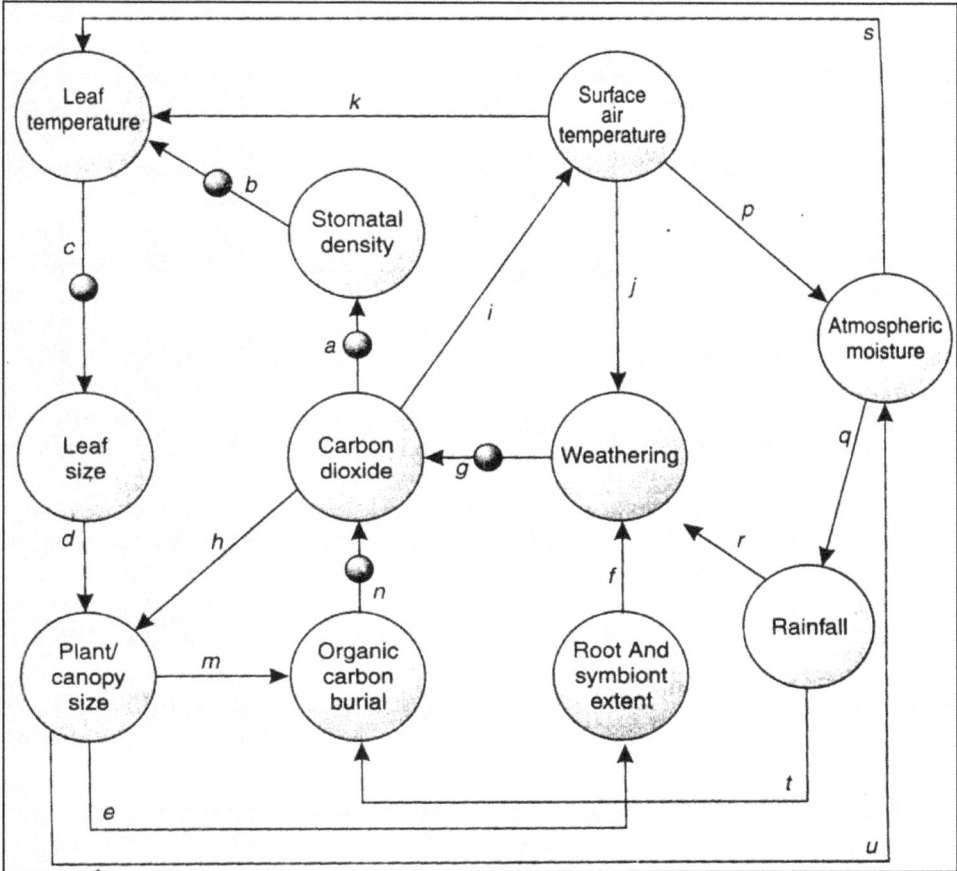

**Figure 8.3: A systems analysis diagram of the geophysiological feedbacks between plants and carbon dioxide, applicable only at times of potential overheating of leaves due to high carbon dioxide levels. Arrows originate at causes and end at effects. Arrows with bull's eye represent inverse responses. Those without bull's eyes represent direct responses. Letters adjacent to arrows designate paths flowed by feedback loops. The time-scale over which the paths operate are as follows: a, b, i, k, p, q, s and u, $10^0$ to $10^1$ years, c and d, $10^1$ years: f, h, j, m, n, r and t, $10^3$ to $10^6$ years, e, $>10^6$ years (after Beerling and Berner, 2005).**

carbon dioxide. Thos described by pathway a-b-c-d-e-f-g and its counterpart, j-k-c-d-e-f-g, involve the action of carbon dioxide on plant evolution and the feeback of plants on chemical weathering rates (Figure 8.3). Three other PFLs emerge from the effect of terrestrial ecosystem evolution on sedimentary organic carbon burial (a-b-c-d-m-n and i-k-c-d-m-n) and the intensity of the hydrological cycle (a-b-c-d-u-q-r-g). All five pathways lead to positive feedbacks, whether carbon dioxide is rising or falling, but only if the paths at some stage involve very warm global climate states that could induce lethal overheating of leaves (Berling and Berner, 2005).

Positive feedback is initiated by any change in the global concentration of atmospheric carbon dioxide that inversely influences the density of stomatal proeson the leaves of vascular land plants. Falling carbon dioxide accompanies higher stomatal density which, in turn, lowers leaf temperatures by increasing latent heat losses due to higher evapotranspiration rates (path a-b). Falling carbon dioxide also lowers ambient temperatures, because of the atmospheric greenhouse effect, and humidity, because of the exponential effect of temperature on the saturation content of water in air. These environmental effects reduce the leaf to air water vapour deficit, and so allow stomal conductance to water vapour to increase and cause further reductions in leaf temperature (path i-p-s). Because of the capacity for more efficient cooling, new trees develop with larger leaves that intercept more solar radiation (path c) without the attendant risks of lethal overheating. Larger leaves increase canopy size and represent an optimal tradeoff between investment in woody supporting tissue and leaf area for photosynthetic carbon gain. Higher stomatal densities reduce the diffusional limitation on photosynthesis through increased rates of carboxylation in the primary photosynthetic enzyme of $C_3$ plants, rubisco. Increased photosynthetic capacity, coupled with greater interception of solar radiation by large leaves, facilitates the evolution of leafier, more productive plants (path d). Higher stomatal densities also permit taller plants by providing improved fine-scale control of transpiration to protect the increasing length of the xylem water pathway from cavitation (Tyree and Zimmerman, 2002) and distribute water and nutrients in the transpiration stream.

Bigger plants demand more nutrients and water than smaller ones. Meeting these enhanced demands required more root (plus mycorrhizal) biomass and/or deeper rooting systems. An expanding root systems, in turn, increases weathering (path f).

The Earth systems' analysis made by Beerling and Burner (2005) shows that plant responses to carbon dioxide across a wide spectrum of timescales have had strong consequences for the evolution of climate and the biota. Indeed, it throws doubt on the widespread belief that establishing positive geophysiological feedback between plants, carbon dioxide, and climate on geological timescales are rare when compared with negative stabilizing feedbacks.

According to Beerling and Berner, the PFLs discussed here (Figure 8.3) led to reinforcement (Odling-Smee *et al.*, 2003), and accelerated rates of evolution on a grand scale. Falling carbon dioxide concentrations are linked to the unparalleled innovation of land plants in the Paleozoic (Osborne *et al.*, 2004) that, in turn, promoted the diversification to terrestrial tetrapod and insect faunas and markedly increased atmospheric oxygen, fuelling the spectacular ecological radiation of gigantism in insects (Graham *et al.*, 1995). These consequences are in sharp contras with positive feedbacks between plants and carbon dioxide in the short-term carbon cycle that are likely to accelerate anthropogenic climate change over the 21st century (Cox *et al.*, 2000).

# Carbon Dioxide Starvation and the Development of C$_4$ Ecosystems

It was the decline of atmospheric carbon dioxide over the last 65 million years (ma) that resulted in the 'CO$_2$-starvation' of terrestrial ecosystems, leading to the widespread distribution of C$_4$ plants, which are less sensitive to carbon dioxide level than are C$_3$ plants. Global expansion of C$_4$ biomass has been recorded in the diets of mammals from Asia, Africa, North America and South America during the interval from about 8 to 5 Ma. This was accompanied by the most significant Cenozoic faunal turnover on each of these continents, suggesting that ecological changes at this time were an important factor in mammalian extinction. Further expansion of tropical C$_4$ biomass in Africa also occurred during the last glacial interval confirming the link between atmospheric carbon dioxide levels and C$_4$ biomass responses (Cerling *et al.*, 1998).

Changes in fauna and flora at the end of the Miocene, and between the last glacial and interglacial, have previously been attributed to changes in aridity. An alternative explanation for a global expansion of C$_4$ biomass is carbon dioxide starvation of C$_3$ plants when atmospheric carbon dioxide levels dropped below a threshold significant to C$_3$ plants. Of course, aridity also may have been a factor in the expansion of C$_4$ ecosystems, but if so, then it was secondary to and perhaps because of gradually decreasing carbon dioxide concentration in the atmosphere. Mammalian evolution in the late Neogene, then may be related to the carbon dioxide starvation of C$_3$ ecosystems (Cerling *et al.*, 1998).

The three photosynthetic pathway used by plants are knows as the C$_3$, the C$_4$ and the CAM pathways. The C$_3$ pathway is the least advanced route and was used by the earliest plants, from the early history of the Earth when carbon dioxide was the most abundant gas in the atmosphere to the present day. The C$_4$ and CAM pathways evolved more recently, apparently in response to lower atmospheric carbon dioxide levels (Ehleringer *et al.*, 1991, 1997). The C$_3$ plants make up most of the global biomass; the C$_4$ plants make up about 18 per cent of the global terrestrial productivity (see Cerling, 1997) and the CAM plants have a lower global productivity than both C$_2$ and C$_4$ plants.

Figure 8.4 shows a model suggested by Cerling *et al.* (1997) and Ehleringer *et al.* (1997) that has important implications about ecosystems when it is considered in the light of the history of atmospheric CO$_2$ on Earth. The 'natural' level of CO$_2$ in Holocene interglacial conditions was about 270 ppmv (the Preindustrial Revolution concentration). Which falls in the range that represents CO$_2$ starvation for C$_3$ plants in the warmer parts of the planet. Such condition favour the C$_4$ monocots; C$_4$ dicots are generally not favoured over C$_3$ dicots except during the interglacial CO$_2$ lows.

The Earth has been in this 'CO$_2$-starved' mode, where C$_4$ monocots make up a significant fraction of the Earth's total biomass, for some 7 Ma, a condition rarely (if ever) attained in the earlier history of the Earth's atmosphere. This paucity of CO$_2$ had strong consequences on global ecosystems and evolution (Cerling *et al.*, 1998).

Between 8 and 6 Ma there was a global expansion of C$_4$ ecosystems. There is no evidence for the presence of C$_4$ biomass in the diets of mammals before 8 Ma (Cerling

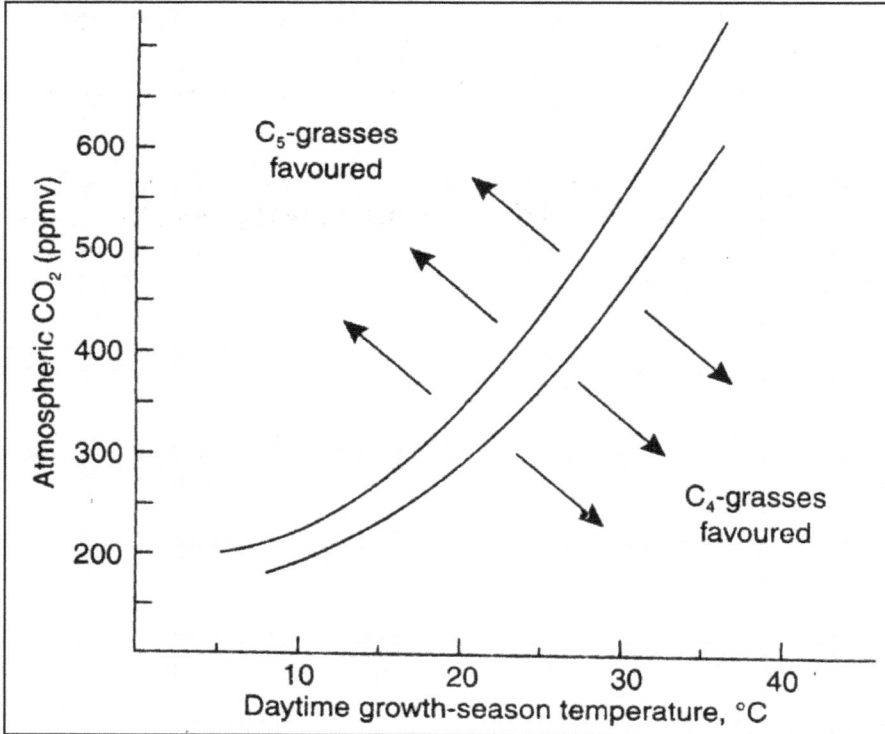

**Figure 8.4: Crossover model of C$_3$/C$_4$ photosynthesis based on quantum yield of C$_3$ and C$_4$ plants (after Cerling *et al.*, 1998).**

*et al.*, 1998). By 6 Ma, there is abundant evidence for significant C$_4$ biomass in Asia, but not in Europe. According to Cerling *et al.* (1998), atmospheric CO$_2$ levels have been generally declining over the last 65 Ma or longer from values more than 1000 ppmv at the end of the cretaceous to glacial and interglacial levels between about 200 and 300 ppmv, respectively, in the last 2 Myr. Physiological studies of plants suggest that C$_3$ plants become increasingly less efficient under low atmospheric CO$_2$ levels, especially when accompanied by high temperatures, leading to conditions where C$_4$ plants have a higher efficiency than C$_3$ plants. Likewise, stable isotope studies of mammalian diets have revealed that there was a global increase of C$_4$ biomass in the latest Miocene to earliest Pliocene (8-5 Ma), a time when faunas similar to the modern type replaced those adapted to more closed habitats in many part of the world. Carbon isotope studies of lake sediments and bogs from tropical regions indicate that in some tropical environments, C$_4$ plants were more abundant under the cooler glacial conditions than under warmer interglacial conditions. These observations are consistent with a model of CO$_2$ starvation of C$_3$ biomass in the late Cenozoic up to the present day.

# Carbon dioxide and Nitrogen Fixation Decline

Increasing atmospheric carbon dioxide ($C_a$) is mostly a product of fossil-fuel burning, land use change, and cement manufacture. It is expected to cause a large carbon sink in terrestrial ecosystems, partly mitigating human-driven climate change (Houghton *et al.*, 2001). Increasing biological nitrogen fixation with rising $C_a$ has been considered as a means to provide the N necessary to support C accumulation (Gifford *et al.*, 1996). Hungate *et al.* (2004) found that $C_a$ enrichment increased N fixation during the first year of treatment in an oak woodland, but the effect declined and vanished by the third year. The Ca enrichment consistently depressed N fixation during the $5^{th}$ to $7^{th}$ years of treatment probably as result of reduced availability of molybdenum, a key constituent of nitrogenase. Their results show how multiple element interactions influence ecosystem response to atmospheric change. They caution against expecting increased biological N fixation to fuel terrestrial C accumulation.

# Reversal of Warming Effect of Ecosystem Water Balance by Plants

Global warming is thought to increase aridity in water-limited ecosystems by accelerating evapotranspiration. This reduces soil moisture which could offset modest increases in continental precipitation and lead to greater aridity in water limited systems around the world (Manabe and Wetherald, 1986). However, certain models that predict this drying do not consider direct biotic responses to warming. Various kinds of vegetation responses to warming, including changes in phenology, leaf area index and rooting depth, may influence evapotranspiration rates and soil moisture availability (Schulze *et al.*, 1994).

Very few experimental observations of the effects of warming on ecosystem water balance have been recorded. In most snow-free ecosystem, vegetation interacts directly with the atmosphere for a considerable part of a year; biotic mechanisms seem to play important roles in mediating water balance. Warming effects on plant production or phenology could influence soil moisture losses by altering the rate or annual duration of plant transpiration. Zavaleta *et al.* (2003) determined the responses of soil moisture availability and plant canopy phenology in a temperate grassland to 2 years of simulated warming under both ambient and elevated $CO_2$. They have presented evidence of a mechanism controlled by temperature responses of grassland vegetation through which warming increases moisture availability at a critical period in a seasonally dry ecosystem. In a 2- year field experiment, simulated warming increased spring soil moisture by 5 to 10 per cent under both ambient and elevated $CO_2$. Warming also accelerated the decline of canopy greenness (normalized difference vegetation index) each spring by 11 to 17 per cent by inducing earlier plant senescence. Lower transpiration water losses resulting from this earlier senescence provide a mechanism for the unexpected rise in soil moisture. Their findings illustrate the potential for organism environment interactions to modify the direction as well as the magnitude of global change effects on ecosystem functioning.

The work of Zevaleta *et al.* (2003) illustrates the potential for organisms-environment interactions to strongly modify global change effects on ecosystem function. In at least some ecosystems, declines in plant transpiration mediated by changes in phenology may offset direct increases in evaporative water losses under future warming. In many grassland and savannas, subtle changes in resource availability at critical times are known to dramatically alter species composition. Extra moisture during the critical time of drought onset promotes establishment of both late-season and woody plant species in California grasslands. A biotic link between warming and water balance thus could influence community responses to climate change in grasslands and savannas.

## Agriculture and Climatic Constraints

The genetic foundation of today's world agriculture is surprisingly narrow. Since times immemorial, the domestication of plants and animals has been achieved by a selective filtering of the species (and a reduction in the number of their varieties) upon which humankind depends for agricultu5ral production. Worldwide, there are only 30 crop species whose individual production exceeds 10 million tones (Mt.) annually. Cereals account for over half of the arable land use, and only three crops-wheat, maize and rice-account for 80 per cent of total cereal production. Similarly, only two animal products (beef and pork) make up approximately three quarters of the world's total animal production.

At local and regional scales, we do not know whether this 'simplification' increases the vulnerability of agricultural production to climatic change. It is prudent to make earnest efforts to introduce species or varieties that are better suited to local earnest efforts to introduce species or varieties that are better suited to local environments as among other things, a hedge against climatic change and variability.

At the global scale, the differences between agricultural regions are immense, which is why total world food production is remarkably stable from one year to the next. The differences pertain not only to environmental and climatic characteristics but to levels of economic development, technology, and human living standard as well. The same major cereal crops are grown and/or consumed in countries as diverse as the US. Australia, and Western Europe and differential regional effects of increased $CO_2$ and climatic change could tip the balance with worldwide repercussions. At the broadest scale, the regional differences are most pronounced between the tropics and the temperate regions and their semi-arid and humid zones.

Within many areas of the tropics, agricultural production is fairly low and unstable as compared to temperate regions. Largely, this is due to the broad problems of underdevelopment, struggling economies and low levels of agricultural technology prevalent in the low latitudes.

The climate of the tropics influences the patterns of agricultural activity and contributes considerably to the persistent agricultural problems. In general, rainfall is the climate variable of primary importance in shaping the spatial and temporal variations of agriculture in the tropics. Temperature is secondary but becomes increasingly important as we move pole wards. In the humid tropics, rainfall is also a major growth-limiting factor due to its variability and the high potential evapotranspiration. There is a close dependence of the growing season on rainfall throughout the entire tropics.

Food production depends not only on the timely appearance of the monsoon but also on its strength and reliability throughout the growing season. Indeed, late heavy rains are often just as disruptive as the late arrival of rain, causing flood damage to established crops at a time when it is too late to replant. This applies even to rice which has high water requirements.

Rice is the staple food for about 60 per cent of the world's population. Production of rice in lowland locations is suited to the humid tropics, particularly if well developed water control and/or irrigation networks are in place. Nevertheless, even lowland rice is susceptible to rainfall variability. There is a positive relationship between rice yield and rainfall throughout monsoon Asia. The relationship is strongest in areas where water control and irrigation are least developed-India. Myanmar, Bangladesh, and Republic of Korea. The effects of rainfall variability are seen clearly in Sri Lanka where marked differences in rice production occur over short distances between the wetter and drier zones.

Under optimally irrigated conditions, radiation and temperature influence rice yields through their effects on photosynthesis and respiration rates, respectively. In Japan (a temperate region), irrigation has greatly reduced the susceptibility of lowland rice to rainfall variation.

It appears that in South Asia, there are three types of agricultural pursuits involving annual crops other than lowland rice that are particularly susceptible to climatic variations. These are currently widespread in the humid tropics and are expanding and include:

1. Shifting cultivation in which traditional fallow cycles are shortened, leading to declines in soil fertility and erosion;
2. Continuous cropping (of maize, sorghum, upland rice, peas, beans etc. primarily for subsistence); and
3. Cultivation of feed crop for export maize production in Thailand.

These activities are typical relegated to upland areas that are subject to soil deterioration and that are hydrologically marginal when intensively utilized for agricultural production.

In several ways, the world's temperate regions differ sharply from the tropics. The mid-and high latitude zones are the centres of grain production other than rice. The potential climatic range of most of the important crops grown in the temperate regions is quite large.

Wheat and potatoes, for instance, can be grown in any state of the U.S. In the mid and high latitudes, temperature is important in shaping the spatial and temporal variations in agriculture. In contrast to the tropics, the growing season in the temperate zones is generally defined by temperature. Polewards, the geographic extent of crop production is ultimately set by temperature, and the length of the frost-free growing season determines the spatial limits of various agricultural activities.

The role of climate in the interannual variability of grain production is a major concern. There are some major differences between regions. In contrast to the tropics, the large increases in total production during recent decades in the temperate, grain-producing regions have resulted largely from intensification, rather than from expansion of cropped area.

## Global Forest Ecosystems and Climatic Constraints

Many forested landscapes deceptively appear natural and largely unmanaged. In fact, the world's forests are subjected to a wide range of management levels. Almost all forests are managed to some degree for purposes ranging from intensive commercial extraction to extensive resource conservation. Therefore, as in agriculture, the potential impacts of increasing concentrations of atmospheric $CO_2$ and climatic change need to be examined in the context of human use and manipulation of the natural system.

In the most extreme case, there has been some experimentation with intensively managed biomass plantations in which trees are irrigated, fertilized and harvested in short (2-5 years) rotations. This form of cellulose production is the type of forestry that most closely resembles intensive agriculture. In more traditional forestry, forest management involves the regeneration of a commercially valuable tree species by altering sites, planting seedling trees at suitable spacing, thinning the trees and harvesting the tree crop. In favourable environments, some of these activities are left to natural processes. Thus, if a commercial tree species regenerates fast in a given environment, the site preparation or planting management steps are eliminated, so that thinning and harvesting of trees become the only concern. In less intensive forestry, trees are periodically harvested, but the thinning of trees to optimize the forest productivity is dispensed with.

Because of this gradient of management intensity, global environmental change could manifest itself in radically different ways. In more intensive forms of forestry, a change in growth and regeneration rates affects management costs or the techniques used to extract wood products. In the less intensively managed forests, an environmental change might actually change the structure, composition, area distribution and the function of the forest ecosystem.

There are two aspects of the behaviour of forest systems that need to be kept in mind while assessing the impacts of environmental change. First, there is some spatial heterogeneity in the potential response of the world's forests to change in climate. Second, at any given place there is wide range of temporal scales over which forests will respond dynamically. Unlike the vast majority of agricultural systems, forests are dominated by long-lived trees that respond to stress or change at several different timescales.

## Climatic Change and the Areal (Macroscale) Response of Forests

The readily discernible correspondence between the distributions of global climates and the spatial patterns of vegetation leads one to expert that a change in the former should eventually produce a response in the latter. The Holdridge life zone system classifies the expected vegetation under differing temperature and moisture conditions. It is similar to other climate/vegetation mapping systems in that it explicitly recognizes the variables of temperature (expressed in this case as 'biotemperature' which is computed as a heat index for periods during which plants can be photosynthetically active) and moisture (expressed as either rainfall or evaportranspiration). It illustrates several relationships that provide perspective for understanding the response of the global vegetation to climatic change. First, there is a parallel between the latitudinal zonation of the Earth (boreal, tropical, etc.) and the zonation of vegetation at different altitudes on mountains (montane, alpine, etc.) second, the responses to temperature and moisture or precipitation changes depend on relative, rather than absolute, changes. A small absolute increase in temperature could be expected to cause a large response in the ecosystems of the cooler climates of high altitudes or latitudes. Similarly, a small absolute increase in moisture may exert a marked effect in an arid region. To cause a vegetational change of comparable magnitude in a wet, warm region, the environmental changes would have to be much greater.

When the global forest cover is plotted as a function of latitude (Figure 8.5) the forested regions of the world resolve into two major forest systems. First there is a considerable extent of forests in the higher latitudes that is dominated by evergreen coniferous trees the circumpolar boreal forests. Second , at allow latitudes, a variety of evergreen tropical forests form a second great forest system. The deciduous forests of the middle latitudes which once covered large areas in Europe, China and the United States have been reduced greatly in area by land conversion. Of the two great

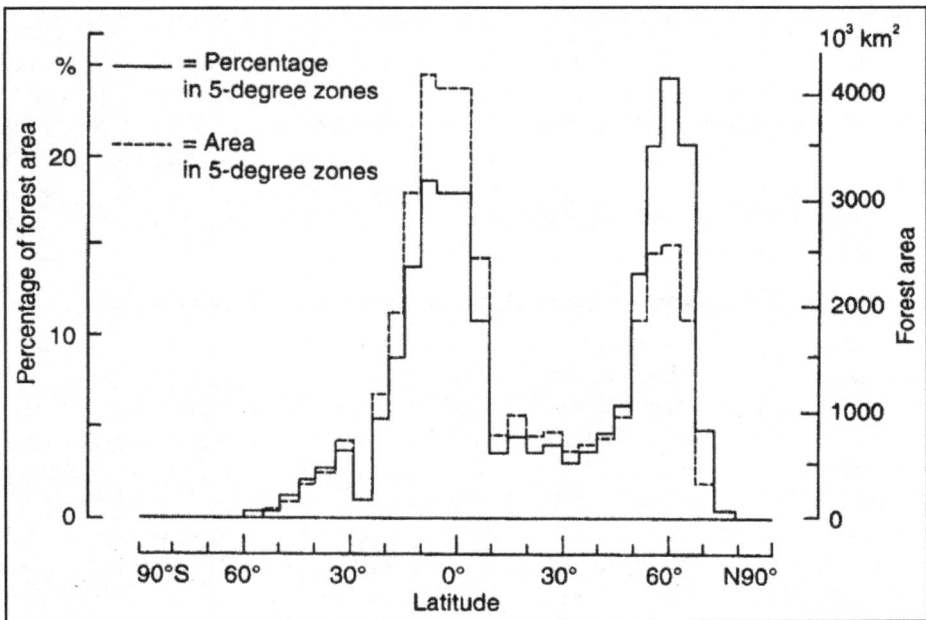

**Figure 8.5: Latitudinal distribution of forested land of the Earth in 5 degree zones. Percentages are related to the total area of each zone. The distinct bimodality of the distribution corresponds with the boreal forests in the higher northern latitudes and the tropical forests in the equatorial zones (from Baumgartner, 1989).**

forests now covering the Earth, the boreal forests may be the most sensitive to a warming.

The immediate responses that one might expect to occur from increased $CO_2$ or climatic change involve modification of forest productivity. Again, there are several differences in productivity and in the factors that modulate the effects of climatic change from one location to another. For example, a warming may do little to increase the productivity of a nutrient limited forest system. Nonetheless, across a broad range of forests, there are positive relationships between the temperature and either the total biomass or the net productivity of forests. Given an adequate supply of water and nutrients, one would expect a global warming generally to enhance the forest productivity.

## Sensitivity of Forests to Climate Change

The sensitivity of forests to climatic change has been estimated by means of forest simulation models which suggest that temperature increases of the size indicated by current climate models for a doubling of atmospheric $CO_2$ are potentially sufficient to produce substantial intermediate and long-term responses in the composition, size and location of forest ecosystems. These climate models predict the highest warming to occur at high latitudes as result of increased concentrations of GHGs with smaller rises in temperature in the lower latitudes. The natural forests of the high latitudes in general, and the boreal forests in particular, may be most sensitive to temperature changes. Warmer conditions could thus possibly lead to large reductions in the area extent of boreal forests and a poleward shift in their boundaries.

The possible problem of a change in climate due to the emission of GHGs should be viewed as one of today's most important long-term environmental problems. It needs to be considered in the context of other ongoing changes of our environment also caused by human activities, such as air pollution, acid rain and deforestation. Only in this way, we can achieve a realistic integrated view of the interplay between the environment as a whole and the global society that is required for meaningful consideration of options and policies for avoiding long-term negative consequences.

## Timescales and the Response of Forests

Forest ecosystem contain a complex web on interactions among physical, chemical and biological processes which can attenuate or amplify direct change in a given process so that the response elicited from a forest are manifested on many different timescales. Several of the important processes in forest ecosystems that have very rapid response times are influenced by carbon dioxide and climatic variables. Among the most important are those involving the exchange of water, heat and carbon dioxide between the leaf surfaces and the environment. An increase in the ambient carbon dioxide concentration could reduce the opening of the stomata required to allow a given amount of carbon dioxide to enter the plant and might thus reduce the loss of water from a tree. This could increase the efficiency of water use and raise the productivity of forests. Changes in the radiation input, temperature or humidity above a forest canopy can also produce almost instantaneous responses in carbon dioxide uptake and water use by the forest.

Tree growth results form the amount of photosynthate produced and the allocation of this photosynthate within the tree. The complex problem of tree growth has been modeled using mechanistic representations of physiological processes, but these models have rarely been used to predict responses over period longer than a year because opposing environmental conditions (high temperature and low precipitation; low temperature and high precipitation) can produce the same response; the production of a narrow annual ring in trees growing in arid sites.

## Genetically-Engineered Forests

While the Kyoto treaty demands reductions in GHGs, the ongoing loss of forests is outpacing their renewal by natural means., This has prompted the possibility of addressing the problem by planting genetically engineered forests. But this can generate another anti-GMO crusade like the well known campaign against GM-food. Genetically modified (GM) trees which, under the Kyoto Protocol already in force, are viewed as a viable, clean mechanism for sequestering carbon in the environment, which contribute to global warming (Herrera, 2005).

Transgenic poplars have already been planted in China. Activists have vowed to launch a global campaign to ban GM trees which, according to them, are hazardous to the environment. It is feared that GM trees are antithetical to Kyoto. Some of the traits being engineered into trees are insect resistance, herbicide resistance, sterility and faster growth. If these were to escape into native forests, the forests may suffer contamination or destruction-worsening global warming.

But proponents feel that the GM trees should be planted without delay. It is a proven fact that trees effectively contain or 'sequester' the GHG's *i.e.* carbon dioxide, carbon monoxide and methane.

Unfortunately, because of growing demand for pulp, paper, building materials, farmlands, and heating and fuel, trees are being replaced. A net loss of trees on the planet is expected within the next two decades, possibly causing havoc in industries, economics, and the environment (Herrera, 2005).

Even today, there are not enough trees on Earth to deal with rising levels of GHG emissions. Most trees take decades to reach maturity. This makes replacing harvested trees a long-term objective. But biotechnology offers much help to nations in bridging the gap that traditional tree breeding cannot. Thanks to forest biotechnology, genetically modified trees are ready to be planted. The only issue is should they be ?

The model species for a GM tree is *Populus trichocarpa* (poplar), whose genome has been sequenced, Its genetic blueprint can provide insights to spur the development of faster growing trees-trees that produce more biomass for conversion to fuels, while also sequestering carbon from the atmosphere, also having unique phytoremediation traits that may be gainfully exploited to clean up hazardour waste sites.

Opponents of genetically engineered forests fear that nearby 'wild' trees and various other wild organisms could become unwitting crossbreeds by coming in contact with the fallen branches and bark, pollens, leaves or sprawling roots from a GM tree bred with foreign genes to yield traits not naturally found in woods. They warn that GM trees could theoretically trigger the emergence of resistant pests and fungi. They advocate reforestation with natural breeds. Advocate counter the above argument by saying that natural forests contain unpredictable cross breeding bugs and fungi that know how to remain one step ahead of our ability to control them.

The new interest in GM trees is driven by continuing advances in genomics and genetic engineering, and by preliminary field trial data that points to the safety of this technology which could benefit the community, industry and the environment in ways that traditional breeds just cannot. Although some improvements to trees can be made through conventional breeding or selection from among the natural breeds because of the 20 to 30 years breeding cycle, it takes too long to breed and select for specific traits. Further, many of the traits sought by industry-low lignin content or altered structure for instance just do not exist either in nature or in breeding populations (see Pilate *et al.*, 2002). It is here that biotechnology proves helpful in accelerating and rationalizing the development process. The world's wood and paper firms are realizing that innovations in wood quality such as fiber length, lignin content, colour, texture, density, grain and energy coefficients, cannot be achieved without the aid of biotechnology.

Many countries intend to develop GM trees not merely as 'carbon sinks' for speeding up carbon sequestration, but also as a source of alternative fuel, erosion and desertification control, industrial waste absorption, and even medicines (Herrera, 2005). Most product-oriented research has a focus on developing trees having reduced or altered lignin content and cellulose composition that is more suitable for the pulp and paper industries; pest and fungal resistance for the fruit industry; growth factors for both the furniture and nursery industries and the conservation-environmental efforts. Many field trials have been made but so far no product has been commercialized.

The amount of timbre that can be sustainably harvested from mature forests in only about 2m$^3$ per hectare per year. Forestry experts have estimated that the current stock of 3.9 billion hectares of global forests cannot supply the average annual demand of 3.4 billion m$^3$ for much longer (Charity, 2003). According to Walter and Fenning (2004), 90 per cent of the trees being felled to meet global demand come from natural forests.

But humans are not the only threats to forests-novel tree running pests and fungi are emerging in forests faster than trees are developing resistance to them, leaving badly damaged and deforested national forests and state parks all over the world (Herrera, 2005). Air pollution and acid rain contribute to soil acidification, which slowly kills trees by robbing the soil of such nutrients as potassium and magnesium.

Traditional breeding has failed to resolve threats to our forests. But, conceivably, GM trees may be made to grow to peak height in 5 to 10 years rather than 20-30 as is the case today. This would be a boon not only for industry but also for environment. The GM trees will be developed that do not produce blossoms, seeds or pollens-or debilitating allergies for such of the world.

# Chapter 9
# Climate Change and India

## General Description

Climate change is one of the most important environmental challenges with profound implications for several different sectors such as food production, natural ecosystems, freshwater supply and health. The Earth's climate system has changed both globally and regionally since the reindustrialize era. Climate change is widely viewed as a major environmental issue of greater concern than, for instance, freshwater scarcity, deforestation and desertification, freshwater pollution, and loss of biodiversity.

The United Nations Framework Convention on Climate Change (UNFC) was prompted by the seriousness of the problem. The Kyoto Protocol proposed guidelines about the extent to which industrialized countries should reduce their greenhouse gas (GHG) emission. The Protocol also suggested mechanisms and instruments not only to promote adoption of climate-friendly mitigation technologies about also facilitate adoption to the negative impacts of climate change.

The proposed reduction of GHGs under this Protocol is a welcome first step but the commitments continue to remain far from the mitigation needs needed to achieve stabilization of the concentrations of GHGs that could prevent the dangerous impact from anthropogenic climate interference (Sathaye *et al.*, 2006).

There are several uncertainties in projections of climate change to the year 2100 and beyond (Mitra, 2004). Nevertheless, most global circulation models (GCMs) do predict global warming. The CGMs are not very strong in projecting changes, or increases and decreases in regional rainfall.

Although India does not contribute much to global GHC emissions, it will suffer from the projected climate change because of significant dependence of the population and economy on such climate-sensitive sectors as farming, forestry and fisheries.

Historically, the responsibility for rising GHG concentrations lies directly with the industrialized worlds but the developing countries are likely to have an increasing share in future emissions. The best way to address climate change may be to orient national development action towards a sustainable development pathway by switching over to environmentally sustainable technologies.

Developing countries like India need to reduce the vulnerability of their GHG inventory estimation reported in the National Communication with respect to the earlier published estimates and highlight the strengths, the existing gaps, and the future challenges. Recent assessment (Sharma *et al.*, 2006) of the trends of GHG emissions from India and a few other countries shows that though Indian emissions grew at the rate of 4 per cent per annum during 1990-2000 and are projected to grow further to meet the national developmental needs, the absolute level of GHG emissions in the year 2020 will still be less than 5 per cent of global emissions and the per capita emissions will still be lower than for many developed countries as well as the global average.

Ravindranath *et al.* (2006) assessed the impact of projected climate change on Indian forests. Under the climate projection for the year in the year 2085, 77 per cent and 68 per cent of the forested grids in India may experience shift in forest types under the A2 (with increasing population, inequity, technological change more fragmented and the heterogeneous world) and B2 (with moderate population growth and economic development, environmental protection and social equity) scenario, respectively, with serious implications for biodiversity loss. Increasing atmospheric carbon dioxide level and climate warming may cause a doubling of net primary productivity under the A2 scenario and nearly 70 per cent increase under B2.

Shukla (2006) presented some emissions scenarios which show that India's per capita emission during this century world rank amongst the lowest. Nevertheless, India's participation in stabilization regime, such as the a 550 parts per million by volume (ppmv) carbon dioxide level, would induce significant changes in energy and technology-mix with a consequent economic burden.

Stabilisation burden would be lesser in those scenarios where underlying development paths are sustainable. Some near-term energy choices might well give sustained development and climate benefits. Development and climate actions should, therefore, be aligned.

Although the science of climate change has progressed I the past decade, it still suffers from significant uncertainties, especially relating to emissions estimates, climate projections, and impact assessments (see Mitra, 2004).

Some priority issues to be addressed are:

1. Improvement in understanding of the exposure, sensitivity, adaptability and vulnerability of physical, ecological and social systems to climate change at regional and local levels.

2. Evaluation of climate mitigation options in the context of development, sustainability and equity at regional, national and global levels indifferent sectors.

3. Development of sustainable and equitable international protocols, mechanisms and financial arrangements to promote mitigation and adaptation to climate change (Ravindranath *et al.*, 2006).

## Greenhouse Gas Emissions

The parties to the United Nations Framework Convention on Climate Change (UNFCCC) are required to report to the Convention on a regular basis a comprehensive and comparable inventory of anthropogenic (GHGs) and the steps take to protect the climate. India submitted its initial national communication to the UNFCCC in 2004. Some improvements were made in GHG inventory estimation reported in the Initial National Communication with respect to the earlier published estimates, but gaps still exist and there are future challenges for inventory refinement (Sharma *et al.*, 2006). Figure 9.1 shows region-wise (world) GHG emission in the years 1990 and 2000. Tables 9.1–9.3 Summarise the GHG emissions from India in the year 1994.

## Emission Scenarios

Shukla (2006) has constructed emission scenarios for India. His analysis is centered on energy sector $CO_2$ emission and spans the 21st century. Across scenarios, aggregate emission trajectories vary significantly, showing that endogenous development choices are key determinants of emission paths.

The scenario building work suggest that India's per capita emission during the century would rank amongst the lowest. Stabilisation at a 550 ppmv $CO_2$ concentration would induce significant changes in energy and technology mix and economic losses.

Stabilisation burden would be lower in those scenarios where the underlying development paths are sustainable. The short-term energy choices given their path dependence, could deliver sustained development and climate benefits.

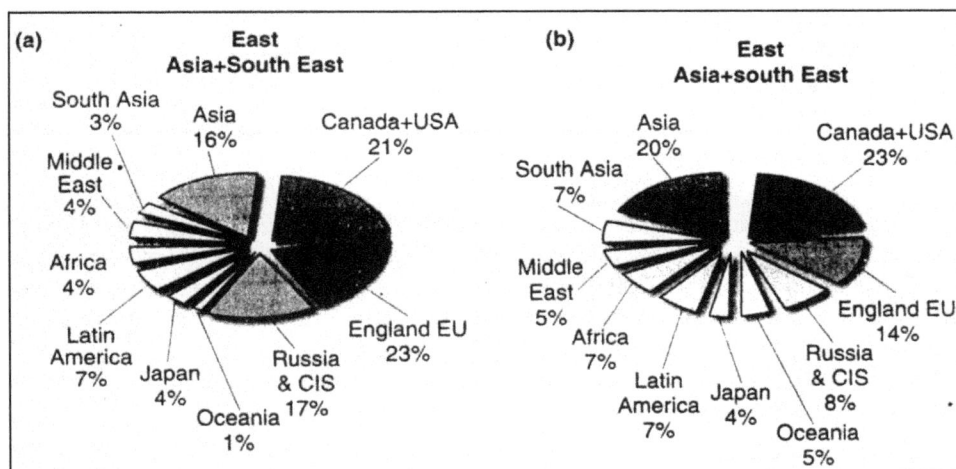

**Figure 9.1: Regionwise (global)GHG emission in (a) 1990 and (b) 2000.**

Source:http://www.rivm.nl/edgar/model/ghg/index.jsp;http://europa.eu.int/comm/ environment/climate staff_work_paper_sec_2005_180_3.pdf.

**Table 9.1: Summary of greenhouse gas emissions in Gg (thousand tons) from India in 1994 by sources and sinks (after MOEF, 2004).**

| Greenhouse Gas Source and Sink Categories | $CO_2$ (Emissions) | $CO_2$ (Removals) | $CH_4$ Emissions | $N_2O$ Emissions | $CO_2$ Equivalent Emissions* |
|---|---|---|---|---|---|
| All energy | 679470 | - | 2896 | 11.4 | 743820 |
| Industrial processes | 99878 | - | 2 | 9 | 102710 |
| Agriculture | - | - | 14175 | 151 | 379723 |
| Land use, land use change and forestry waste | 37675 | 23533 1003 | 6.5 7 | 0.04 23233 | 14292 |
| Total national emission (Giga gram per year) | 817023 | 23533 | 18083 | 178 | 1228540 |

*:  Converted by using global warming potential (GWP) indexed multipliers of 21 and 310 for converting $CH_4$ and N2O respectively.

**Table 9.2: GHG emission trends in India (after Sharma *et al.*, 2006).**

| GHG Sources and Sink (Gg) | 1990 ($CO_2$ eq-Mt) | 1994 ($CO_2$ eq. Mt) | 2000* ($CO_2$ eq. Mt.) | CAGR* in % (1990-2000) |
|---|---|---|---|---|
| All energy | 622587 | 743820 | 959527 | 4.4 |
| Industrial processes | 24510 | 102710 | 168378 | 21.3 |
| Agriculture | 325188 | 344485 | 328080 | 0.1 |
| Land use, land use change and forestry | 1467 | 14291 | - | - |
| Waste management | 14133 | 23233 | 28637 | 7.3 |
| Total emissions (Gg) | 987885 | 1228539 | 1484622 | 4.2 |
| Population (million) | 853 | 914 | 1000 | - |
| Per capita emissions (tones/capita) | 1.2 | 1.3 | 1.5 | - |

* CAGR = compound annual growth rate.

The best-known emissions scenario exercise is the IPCC SRES (Special Report on Emissions Scenarios) (see Nakicenovic *et al.*, 2000). The IPCC SRES methodology (Shukla *et al.*, 2003; Shikla, 2006). Four scenario families could be classified by 2x2 matrix whose one dimension is governance (A: Centralisation or B. decentralization) and the second is market integration (*i.e.* integration with global markets; 1: high and 2: fragmented). Four Indian (prefix I) scenarios are named IA1, IA2, IB1 and IB2, as per IPCC scenarios (Figure 9.2).

As in the IPCC scenarios, the key driving forces of Indian scenarios are also the economic growth, demographic profile, technological change, energy resources, geographic integration of markets, institutions and policies. These drivers can be quantified. The conclusions from comparisons among and across scenarios are reliable and robust.

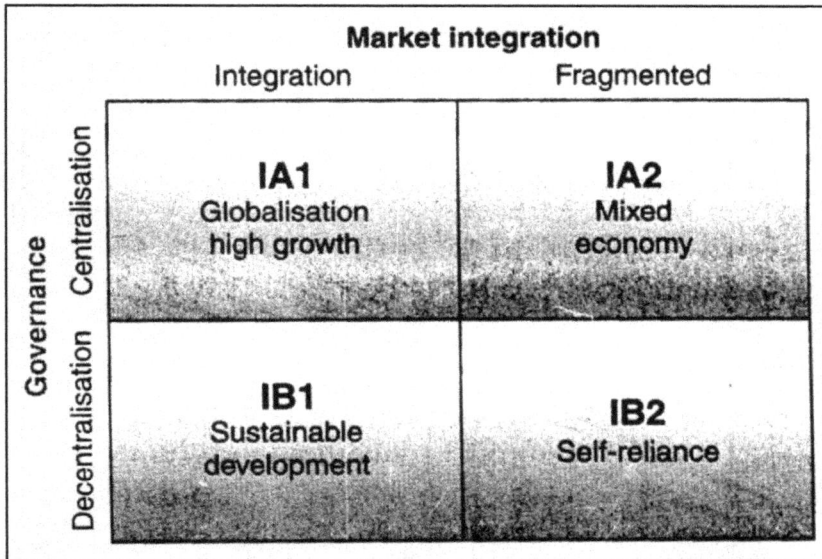

Figure 9.2: Classification of India's emissions scenario families (after Shukla, 2006).

Table 9.3: Comparative trends of GHG emission for a few selected countries (after MOEF, 2004).

| Country | $CO_2$ eq. Emissions in mmt | | |
|---|---|---|---|
| | (1990) | (2000) | CAGR (per cent) |
| Russian Federation | 3208 | 1833 | -3 |
| Germany | 1246 | 1019 | -2 |
| United Kingdom | 738 | 640 | -1 |
| Japan | 1103 | 1297 | 2 |
| USA | 5080 | 6209 | 2 |
| India | 988 | 1458 | 4 |
| China | 3837 | 4820* | 5 |
| Brazil | 1187 | 1477** | 6 |

*: Data available up to 1994; **: Data available up to 1995.

Medium-term carbon emissions (Figure 9.3) show general increase in emission in all scenarios. The emissions range varies significantly across scenarios in the year 2030; the IB1 scenario that follows classic sustainable development pattern with medium economic growth shows emission trajectory that is significantly less than the high growth scenario 1A1 accompanying higher growth and advanced technologies. Carbon emission in IA1 scenario in the year 2030 is nearly 30 per cent higher than 1B1 scenario, even though the carbon intensity (*i.e.* emissions per unit of GDP) of both scenarios is about the same. The other scenarios and IB2 (1A2 see Table 9.4) which assume medium and low economic growth rates respectively point to

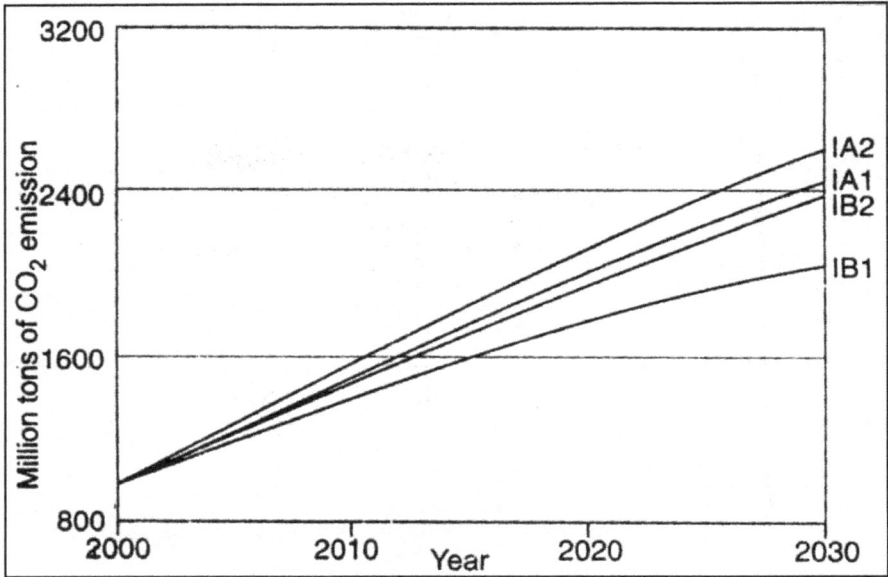

Figure 9.3: **Carbon dioxide emissions scenarios for India (after Shukla, 2006).**

high emissions intensities. Development along a sustainable pattern (*e.g.* IB1 scenario) is inherently more climate friendly.

Table 9.4: **Emission trend for 1A2 scenarios for India (after Shukla, 2006).**

| Emissions (Tg or million tones) | 2000 | 2010 | 2020 | 2030 |
|---|---|---|---|---|
| $CO_2$ | 956 | 1507 | 2080 | 2572 |
| $CH_4$ | 18.63 | 20.08 | 21.73 | 24.36 |
| $N_2O$ | 0.308 | 0.505 | 0.689 | 0.807 |
| $CO_2$ equivalent GHG* | 1554 | 2115 | 2839 | 3507 |

*: Global warming potentials used for conversion to $CO_2$ equivalent GHG emission are: $CO_2$(1), $CH_4$ (21) and $N_2O$ (310).

The integrated market reforms scenario (*e.g.* 1A1), which enhances the fuel and technology choices has better emissions intensity than those which assume a fragmented market. This means that emissions intensities can be brought down through alternate development approaches.

In all the scenarios, domestic coal attracts high share in primary energy demand. Although the $CO_2$ emissions rise in all scenarios, the per capita emissions in the 2030 for all Indian scenarios are much below the global average reported for comparable IPCC SRES non-climate intervention scenarios (Shukla, 2006).

## Climate Change Scenarios

Global atmosphere-ocean coupled models have generally provided valuable and fairly reliable representations of the planetary scale features, but their application to regional studies has been limited by their coarse resolution (-300 km). These models do not contain some topographical features such as the Western Ghats along the west coast of India, and so fail to reproduce their predominant influence on the peninsular monsoon rainfall patterns (Pant and Rupa Kumar, 1997; Rupa Kumar and Ashrit, 2001; Rupa Kumar *et al.*, 2003).

In contrast, regional climate models (RCMs) facilitate dynamical downscaling of global model simulations to superimpose the regional detail of specified regions. Developing high-resolution climate change scenarios also helps in:

1. A realistic simulation of the current climate by considering fine-scale features of the terrain;
2. Better and more reliable predictions of future climate change based on local features and responses;
3. Representation of the smaller islands and their unique features;
4. Better simulation and prediction of extreme climatic events; and
5. Generation of detailed regional data to drive other region-specific models analyzing local-scale impacts (Noguer *et al.*, 2002).

Rupa Kumar *et al.* (2006) applied a sate-of the art regional climate modeling system, known as PRECIS (Providing Regional Climates for Impacts Studies) to develop high resolution climate change scenarios for India. Evaluation of simulation (1961-1990) with PRECIS included examination of the impact of enhanced resolution and an identification of basis.

According to Rupa Kumar *et al.*, the RCM can resolve features on finer scales (50 km x 50 km) than those resolved by the GCM, particularly those related to improved resolution of the topography. The RCM gives a more realistic representation of the spatial patterns of summer monsoon rainfall such as the maximum along the windward side of the Western Ghats.

The PRECIS simulations under scenarios of increasing GHG concentrations and sulphate aerosols show marked increase in both rainfall and temperature towards the end of the 21$^{st}$ century. Surface air temperature and rainfall show generally similar patterns of projected changes under A2 and B2 scenarios, but the B2 scenario shows slightly lower magnitudes of the projected change.

The warming is monotonously widespread over the country, but substantial spatial differences occur in the projected rainfall changes. The west central India shows maximum expected increase in rainfall. Extremes in maximum and minimum temperatures are also likely to increase in the future, but the night temperatures increase faster than the day temperatures. Extreme precipitation shows substantial increases over a large area, and particularly over the west coast of India and west central India (Rupa Kumar *et al.*, 2006).

Although the scenarios presented by Rupa Kumar *et al.* (2006) point to the expected range of rainfall and temperature changes, it needs to be borne in mind that the quantitative estimates do suffer from some uncertainties. Further work will, however, minimize or remove the uncertainties.

# Climate Change Economics

India can potentially achieve substantial mitigation at fairly low price, both on the supply and demand side of energy, for carbon emission. Methane and nitrous oxide also can be mitigated at low cost. In the short term, under the Kyoto Protocol, carbon, methane and nitrous oxides can be mitigated at costs less than $ 30 per tons of carbon equivalent (or $8 per tons of $CO_2$ equivalent)- less than the prevailing price of traded carbon in European market. For the long term, modeling exercises show that between 2005 and 2035, India could supply a cumulative 5 billion tones of carbon equivalent mitigation from the energy options at less than $10 per tonne of carbon equivalent (Figure 9.4 and Table 9.5). Adaptation has attracted lesser attention than mitigation. Adaptation happens to be a private or local public good while mitigation is a global public good.

As India is a large developing country having several different climates zones, the livelihood of many people depends on such climate-sensitive economic sectors as agriculture, forestry and fisheries. The costs of not addressing climate change or failure to adapt to it are highly uncertain, but their welfare consequences are very high. Early actions on adoption, therefore, are sensible and desirable as per the 'precautionary principle' (Sathaye *et al.*, 2006).

**Table 9.5: Mitigation options, potential and costs (after Chamber *et al.*, 2002).**

| Mitigation Options | Mitigation Potential 2002-2012 (million tones) | Long-term Marginal (s/tonne of carbon equivalent) |
|---|---|---|
| **Carbon** | | |
| Demand-side energy efficiency | 45 | 0-15 |
| Supply-side energy efficiency | 32 | 0-12 |
| Electricity T&D | 12 | 5-30 |
| Renewable electricity technologies | 23 | 3-15 |
| Fuel switching-gas for coal | 8 | 5-20 |
| Forestry | 18 | 5-10 |
| **Methane** | | |
| Enhanced cattle feed | 0.66 | 5-30 |
| Anaerobic manure digesters | 0.38 | 3-10 |
| Low methane rice varieties | Marginal | 5-20 |
| Cultivar practices | Marginal | 0-20 |
| **Nitrous oxide** | | |
| Improved fertilizer application | Marginal | 0-20 |
| Nitrification inhabitors | Marginal | 20-40 |
| Climate Change and Indian Agriculture | | |

For the Indian subcontinent, the mean atmospheric temperature is predicted to increase by 1 to 4°C (Sinha and Swaminathan, 1991). The solar radiation received at

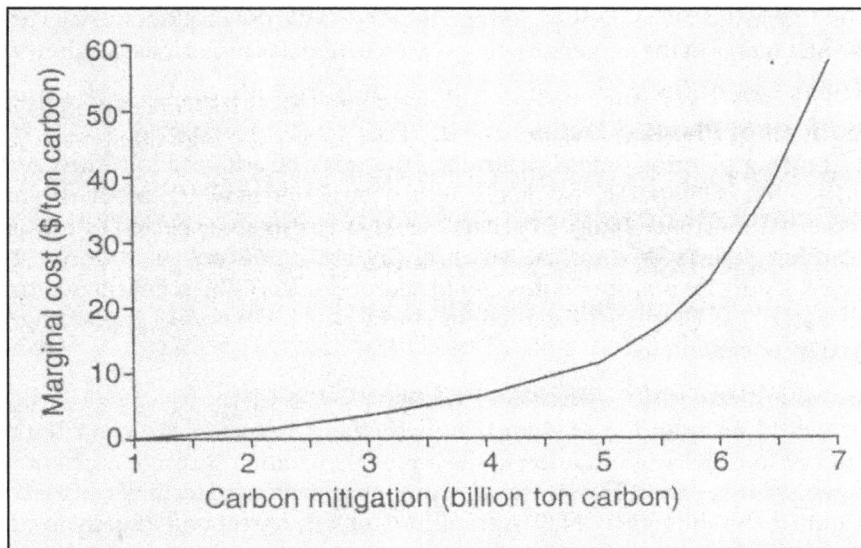

**Figure 9.4: Carbon mitigation supply curve for India for the period 2005-2035 (after Shukla *et al.*, 2004 and Sathaye *et al.*, 2006).**

the surface is Variable geographically; on average, it is expected to decrease by about 1 per cent Rice and wheat are the two most important cereals. While rice and wheat constitute the major cropping system of Indo-Gangetic plains of northern India, a the southern peninsula comprising the Godavari and Cauvery delta exclusively depends on rice. For India as a while, rice may become even more important in the future, because it car give high yields under a wider range of growing conditions than wheat (see Kumar 2006). Rice is grown worldwide over a wide geographic range from 45°N to 40°S to elevations of more than 2500 m but with average daily temperature in the range of 20° to 30°C. The impact of climate change on rice production is of paramount importance in planning strategies to meet its increasing demand. Most present varieties are highly sensitive to daytime temperatures with yield decreasing linearly with increase in daytime temperatures above 33°C. It is conceivable that high night temperatures may also reduce the potential yield of rice, primarily as a result of sterility.

## Options to Address the Negative Effects of Climate Change

Adhya *et al.*, suggested the following options to address the negative effects of climate change on rice production for assuring food security of the Indian populace.

### Varietals adaptation

There is considerable variation between rice varieties in tolerance to high temperatures. If the sensitivity of spike let sterility to temperature is shifted by 2°C higher for the new varieties, it can offset the determinental effect. Two possible adaptations are likely to occur. One could be the use of varieties more tolerant to temperature in the low latitude region and the other is the use of late maturing varieties to take advantage of the longer growing season in high latitude areas (Mathews *et al.*, 1995).

## Adjustment of Planting Dates

Adjusting planting dates is a strategy that may be adopted in future. At high latitudes, a rise of temperature would lengthen the period in which rice can be grown. In northern China (Shenyang), 43 per cent yield increases are expected by advancing the planting date by 30 days. At Madurai (India), significant yield decrease was predicted if current planting dates were used under the GISS scenario due to high spikelet sterility. These yield reductions could possibly be offset if planting were delayed by one month.

## Environment-friendly Cultivation Practice

It would be useful to standardize methods of reducing the contribution of agriculture to GHG accumulation in the atmosphere either through efficient water management to control $CH_4$ release or by proper utilization of fertilizer-N that could contribute to the emission of $N_2O$. Agriculture is likely to respond initially to climate change through a series of automatic mechanisms. Some of these mechanisms are biological whereas others are routine adjustments by farmers and markets. Climate change will impact agriculture by causing damage or gain at scale ranging from individual plants or animals to global trade networks.

At the plant or field scale, climate change can interact with rising $CO_2$ concentrations and other environmental changes to affect crop and animal physiology (Figure 9.5). Climate change involving alterations in temperature, precipitation and sea level as well as increased incidence of ultraviolet B radiation (280-320 nm) are distinct possibilities in the not too distant future (see Sen *et al.*, 2003).

Emissions of GHGs and particularly the non-$CO_2$ ones from the agriculture sector are quite significant in India. The primary sources are the large agricultural areas under paddy cultivation and high cattle population in India. The overall budget of atmospheric $CH_4$ emission amounts to 500 Tg year$^{-1}$ of which 60 Tg year$^{-1}$ come from paddy fields worldwide. Methane emissions from the agricultural sector in Tg are as follows (see Sen *et al.*, 2003).

Livestock approximately 0.3 Tg/y (largest contribution is from non dairy, followed by buffaloes); rice paddy cultivation approximately $4.07 \pm 1.25$ Tg $CH_4/y_4$; animal manure ~ 0.9 Tg/yr; field burning of agricultural residues approximately 0.116 Tg/yr.

1. The Indian domestic livestock population increased 456 million in 1987 to 467 million in 1992 and is expected to increase to 625 million in 2020.
2. The paddy cultivation area of 42.32m ha in India is the largest in Asia. The global emission of methane from paddy cultivation globally is 60 Tg.
3. Field burning of agricultural residues also releases: CO=2531 Gg; $N_2O = 3$ Gg and agricultural soils also release $N_2O$=0.24 Tg/yr (1 Tg=$10^{12}$ g or 1 million tons).

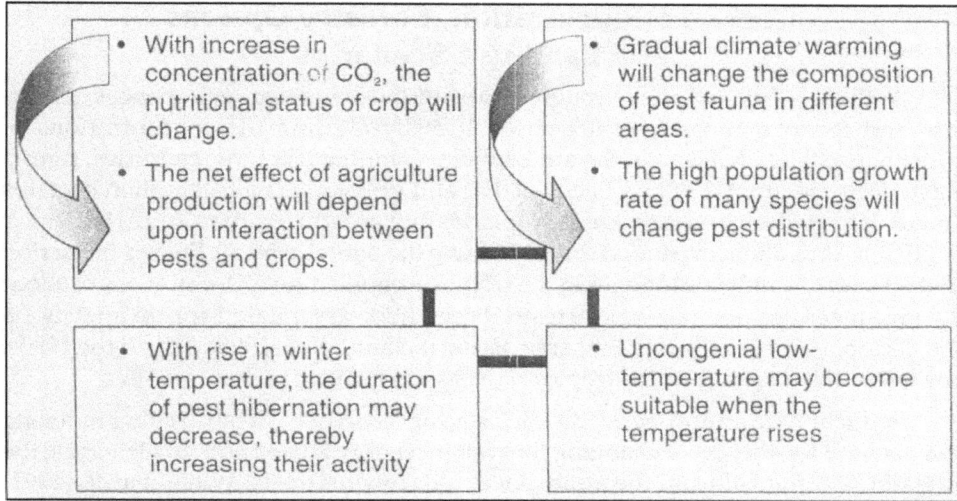

Figure 9.5: Contacts between climate and agriculture (after Sen *et al.*, 2003).

## Jute

Jute is an effective fixer of atmospheric carbon into high=energy carbon in the form of wood and hence can be utilized as energy plantation. It is a fast-growing field crop. Jute plants clean the atmosphere by consuming large quantities of carbon dioxide. Theoretically, 1 ha of jute plants can consume about 15 tons of carbon dioxide from atmosphere and release about 11 tons of oxygen in the 100 days of the jute-growing season. The carbon dioxide assimilation rate of cute is several times higher than that of some trees (Sen *et al.*, 2003).

Jute cultivation and retting emits only a negligible amount of methane. The rate of methane emission from jute retting ponds and ditches under typical agriculture condition of southern West Bengal was found to vary considerably from 8 to 779 mg $m^{-2} h^{-1}$. The annual addition of methane to the atmosphere is of the order of 0.01306 Tg, and India's contribution would be about 0.008541 Tg $yr^{-1}$. This is only a small fraction of the current estimates of global atmospheric methane increases from waterlogged fields, animals and other sources. Unlike paddy cultivation, jute is retted for only a short period in a limited area. All these factors suggest that the traditional jute retting practice in natural conditions is a negligible contributor to atmospheric methane. According to Sen *et al.* (2003), agriculture may well adjust to the global climatic change. These adjustments may involve changes in crop varieties and management practices. On an average global scale, the food production may not alter too much and harmful effect of increase in temperature and depletion of soil moisture could also be compensated by some other factors. Yet, we must enhance our knowledge of the changed climate and its agricultural implications so as to adjust our crop management practices to cope with global climate change. These authors visualize the following scenario concerning the impact of climate change on pest dynamics in agriculture.

# Carbon Sequestration, Forestry Options and Land-use Scenarios

Anthropogenic activities strongly impact the biosphere through changes in land-use and forest management activities, thereby altering the concentrations of greenhouse gases (GHGs) in the atmosphere. Over the last three centuries, forests have decreased by 1.2 billion hectares (ha) and grasslands by 560 million hectares (mha), largely due to an increase in croplands and growth of urban areas (Watson *et al.*, 1995). Agriculture expanded sharply during the period 1950-80. During the period 1981-90, while land- use changes in the tropics accounted for $CO_2$ emissions of about 1.6 giga tones per year (Gt/yr), terrestrial vegetation assimilated approximately 1.8 Gt. of carbon per year during the same period (Schimel *et al.*, 1995). Thus, I the 1980s, the terrestrial vegetation I the tropics acted as a net sink of carbon.

Reductions in atmospheric $CO_2$ can be achieved in two ways: (a) either reducing the demand for energy, or changing the way theenergy is used and (b) increasing the rate of $CO_2$ removal from the atmosphere by growing frosts. Winjum *et al.* (1992) stated that the best management practices for $CO_2$ mitigation are reforestation in the temperate latitudes, and agroforestry and natural reforestation I the tropics. Indeed, drawing $CO_2$ from the air into the biomass seems to be the best practical way for mitigation of this GHG from the atmosphere (Bhadwal and Singh, 2002).

Land resources have faced severe pressure for food and forest products from the very beginning of civilization. For proper land-use management, land resources must be used sustainable. To manage these resources effectively while safeguarding future food security, the demands and potential for our land need to be kept in view. Bhadwal and Singh (2002) made a comparative estimate of land-use and carbon sequestration potential for different forestry options for India, using the Land Use and Carbon sequestration potential for different forestry options for India, using the Land Use and Carbon Sequestration (LUCS) model. This model systematically incorporates the Indian agricultural and forest statistics as also relevant geographical and demographic data. It also considers various pressures exerted on land-use system from the existing growth rates of population, fuelwood requirements, export of agricultural commodities and probable transfer of lands from one category to the other. Bhadwal and Singh (2002) generated three scenarios (LUCS-1, LUCS-II and LUCS-III) with different land-use options following the demands and present land-use pattern prevailing in the country. The scenario LUCS-1 puts maximum amount of land into the forestry sector and is economically attractive. The LUCS-II is a 'business as usual' scenario, projected according to the current five-year plan. LUCS-III assigns maximum area of land to plantations. These scenarios were generated for a time period of fifty years starting from 2000. The LUCS model estimates the amount of carbon sequestered by approximating land-use and relative biomass changes I the landscape over time. The amount of carbon sequestered in scenario LUCS-III is estimated to be 6.937 billion tones, which is the highest among those sequestered in all the three scenarios. According to this scenario, the carbon sequestered in the aboveground vegetation of India will more than double by the year 2050 (Bhadwal and Singh, 2002).

In the 1950s, the area under the forests in India was approximately 40.48 mha; it increased to 63.92 mha during 1970s, 64 mha during 1980s and has been more or less constant since then (see FSI, 1999). Indian has about 19 per cent of its land area under forests (excluding land under three plantations). About 43 per cent of the land area in India is devoted to croplands. The remaining 113.535 mha accounts for land in forest fallows, restored forest and degraded land. Table 9.6 gives the estimated area under each of various land use classes for the year 2000.

**Table 9.6: Total area in each land-use class and their total biomass for the year 2000 (after Bhadwal and Singh, 2002).**

| Land-use class | Area (mha) | Biomass (t/ha) |
|---|---|---|
| Closed forest | 36.72 | 129.00 |
| Degraded land | 84.90 | 2.00 |
| Forest fallow | 7.21 | 16.00 |
| Open forest | 26.13 | 66.55 |
| Permanent agriculture | 141.73 | 16.00 |

Bhadwal and Singh (2002) divided the geographical area of India into nine categories. Changes brought about in the land-use pattern after a period of fifty years for the three scenarios considered by Bhadwal and Singh are shown in Table 9.7.

**Table 9.7: Change in land-use categories and associated carbon uptake during the time period 2000-2050 (after Bhadwal and Singh, 2002).**

| Land Category (mha) | Year 2000 | Year 2050 | | |
|---|---|---|---|---|
| | | LUCS-I | LUCS-II | LUCS-III |
| Closed forest | 36.72 | 92.63 | 93.88 | 91.64 |
| Degraded land | 84.9 | 0.10 | 0.38 | 1.10 |
| Open forest | 26.13 | 6.20 | 6.40 | 6.20 |
| Permanent agriculture | 141.73 | 130.18 | 129.88 | 130.18 |
| Forest fallow | 7.21 | 1.19 | 1.12 | 1.19 |
| Net carbon uptake (bt) | 5.27 | 11.92 | 10.50 | 12.21 |

Land-use is, of course, constrained by such biophysical factors as soil, climate, relief and vegetation. Most land-cover modifications and conversions happen to be anthropogenic rather than natural. Human activities change land attributes and act as proximate sources of land-use/cover change which disturbs the ecological balance, in turn reducing the potential productivity of land resources (Bhadwal and Singh, 2002).

# Chapter 10

# Base of Forest Resource

## General Description

The world's forest resource base is mind-boggling in its extent,complexity and diversity. Much of the earth surface is or has been capable of supporting woods or forests, and despite centuries or even millennia of clearance perhaps as uch as one-third retains a tree cover. Perhaps as much as 80 per cent of the pre-agricultural forest area survives, although the pattern of clearance has been very variable. Forests and woodlands still occupy around 4,000 million hectares, and in some areas such as parts of Canada and Siberia they still give the impression of being almost limitless. On the other hand, the impression created when the forest area is related to the human population is rather different. The average forst area per paerson has now shrunk to around three-quarters of a hectares, or an area similar to that of a soccer pitch. The average forest area per person is of course an abstraction: the distribution of dense population and densely forested areas are almost mutually exclusive at all scales from the global to the local. And the distributions of forest areas and forest types are remarkably uneven. This is one of the key characteristics of the forest resource base,and it is a factor of fundamental significance for the use of the resource. Another fundamental factor is the ecological character and type of the forest. Forest ecosystems vary greatly in their structure and dynamics. The usefulness of a forest, in terms of the products and services it provides, depends in part on ecological characteristics. In addition,these characteristics may determine the way in which the forest responds to use and is modified by it. In short, forest ecosystems provide the base on which resource use, and its consequences, take place.

The purpose of this chapter is to present a brief review of the forest resource base and its distribution pattern, in order to provide context and background for the ensuing discussion of resource use. The focus is on the forest resource base, rathr than on forests per se. For more detailed reviews of forst vegetation and its various types, the reader is referred to texts such as those of Eyre (1963), Richards (1952) and Waltere (1985) and the volumes in the series 'Ecosystems of the world' (*e.g.* Ovington, 1983).

# Distribution

At the global scale, the ultimate limiting factor inforest growth and distribution is climate, while climate and geographical location (in relation to the origin and dispersal of tree species) are the main natural determinants of the distribution of forest types. Two climatic elements are of fundamental importance: temperature and rainfall. Temperature in conjunction with wind exposure, determines the northern latitudinal limits of the forest as well as its altitudinal limits. Rainfall is the main control on the mid-latitude distribution of lowland forests, and its relative absence accounts for the desert and steppe zones which separate tropical and temperate forests.

Much effort has been devoted to identifying the climatic values which coincide with and therefore appear to determine natural forests distributions. The search for perfrect fits between forest margins and specific lines of temperature or rainfall values is somewhat futile: local variations in soil type, drainage and slopoe will always be complicating factors. Nevertheless, the isotherm of 10°C for the warmest month has long been regarded as coinciding approximately with the poleward limit of forest growth. The fit is far from perfect, with the forest margin lying polewards of the isotherm in maritime regions and equatorwards in more continental areas, and attempted refinements have sought to make use of data on other variables such as length of growing season.

If temperature is the determining influence on the poleward extent of forests, rainfall is the limiting factor in mid-latitudes,. The effectiveness of rainfall for plant or tree growth depends on its seasonal distribution and on the amount of evaporation, which is turn is related to temperature. Again, perfect coincidence between lines of equal rainfall (isohyets) and forest limits cannot be expected, but in practice little woodland is found in areas of less than 400 to 500 millimeters annual rainfall.

## Forest Types and Distribution

Figure 10.1 illustrates the broad distribution of the natural forest within its basic climatic controls of temperature and rainfall, and also indicates something of the diversity of forests types. It is important to emphasise that this is merely a highly generalized version of the pattern that would exist under purely naturala conditions, in the absence of human interference. It also omits local variations arising from different soil types and mountain ranges, for examples. The pattern portrayed, therefore, it not that of existing forests: indeed that pattern is very different, as will be discussed later. Much of the mid-latitude forest has long since been cleared,and the boreal (northern) coniferous forests and tropical moist forest together account for most of the remaining area of the global forest. Tropical forests, including evergreen and deciduous moist forest and dry open woodlands, make up about half of the world area of forest and woodland, and the high-latitude (mainly) coniferous forest around one-third (Brunig,

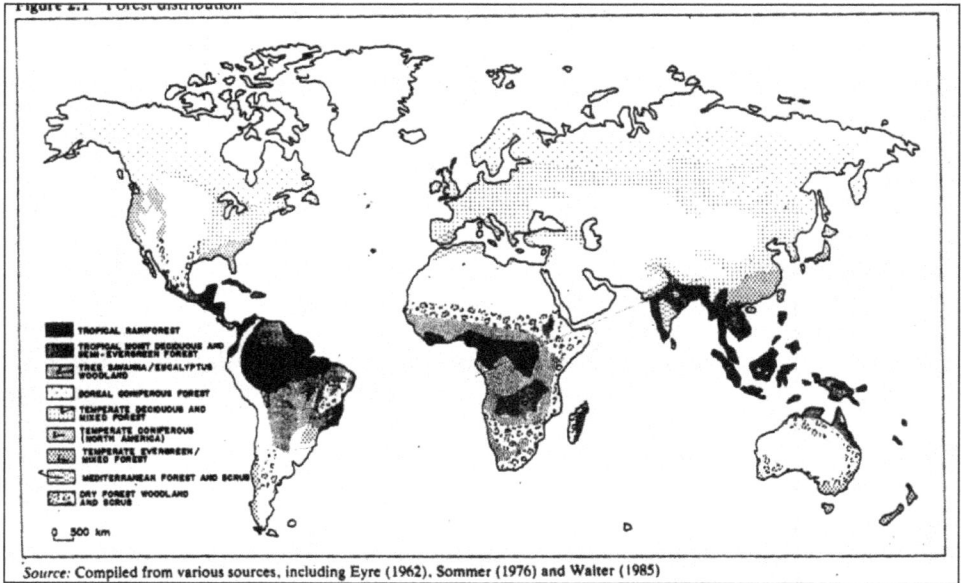

Source: Compiled from various sources, including Eyre (1962), Sommer (1976) and Walter (1985)

**Figure 10.1: Forest distribution.**

1987a). Subtropical and mid-latitude forests are now therefore relatively insignificant in areal extent at the global scale.

## The Tropical Forest

The tropical forests varies in its character and composition along latitudinal and altitudinal gradients. The key variable on which it depends latitudinally is rainfall, which tends to decrease with increasing distance from the equator. There is therefore a vegetational gradation from moist evergreen (rain) forest in the equatorial zone to drier and more openwoodland merging into savannah grasslands as latitude increases.

## Tropical Rain Forest

The tropical rain forest is found in three main areas of low latitudes, in South America, Africa and south-east Asia. There are important differences between these widely separated areas, especially in terms of floristics,but perhaps the main characteristics of the tropical rain forest as a whole is the diversity of tree species. The Brazilian Amazone region contains around 6,000 tree species, many of them endemic to specific areas (Correa de Lima and Mercado, 1985). As many as fifty or a hundred species per hectare may occur in the tree layer of the tropical rain forest, and many of them belong to different plant families. Forests in the Amazon basin have on aveage 87 species per hectare (Ramade, 1984). As many as 300 tre species have been recorded in each of two one-hectare plots in Iquitos, Peru, while ten similar plots in Kalimantan, Indonesia, were found to contain 700 tree species: the number of tree species native to the whole of North America is about the same (Wilson, 1989). Species diversity is not equally high throughout othe tropical rain forest area: for example the forests of parts of south-east Asia contain species of only a few families. Furthermore, African forests are species-poor compared with those of Amazonia and parts of south-east Asia, partly because of long- continued human influence (Jacobs, 1988). Nevertheless, the generalization of high diversity is vaoid, and is extremely important from the viewpoint of the utilization and management of the forest.

The tallest trees reach to over 50 metres, and sometimes a stratified structure is recognizable. The highest level or storey consists of a relatively small number of giants forming an open,discontinuous canopy well abaove the denser middle and lower levels. The tallest trees are characterized by enormous buttress roots radiating outwars from the base of the trunk. When one of these upper-storey trees eventually falls, a gap is opened in the canopy and ultimately may be filled by a replacement of a different species, the original tree probably having no saplings of its own species immediately below it because of the lack of light under the dense canopies. Perhaps several tree generations may pass before the original species returns to the site. Since the light intensities below the main tree storeys are usually poor, forests-floor vegetationis very limited, although most of the trees themselves are festooned with climbing plants such as lianas.

Climatic conditions are warm and moist throughout the year, and there is no climatically limited gorwing season. Individual trees shed their leaves at different times, and the forest as a whole is evergreen. Fallen leaves and other plant debris accumulate on the relatively bare forest floor, where a litter layers forms. Decompostion of the plant litter proceeds very rapidly under the continuously hot and damp conditions, and the plant nutrients are rapidly released for subsequent uptake. Despite the apparent luxuriance of the vegetation, the soils in many equatorial areas are usually very old, weathered and leached. They are often very poor in nutrients and acid in reaction. Most of the plant nutrients are contained in plant material, rather than in the soil, and these nutrients are rapidly cycled from plant to forest floor. If the tree layers are removed for example by burning, the plant-nutrients cycle is disrupted, and the nutrient reserves are suddenly and abruptly leached. A major loss of nutrients from the local ecosystem results, and the improverished soils may be capable of supporting only grasses or other non-tree species. The pattern of nutrient cycling in the tropical rain forest is therefore of crucial significance and has profound implementation both for the utilization and management of the forest itself, and also for the conversion of the forest into agricultural land.

As in other latitudinal zones, the nature of the forest changes in mountainous areas, where characteristics of both temperature and rainfall differ from those in the lowlands. Rainfall is higher at normal cloud level, which oftenlies between 1,000 and 2,500 metres, and the cloud forests which occur in this zone are characterized by ferns and mosses which drape the branches of the trees. Above the cloud level, rainfall decreases rapidly, and tree species suited to the drier environment (for example Podocarpus conifers) may be found. With increasing exposure to wind, the trees become shorter and more gnarled and stunted, and eventually the forest gives way to a shrub zone. Well before that stage is reached, however, and at altitudes of only a few hundred mitres, clear contrasts between the lowland and mountain forest can become apparent. The number of tree storeys may drop from three to two and their average height rapidly decreases. Average trunk diameter also decreases, as does the timber volume per unit area. At the same time, the number of tree species per unit are decreases. Table 10.1 summarises these characteristics, while Figure 10.2 illustrates the relationship between species richness and altitude for the example of Brunei.

**Table 10.1: Lowland mountain tropical rain forest contrasts.**

| *a) Mount Maquilingk, Philippine Islands* | | | |
|---|---|---|---|
| Forest | Dipterocarp | Mid-mountain | Mossy |
| Altitude | (450m) | (700m) | (1020m) |
| Number of tree storeys | 3 | 2 | 1 |
| Average height of storeys (m) | 27,16,10 | 17.4 | 6 |
| No. of species of woody plants | 92 | 70 | 21 |
| *b) Eastern Zairel Belgian Congo* | *Tropical Rain Forest* | *Transition Forest* | *Montane Rain Forest* |
| No. of trunks (> 20cm diam.)/ha | 115 | 180 | 200 |
| Mean height of boles (m) | 13 | 12 | 10 |
| Mean diameter of trunks (cm) | 60 | 40 | 35 |
| Timber volume per ha (m³) | 400-600 | 300 | 200 |

*Source*: Based on data in Richares (1952).

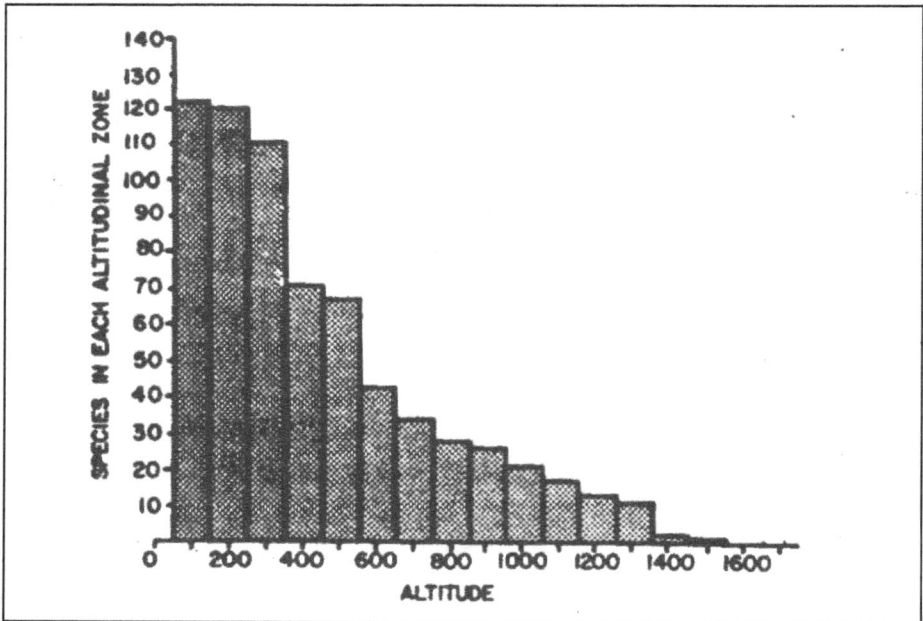

**Figure 10.2: Tropical rainforest: specis richness and latitude (metres) (Brunei).**
*Source*: Modified after Ashton (1964).

In short, the mountain forests are less rich both in timber and in diversity of species than their lowland counterparts. The more inaccessible mountain forests are amongst hose least likely to be cleared or converted; it is unfortunate that they are also the elast diverse in terms of species.

## Tropical Seasonal Forests

Outwards from the the equator, the seasonality of rainfall becomes more pronounced than in the uniformly wet equatorial zone, and the character of the forest changes as dry seasons become longer and clearer. Semi-evergreen forests are widespread in the northern part of South America and in the form of monsoon forests in parts of the area stretching from north-east India to nortern Australia (Indo-Malaysia). They are less well development in Africa. Comapred with the tropical rain forest, their structure is generally simpler and lower. There are usually only two tree storeys, with the upper one reaching to between 20 and 30 metres. Most of the tree species in the lower layer are evergreen, but a proportion of those making up the upper storey are deciduous. This forest type is represented by the so-called moist teak foests of Burma and neighbouring areas; teak, however, is only one of a number of tree species found in these areas of seasonal rainfall, and rarely occupies more than 10 per cent of a stand.

Where the patter of rainfall becomes more seasonal, the nutrient cycle differs from that of the tropical rain forest. The leaf fall at the beginning of the dry season forms a deep layer of litter which does not begin to decay rapidly until the following wet season. Similarly, with higher light intensities during the 'winter', more profuse and diverse undergrowth and forest floor vegetation may develop.

With longer dry seasons and lower rainfall, the semi-evergreen forest gives way to one more correctly designated 'tropical deciduous'. In Central and South America and in Indo-Malaysia, this forest typically has a two-storey structure, with an upper storey at around 20 metres composed mainly of deciduous species, below which may lie a lower, mainly evergreen layer. In Indo-Malaysia, this type is represented by the so-called dry teak forest, where the associated species differ from those of the moist variant.

*Tropical deciduous forest* is less well represented in Africa. It is typically discontinuous, and usually takes the form of stands of tall deciduous trees separated by open grassland or parkland. The relative proporition of the former is greatest in the belt of miombo forest which stretches across Africa from Angola to Tanzania. Here, and in a number of similar but smaller areas to the north of the equator, the dominant trees are usually flat-topped and under 20 metres in height. Tall grasses, and shrubs in some areas, dominate the undergrowth. Miombo forest grades into more open paprkland or savannh where the trees may have similar form but become more scattered and the woodland is more open. *Acacia* spp. Are usually the most common species in this zone. With decreasing rainfall the tree savannah becomes increasingly open and mergres into thorn scrub and grassland. In terms of industrial timber production, the native woodlands of the more open savannahs and dry forests are unimportant, but in relation to the supply of fuelwood their significance is much greater.

Climate, especially in the form of low rainfall, and soil conditions may both be limiting factors to forest growth over large areas of the savannah zone, but fire is a major ecological factor. During the dry season of several months, the plant litter is highly combustible, and frequent fires over a long period of time are probably responsible for the present character of much of the dry forest zone. The chances of survival of tree seedlings or samplings are low, and most of the trees making up the open woodland are of fire-resistant species such as the baobab (*Adansonia digitaja*). Where fire has been excluded in experimental areas, increases in woody growth have taken place, and commercial plantations have been established in many parts of the savannah zone.

## Boreal Coniferous Forest

Outside the tropics, the boreal coniferous forest is by far the most extensive surviving forest type. It occupies two great zones, one stretching across North America from Alaska to Newfoundland and the other from the Atlantic coast of Scandinavia to the Pacific coast of Siberia. In these zones, the climate is characterized by short growing seasons and long, could winters. The forests are dominated by conifers, most of which are evergreen. Their dominance is ensured by an ability to survive the harsh winter conditions by minimizing water-loss through their needle-like leaves and to begin photosynthesis and growth rapidly during the hsort growing season.

Although there are similarities of form and structure between the North American and Eurasian parts of the boreal coniferous forest, there are important differences in species composition both between and within these areas. The European section is much poorer in species than its counterparts in North America and eastern Asia. Scots pine and Norway spruce dominate in the European section, on the drier and wetter soils respectively. Siberian species gradually increase eastwards. Spruce gradually decreases towards the most continental areas of eastern Siberia, and its place is taken by larch, a deciduous confier. The shallow-rooted Dahurian larch is the dominant species over huge areas of permafrost. Larch forests occupy some 2.5 million square kilometers in Siberia (Walter, 1985). Further east again, towards the Pacific coast, fires and spruce again become dominant although larch remains common in northern Japan and Manchuria.

In terms of species diversity, the North American section more closely resembles the east Asian part of the boreal forest than the European area. As in Eurasia, however, there is a west-east gradation. In eastern Canada the white and black spruces are commonly found on areas of better and poorer drainage respectively and give way to species such as jack pine on poorer soils. The species pattern in the west is complicated by mountain chains, but the eastern dominants give way to lodgepole pine and western species of spruce and firt.

A distinctive feature of the boreal forest is the frequent occurrence of almost pure, single-species stands. This facilitates exploitation on a large scale, but when extensive areas of pure stands are removed by felling or fire they may not regenerate directly. Deciduous species such as birch and aspen may be the first trees to invade, and may in turn themselves form even-aged stands. Conifers such as jack pine may eventually displace the birch or aspen, and in turn may themselves be followed by the probable climatic-climax dominants such as white spruce and balsam fir.

Borth the structural and functional characteristics of the boreal forest ecosystem mean that utilization and dmanagement face problems quite different from those of the tropical rainforest, for example. In contrast to the rainforest, the plant litter on the forest floor decomposes only slowly in the cool conditions. The meman residence time for organic matter on the forest floor is quoted by Cole (1986) as 350 years, compared with four years and 0.7 years respectively for temperature deciduous and tropical forests. In the boreal forests most of the nutrients are found in the litter and in the tropical rain forest in the vegetation. Indeed the for4est-floor litter contains as much as 84 per cent of the above-ground nitrogen and 71 per cent of the potassium, whereas for the tropical rainforest the corresponding figures are only 6 per cent and 1 per cent respectively. Nutrients reserves in the leached podzolic soils of the boreal forest are poor, but the tree roots generally possess mycorrhizal fungi which may help to make the nutrients contained in the raw litter more easily available. In addition, the boreal forest appears to be able to make more efficient use of available nitrogen (as indicated by the amount of biomass produced per unit nitrogen uptake) than most other forest types.

## Other Forests

Although the greater part of the remaining forest area is contained within the boreal and tropical forest belts, forests in intermediate latitudes have been or primary importance as the hearths of modern silviculture and forest management.

## Temperate Deciduous Forest

To the south of the boreal forest zone lies temperate mixed and deciduous forest. Wide belts of transition occur in both Eurasia and North America, with the detailed pattern of distribution of conifers depending on local conditions of relief, soil and drainage. The deciduous forests of Europe stretched from the Atlantic to the Urals and were dominated by the pedunculate oak over vast reas. Depending on soils, climate and drainage, the oak was usually associated with a number of species including elm, as and beech. Beech, for example, was prominent or dominant on some calcareous areas and on areas of well-drained and rich soils as far north as the south of Sweden. It also occurred widely at intermediate altitudes in the Alps and the Mediterranean peninsulas, where it frequently occupied a zone below the coniferous mountain forest. In the southern part of the temperate belt of Europe, oak often remained dominant (for example in areas such as norther Spain and Italy), but was represented by species different from those of north-west Europe.

In North America prior to the arrival of European settlers, the deciduous forest occupied a block of country between the Atlalntic and the Mississippi. Much of this forest has now been removed or modified, as in Europe, but it seems that many structural similarities existed between the two continents. The upper storey of trees usually formed a canopy suffiently open to allow sunlight to penetrate and support a rich shrub layer. On the dother hand the American formation is characterized by a wider range of species. In New England and much of the Appalachians beech and maple are usually dominant, and are found in association with a range of other broaodleaf species as well as conifers such as hemlock in some areas. Further south, the beech-maple forest gives way to one of oak and chestnut, while in the drier areas to the west oak-hickory forest predominated. Similar deciduous forest, usually dominated by oak pseices, occupied large areas in northern China, Korea and the south-eastern part of the Soviet Union. The oaks of Shantung at one time provided food for the silkworms which produced local 'wild silk'.

In contrast to the extensive blocks of deciduous forests in the eastern parts of North America and Asia, narrow belts are found to the south of the boreal forest in the interiors of both continents. In Canada this belt is only 75-150 kilometres wide and is dominated by balsam, poplar and aspen. Aspen is also often dominant in the Siberian belt, where it is usually associated with birch. In both continents these deciduous belts have proved to be more attractive for settlement than either the boreal forest to the north or the grasslands to the south. The Trans-Siberian railway, for example, closely follows the narrow belt of deciduous forest. In both continental and maritime deciduous forests, soils are usually bettern than those in the boreal forest zone, and the more rapid decomposition of the plant litter in the waimer climate allows faster recycling of nutrients. Throughout the zone, therefore, the deciduous forest has been extensively cleared for agriculture, and most areas of forest that have not been cleared have been heavily modified.

## Temperate Coniferous Forests

Extensive areas of coniferous forest are also found outside the boreal forest, notably in North America. The so-called 'take forest' occupied a huge area stretching from Minnesota to New England; but like the deciduous forest has been severely modified by clearance and logging. White and red pine and eastern hemlock were the predominant species. To the south and east, forests dominated by loblolly and other species of pine occupied much of the coastal plain from New Jersey to nortern Florida and westwards to Texas. This forest is characterized by pure stands of mainly pine speices on low-lying sandy or marshy soil along the coastlands. The uncompromising nature of the typical soils meant that there was little incentive to clear the land for agricultural purposes, and this forest has survived better than much of the deciduous and lake forests to the north. The western forests of North American form a third type of non-boreal coniferous forest. The 'coast forest' is dominated by Sitka sspruce from the coastal lawlands of Alaska south to British Columbia, and from there southwards is gradually replaced by western hemlock and western cedar, which attain heights of 60-80 metres. From Oregon soutwards these species in turn are challenged by the even larger coastal redwood. Douglas fir occurs extensively both in coast forest of British Columbia and Washington, and in the drier mountain forest inland from the coast. In the former areas, its presence mau be related to successions following extensive fires rather than directly to the nature of the climate.

## Mediterranean Forest

Each of these types of coniferous forests in North America, as well as the other forest types previously outlined, have posed their own problems for foret management and have provided the settings for successive phases of forest utilization and development of the lumber industry. At a much earlier stage of human history,the forests around the Mediterranean did likewise. Most of these Mediterranean forests wer typically mixed and evergreen. Evergreen oak species such as the holm oak were commonand widespread, while pines such as the stone, maritime and Aleppo pines also occurred widely around the Mediterranean basis. As long ago as classical times it was known that a continual supplyof small timbers could be obtained from stump growhs (coppicing) of oak, but this knowledge did not safeguard the forest resource base. Extensive deforestation and forest destruction or modification are long established, and with them came the hydrological changes and soil erosion that are still associated with modern forest removal inother parts of the world. Much of the original forest has now been totally destroyed or reduced to scrub by centuries of grazing and burning, especially in the drier areas. Today's forests in Mediterranean Europe and Turkey occupy only 5 per cent of their original area (Ramade, 1984). On the other hand in the cooler and moister mountains, the mixed forest fo the lowlands typically gave way to coniferous forests, including the well-known cedar forests of parts of North Africa and Lebanon.

Areas of Mediterranean-type climate outside the European area-for example in parts of California, South Africa and Australia-have been less severely modified by humans and are characterized by general and species different from those of the Old World Mediterranean area. In these areas of dry woodland and scrub, summer drought is a major limiting factor, and the ttee and plant species making up the natural vegetation of open 'sclerophylous' woodland or scrub are characterized by small leaves with thick cuticles, adapted to minimize transpiration.

## Other Temperate Evergreen and Mixed Forests

Mixed evergreen forests are also found in the Southern Hemisphere, in South America, South Africa, and Tasmania and New Zealand. The tree species comprising these forsts are usually quite different from those in the Northern Hemisphere. In Chile, for example, the dominant species is Araucaria pine, which is often associated with beech (Nothofagus spp.). Forests of similar structure are found in New Zealand, but with different species: for example the kauri pine (Agathis australis) is associated with the northern part of the country, while various pines of Podocarpus spp. Are found along with broad leaved evergreens elsewhere in the islands.

Evergreen forests dominated mainly by broad-leaved species occur in restricted areas in both hemispheres. In the Northern Hemisphere the main occurrence was in central China and Southern Japan, where a range of evergreen oaks was predominant, often occurring in association with laurel, magnolias and some conifers. Forests of this type also covered much of south-east and south-west Australia prior to European settlement. These forests were dominated by drought-resistant species of *Eucalyptus*. Trees of this species are typically very tall, but cast only a light shade. They are therefore associated with a well-developed undergrowth of grass and scrub. Because of these characteristics of droght resistance and light shade, *Eucalyptus* species have been widely used in agro-forestry projects and in other forms of planting in many parts of the world far removed from their native Australia. Evergreen broad-leaved forests also occupy parts of New Zealand with much wetter climates than those found in most of Australia: for example many of the slopes of the Southern beech (*Nothofagus* spp.).

The forest resource base is therefore highly varied in its structure, composition and ecological characteristics. Although climate is a determining factor in forest type and distribution, it is not the only one. Variations in soil type, for example, may complicate the simple relationship between type of forest and type of climate, as for example in the case of the pine forests of the eastern coastlands of the United States. Furthermore, the history of changing climate over the last few thousand years is a major complicating factor. Forest type and distribution may not yet have adjusted to an equilibrium state in some areas. Locational or plant-geographical factors are obvisous: although laife forms may be similar inforests of both hemispheres the tree specis are often quite different. Human interference has been a major factor, selectively removing some species or causing indirect modification by practices such as grazing. The pre-human character of some forest types remains uncertain or unknown, simply because few completely untouched areas have survived. Forests in mountains and in areas of poor soils are more likely to have escaped clearance for agricultural purposes than forests in lowlands and more fertile areas.

With a selective pattern of forest survival, the basis for our understanding of forest ecology may be somewhat biased, especially in relation to the former forsts of the now densely populated and severely modified parts of the world. Furthermore,the concepts and systems of management that evolved in response to the ecology of one type of forest may be less appropriate in other settings. For example the tenets of silviculture that evolved I the temperate deciduous forest of west-central Europe were not necessarily appropriate for the coniferous forests of North America. At the same time humans have themselves become a major factor in the distribution of tree species as well as of forests. Tree species for use in forest plantations have been transported from continent to continent and even from hemisphere to ohemisphere. The most usual direction of movement in temperate latitudes has beenfrom the west coast of North America to Europe and Asia: movement from Europe to North America oahs been less successful (Zobel *et al.*, 1987). The extensive use of Monterey pine (*Pinus radiate*), a native of California, in plantations in Chile, Australia and New Zealand is a major example, as is the use of Sitka spruce (*Picea sitchensis*) from the Pacific coast of North America in the man-made forests of Britain and Ireland. The expansion of eucalyptus outwards from Australia to many low-latitude lands is another major example. But while species can be taken from area to area and while a certain amount can be done to modify drainage and soil factors locally the ultimate control of climate remains as an unyielding limiting factor. It is this factor above all others that determines forest growth and productivity.

## Forest Ecosystems: Growth and Volume

The forest resource-base is perhaps even more variable in its timber volume and growth rates than it is in its structure and composition. Tremendous contrasts in volume per hectare and rates of growth exist between different types of forests. Rainfall and length of growing season are major determinants of these variables. The volume of standing timber varies from around 350 cubic metres per hectare in the tropical rainforest to 50 cubic metres in savannah woodland and perhaps 150 cubic metres in the temperate forest (Persson, 1974). It is estimated that around 20 to 25 per cent of the photosynthetic matter produced on earth is in the form of wood, as is about half of the total biomass produced by the forest (Spurr and Vaux, 1976).

There is a voluminous literature on the amounts of living matter found in the various ecological zones, and on its rate of growth. Net primary productivity (NPP) is the rate of increment of plant material, and is usually expressed in terms of dry matter per square metre per year. One measure of the extent of human use of ecological resources is the proportion of terrestrial NPP used or directed by humans. This amounts to around 40 per cent.

Whilst the use of the forest resource accounts for over one quarter of the 'human' proportion, much of thisfraction arises fromforest clearing for cultivation and other forms of forest destruction without beneficial use (for example during forest harvesting). The part represented by wood actually used as lumber or for paper or firewood is equivalent to little more than one-twentieth of the NPP used or directed by humans, and to around 2 per cent of all terrestrial NPP. (Diamond, 1987).

Table 10.2 shows illustrative figures for productivity, production and biomass in the main forest types. A wide range of NPP values is indicated for most of the divisions, and different sources quote different estimates for mean values. For example, Lieth (1976) quots a figure of 2.8 kilograms per square metre per year for the tropical rain forest, while Eyre (1978) estimates the mean value at 2,500 grams per square metre per year. (See also Cannell (1982) for a compilationof onumerous reports of net production and biomass relating to different tree species and conditions world wide.)

**Table 10.2: Net primary productivity.**

| | Area (10⁶ km²) | Net Primary Productivity | | Total production (10⁹t) | Mean Biomass (t/ha) |
|---|---|---|---|---|---|
| | | Range (g/m²/year) | Approx. (g/m²/year) | | |
| Forests | 50.0 | | 1290 | 64.5 | |
| Tropical rain forest | 17.0 | 1000-3500 | 2000 | 34.0 | 450 |
| Raingreen forest | 7.5 | 600-3500 | 1500 | 11.3 | |
| Summergreen forest | 7.0 | 400-2500 | 1000 | 7.0 | |
| Mediterranean forest | 1.5 | 250-1500 | 800 | 1.2 | |
| Warm temp. mixed forest | 5.0 | 1000-1500 | 1000 | 5.0 | 300 |
| Boreal forest | 12.0 | 200-1500 | 500 | 6.0 | 200 |
| Woodland | 7.0 | 200-1000 | 600 | 4.2 | |

*Source*: Compiled from data in Lieth (1975) and Jones (1979).

As Table 10.2 clearly indicates, the torpical rainforest is the most productive forest type, and has the highest biomass. Net primary productivity is up to four times greater than that in the boreal forest, and twice that I the warm temperatre mixed forest. As Jordan (1983) observes, there is almost unanimous agreement amongst professional ecologists that productivity in tropical forests is higher than in any other forest type. Productivity tends to decline along gradients toward zones with seasonal humid and semi-arid climates, and also with altitude, on poorer soils and in areas of very high rainfall (Brunig, 1987). In this latter respect the gradient of a primary productivity resembles those of forest structure and complexity, and of species richness.

The tropical rain forest is also dominant in terms of total production and biomass per ounit area. It will be apparent from the earlier part of the chapter that the composition of the biomass varies greatly between different forest types, in terms of both therelative role of wood and the composition of tree species. A distinction needs to be drawn between total net primary productivity and wood production. Little more than half of the biomas may consist of the stem of trees in the temperate forest: for example these account for 52 per cent of the plant biomass in Russian oak forests (Walter, 1985). (Stems, of course, comprise a greater proportion of tree biomass. For example Parde [1080] quoate a range of 65-70 per cent for mature spruce forest in Canada). Although the tropical rain forest is characterized by high rates of NPP, these do not necessarily extgend to wood production. In the words of Wadsworth (1983), 'The relevance of the level of primary productivity to that of usable wood is apparently no greater than that of primary forest luxuriance to soil fertility in the tropics'. In the same way that explorers and colonizers have often mistakenly assumed that luxuriance of troical vegetation meant fertile soils, so also have politicians, planners and entrepreneurs often assumed that high NPPs meant rapid growth of useful timber. Furthermore, timber growth is spread across numerous tree species, whereas in the boreal forest, most of the tree biomass is composed of only one ot two species. Commerical harvesting (for timber) of the former is more difficult and more costly for the former, especially since only a few of the species may be useful or valuable, than it is for the latter. In short, while total NPP is much higher in tropical forsts than in those in the temperate zone, those for wood only may be much more comparable. Considerable variations exist for the latter in both zones, and especially in the temperate zone. These variations are related especially to climate, and in particular to radiation balance and rainfall.

These variations are related especially to climate, and in particular to radiation blance and rainfall.

Estimates of timber volume per hectare in different forest types have been assembled by Persson (1974) and have been used, in conjunction with his estimates of forest areas, to obtain estimates of the total volume of standing timber. These estimates are illustrated in Table 10.3, which relates to closed forest only. In the table, the dominant role of tropical and boreal zones is clearly indicated: together they account for over three-quarters of the forst area and nearly 80 per cent of the volume of standing timber.

In natural forests and other stable, climatic- climax communities, both NPP and biomass are constant: new growth is matched by death and decay. When management is introduced, however, changes may occur. Tree growth rates vary with age, and after reaching a peak begin to slow down. If management aims at maximizing timber yield, therefore, it may adjust the length of rotation to the pattern of growth. In this way the productive characteristics of the managed forst may differ from those of the natural forest. In the mainly managed forests of temperate regions usch as Europe, North America and Japan, annual growth rates or net annual increments are usually equivalent to between 1 and 6 per cent of growing stock, and are usually of the order of a few cubic metres per hectare per annum, with gradients of variation paralleling those of temperature and rainfall. In the Soviet Union, Canada and the United States for example,annual increments per hectare are respectively 1.2, 1.5 and 3.1 cubic metres. King (1975) has suggested a global average increment of around 1.1 cubic metres per hectare, equivalent to over 3,000 million cubic metres per year or around 1 per cent of the growing stock.

**Table 10.3: Distribution of forest area by type and timber volume.**

| Type | Area (million ha) | Mean Volume (m³/ha) | Total volume (1000 million m³) |
|------|------|------|------|
| Tropical wet evergreen | 560 (20.0) | 350 | 196 (49.5) |
| Tropical moist deciduous | 308 (11.0) | 160 | 49 (12.3) |
| Tropical and sub-tropical dry | 784 (28.0) | 50 | 39 (9.8) |
| Other sub-troopical | 28 (1.0) | 80-200 | 4.6 (1.6) |
| Temperate | 448 (16.0) | 150 | 67 (16.9) |
| Boreal | 672 (24.0) | 60 | 40 (10.1) |
| Total | 2800 | 157 | 396 |

*Note:*Figures in parentheses indicate column percentage.

*Source*: Compiled from data in Person (1974).

## Potential Productivity

In order to estimate the potential productivity of the world's forests, Paterson (1956) attempted to relat ideal site class, or poroductive capacity of forested lands under conditions where rotation length and management are geared to maximizing yield, to climate. He derived a CVP (climate vegetation productivity) index in which data on temperature, rainfall and growing season were incorporated. This index was closely and positively correlated ($r=+0.90$) with values for ideal site class. Difficulty was experienced in obtaining acceptable values for ideal site class, and although few of his values came from Africa and North America, the majority were taken from Sweden. From the statistical relationship that he established between the CVP index and ideal site class, he proceeded to estimate the potential productivity of the world's forests on the basis of available climatic statistics. The resulting pattern of variation is shown in Figure 10.3.

Since the map is based on a climatic index, it will be apparent that influences from factors such as soils are not portrayed, and furthermore, the climatic index itself does not incorporate certain elements such as wind exposure. Nevertheless, even

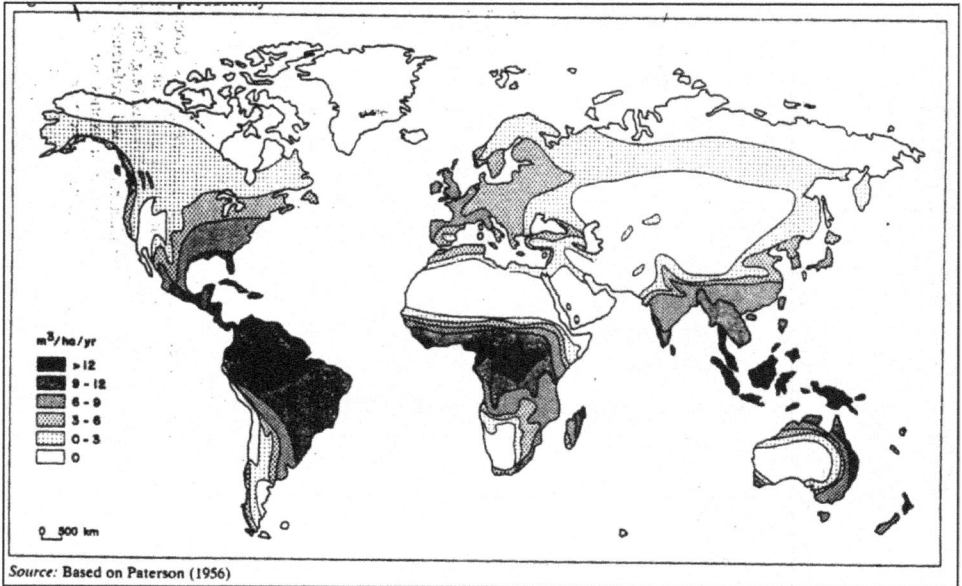

**Figurre 10.3: Potential productivity.**

though it is an oversimplification, the map shows an interesting pattern of varation in estimated potential productivity. A degree of correlation exists, as might be expected, with the map of forst distribution and type, with high values in much of the tropical zone and low values in the boreal forest zone. In most of the equatorial zone and the humid tropics, the calculated potential productivity is in excess of 12 cubic metres per hectare per year, compared with under 3 cubic metres in the boreal forests. Potential porudctivity is relatively high in the humid temperate climates of north-west Europe and eastern North America,and in parts of eastern Australia and New Zealand. On the other hand the limiting factor of moisture means that potential productivity in much of Africa and western Asia is zero or negligible.

**Plantations**

In practice much higher growth rates than those indicated in Figure 3 can be obtained in plantations, and these may have the additional advancatge of higher proportions of potentially useful stemwood than in natural forests. Their real comparative advantage, in practical terms, may therefore be greater than the apparent one. Caution needs to be exercised in comparing plantations with each other and with natural forests. Plantation records usually refer only to yield, and indicate nothing of the inputs (Jordan, 1983). Furthermore, plantation figures are often quoted over-bark, which may be thick (up to 45 per cent by volume) in young pines in particular, whereas in the temperate zone the usual practice is to quote under-bark (Zobel *et al.,* 1987). These points need to be borne in mind in considering Table 4, which illustrates some examples of annual growth rates achieved in tropical plantations. (Zobel *et al.,* quote some exceptionally high rates of up to 60 or even 100 cubic metres per hectare per year). Some figures for non-tropical plantations are included for comparison as well as some data for other managed forests.

The table suggests that growth rates in tropical plantations are much greater than those in temperate zones, and it is usually assumed that this is so. Some authorities, however, have cautioned tha this assumption needs to be viewed with care. Jordan (1983) concludes form a review of published figures that the *wood only* productivities of temperate plantations are similar to (and sometimes slightly higher than) those in the tropics and substropics. Whether this is simply a reflection of length of experience with silviculture in the respective zones remains to be seen.

**Table 10.4: Examples of average growth rats (mean annual increments (MAI) achieved in plantations**

| Country | Species | Rotation Length (yr) | MAI (m³/ha/yr) |
|---|---|---|---|
| Brazil Amazonia | Gmelina arborea | 10 | 35 |
| | Pinus caribaea | 16 | 27 |
| Central | Eucalyptus spp | - | 25 |
| Costa Rica | Pinus caribaea | 8 | 40 |
| Chile | Pinus radiate | 25 | 22 |
| New Zealand | Pinus radiate | 18 | 25 |
| Swaziland | Pinus patula | 15 | 19 |
| Malawi | Gmelina spp | 16 | 18 |
| Gambia/Senegal | Albizzia falcataria | 10 | 15 |
| Philippines | Pinus taeda | 10 | 28 |
| South USA | Pseudotsuga menziesii | 30 | 12 |
| NW USA | Picea abies | 40 | 13 |
| Scandinavia | | 50 | 5 |
| Canada (average) | | - | 1 |
| Sweden (average) | | 60-100 | 3 |
| Tropical high forest (managed) | | - | 0.5-7 |

*Source*: Compiled from data in Evans (1982) and Sedjo (1984).

Productivity or growth rat ein plantations varies with species type, management (including the nature and intensity of inputs), soil type and climae. In Europe, for example, gorwht rates vary with climate in two different dimensions. One gradient, determined mainly bytemperature, slopes upwares from north to south.The other is from extreme oceanicity of climate in western Ireland to extreme continentality in the Soviet Union. This gradient operates differentially between different species.Oceanic species such as Sitka spruce attain their maximum growth in the west. More continental species such as Scots pine grow well in east (*e.g.* Christie and Lines, 1979). Knowledge of the characteristics and preferences of species (and provenances) and experience in use of exotics are rapidly increasing, and hence growth rates can be expected to increase.

The area undr plantations in general and tropical plantation in particular is increasing rapidly, and this expansion combined with high productivities is likely to mean tha an increasing proportion of industrial wood requirements will be supplied from this source. On the other hand some controversy exists over the sustainability of high productivities beyond the first rotation in tropical plantations and pests and diseases may yet be problems. It should perhaps also be borne in mind that the expansion of plantations itself reflects the existence of some pressure on the forest resource base.

# Chapter 11

# Historical Perspectives on Forest Resource Use

## General Description

The management and use of the forest resource are sometimes seen as problems peculiar to the late twentieth century. We read much about shrinking forests and shortages of wood, and about the floods and accelerated soil erosion that are alleged to follow the removal of the forest. Topical as these problems are, however, they are not new. Although the scale of some of them may be greater than in the past, they affected our forefathers as well as ourselves. In his arrogance, modern 'technical' man may assume that both his perception of modern problems and his responses to them are (and must be) unique, and that he has little to learn from the past. If he takes this view, he is limiting his vision as effectively as if he wore blinkers.

In many parts of the world a sequential pattern of use of the forest resource can be demonstrated (Table 11.1). Initially, the forest is seen as an almost unlimited resource, with little danger of exhaustion and little need of conservation. Trends in forest use and area are perceived in neutral or positive terms: a reduction in forest area may be welcomed in allowing an expansion of the agricultural area. As this phase of resource destruction proceeds, some voice begin to call for conservation. Initially they are largely ignored, but from the faint stirrings a clamour may arise and may eventually lead to legislation or other government action aimed at halting the trends. The effectiveness of these voices has been variable. In some countries forest destruction was halted while significant forest areas still survived. In others it continued to a point where the forest resource was all but exhausted. In some instances, near-exhaustion was followed by attempts at the re-creation of the resource, and in a few of these cases the expansion of the forest area has recently attracted adverse reaction. Perhaps there may also (at least in theory) be a phase of equilibrium, when the resource is neither expanding nor contracting. Such phases, however, have been in reality few and far between.

**Table 11.1: Sequential model of forest resource trends.**

| Sl.No. | Stage | Trend in Resource Area | Perception of Trend |
|--------|-------|------------------------|---------------------|
| 1. | 'Unlimited' resource | Contraction | Positive or neutral |
| 2. | Depleting resource | Contraction | Negative |
| – | | Forest transition | – |
| 3. | Expanding resource | Re-creation/expansion | Neutral/negative |
| 4. | Equilibrium | (stability) | NA |

This three or four-phase cyclical model can be demonstrated at various scales. For many centuries, the forests of Britain suffered contraction before a dramatic turn-round in the early twentieth century led to expansion which has continued ever since. In detail the model is not quite so simple as the outline may suggest: the late eighteenth and early nineteenth centuries saw a 'false dawn' of forest expansion which was not sustained. Nevertheless in general terms it appears to be as valid as it is simple. Different parts of the world have reached different stages in the model, while at the global scale we are still firmly in the phase of resource destruction.

At the same time we can perhaps conceive of management of the forest resource in terms of the model or analogy of hunting/gathering and agriculture. Initially, utilization of the forest resource resembles hunting or gathering rather than farming. It involves the direct use of an ecological resource with little or no management or manipulation. Only later is management applied, and later still trees are grown under conditions as 'artificial' as those under which crops such as wheat or rice are produced. Again the model is perhaps not simple or linear: there may be deviations and reversals. Nevertheless, the transition from hunting/gathering to farming the forest has been made in many parts of the world, although in many others it has still to begin.

An important contextual factor is the incorporation of an area for the first time into the wider world economy. This process brings new demands for forest products, as well as new forms of control. It often coincides in time with rapid population growth, and severe pressures may be exerted on the forest resource. These may eventually be followed by a shift from one exploitation phase to the next, and by a transition from hunting/gathering to farming. Incorporation may be an important trigger factor, but the time-lag between it and the response appears to be as variable in length as it is critical for the condition of the forest resource.

While the sequential and hunting/gathering models both provide a background against which current trends in the forest resource can be viewed, they are perhaps of more value in combination than they are separately. The driving-force in the shift from hunting/gathering to farming the forest also provides the dynamic for the progression from the phase of destruction to that of conservation or re-establishment.

It may thus be both meaningful and helpful to think in terms of a forest transition, from a 'hunted', dwindling forest resource to one which is 'farmed' and stable or expanding in area. Two separate transitions may occur–the one from hunting/ gathering to farming and the other from contraction to stability or expansion–but in practice they often coincide in time.

In this chapter, forest resource use is initially considered in two contrasting areas–the Mediterranean basin in the Old World, and the United States and neighbouring areas in the New World. These histories are then considered in relation to other parts of the world, and in particular to the expanding world economy.

## The Mediterranean Basin

The Mediterranean region has been described as the 'type situation' of an unhappy history of forest management and its consequences (Thirgood, 1981, p. 163). Its degraded scrub vegetation, denuded slopes and silted river mouths are widely perceived as symptoms of mismanagement of the environment in general and of the forest in particular. Some have gone so far as to implicate this environmental mismanagement in the decline of classical civilizations (Hughes and Thirgood, 1982). The long and troubled relationship between man and the Mediterranean forest gives ample scope for thought. Several classical writers refer to the rate at which forests were being replaced by fields, pastures or scrub. Perhaps there is a parallel between the Mediterranean forest of 2,000 years ago and the topical forest of today.

### Resource Demands

In classical Greece and Rome, the forest was the source of fuel, building material and war material. Furthermore, the forest had to be cleared before agriculture could be practiced, and in areas remaining uncleared its shrub and field layers offered grazing and browsing. It is not surprising, therefore, that the forest resource was subjected to severe pressure: perhaps it is more surprising that so much of it survived the classical civilizations.

With growing populations in the city-states, more and more forest land had to be cleared for agriculture, and this was perhaps the main cause of contraction of the forest area as indeed it is today. Forest clearance occurred not only around the cities themselves, but also further a field in Greek colonies and in Africa and in other Roman provinces. An even if direct clearance did not occur, the same result was produced by grazing, which also prevented the recovery or regeneration of forests previously cleared or degraded.

The biggest single demand for wood was for fuel. In the absence of significant use of coal and oil, the Greeks and Romans had to look to the forest for fuel for both domestic and industrial purposes. According to Hughes and Thirgood (1982), probably close to 90 per cent of wood consumption was as fuel, as it is in man developing countries today. Charcoal, rather than wood, probably accounted for much of this consumption, and its production employed thousands. As a more easily transportable material than timber, it could be economically produced at greater distances from the cities and allowed urban demand to be met from some of the remoter mountain forests. Timber for building material, on the other hand, was less easily transported though more valuable. Even, when timber scarcities led to the more extensive use of stone as a building materials, with the accompanying flowering of classical architecture, some timber was still required for purposes such as scaffolding as well as for fitting our the buildings. It was also required for shipbuilding, and became a strategic raw material of primary importance. In short, with the growth of city-states of substantial populations numbered in hundreds of thousands, demand for forest products escalated at an unprecedented rate.

It is not surprising that this demand resulted in the local destruction of the forest resource around the main cities. Many classical writers refer to deforestation, and indeed it seems that the area around Athens was already bare by the time of Plato. Many also refer to the difficulty of obtaining timber suitable for shipbuilding and construction, and Plato describes the permanent damage to the environment that could result from deforestation. The loss of soil from the deforested slopes was a particular problem in both Greek and Roman times. Post-deforestation erosion products now form deposits 10 meters thick in parts of the Roman provinces of Syria and Africa (Simmons, 1989). Indeed most of the environmental concerns of the present day, with the possible exception of fears about the loss of genetic diversity, were voiced two thousand or more years ago.

Yet although the areas around the main cities may have been extensively deforested, the regional forest resource was not completely destroyed. Although some writers have assumed that Plato's description of deforestation in Attica applied also to other parts of classical Greece, it is unlikely that this was so. In the words of Thirgood (1981, p. 46). 'The overall conclusion is that, despite considerable inroads, and with the exception of the more arid, thin-soiled conditions, extensive timber forests, though often depleted, still remained at the end of the classical period, and that natural regeneration was able to maintain the forests in being'. That some of the forest resource was able to survive these unprecedented pressures was probably at least partly due to a combination of technical and organizational ability. Knowledge of coppicing could ensure a continuing supply of timber without the elimination of the forest. Methods of sowing and planting were understood, as was the practice of thinning. Without an appropriate institutional framework, however, such technical knowledge would have been of little practical use. In the past, as at present, the major problem was organizational rather than technical.

## Resource Management

Government controls on forest use were imposed in both Greece and Rome. Since ship timber in particular was an important strategic material. Athens banned the export and re-export of all such material, and indeed strict export control characterized states with significant forest resources. Those lacking these resources often sought to enter into treaty arrangements with peoples still in possession of forests. Athens and the other powers that controlled the Macedonian forests at various times tried to manage them for national purposes and applied controls to this end, Although both then and in more recent times short-term national security overshadowed long-term conservation as a policy objective.

The association of state control with forest management was even more clearly demonstrated in the case of Rome. Forests were in effect under state ownership and administration. In both unoccupied and conquered areas, government ownership was assumed, but elsewhere some areas were leased to syndicates of businessmen for commercial development and in some of the settled areas forests were included in the larger *latifundia*. The potentially damaging effects of grazing animals on forest regeneration were understood and reflected in at least some colony charters by clauses prohibiting grazing on land where young trees were growing (Meiggs, 1982). In theory, then, an effective framework existed for the management and conservation of the forest resource. A forest guard service gradually evolved and became responsible for various aspects of forest management, including what would now be termed watershed protection.

Both in classical times and more recently, however, this framework was perhaps necessary but not sufficient for sound management. Then, as now, governments from time to time sought to raise revenue by selling or leasing forest land to private owners, who would proceed to clear it or exploit it by producing timber or other forest products for the market, or even by turning it into residential subdivisions. Furthermore, a Roman law of 111 BC confirmed that anyone who had occupied public land of up to 30 acres (12 hectares) for purposes of bringing it under cultivation had a full legal right to ownership (Meiggs, 1982). Sizeable areas of woodland might be retained (or even planted) on the larger farms and products such as honey, nuts and resin produced, but on the smaller farms and residential subdivisions significant areas of woodland were less likely to be found. Then, as now, the nature of the rural population and settlement pattern helped to determine the nature of the forest.

### Forest Survival and Forest Destruction

The survival of the forest, therefore, depended on several factors. Consistent government control was one of the most important. Location was another. After the exhaustion of resources in the immediate vicinity of the city-states, those forests along rivers where logs could be floated and those around potential seaports were the next to be utilized. Conversely, the more inaccessible mountain forests were more likely to remain intact. The third major variable was the nature of the physical environment, and especially of climate. In the drier areas of the eastern Mediterranean, the forest was initially sparser and subsequently less resilient than in the better-watered areas. The drier limestone areas were especially vulnerable, and parts of eastern Greece and the neighbouring Aegean islands and coasts were amongst the fist to experience complete loss of forest and the associated loss of soil. The end stage of the process was the reduction of the forest to little more than bare rock (Figure 11.1).

The fate of the forest, therefore, depended on a combination of climate, location and control, as well as on demands for agricultural land and forest products. These were, of course, related to population trends. If one of These factors were favourable, it could to a large degree mitigate the adverse effects of the others. Even in the unlikely desert climate of Egypt, forests of considerable extent survived as late as the eleventh or twelfth centuries along the Nile Valley and in the desert fringes and in some of the wadis (Thirgood, 1981). As early as the Ptolemaic period (332-30 BC), there was

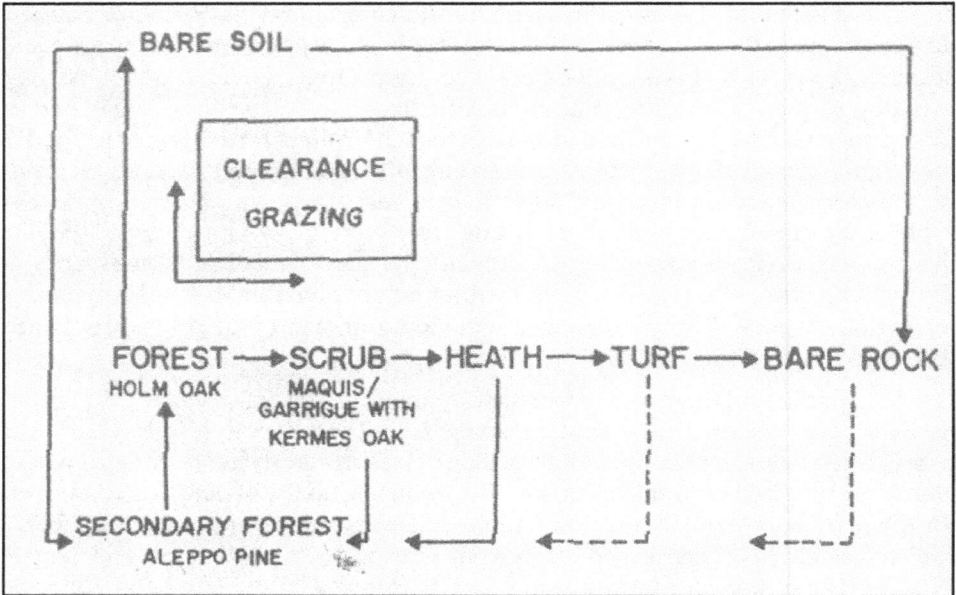

**Figure 11.1: Forest degeneration resulting from clearance and grazing in the Medirerranean world.**

government control of felling and programmes of tree-planting on wasteland and long the banks of rivers and canals. Later all forests became state property, and an efficient central government strove to conserve the limited forest estate. Only with the breakdown of this control, and with the extension of the cultivated area in the Nile Valley in the face of population growth, did forest depletion occur. Indeed remnants of the ancient Egyptian forest survived as late as 1880, when according to Thirgood, contemporary writers were predicting its exhaustion within twenty years.

## The Forest and Political Instability

Perhaps the significance of control and management is best illustrated by what happened during periods of strife, instability and power vacuums. From the Persian invasion of Greece in the sixth century BC through the Roman siege of Jerusalem in AD 70 to the depletion of the surviving forest of the Levant coast by the Turks during the First World War, there are numerous reports to the effect that the forest was, if not the fist, then a significant casualty of war. It is claimed, for example, that Lebanon lost 60 per cent of its remaining trees as railway fuel during the fist three years of the First World War. But instability did not have to develop to such extreme form as warfare and conquest for the forest to b threatened. With the decline in the classical civilizations, two opposing tends were set in motion. On the one hand, large tracts of cultivated land reverted to forest, at least in areas where the effects of the removal of the forest had not been so devastating as to prevent regeneration. Subsequently, fluctuations in population and in the rate of population growth were reflected in the waxing and more usually in the waning of the forest. For many centuries population pressures were generally light: not until the Renaissance Period in Italy, for example, was there rapid growth in population and demand for agricultural land, and hence the stripping of the forests.

On the other hand, with the decline of the classical civilization there was a decline in administration and in the enforcement of such forest laws as were still extant. Laws were now enforced only where there was strong rule: repeatedly over the centuries political weakness, uncertainly or change was accompanied by inroads into the forest area. In the words of Thirgood (1981, p. 59), 'disruption of settled government has almost inevitably led to an increase of pastoralism', and an increase in pastoralism has equally inevitably led to an increase in grazing pressures and in further attenuation of the forest. Under pressures of wars or invasions, populations might seek refuge in previously unoccupied mountains and the forests rapidly converted to crops or pastureland. Major upheavals such as the Arab Conquest of the Maghreb and subsequent mass immigration set in motion what Thirgood (1986a) describes as a process of inexorable decay of the forest.

The adverse effects of political instability continued until the nineteenth and twentieth centuries. In the upheaval that followed the Greek War of Independence in 1821, for example, many forest areas that had previously been protected by their inaccessibility were now destroyed or degraded. In North Africa, the coming of French colonists led to the partial displacement of the native population to the hills, bringing new demands for fuel wood and new intensities of gazing in the forests. Loss of traditional grazing grounds in the plains gave rise to bitterness expressed in the deliberate burning of lowland forests (Meiggs, 1982). More recently still, the political unrest that accompanied the Independence movement in Cyprus in the late 1950s and 1960s was itself accompanied by an increase in fire damage to the remaining forests, and indeed the island suffered the worst fires in its known history at the time of the Turkish invasion in 1974 (Thirgood, 1981, 1986b, 1987).

A hundred years earlier, the forests of Cyprus were in a sorry state: in the third quarter of the nineteenth century it was estimated that in the previous twenty years they had decreased in area by one-third and in volume by one-half. This spectacular degradation of the Cyprus forests during the latter phase of Ottoman rule was caused mainly by the depredations of the goat, that well-known enemy of the Mediterranean forest. Both in Cyprus and in many other parts of the Mediterranean, the traditional woodland grazing economy had begun to break down in the face of rapidly growing population and growing (and uncontrolled) grazing pressures from increasing number of nomadic shepherds and their herds and flocks. Pastoralism was of course not a new activity nor were its damaging effects on the forest previously unknown. In Spain, for example, the large scale annual migrations of sheep flocks had done much to reduce the forest area over the centuries (Darby, 1956). By the nineteenth century, however, the scale, extent and intensity had greatly increased.

To these pressures were added in many parts of the Mediterranean those arising from changing accessibility. In the age of the railway, previously inaccessible forests could now be exploited. The railway age witnessed the destruction of the forest resource in some of the more mountainous and previously inaccessible areas. During the same period changing political structures probably contributed to the destruction. Since 1870 the modern Italian state had the unenviable task of unifying the forest organizations (such as they were) of the many small states and principalities which it succeeded. The finances of the new state were desperately weak. Many estates on former papal lands were sold, and the purchasers proceeded to clear-fell the timber they carried (Meiggs, 1982). Earlier in the century, huge areas of forest in the public domain in Spain were sold to private owners and obligations for woodland protection and management in privately owned forests wee removed (Thirgood, 1981). AT the other end of the Mediterranean, nineteenth-century travelers in the Levant repeatedly referred to contemporary and extensive forest destruction.

There is, therefore, considerable evidence to suggest that the nineteenth century was a decisive period for much of the Mediterranean forest. Previous periods had witnessed resource destruction (and partial recovery) in relatively localized areas, but the new pressures meant destruction on a wider scale, and extended to better-watered, more resilient areas that had previously escaped destruction or that had regenerated. These pressures included spectacular growth in population, greatly increased demand for timber for fuelwood and building, and increased numbers of nomadic shepherds and heavier grazing pressures in the woods. They also resulted from the arrival of the railway, and the transformation of transport and accessibility. Forests previously protected by virtue of remoteness and inaccessibility were now vulnerable. Finally, the century saw great changes in political organization, and at a time of political change forest laws and efficient management practices could not easily be enforced. With the pressures and the resulting deforestation (especially on vulnerable mountain slopes) there came the inevitable consequences of soils erosion and environmental degradation, epitomized in the bare slopes of areas such as Calabria.

## A Model for Modern Forest Resource Use?

This Mediterranean combination of pressures is perhaps unique in detail, but in more general terms is matched in much of the developing world in the second half of the twentieth century. If the phase of resource destruction reached a climax in the Mediterranean in the nineteenth century (and locally in small areas much earlier), then perhaps much of the developing world is experiencing similar changes today. The Mediterranean experience in the present century would suggest that the turn-round from the phase of destruction to the phase of re-creation (or at least stabilization) is difficult but not impossible. Forestry histories such as that of Cyprus (Thirgood, 1987: Dunbar, 1983) related the setbacks and dis-appointments as well as the successes, but in most of the region the forest area has once more expanded or at least stabilized, despite continuing problems of fire and other hazards. At the very least, the rate of forest destruction has slowed down. Tree 'farming' in the shape of the establishment of forest plantations has expanded far beyond its very limited extent of classical and medieval times. Nineteenth-century projections of the Mediterranean forest would in all probability have indicated the complete exhaustion of the forest resource long before the late twentieth century. The fact that this has happened neither here nor in North America, where projections were similar, is significant, and offers grounds for some optimism about trends in the tropical forest today. On the other hand, the Mediterranean experience is also acutely depressing. Technical knowledge about forest management and conservation was available from classical times. There were some technical triumphs, such as the silver fir plantations in the Tuscan hills, managed initially to ensure a supply of ship masts for the Florentine state sand still maintained today. But consistent application of that technology was sadly lacking, at least until very recent times. The result was that all but a tiny fraction of the forest was destroyed or degraded. This history of the Mediterranean forest is characterized by destructive exploitation, spread over several eras in some areas. This exploitation has at best been followed by only partial transition to forest conservation and more efficient use. It has been accompanied by changing perceptions of what constituted potentially useful resources, especially as transport and accessibility underwent fundamental change. It has also been accompanied by severe resource shortages and by serious environmental damage. In short, the Mediterranean forest resource model does not bode well for the future of the world forest.

## North America

If the forest history of the Mediterranean area is a rather depressing one of deforestation and environmental degradation, that of North America perhaps gives more grounds for optimism. There the turn-round from deforestation to forest stability or expansion occurred more abruptly and at a less advanced stage in the process of forest removal. Despite forecasts of timber famines dating from more than a hundred years ago, there have in fact been increases in forest volume during the present century. Many of the worst environmental effects arising from the use of the forest resource have been curbed, and while it would be wrong to suggest that the problems of sue of the forest resource have been solved, the present situation is less gloomy than might have been predicted at the end of the last century. It is debatable whether the American model of forest resource use can be applied to the developing world or its constituent parts today, but the model certainly offers a thought provoking perspective for modern problems of global forest resources.

### The Pre-Industrial Forest

The history of use of the American forest is more concentrated an more dramatic than that of the Mediterranean. Indigenous peoples for centuries had used the forest for hunting, and as a source of numerous domestic and medicinal products, as well as for limited agriculture. Human impact on the forest was not absent: some timber was exported as early as Viking times (Cox *et al.*, 1985), and cops such as beans, corn and pumpkins were produced in forest clearings in New England when English colonists arrived in the seventeenth century (Carroll, 1973). The use of the forest by the Indians resembled the shifting agriculture of the tropical forest of today. Further south, Mayan exploitation of the topical fruit of south-east Mexico actually increased the number of useful tree species (Edwards, 1986). Human impact was therefore not absent. In addition to the effects of gathering and of agriculture, those of fire were also widespread. When the first Europeans arrived, portions of forest had already been changed to more open, park like vegetation, and forest composition in some areas is also likely to have been altered (Williams, 1989a). Nor did the fist European settlers in other parts of the world always encounter completely natural environments. In

New Zealand, for example, the introduction of the potato at the end of the eighteenth century caused a great expansion of shifting cultivation over forest land, and an increased use of fire (Cameron, 1964). There is little doubt, however, that the arrival of Europeans in North America was soon followed by increased impact.

## The Quickening Pace of Destruction

From the seventeenth to the nineteenth centuries the North American forest shrank at an accelerating rate. Initially the forest was both a resource and an obstacle. It provided ample fuel and building materials, and it had to be cleared to provide agricultural land. It was also perceived as a threat, in both physical terms in harbouring dangerous wild animals and hostile Indians and in psychological or spiritual terms as an untamed wilderness or domain of Satan (Cox *et al.*, 1985). The clearing of the forest had an almost sacred motive as well as a secular one: it was the wild, dark and horrible abode of the supernatural and the fantastic (Williams, 1989a). The attitude of early settlers in Ontario, for example, was one of outright antagonism: 'They attacked the forest with a savagery greater than that justified by the need to clear land for cultivation, for the forest smothered, threatened and oppressed them' (Kelly, 1974).

The apparently endless forest seemed inexhaustible, and its removal was as much a key to agricultural expansion as it was a symbol of control or dominion in the erstwhile wilderness: removal of the forest was the first step towards civilization. Under such conditions, therefore, it is not surprising that the conservation ethic was not at the forefront of forest management. At the close of the eighteenth century. British North America was largely a wilderness covered by forest. In the words of Lower (1973, p. 31), 'no one could possibly think of forest conservation, the inscrutable forest. Was itself the enemy. Against it settles would wage a century long war.'

Even in this unpromising climate, however, the relationship between the settlers and the forest was not wholly destructive. Although 200,000 hectares of woodland had been cleared by the end of the seventeenth century, the cutting of timber was restricted (at least in certain areas) by various laws. Perhaps these were motivated in part by memories of the timber shortage in England: in addition at least some of the settlers considered that wanton destruction of timberland was 'displeasing to Almightie God, who abhorreth all willful waste and spoil of his good Creatures' (Carroll, 1973).

This early chapter of the history of the use of the American forest is not without its paradoxes. On the one hand there is the contradiction between expressed attitudes towards this part of God's creation and the exhibited behaviour of forest removal. On the other hand, it has been suggested that migrants from the better-wooded areas of Europe (such as German settlers in Pennsykyania) were more appreciative of the forest and more careful of its use than those from England, where the near-exhaustion of the forest had already resulted in a timber shortage (Cox *et al.*, 1985).

Although most of the early forest clearance was related to agricultural expansion and to procurement of firewood and building materials, commercial exploitation for export markets began very early. As early as the seventeenth century, some timber was shipped to the colonies in the West Indies, to southern Europe, and to England

(Carroll, 1973). This trade, however, was on a very small scale, and consisted largely of valuable products such as ships' masts, by now in short supply in much of maritime Europe. (See for example, Albion [1926] and Bamford [1956].) Indeed the British Crown attempted to reserve the best of the pines along the New England rivers, by means of the broad arrow' marked by the surveyors of the 'Kings Woods'. Even with an acute shortage in England, the export trade was largely restricted to the most valuable timber such as masts, and the stormy Atlantic remained a formidable obstacle for bulky, low-value cargoes of timber for other purposes.

In what is now Canada, the timber export trade took off at the beginning of the nineteenth century. Up until that time, the British Navy relied strongly on timber from the Baltic lands, and with supplies threatened during the Napoleonic Wars it began to look more intently across the Atlantic. In 1800 British imports of timber from the colonies were negligible. In 1803 they amounted to 10,000 loads and in 1811 to 175,000 (Lower, 1973). By 1810 timber had replaced fur as Canada's leading export (Mackay, 1979). The Atlantic trade in timber continued after Napoleon's blockade of the Baltic was lifted in 1812. By then, Canadian timber merchants had become established in British markets. Furthermore, in order to encourage secure imports from North America, the British government in 1810 doubled duty on timber imports from northern Europe, while imports from the colonies were duty free and those from America attracted only a low duty. This preference lasted until 1866, and stimulated a rapid expansion of timber extraction in coastal areas such as New Brunswick, where exports multiplied nearly twenty fold between 1805 and 1812 (Wynn, 1981). In that province, the next few decades saw the unfolding of a story which has been often repeated in more recent times. A first, only the finest and soundest pines were taken for masts, but a more general demand for timber meant that progressively smaller timber was taken. Initially, squared timber was produced, and this form of exploitation utilized relatively few trees in a stand. Around mid-century, demand for lumber grew rapidly from the rapidly growing American cities, and far more trees were utilized. Furthermore, large amounts of slash were produced in the course of logging, and became a major fire hazard (Head, 1975).

This progressive use of increasing proportions of trees per stand and of small trees has been accompanied in many areas by a progressive widening of the range of utilized species. In Quebec, for example, pine and oak were selectively lumbered at the beginning of the nineteenth century. Subsequently other species, including hemlock, spruce, maple, yellow birch and beech were lumbered in a second generation of forest exploitation (Bouchard *et al.*, 1989). A similar sequence was witnessed in the Lake States half a century later: the exploitation of pine was followed by that of hemlock and hardwoods, which were initially perceived not to be of value (Whitney, 1987; Williams, 1989a). With passing time and changing technology, the Zimmerman 'wedge' of resources (Chapter 1) occupied an increasing proportion of the forest.

In the early years of exploitation in areas such as New Brunswick, most, most of the timber was cut by small operators, combining the seasonal activity with farming. Gradually, however, operations increased in scale, and a Glasgow-based timber company dominated much of the north-east of the province. The growth of capitalist organization accompanied the incorporation of the area into the expanding world economy. With this development came the emergence of a powerful entrepreneurial class and a growing proletariat (Wynn, 1981). The forms of control and use of the forest were being transformed. Whereas in earlier times control and use were local, both were now geared to the much larger scale of the North Atlantic. In practice, control was now external, as was the destination of both products and profits. In the words of an early historian of New Brunswick.

> The wealth that has come into it, has passed as through a thoroughfare. The persons principally engaged in shipping the timber have been strangers who have taken no interest in the welfare of the country, but have merely occupied a spot to make what they could in the shortest possible time. The capital of the country has been wasted. The forests are stripped and nothing is left in prospect, but the gloomy apprehension when the timber is gone, of sinking into insignificance and poverty. (Lower, 1973, p. 32).

White the colonial exploitation of the forests of New Brunswick and other areas was proceeding apace during the first half of the nineteenth century, the forests of the United States were being cleared primarily for agriculture. It is estimated that at the very least 40 million hectares of forest had been cleared, and a further 15 million by 1860 (Williams, 1982, 1989a). Perhaps 60 million hectares had been cleared by 1860, compared with about one-twenty of that amount cleared as a result of industrial lumber, mining and urban development (Cox et al., 1985). The consumption of timber was enormous; for example farm fencing in the state of Kentucky alone during the 1870s was estimated to consume 10 million trees annually (Clark, 1984). In some Midwest states such as Ohio, clearing reached a peak during the 1870s, when 1.46 million hectares were removed. Extrapolation of such rates of forest removal caused local concern and alarm, pointing as it did to timber shortages or famines by the early decades of the twentieth century. These fears prompted a questionnaire survey of 'all township officials and intelligent farmers': agricultural clearance was perceived as the primary cause in 90 per cent of the state's eight-eight counties, compared with use as fuel and construction in 9 per cent, and gazing by cattle and sheep in one county (Williams, 1983). By the end of the century, the formerly continuous forest of many states had been reduced to a few isolated patches (Figure 11.2).

After 1860, however, the rate of clearing for agriculture declined sharply, while the rate of cutting for industrial purposes rose. The next few decades witnessed the quasi-colonial exploitation of successive forest areas in New England, the Lake States, the South and the Pacific Northwest. With the exhaustion of the forest resources of one area, the lumber industry simply moved on to a new one. Successive areas waxed and waned (Table 11.2). Between 1870 and 1910, the industry went through its period of greatest growth, production and destructiveness': the cutover land, great fires, and big mills 'assured to the Lake States an indelible image as the epitome of destructive lumber exploitation' (Williams, 1989a, p. 237). The legacy was not only of devastated forest lands but also of unfettered capitalist exploitation, and of reactions, attitudes and ideals that have continued to influence American Forest Policy to this day.

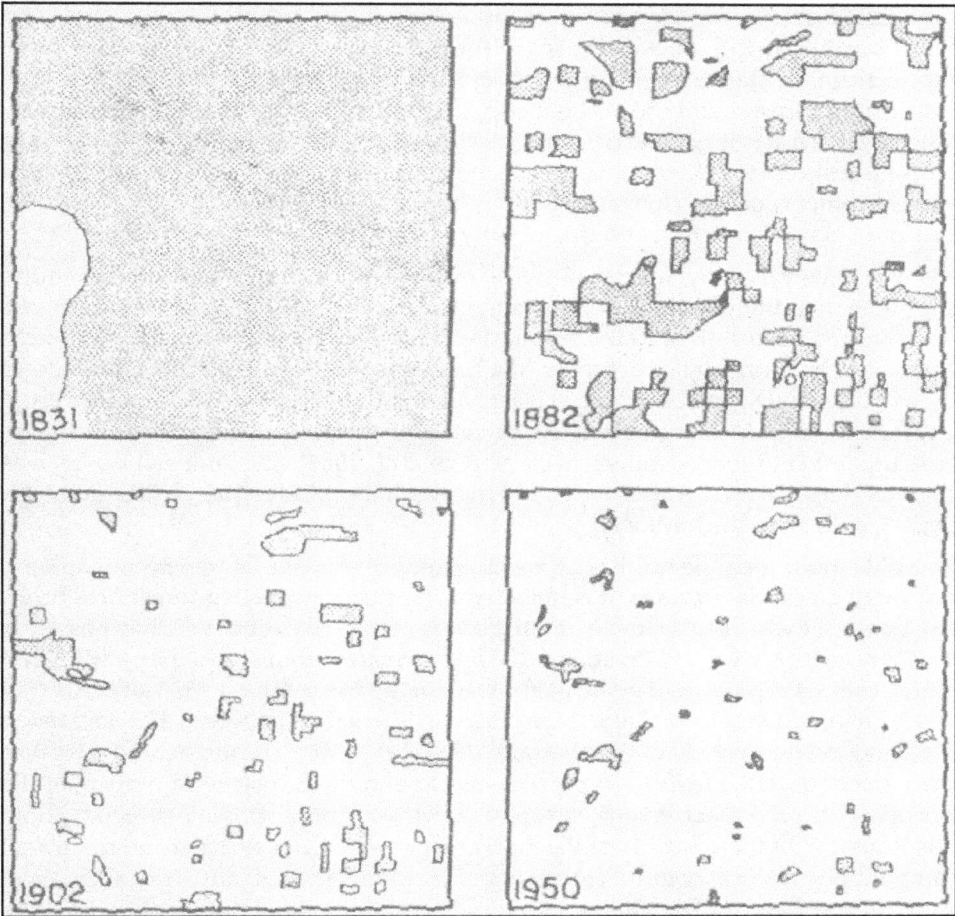

**Figure 11.2: Changing forest area (stippled). Cadiz Townshiop, Green Country, Wisconsin, 1831,1882, 1902 and 1950.**

**Table 11.2: Lumber production in parts of the United States 1800-20**

|  | Lake States (Billion board feet) | Gulf South | Pacific States |
|---|---|---|---|
| 1900 | 9 | 5 | 5 |
| 1910 | 5 | 9.5 | 7.5 |
| 1920 | 2.4 | 8 | 10 |

*Source*: Compiled from data in Cow *et al.* (1985).

## Reaction

From the beginning of this period of almost unbridled exploitation, calls for more careful husbanding of resources and fears of timber famines were voiced. These came first from New England, the earliest area to suffer, but while new timberland could be obtained in the west or south for as little as $1.25 per acre, it was more profitable for operators in Maine or Pennsylyania to relocate than to stay where they were and reforest their cutover lands (Cox *et al.,* 1985), By the turn of the century, however, the last frontier had been reached in the Pacific Northwest, No new areas of virgin forest remained, and in any case the increasing capital investments in mills and equipment could no longer be lightly abandoned. The relationship between the 'cut and get out' approach and that of 'stay put and reforest' was now changing.

With many men and much equipment in the forests in the late nineteenth century, and huge quantities of slash, it is not surprising that devastating forest fires broke out. Some of these destroyed the organic layers of the soil over which they passed as well as neighbouring stands of timber. The lumberman's promise that the plow follows the ax' did not always hold good: at least half of the cutover land was not converted into farms, and a quarter of that was left barren, without trees or people. The devastated area amounted to over 30 million hectares (Cox *et al.,* 1983). The lumber industry was characterized by the features typical of the expanding industrial capitalism of the period, with price fluctuations.

With many men and much equipments in the forests in the late nineteenth century, and huge quantities of slash, it is not surprising that devastating forest fires broke out. Some of these destroyed the organic layers of the soil over which they passed as well as neighbouring stands of timber. The lumberman's promise that the plow follows the ax' did not always hold good: at least half of the cutover land was not converted into farms, and a quarter of that was left barren, without trees or people. The devastated area amounted to over 30 million hectares (Cox *et al.,* 1983). The lumber industry was characterized by the features typical of the expanding industrial capitalism of the period, with price fluctuations, severe competition and periodic overproduction. Associated with the volatility of the market were waste, devastation and disrupted communities-features common to a number of resource based industries at the time. Agricultural clearing, as well as industrial lumbering, was now recognized as a source of environmental damage. Over 90 per cent of the respondents to an Ohio survey in the 880s perceived it as having damaging effects (Williams, 1989a). The perceptions of George Perkins Marsh (Chapter 1) were now rapidly spreading.

By the turn of the century, the use of the American forest resource epitomized all that was inharmonious in the relationship between man and environment, and reaction was beginning to set in. Devastation of the forest became a powerful symbol for the first American conservation movement during Theodore Roosevelt's presidency in the early years of the century. The leading figure in that movement, Gifford Pinchot, was head of the new forest service, whose creation was in itself a reflection of growing unease about the use of the forest resource.

During the nineteenth century, people to whom the American forest had seemed inexhaustible began to realize that it was, after all, finite. This dawning realization gave rise to the Forest Reserve Act of 1891, under which forest reserves, later to become national forests, were first established. Although initially there was uncertainty about the role and management of these forests, in practice they were safeguarded, under federal control, from private exploitation. Even before the peak of lumber production in 1909, it was becoming clear that attitudes towards the forest resource were undergoing a fundamental change. From the 1870s onwards, various government officials were referring to prospects of timber scarcity or famine. Thirty years later, the president himself publicly espoused such fears. Addressing the American Forest Congress in 1905. President Roosevelt declared that 'if the present rate of forest destruction is allowed to continue, with nothing to offset it, a timber famine in the future is inevitable' (quoted in Olson, 1971, p. 1).

## The Forest Turn-round

By the end of the nineteenth century, professional foresters, recreationists and wilderness preservationists, government agencies and even the lumber industry itself were advocating change (Rakestraw, 1955; Williams, 1989a). The first step towards a fundamental change in forest resource use was an awareness of malaise in the existing pattern of use. From this growing awareness, which as itself reinforced by the apparently unsuitable demand for lumber during the First World War and by the devastation wrought by the forest industry in the South at this time (Clark, 1984), there emerged a radical change in the fortunes of American forests. Whereas forest area and timber volume had progressively decreased for many decades, a turn-round towards, stability and expansion now began to occur. This happened over a period of decades rather than overnight, and the transition was not effected without difficulties and setbacks. Nevertheless, a fundamental transition did occur, and the predicted timber famines failed to materialize.

If the historical model of American forest-resource use is applicable to the forests of the developing world today, the factors underlying the transition are of crucial significance. Several are important, one of which is changing demand. In the nineteenth century, timber consumption rose steadily and apparently exponentially (Figure 11.3).

If the historical model of American forest-resource use is applicable to the forests of the developing world today, the factors underlying the transition are of crucial significance. Several are important, one of which is changing demand. In the nineteenth century, timber consumption rose steadily and apparently exponentially (Figure 11.3).

Timber was plentiful and cheap: it was a major source of fuel as well as constructional material. The extrapolation of these nineteenth-century trends must have indeed been worrying, and if that extrapolation had been valid there is little doubt that the consequences would have been serious. But the pattern of demand changed radically after the early years of the century. Electricity and gas supplied growing proportions of domestic fuel, while demand for constructional and industrial timber declined. The railways, for example, consumed about one-fifth of the timber harvest at the beginning of the century but only 3-4 per cent by the 1960s (Olson, 1971). Substitutes such as steel and concrete were increasingly employed: predictions of timber famine themselves encouraged the search for substitute materials. Consumption for the basic domestic and industrial uses steadily decreased during the fist few decades of the century (Figures 11.3 and 11.4), and were only very partially

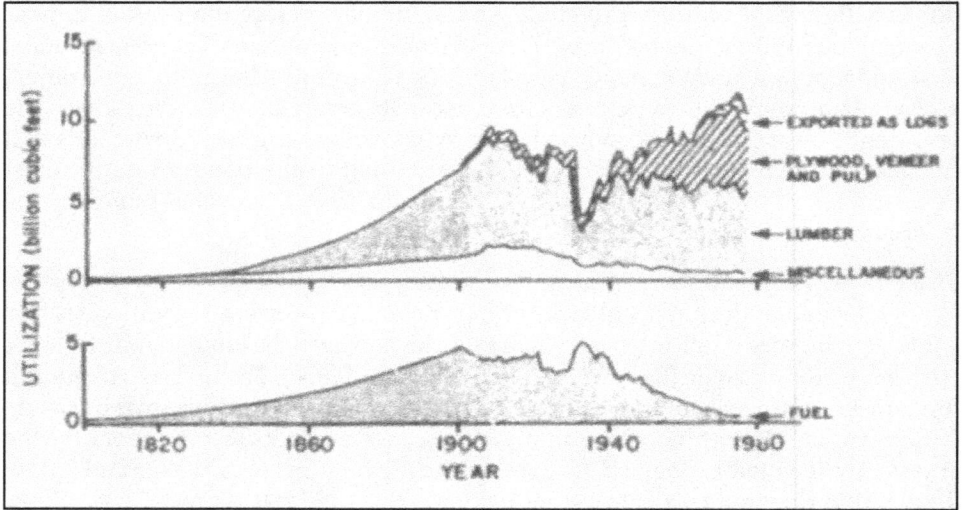

**Figure 11.3: Total utilization of US grown wood (in roundwood equivalent) by major form of use.**

offset by rising consumption of pulp and plywood. In short, the nature and magnitude of demand underwent a fundamental change, which completely undermined the validity of contemporary projections.

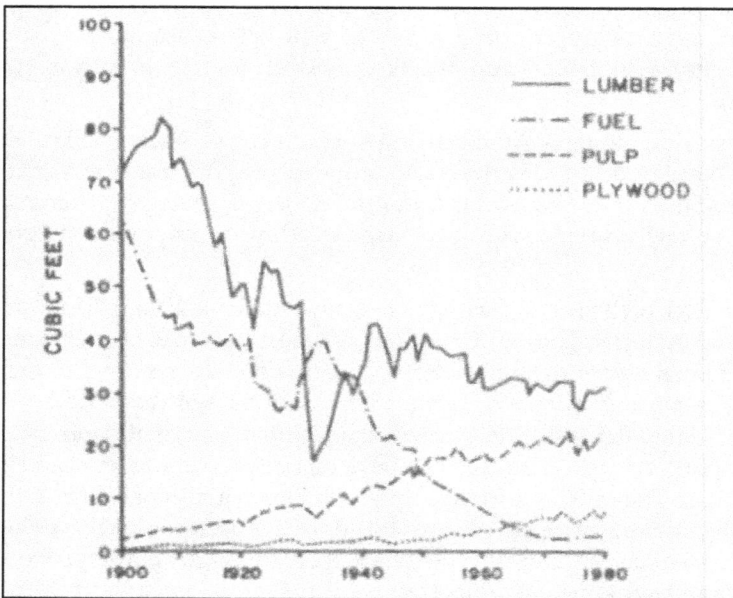

**Figure 11.4: Per capita consumption of timber products, United States.**
*Source*: **Based on Clawson, 1979.**

Major changes were also taking place in the management of the resource, and hence in potential supply. Predictions of timber famine were based on the fact that consumption was outstripping growth: indeed as late as 1920 consumption was estimated at 26 billion cubic feet compared with growth of only 6 billion (Cox *et al.*, 1985). Such a ratio was anathema to proponents of sustained-yield policy: simple logic seemed to indicate that this state of affairs could not continue indefinitely.

Professional and technical forestry in the United States had its roots in the managed forests of central Europe, where the concept of sustained yield was well established. The concept, however, could not be easily transplanted to the essentially natural forests of North America. In natural forests, timber volume is relatively constant, and no net growth takes place as long as a state of equilibrium is maintained. Growth is balanced by loss through age, disease and fire. In this state, timber volume is at a maximum, but net growth is much higher in younger stands of trees (Figure 11.5).

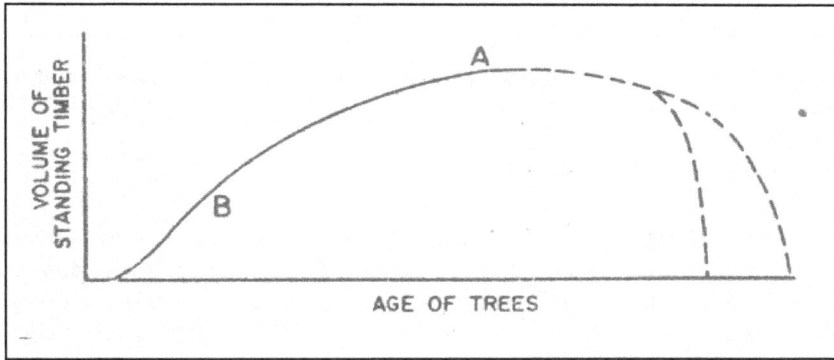

**Figure 11.5: Timber age-volume relationship (A maximum stand volume, B maximum mean annual growth).**
*Source*: **Based on Clawson, 1979.**

In other words, exploitation of the 'old growth' of mature natural forests resulted in a reduction in timber volume, as it must inevitably do whenever such forests are harvested, but not in annual growth. The application of the sustained-yield principle, in its strict sense of matching growth and removals, presents special problems when management is first introduced into natural or virgin forests (Eckmullner and Madas, 1984). Initially, removals must exceed growth and cause a reduction in volume, and yield appears not be sustained. If the old, mature stands are replaced, however, then growth will occur, and eventually growth and removals can be matched. There would be a time lag between the cutting of the old growth and the time when the new forest would be producing at its optimal rate. This time lag was inevitable in the transition from the unmanaged to the managed forest.

Since the implications of this basic principle were not widely appreciated at the time, it is not surprising that commentators of the day were concerned at the disparity between consumption and growth. In practice, annual growth has steadily increased from around 6 billion cubic feet at the beginning of the century to almost 22 billion by 1977 (Clawson 1979). The downward trend to famine predicted at the beginning of the century has not materialised (Figure 11.6).

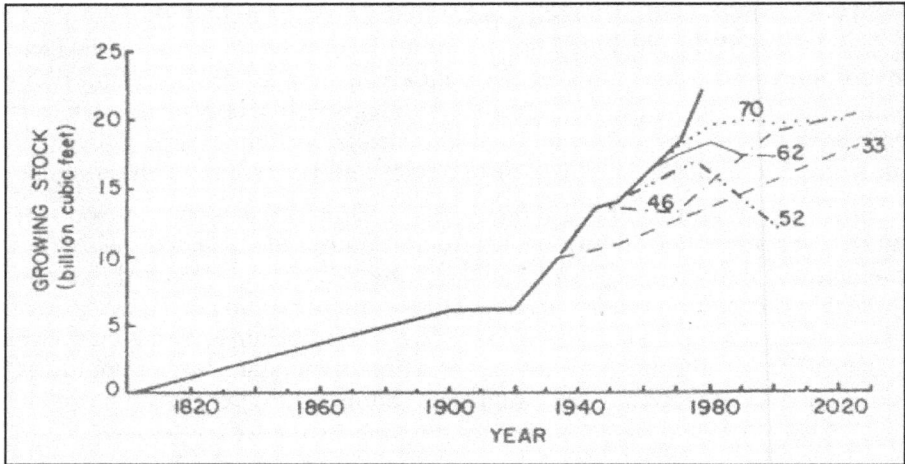

Figure 11.6: Annual net growth of timber in the United State, and US Forest Service projections of future growth made in 1933 (33), 1946 (46), 1952 (52), 1962 (62) and 1970 (70).
*Source*: Based on Clawson, 1979.

The trendlines of growth and cut gradually converged during the present century: whereas the US Forest Service pointed out in 1920 that logging was proceeding at nine times the rate that new wood was being grown (Cox, 1983), by 1963, annual growth of timber exceeded the annual cut for the first time since record keeping began (Cox *et al.*, 1985).

This spectacular reversal of resource trends reflects a major transition from 'hunting and gathering' the forest resource to 'farming' it (*e.g.* Sedjo and Clawson, 1984; Sedjo, 1987). It was accompanied by another major transition, from a resource that was shrinking in a real terms, to one that was stable or expanding in area as well as in volume and growth rates. In addition to changing demand and supply, various other factors contributed to the transition. They included a changing perception of the forest resource, with increasing value placed on recreation and wilderness qualities. They also included a reaction against the concentrated pattern of capitalist control and use of the forest resource, and a climate of pinion that favoured increased government intervention and involvement in forestry matters.

This transition from a shrinking forest resource, apparently heading for exhaustion, to one which is now expanding, has taken place against a background of forest ownership divided between government, industry and private individuals. From the early years of the century, government has sought to influence the private sector in its forest management. Initially there was much debate about whether government policy should be based on compulsion or co-operation, but the theme of co-operation won the day, and from the early 1920s federal government assistance for tee-planting has been provided in a series of acts. Co-operation between government and private owners has also been effective in greatly reducing timber losses through fire. The timber industry itself has undergone radical change. At the beginning of the3 century few of the large mill operators in the South invested in forest research or supported reforestation (Clark, 1984); then the tradition, perhaps understandably in an age of apparently limitless forest resources, was simply to move on to a new area. Today with no such areas remaining and huge investments in mills and processing plants, a policy of 'cut and run' makes less sense in the United States (if not in other parts of the world). This is not to say that all the problems associated with the exploitation of the forest resource are problems of the past. In recent decades, the depletion of the timber stocks in the last 'forest frontier' of the Pacific Northwest has led to the old pattern of mill closures and high unemployment in that region (Robbins, 1985) and a shift in industrial forestry back to the regrowth forests of the South. Nevertheless, at the national scale the transition has been made. The forest area is now relatively stable, and much of it is in effect 'farmed'. Gloomy predictions of timber famine have not materialized, although the fact that these predictions were made may well in itself have helped to ensure that they did not come is pass. What was at risk were the virgin forests of the Pacific Northwest (and of other areas previously) and their aesthetic, amenity and conservation values, rather than their timber-producing potential. Indeed these values were extensively and permanently destroyed, while the timber-producing potential was not harmed.

There are several implications for other parts of the world at the present. The fact that a clear transition took place after a phase of destructive exploitation is one. The fact that a growth in recreational and preservational values contributed to the transition is another. The fact that the lat frontier ha been reached is perhaps a third. Perhaps more serious and ominous is the fact that a century elapsed between the voicing of the first concerns and the eventual turn-round, and almost half a century intervened between the first popular warnings of environmental consequences (such as those of Marsh) and the peak of the forest conservation movement around the turn of the century. A speedier transition will be required if the tropical forest is to survive.

## Models of Forest Resource Use and the Expanding World Economy

The contras6ting histories of the Mediterranean and American forests offer different model of forest resource use, the one leading to the virtual exhaustion of the resource and the other suggesting that a fundamental transition (in area as well as in management) can be made while substantial areas of forest remain. In the former case, the transition was long-delayed and was slow, hesitant, and incomplete. In the latter, a transition was effected not only relatively earlier in the history of forest depletion, but also more quickly and completely.

If the models are to have validity or utility as perspective on modern trends in forest resource use and depletion would-wide, they need to be evaluated against the experience of other parts of the world. The remaining part of this chapter attempts such as evaluation and also considers the dynamic of expansion of the world economy (and incorporation of new areas into it) as a driving-force or trigger for destructive exploitation.

Many European countries conform to a greater or lesser degree to one or other of these models. For example, in the case of the United Kingdom and the Republic of

Ireland, the resemblance is to the Mediterranean model, even though the original forest and the details of its removal are very different. In these countries, the native forest all but disappeared before a transition (in area) occurred in the twentieth century. On the other hand most central European countries conform more closely to the American model, even if once more there are many differences in detail and timing.

A basic outline of forest trends can be roughly sketched for many parts of the world, but detailed time series data are rarely available. This deficiency is understandable: as Chapter 4 indicates, there are conflicting estimates of the forest area even today. It seriously limits the extent to which the models or generalization constructed for one area can be evaluated for more general applicability. In the words of Richards (1986). 'More precise, better documented data on the extent and rate of depletion for each individual forest would do much to correct our perspectives of the global process [of deforestation].

Despite the deficiency of data, there is clear evidence of a transition in area (as well as in management) in many countries during the nineteenth century. For example, forests covered 18.6 per cent of Swiss territory in 1863 and 31.8 per cent in 1983 (de Saussay, 1987), while in Denmark the forest cover is now 11 per cent, compared with 4 per cent in 1800 (Knudsen, 1987). Forest trends are illustrated for the examples of France and Hungary in Table 11.3 (a and b).

**Table 11.3: The changing extent of forest.**

|                                   | Date    | Wooded Area as Percentage of Total |
|-----------------------------------|---------|-----------------------------------|
| (a) France                        | 3000BC  | 80                                |
|                                   | 0       | 50                                |
|                                   | 1400    | 33                                |
|                                   | 1650    | 25                                |
|                                   | 1789    | 14                                |
|                                   | 1862    | 17                                |
|                                   | 1912    | 19                                |
|                                   | 1963    | 21                                |
|                                   | 1970    | 23                                |
|                                   | 1977    | 24                                |
|                                   | **Date**| **Area (ha)**                     |
| (b) Hungary (present boundaries)  | 1800    | 2765.8                            |
|                                   | 1925    | 1090.8                            |
|                                   | 1938    | 1106.0                            |
|                                   | 1946    | 1124.2                            |
|                                   | 1950    | 1165.9                            |
|                                   | 1960    | 1306.2                            |
|                                   | 1970    | 1470.7                            |
|                                   | 1980    | 1610.2                            |

*Contd...*

**Table 11.3–Contd...**

| | Date | Percentage of land surface |
|---|---|---|
| (c) Central America | 1700 | 92.0 |
| | 1800 | 91.8 |
| | 1850 | 85.0 |
| | 1900 | 77.8 |
| | 1940 | 70.7 |
| | 1950 | 67.7 |
| | 1960 | 62.4 |
| | 1970 | 55.4 |
| | 1977 | 50.9 |

| | Date | Area (ha) |
|---|---|---|
| (d) Ivory Coast | 1900 | 14 500 |
| | 1955 | 11 800 |
| | 1965 | 8 983 |
| | 1973 | 6 200 |
| | 1980 | 3 945 |

| | Date | Percentage of land area |
|---|---|---|
| (e) Thailand | 1913 | 75 |
| | 1930 | 70 |
| | 1949 | 69 |
| | 1959 | 58 |
| | 1969/70 | 52 |
| | 1978 | 25 |

| | Date | Percentage of land area |
|---|---|---|
| (f) Northern India | 1870 | 24.6 |
| | 1890 | 21.5 |
| | 1910 | 19.2 |
| | 1930 | 18.2 |
| | 1950 | 16.2 |
| | 1970 | 13.7 |

(combined total for forest woods' and 'interrupted woods)

| | Date | Area (ha) |
|---|---|---|
| (g) Liberia | 1920 | 6475 |
| | c.1950 | 5520 |
| | 1968 | 2500 |
| | 1980 | 2000 |

*Contd...*

**Table 11.3–Contd...**

| (h) Ghana | Date | Area (ha) |
|---|---|---|
| | 1920 | 9871 |
| | 1937/8 | 4789 |
| | 1948/9 | 4236 |
| | 1953 | 2810 |
| | 1958 | 2493 |
| | 1960/1 | 2424 |
| | 1968 | 2207 |
| | 1980 | 1718 |

*Source*: Complied from data in Prieur (1988) (a), Kereztesi (1984) (b), Keogh (1986) (c), Bertrand (1983) (d), Feeny* (1988) (e), Richards *et al.* (1985) (f) and Gronitz (1985) (g and h).

*: Feeny tabulates numerous estimates of the forest area made by different commentators using different bases. This is a selection from his data.

In countries such as Austria and Finland, for example, legislation to safeguard the remaining forest area was enacted in response to the growing demand for timber (in both the local and wider economies) during the second half of the nineteenth century (Raumolin, 1984; Johann, 1984), and in effect marks the beginning of a relatively abrupt and complete transition.

Perhaps the earliest forest transition was in Japan suffered a timber shortage in the early modern period, as a result of a combination of factors including warfare, urbanization, intensified agriculture and growth in population. Prior to the Tokuwaga era (1600-1868), deforestation was not a significant problem. According to Osaka (1983), villagers had free access to the forest but exercised restraint in their use of forest products because of their respect for fellow villagers and for nature. From the seventeenth century, however, pressures on the forest increased, not least because of the huge quantities of timber used in the construction of castles and cities. In addition, demand was created by the growing urban population, and increasing areas of land were cleared for agricultural purposes (Totman, 1984). Indeed during the seventeenth century the new pressures on the forest gave rise to a set of problems similar to those in many parts of the world today. Amongst these problems were flooding, soil erosion, shortages of timber and fuel wood, and endless wrangles over rights of forest use (Totman, 1986). In response the daimyos or lords took over the village forests. Conservation measures based on complete or partial prohibitions on the use of forest products were introduced and rigorously enforced, along with regulation of use in other areas and with compulsory reforestation (Osaka, 1983). From the time of the Meiji Restoration in 1868, most of the forests owned by the daimyos became in effect government property, and regulation of production was maintained despite peasant resistance. Japan remained self-sufficient in fuel and other forest products until late in the nineteenth century, and still retains an unusually high degree of forest cover .

The case of Japan is instructive for at least two reasons. First, it reveals a forest transition at a much earlier date than that in America, and probably one at an earlier stage of forest depletion. Second, it shows that deforestation was experienced, and was perceived as a problem, during the early stages of population growth and even before industrialization or incorporation into the world economy, which are so often associated with environmental deterioration in general and forest depletion in particular.

Nevertheless, this is not to invalidate the association between these factors and forest depletion. Numerous examples could be cited of how the expansion of the world economy though the imposition of colonial control and development in the nineteenth and twentieth centuries has led to new pressures on forest resources, and indeed also to new regimes of management. In many parts of what is now called the developing world, forests were cleared for the production of cash crops such as tea in Assam (Tucker, 1988), rice in Burma (Adas, 1983) and cotton in the Bombay Deccan (Richards and McAlpin, 1983). The physical integration of such areas with the world economy through the construction of railway systems itself brought new pressures to bear on forest resources, with new demands for fuel and for timber for railway sleepers (Tucker, 1983). Similarly the processing of some of these new cash crops consumed large quantities of timber. The forest in the Philippines, for example, suffered not only direct clearance for sugar-growing, but also for wood to fuel refineries (Roth, 1983).

The displacement of indigenous peoples from areas cleared for each crops meant new pressures on the forest in areas of resettlement, which frequently were in the more fragile mountain environments less coveted for cash crops. For example Thirgood (1986a) records such displacement following the French colonization of Algeria. This is but one specific example of a widespread process identified by Westoby (1989) as a major cause of forest problems. Similarly, the clearing of land for the subsistence needs of peasants who had been plantation workers made inroads into the remaining areas of forest. Tucker and Richards (1983, p. xvii) conclude that 'We may well find that these secondary effects of the international economy were the most important, as well as the most difficult to measure, of all causes of forest degradation until the massively mechanized agriculture of very recent years.

The 'incorporation of new areas into the world economy in the nineteenth and twentieth centuries was often accompanied by devastating effects on the forest, irrespective of whether the colonialism was external (as for example in the nineteenth-century case of New Brunswick) or internal in Brazil, for example, perhaps as much as 30,000 square kilometres of forest were cleared for coffee-planting during the nineteenth century (Dean, 1983), McNeill (1988) has documented the disastrous effects of the incorporation of parts of southern Brazil in the present century. Of the original 20 to 25 million hectares of Araucaria forest, only 445,000 remained in 1980. Deforestation as a result of clearance for agriculture proceeded gradually during the first three decades of the century before accelerating to reach a peak during the period from 1945 to 1970. The waste of timber has been tremendous; perhaps as much as 1 billion cubic metres of Araucaria wood may have simply been burned. The forests have ceased to exist as an economic asset; logging and sawmilling, once the largest source of employment in Parana state, have greatly diminished, and the drainage basin of the Iguacu river has been so disturbed that navigation is no longer possible.

In neighbouring Chile, the twentieth-century history of the forest resource has been similar, if on a smaller scale. In the province of Aysen, the pastoral settlement of the interior was promoted by the government for strategic reasons in the 1920s. Massive clearance of *lenga* forest (dominated by *Nothofagus pumilio*) took place. Half of the original forest cover was burned to make way for cattle and sheep grazing with the familiar result of accelerated erosion and estabilised slopes. Ironically, the value of *lenga* was not perceived until the late 1970s, when exports of its decorative hardwood to Europe began (Veblen, 1984).

Perhaps the story of southern Brazil and similar areas in the twentieth century is no more than a repeat of that of many other areas of expanding frontiers in the nineteenth century. It reminds us that external colonial powers need not be directly involved for this type of development to occur, any more than they were in the case of nineteenth-century America. The effects in terms of displacement of indigenous population and forest clearance seem to be similar whether the colonial process is driven internally or externally. Characteristically, pioneers have tended to have exploitative or antagonistic attitudes towards the forest. Perhaps these attitudes are at least in part a reaction against the hardships and dangers of life on the frontier, and to the insecurity that they engendered. It is suggested by Wynn (1979) that the unusually early conservation response in New Zealand, in the form of the Forest Bill of 1974, may b related to the presence of a significant number of emigrants from the middle and upper ranks of British society. Their reserves of capital insulated them from some of the rigours and uncertainties of the pioneer life, and permitted a more benign and concerned attitude towards the forest than was common in similar-territories. Perhaps New Zealand was also unusual in that much of the timber felled in the course of clearing land for agriculture was turned to profit (Arnold, 1976). In the case of southern Brazil, as in many other areas in the nineteenth and twentieth centuries, clearing was associated with tremendous waste of timber. Exploitation was often inefficient and wasteful, as well as destructive.

With the arrival of colonial power and the expansion of cash crops, there might come the first positive measures to protect the remaining forest area. Colonial powers might seek to reserve forest areas in order that forest conservation might be achieved for both environmental and commercial reasons,; for the avoidance of soil erosion and other adverse processes as well as for safeguarding timber resources. In parts of India, for example, trends in forest resources began to worry the colonial authorities as early as the 1850s, and in 1864 Dr Dietrich Brandis, a German forester, was appointed Inspector-General of Forests. The enactment of forest laws in 1878 provided the legal basis for government ownership and management of parts of the remaining forest area, through the creation of forest reserves. In some areas, policies of forest preservation were accompanied by early attempts at planting exotic species. Such attempts began in 1902 in Kenya, for example (Ofcansky, 1984), and by 1914 over 4,000 hectares of forest land had been replanted in Punjab and Uttar Pradesh (Tucker, 1983). Similar attempts at forest regulation and replanting began half a century earlier in New Zealand (Roche, 1984), and within thirty-five years of having become a British colony the 1874 Forest Bill was enacted to 'make provision for future industrial purposes, [and] for subjecting some portion of the native forests to skilled management and proper control' (Wynn, 1979, p. 172).

The imposition of external or colonial control gave rise to many and persistent problems. In Africa and India in particular, these problems have been very long-lasting (*e.g.* Anderson (1987); Shiva and Bandyopadhyay, 1988). In Africa, for example, many colonial forestry departments shared a number of commons aims, which are summarized in Table 11.4.

**Table 11.4: Characteristics of colonial forestry departments in Africa.**

☆ Emphasis on commercial forestry, and on growing timber that could be sold either within the county (primarily for mines and industries) or for export.

☆ Corresponding emphasis on exotic species (especially pines and *Eucalyptus* spp.)

☆ Conversely, little interest in indigenous species, which for the most part were ignored.

☆ Little enthusiasm for ethno-botany of local people, who were usually regarded as ignorant about forestry and as spoilers of the environment.

☆ Most forest officials were dedicated to setting aside certain proportions (*e.g.* 6 per cent) of the total land areas for forest reserves.

☆ Foresters were regarded by local people as a sort of auxiliary police, since one of their functions was to keep people out of reserves and to prosecute them for infringement of regulations.

*Source*: Based on Little and Brokensha, 1987.

In detail the role of a department might be complex: in Kenya, for example, a three-sided political struggle developed between it, the company that had leased the forest for commercial exploitation, and the indigenous peoples (Anderson, 1987). In general terms, however, the new regime reflected Western notions of conservation and management, and alienation was often the inevitable result of the new orientation. The policing of forest reserves, with their symbolism in terms of alien control and regulation of traditional resources, posed many problems and not infrequently met with the response of incendiarism (see, for example, Castro [1988] and Tucker [1988] for discussions of forest reserves in Kenya and India respectively). Nevertheless, such reserves may have helped to reduce the attrition on the forest, even if they failed to reverse the downward trend of forest area. In many developing countries and regions, no real transition in the trend of forest area has yet occurred (Table 11..3 c–h), and indeed the rate of forest contraction in recent decades has been comparable to that in the United States in the nineteenth century.

In some instances the incorporation of new areas into the world economy led to the lessening, rather than intensification of pressures on the forest. For example, Goucher (1988) has shown that the decline of the local iron industry during the colonial period in Togo in West Africa reduced the local demand for charcoal and thus relieved one of the greatest pressures on the remaining forest. More usually, however, this incorporation whether carried out by colonialism or imperialism, was accompanied by fundamental changes in the relationship between man and land, including a revised perception of the forest resource. The usual result was the diminution or degradation of the forest. Exploitation was often destructive, but the phase of destructive exploitation was by no means always followed by a clear shift towards conservation and more efficient use of the forest resource (Chapter 1). Perhaps the existence of external control (political or economic) meant that the transition from *Roubwirtschaft* to resource conservation, as indicated in Friedrich's theory, was far more difficult to achieve in peripheral than in core areas of the world economy. Perhaps the case of Cuba is instructive. At the beginning of the nineteenth century, forests covered 90 per cent of the land area, but by 1946 it had shrunk to 11 per cent. The expansion of sugar-growing and ranching, organized as 'peripheral' activities geared to the American market, was a major factor in that contraction. Following the overthrow of the Batista regime in 1959 and its replacement by Fidel Castro, a form of forest transition was achieved. By the early 1980s, the forest area occupied 26 per cent of the land surface.

The incorporation could apparently be set at various scales,. In seventeenth century Britain, for example, timber shortages in England gave rise to widening searches and to the exploitation of forests in the north of Scotland for naval timber. A century later, British companies expanded their activities to Eastern Europe; one, for example, began to exploit Belorussian forests for timber in 1793 (French, 1983). This exploitation, as in many similar 'expanding frontier' episodes, was typically destructive. This was but one contribution to the widespread clearing of the Russian forest in modern and early modern times. From the end of the seventeenth century to 1914 about 70 million hectares of forest were cleared in European Russia alone, including 3 million hectares between 1888 and 1908 (Barr and Braden, 1988). Growth in population coincided with incorporation into the wider economy to unleash pressures on the forest comparable to those in the United States last century and in the tropical world today. One facet was the growing demand for timber in Sweden in the second half of the nineteenth century, which could not be satisfied from the traditional producing areas in the south of the county and which led to 'frontier' development in the north (Gaunitz, 1984) and a far east as the White Sea (Bjorklund, 1984). On a much larger scale, the nineteenth-century colonial period in Africa and Asia, and the same era in colonial and independent Latin America, witnessed the first impact of the global economy as it began to penetrate the topical forest zone (*e.g.* Tucker, 1986).

This incorporation may be a sufficient condition for forest contraction, but it appears not to be a necessary one. In countries as disparate as Scotland and China, the forest resource has been largely exhausted even without (or before) radical changes in economic structures or organization. Scotland was largely deforested even before its growing incorporation into the wider British economy was followed by the exploitation of some of its surviving forests. On a much larger scale (China was effectively denuded of forest cover by the nineteenth century. This denudation took place over a very long period, beginning as long ago as 3000 BC and accelerating with the diffusion of bronze and then iron tools, but most of all with population pressures in the context of deteriorating economic, social and political orders from the seventeenth century onwards. Most of the deforestation was caused by peasant use of wood for fuel, building and coffins (Murphey, 1983), rather than by 'colonial' exploitation.

Perhaps the experience of counties such as Scotland and China (as well as parts of the Mediterranean world) suggests that when pressures on forest resources build up gradually over long periods, the forest transition is both late and uncertain. On the other hand, the examples of America and Japan may suggest that a relatively sudden build-up of pressures gives rise to perceptions that lead to effective responses and in turn to relatively early and complete forest transitions. A sudden build-up of pressures may lead, in other words, to a literal crisis. In theory the process of incorporation into the world economy may thus be the impulse that ultimately gives rise to the transition, but in practice a long time lag is usually involved. The fact that there is usually little local control or political power in the newly incorporated area means that the government action crucial in effecting the transition is weak, delayed or completely absent.

National forests all but disappeared in countries such as Scotland and China, but nevertheless there has been a form of forest transition even in these countries. In both cases the forest area is now increasing. In Scotland, the forest area has almost doubled since the Second World War, while some reports indicate that at least a partial transition may also have taken place in China. According to Li Jinchang *et al.* (1988), the forest area in China increased from under 9 to 12 per cent between 1950 and 1981, and there are numerous reports of ambitious and grandiose afforestation plans (see also Chapter 4). On the other hand it is suggested by come commentators (*e.g.* Forestier, 1989) that the forest area is still shrinking. It can safely be concluded, however, that strenuous attempts are being made to achieve a transition, even if the transition has not yet been successfully effected.

Perhaps the Mediterranean model can be applied to them, as to other long-settled countries, with some validity. In at least some of the newer lands, there are perhaps grounds for optimism that the American model can be applied despite the vicissitudes of colonial or imperial incorporation into the world economy, and that a turn-round in the trend of the forest area achieved well before forest exhaustion looms. A changing perception of the forest resource, and a growing appreciation of recreational and preservational values, underlay the American transition. Perhaps similar shifts will eventually lead to a comparable transition in the tropical forest.

# Chapter 12

# The Extent and Distribution of the Resource

## General Description

> We know quaite a lot abaout the moon but do not know how much of the world;s surface is covered by forests and woodlands (Persson, 1974).

This was the despairing conclusion reached by Person at the end of a major assessment of world forest resources which he carried out whilst working in the Forestry Department of the Food and Agriculture Organization (FAO) of the United Nations. Since the early 1970s, our knowledge have have improved to some extent, but many uncertainties remain. Since we do not know with accuracy and precision the present extent of other forest resource, we cannot reliably estimate the fraction of the original resource that still remains, nor can we evaluate the current trends of grains and losses in relation to the global extent of forest. Even during the last two decades, estimates of other forest area have varied from under 3000 to over 6000 million hectares, or from 20 to 45 per cent of the global land surface. One of the earliest estimates, made by Zon and Sparhawk as long ago as 1923, is not unlike some modern figures, and is considerably lower than some of those made more recently. It cannot be concluded from this, however, that the forest area has been static or increasing. Since such diverse estimates have been suggested, some of which are illustrated in Table 12.1 it is apparent that reliable time-series data cannot readily be assembled.

**Table 12.1: Estimates of global area of forest and woodland.**

| Source | Category | Area (10⁶)ha |
|---|---|---|
| Zon and Sparhawk (1923) | Forest Area | 3010 |
| FAO (1946/1937) | Forest | 3650 |
| Haden-Guest *et al.* (1956) | Forest | 3914 |
| FAO (1963) | Forest land | 4126 |
| Persson (197) | Forest land | 4030 |
| | Closed forest | 2800 |
| Eyre (1978) | Total forest | 6050 |
| Global 2000 (1975) | Closd forest 2563 | 3763 |
| (Barney, 1980) | Open woodlands 1200 | |
| Matthews (1983) | Forest 3927 | 5237 |
| | Woodland 1310 | |
| World Resource Institute | Closed forest 2865 | 4147 |
| WRI (1986) | Open forest 1282 | |
| | Other wooded land* | 1081 |
| Total wooded area | 5228 | |
| FAO (1987) | Forest and woodland | 4087 |

Differing definitions of forest and woodland are one of the main reasons for the apparently conflicting estimate illustrated in Table 12.1. The forests are extremely varied in their structure and composition and range from dense assemblages of trees to open woodland and scrub. Some definitions are required to provide context for these and other estimates. *Forest and woodland* is land under natural or planted stands of trees, and in FAO statistics includes land from which forests have been cleared but which will be reforested in the foreseeable future. *Closed forest* in land with a forest cover, with tree crown covering more than 20 per cent of the land area. Land under shifting cultivations included if it is expected to return to forest in the foreseeable future. *Open woodland*, on the other hand consists of land with tree-crown cover of 5-20 per cent of other surface area, and includes the savannah belts of the tropics even if crown cover sometimes exceeds 20 per cent. The most convenient and comprehensive source of statistics of the extent of the forest and woodland resource is the FAO Production Yearbook. This publication lists area statistics for forest and woodland on country basis. The apparent degree of comprehensiveness may to some extent be misleading, since the same data are quoted for some countries over periods of several years and on the other hand major changes may occur from year to year. FAO depends largely on the returns made by individual countries, and the degree of precision (and indeed also of accuracy) is variable. Nevertheless, the source remains the most convenient one in terms of availability, completeness and frequency of cover, despite the fact that it quotes figures only for 'forest and woodland' and not for 'closed forests'. (Some other FAO publications do distinguish between closed and other forests).

## Forest Area and Distribution

The total area of forest and woodland in the mid-1980s, as listed in FAO Production Yearbook, is just over 4,000 million hectares, or 31 per cent of the world land surface. Of that area, between 2,500 and 3,000 million hectares (c. 20 per cent) are closed forest. Perhaps the most detailed and authoritative estimate of the closed-forest area, as the global scale, is that made by Person (1974) in the course of work intended to contribute to the never- completed fifth World Forest Inventory. His work is characterized by carefulness and comprehensiveness, but he clearly warns the reader of possible inaccuracies, and indeed indicates the estimated level of accuracy of his data. Few countries fall within his narrowest range of –5 to +5 per cent. Such a range of possible error is a salutary reminder of the uncertainties that still surround our knowledge of the world forest area, and provides a context in which his and other estimates may be considered. Persson's estimate of the closed-forest area around 1970 is 2,800 million hectares; the World Resource Institute (WRI) (1986) estimated the closed forest area to be 2,865 million hectares in 1985, compared with 2,948 million hectares in 1980. Just under 40 per cent of that area consists of coniferous forest: WRI figures suggest that the coniferous forest is increasing in both absolute and relative terms (from 38.0 to 39.8 per cent between 1980 and in 1985), as the broad-leaved forests contract and coniferous ones expand through afforestation and reforestation.

Both the volume of growing stock per hectare and growth rate are vary from area to area, depending on environmental factors and forest type. On the basis of information that he was able to assemble for 70 per cent of the closed forest area. Persson (1974) concluded that the average volume of standing timber was around 110 cubic metres per hectare, and that the total volume was around 3,10,000 million cubic metres, of which perhaps 1,00,000 million cubic metres were coniferous. The former figure compares with one of around 4,00,000 million cubic metres derived from the application of average figures to estimated extents of forest types (Table 12.3). To these closed-forest figures must be added timber in open woodland areas, estimated at around 30,000 million cubic metres with perhaps a similar volume of growing stock was around 3,50,000 million cubic metres. This estimate should be viewed with caution, as Persson himself carefully emphasized: his "guestimate" may be far from the truth'.

The distribution of closed forest and open woodland by area and growing stock is indicated in Table 12.2. Among the main features emerging from the table area the dominant position occupied by the Soviet Union in terms of both ara and volume, and the significance of open woodland in Africa in terms of area and, to a lesser extent, volume.

The distribution of closed forest is strongly concentrated: the Soviet Union, Canada and Brazil account for close on half of the total area, and the Soviet Union alone accounts for over a fifth of the area, nearly a quarter of its growing stock, and 50-60 per cent of its coniferous volume (Sutton, 1975; Barr, 1988). South America stands out as having the most extensive closed forest in the developing world, and Brazil comprises much of that area. Open woodland is even more concentrated in distribution, with Africa accounting for more than half of the area.

**Table 12.2: Distribution by area and growing stock.**

| Region | Closed Forest Area (mill ha) | Closed Forest Volume (100 mill m³) | Open Woodland Area (mill ha) | Open Woodland Volume (100 mill m³) |
|---|---|---|---|---|
| North America | 630 (459) | 585(503 | -(275) | - |
| Central America | 60 | 55 | 2 | 1 |
| South America | 530 | 915 | 150 | 40 |
| Africa | 190 | 250 | 570 | 140 |
| Europe | 140 (145) | 120 (160) | 29 (35) | 8 |
| USSR | 765 (791) | 733 (841) | 115 (138) | 56 |
| Asia | 400 | 380 | 60 | 20 |
| Pacific | 80 | 60 | 105 | 25 |
| World | 2800 | 3100 | (1000) | (300) |

Figures in brackets are from UNECE/FAO (1985) and in columns 1 and 2 are for exploitable closed forest only. Differences compared with Persson's data reflect redefinition rather than physical changes.

*Sources:* Compiled from data in Persson (1974) and UNECE/FAO (1985).

The extent of forest and woodland in relating to total land area is shown in Figure 12.1. Countries with an extent of forest and woodland grater than the world average are to a large degree concentrated in the high and low latitudes, reflecting the area dominance of the boreal and tropical forests. Japan and the Kores stand out as well forested areas in mid-latitudes. Outside these exceptions, the highest percentages for cover of forest and woodland are found on the one hand in Finland and Sweden, which fall largely within the boreal forest belt, and in parts of the equatorial zone such as Surinam, French Guiana and Guyana in Latin America, Congo, Gabon and Zaire in Africa, and Papua New Guinea. It is noticeable from Figure 12.1 that the most poorly endowed countries are not only those of the arid and semi-arid intermediate latitudes: Britain, Ireland and some neighbouring countries in north-west Europe also fall into this group. The controls on the pattern are obviously human as well as climatic.

## Forest Area and Population

In general terms, there is an inverse relationship between forest cover and population density. Densely people countries generally have low degrees of forest cover, while the reverse is true of sparsely populated areas with physical environments suitable for forest growth. For example Palo (1987) reports a correlation coefficient between forest cover and population density of –0.334 for 72 tropical countries, while Palo and Mery (1986) quote much higher coefficients for individual continents (-0.80 for Latin America, –0.96 for Asia) and areas (-0.80 for moist Africa). A highly significant relationship ($r^2 = 0.62$. P = 0.0001) was found for the Greater Caribbean area by Lugo *et al.* (1981).

The area of forest and woodland per head of population at the global scale is now around 0.8 hectares while that of closed forest is little more than 0.5 hectares.

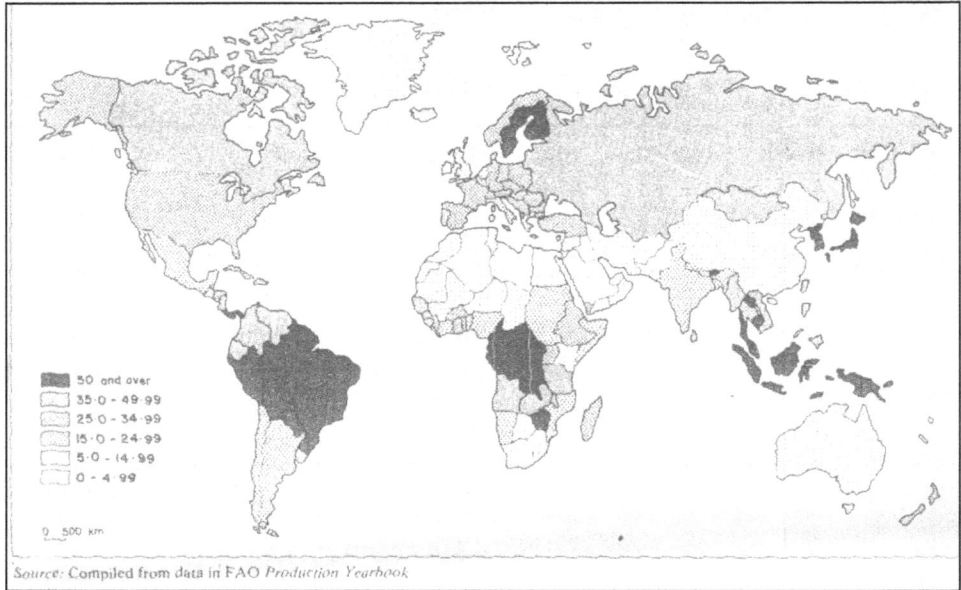

**Figure 12.1: Forest and woodland as percentage of land area, 1985.**

(For comparison, the corresponding figures for arable land (with permanent crops) san permanent pasture are approximately 0.3 and 0.6 hectares respectively). The unevenness of the per capita distribution of both closed forest and of forest and woodland is very pronounced, as Figures 12.2 and 12.3 illustrate.

Figures 12.2 clearly depicts a division into the 'haves' and 'have nots' in terms of the closed forest resource at the level of individual countries. It is based on data assembled by Persson (1974): separate data for closed forest are not published in FAO *Yearbooks*. Ratios of more than one hectare of closed forest per head of population are largely confined to a few low-latitude countries, and to the Soviet Union and North America. On the other hand most of Africa and Asia (outside the Soviet Union) have less than one-quarter of a hectare per head, as indeed also has a group of densely populated countries in north-west Europe. Several significant features emerge from the pattern. One is that some tropical countries that are (or have been) associated with dense forests and active utilization of the forest resource were near of below the world average even around 1970, since when the ratio has further declined with increasing population and decreasing forest area. Several countries in east and West Africa (for example Uganda and Nigeria) fall into this category, as do some in south-east Asia (including for instance the Philippines and Sri Lanka). Closed forest is a scarce resource in both absolute and relative terms throughout much of Africa and Asia, and not just in the areas traditionally associate with treelessness. Huge tracts in Africa, as well as in south-west Asia, have less than one-tenth of a hectare of closed forest per person.

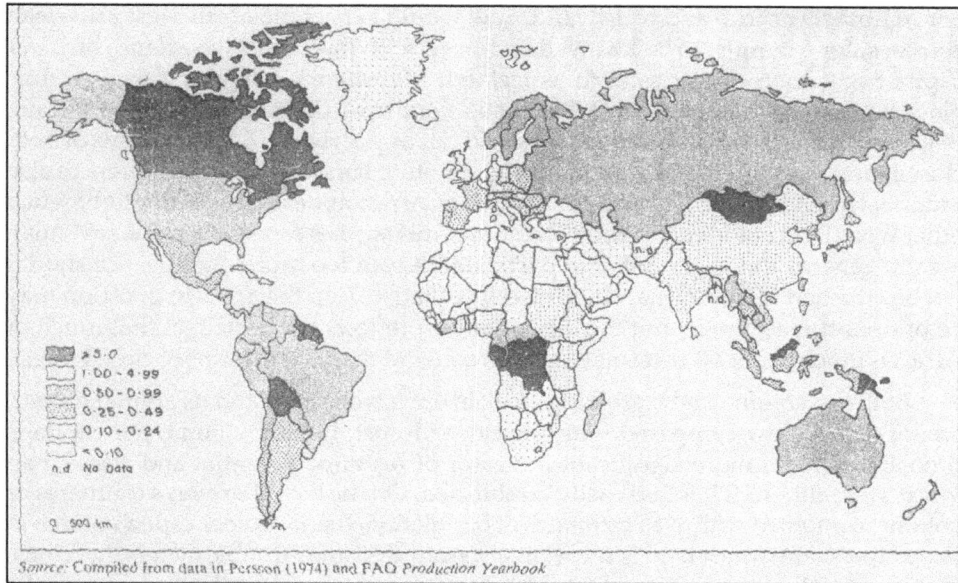

**Figure 12.2: Per capita area of closed forest (early 1970s) (ha).**

Perhaps the picture portrayed in Figure 12.3 is slightly less gloomy, reflecting as it does an apparently more generous per capita endowment. Here other woodland is added to the category of closed forest. Caution is necessary in assessing he pattern, since the definition of 'forest and woodland' employed for some countries is a wide

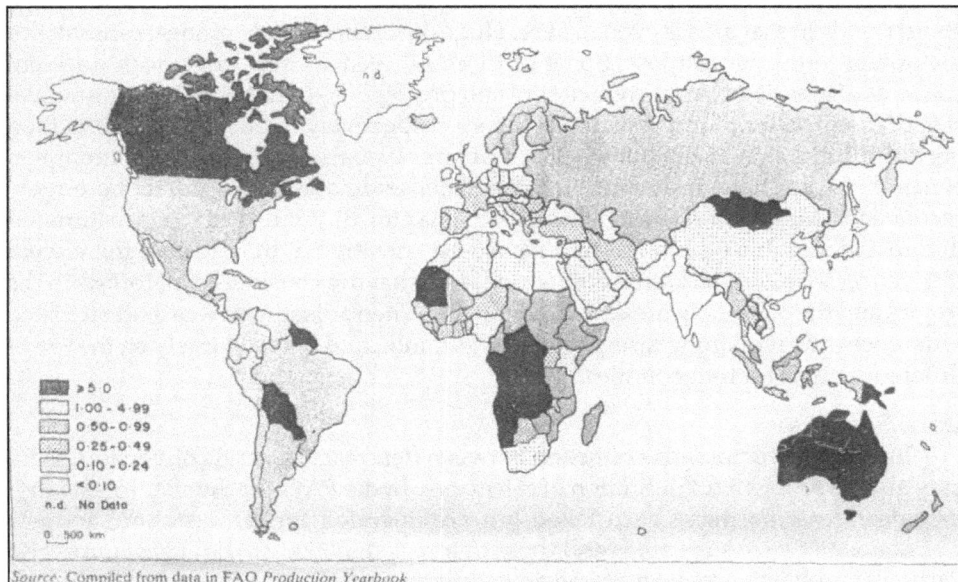

**Figure 12.3: Per capita area of forest and woodland 1985 (ha).**

one, embracing scrub and brush land. Few would expect the north-west European counties, for example, to be less well endowed with forest and woodland, on a per capita basis, than countries such as Iceland, Mauritania, Libya or Saudi Arabia. Nevertheless, these north-west European countries (Ireland, United Kingdom, Belgium, Netherlands and Denmark) stand out as poorly endowed in terms of both closed forest and forest and woodland. On the other hand significant differences also occur between Figures 12.2 and 12.3. Much of Africa appears in a better light when other woodland is added to closed forest, and this applies especially in the savannah area in general and in the Sahel in particular. Again too much significance should not be attached to individual countries since statistics on the areas in question may be of doubtful accuracy, but the broad pattern reflects the basic fact that much of Africa is much better off in terms of open woodland than it is in terms of closed forest.

Both the volume of the growing stock in open woodland and its annual growth are, of course, low compared with the closed forest. Indeed volume per capita is probably a much more significant indicator of resource potential and stress than forest area, although it is less easily established. On the basis Persson's estimates or volume, combined with 1985 estimates of population, the world per capita volume in closed forest and forest land is respectively around 65 and 70 cubic meters. In Africa, the figures fall to about 45 and 70 cubic meters, whilst at the other extreme in the Soviet Union the ratios are around 280 cubic metres per person. Major disparities occur even at the scale of continents and large regions: at the intra-continental scale these disparities are of course much greater.

## The Dynamics of the Forest Area

### Trends and Patterns of Deforestation

If uncertainty exists about the total area of forest and woodland, our knowledge about trends in that area is even unsure. Huge differences exist amongst estimates of the annual rate of forest loss. Some of these differences may arise from different concepts of loss. For example two careful and prestigious studies estimated the annual rate of disturbance in tropical closed forests respectively to be 7.4 and 22.0 million hectares in the late 1970s (Lanly, 1982; Myers, 1980a). A subsequent comparison concluded that these apparently irreconcilable estimates could in fact be largely reconciled (Melillo *et al.*, 1985) (see also Chapter 8). One study had estimated deforestation, whilst the other had measured conversion. In addition, there were differences in the initial assumptions, one study having considered all forests in the tropics and the other only broad-leaved forests. When adjustment was made for such differences, the remaining disagreement was minor, and hinged largely on the use of different figures for four counties.

### Data Sources

In addition to possible confusion between deforestation and conversion, there may also be problems of definition of categories. In the FAO Yearbooks, 'forest land' includes areas that have been felled but are intended for reforestation, and the distinction between such areas and those deforested in likely to be at best imprecise Particular problems are encountered as woodland opens out to grassland. Shortening rotations of shifting cultivation may gradually reduce the forest, whilst on the other hand the gradual encroachment of the forest onto land abandoned by agriculture may similarly give rise to problems of definition and criteria. Neither increases nor reductions in the area of forest and woodland are easily capable of precise measurement, and therefore the net change is the balance between estimates of two quantities which are both likely to be subject to inaccuracies. The data on apparent changes in area of forest and woodland that are readily calculable form the *FAO production Yearbooks* must therefore be viewed with caution. Myers (1980a) suggests that the FAO has been unable to apply vigorous appraisal to nationally reported statistics for fear of threats to funding, and refers to specific cases such as that of Indonesia, for which at his time of writing the published figure for forest cover was more than twenty years out of date and took no account of continuing logging and shifting cultivation. The quality of data has probably increased considerably in recent yeas, but in general terms, estimates of deforestation are usually five to seven years out of date, and while the use of remote-sensing techniques such as the analysis of LANDSAT imagery is a major step forward, their usefulness may be limited by the difficulty o differentiating between original forest and areas of regeneration (Salati and Vose, 1983), as well as by problems of cloudiness.

With this proviso in mind, we can conclude from FAO Production Yearbooks that the total area of forest and woodland appears to have decreased by approximately 2 per cent during the period from 1975 to 1985. During this period the net change has amounted to around 80 million hectares, or an area more than three times the size of Britain. The annual net change during the 1980s is around 5 million hectares, equivalent to around two-third the size of Scotland. The definition of forest land, as discussed earlier, is likely to mean that this figure is an underestimate of the area cleared, as 'forest land' includes those areas where there is an intention of eventual reforestation. Furthermore, data from the Yearbook relate to change in the aggregate forest area rather than to deforestation. Net changes in the forest area result from the b balance between the area 'deforested' and the area 'reforested': the data therefore understate 'gross' deforestation. This important proviso should be borne in mind in considering the ensuing figures and discussion.

Data from the Yearbooks suggest that the annual rate of change increased from around 4 million hectares during the first half of the 1970s to over 10 million during the second half, before slowing to the more recent rate. Gross deforestation is of course very much more extensive, and net deforestation in the tropics is likely to be greater than that in the world as a whole.

Attempts to measure rates of change encounter great problems of source data as well as of definitions. Well-known examples of estimates of rates of tropical deforestation include those of Sommer (1976), Myers (1980a), and Lanly (1982). The later concluded that the annual rate of clearance of tropical forests around 1980 was 11.3 million hectares of which 7.5 million were in closed forests and the remainder in open woodlands (see also Chapter 8). This corresponded to around 0.60 per cent of the tropical closed forest area. Whereas Lanly estimated that the rate was similar in the three major tropical areas, more recently Arnold (1987) has reported annual deforestation rates of 0.77 per cent in Africa, 0.67 per cent in Asia and 0.60 per cent in America.

Of Lanly's total area of 11.3 million hectares, 6.2 million hectares were transferred to other land uses while the remainder retained some tree cover. Myers (1983a) concluded that a minimum of 10 million hectares is eliminated each year, and that the overall conversion rate is around 24.5 million hectares. Some recent work based on remote sensing in the Amazon region suggests that up to 20 million hectares per year may be burned in that area alone (*e.g.* The Times, 6 August 1988), and indicates that the rate of deforestation increased exponentially between 1975 and 1985 (Malingreau and Tucker, 1988). The use of remote sensing in some other parts of the world has also suggested that other sources may have given rise to underestimates. For example, it has been alleged that while the forest cover of India was officially reported to extend to 74 million hectares in the early 1980s, its extent as determined by remote sensing was only 37.9 million hectares (Bowonder *et al.*, 1987). It must be emphasized, therefore, that there are serious limitations to the accuracy o the data in the FAO Production Yearbooks: on the other hand the latter source is unrivalled in terms of its convenience and availability.

## Patterns

The broad pattern of distribution of change is indicated in Table 12.3, from which it clearly emerges that reductions in area of forest land are primarily characteristic of the developing world, which accounts for some 70 million of the 80 million hectares apparently cleared between 1975 and 1985. The net change in Latin American alone is almost two-thirds of the net global figure, and African also registers a large decrease. In Brazil, forest land contracted by over 24 million hectares between 1975 and 1985–an area larger than the whole of Britain. Colombia, Ecuador and Venezuela also appear to have suffered large reductions of the scale of millions of hectares, while in Africa the largest decreases were in Ivory Coast, Nigeria, Sudan and Zaire. The apparent net increase in Asia results from the expansion of forest land in China (by a reported 22 million hectares, but it is alleged by Smil (1983) that much of the claimed afforestation indicates the target rather than the achievement) and to a lesser degree in India: countries such as the Philippines and Malaysia experienced substantial decreases. These trends between 1975 and 1985 merely continue those that have been operating for several decades. For example in Sri Lanka, natural forest cover shrank from 50 to 24.9 per cent between 1950 and 1981 (Erdelen, 1988). In Thailand, forest cover decreased from 70 per cent in 1945 to 53 per cent in 1961 and 18-19 per cent in 1988 (Lohmann, 1989).

## Reforestation and Afforestation

If dwindling areas of forest land are a characteristic of much of the developing world, much of the developed world has undergone expansion. This is particularly true of Europe and the Soviet Union. As Table 12.3 indicates, expanding forest land appears to be a particular feature of the centrally planned economies. The reported increase in the Soviet forest area between 1961 and 1978 amounted to 52 million hectares (Barr, 1984), or more than twice the entire land area of Britain. The pattern of annual cut and replanting, however, indicated that the forest area should have been declining, and at least some of the reported change may be more apparent than real. Barr (1988) suggests that increases in the total forested area and volume of growing stock may reflect changes in survey techniques more than real changes in these parameters. Even if expansion in the Soviet Union and China is overstated, however, perhaps a three-world model can be applied effectively, with expansion and contraction characteristics respectively of the 'Second' and 'Third' worlds, and a mixture of expansion and contraction in the 'First'.

**Table 12.3: Not changes in area of forest and woodland 1975-85.**

|  | *Area (million hectares)* | *Per cent* |
|---|---|---|
| World | -82.9 | -1.99 |
| Developed world | -12.9 | -0.70 |
| Developing world | -70.0 | -3.01 |
| Centrally planned ecs. | +60.7 | +5.55 |
| Latin America | -.54.1 | -5.21 |
| Africa | -29.0 | -3.99 |
| Asia | +13.8 | +2.51 |
| Europe | +2.0 | +1.29 |
| USSR | +38.0 | +4.24 |

## Patterns of Change

A more detailed spatial pattern of change in forest land is depicted in Figure 12.4. Here the note of caution about the validity and reliability of statistics for individual countries needs to be sounded yet again: while the overall pattern is probably valid, over reliance should not be placed on the values shown for small area. It should also be emphasized that the change is net, and is based on data in FAO Yearbooks. It is likely, therefore, to understate gross deforestation.

At the general level, forest contraction is shown to be almost ubiquitous in the tropics. The whole of Central America from the US border to Ecuador, with the sole exception of Panama, experienced a loss of 10 per cent or more during the period form 1975 to 1985, together with large areas in the Sahel and West Africa, and part of south-east Asia. While high percentage rates of change are most likely be found in small countries, the spatial propinquity of the areas in question is a notable feature of Figure 12.4. Furthermore, comparison with data for the decade of the 1970s suggests that he 'high loss' areas are expanding. Mexico, Guatemala and Ecuador have been added to the Latin America group and Malaysia has joined the 'high' group of Thailand and Indonesia in south-east Asia. Outside the tropics, reductions in forest land are most notable in the United States and Australia, where high agricultural prices in the 1970s and early 1980s contributed to forest clearance, and a few other small and isolated areas. In the case of the United States, at least part of the forest loss incurred during the 1970s may soon be replaced under a progamme for the afforestation of erodible cropland.

The association between forest contraction in the tropics and expansion in the higher latitudes is reflected at the continental and national levels in South America and Africa, as well as at the global scale. High rates of increase are mainly confined to countries where the forest area is small. Almost all the centrally planned economies,

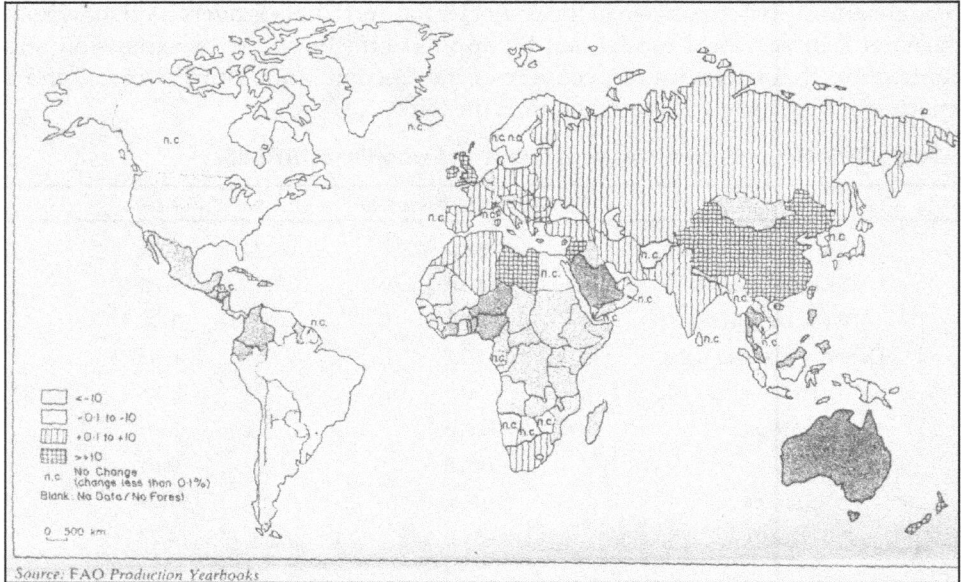

**Figure 12.4: Percentage change in area of forest and woodland, 1975-85.**

**Note: No change relates to the area as listed in the sources: it does not necessarily imply that the area remained constant. Please see text for review of source data.**

including the Soviet Union and Eastern Europe as well as China, stand out as areas of apparent forest expansion. Similarly, almost the whole of Europe, with the exception of Austria and the Benelux countries, has seen expansion even in the period before the agricultural surpluses of the mid-1980s began to encourage a faster transfer of land into forest. This tend applies even in counties such as Iceland, which is not traditionally associated with extensive forests. Its case, incidentally, illustrates the general point about the deficiency of data on which the map is based. The source indicates no change in forest area, but other reports suggest that by 1987, some 41,000 hectares were enclosed for afforestation, of which some 5,000 hectares had already been planted (Blondal, 1987). Previously, the extent of Icelandic birch woods ha decreased by a factor of ten since the early eighteenth century, with the clearing o woodlands for fields, fuel and lumber (Arnalds, 1987).

Tends in forest area clearly do not exist in isolation, but are closely interrelated with trends in agricultural areas. This relationship is depicted in Figure 5, where arable and forest trends are plotted for a large group of countries. Most of these countries are located in the parts of the graph indicating arable expansion combined with forest shrinkage, and vice versa (despite the fact that these two categories of land cover comprise just over 40 per cent of the total land area). Most European countries occupy the top left quadrant of the graph. In France, for example, the forest area has increased from 20 per cent of the land area in 1959 to 27 per cent in the mid-1980s (Devaud, 1987). In the Netherlands, the forest area has grown from 250,000 hectares during the Second World War to 330,000 hectares by the mid-1980s

(Grandjean, 1987). Most developing countries, on the other hand, are in the lower-right part, indicating a combination of agricultural expansion and forest contraction.

## Population Trends and Deforestation

The patterns indicated in Figures 12.4 and 12.5 suggest a relationship between forest trends and population trends at the global scale, and an example of a similar relationship at the national scale is contained in Figure 12.6. In countries with rapidly expanding populations the forest land area is rapidly contracting. Conversely, where the population is stagnating or growing only slowly, forest land is expanding. Probable errors in the published data values for both population and forest land area complicate the measurement of the statistical correlation between these two trends, but significant relationships between population growth, expansion of the arable area and contraction of the forest land area have been confirmed for developing counties by Allen and Barnes (1985) on the basis of data from FAO Production Yearbooks. They conclude that recent population growth an change in arable a5rea are associated with deforestation, and also show that the relationship is stronger in Asia and Africa than it is in Latin America, where presumably other factors play relatively stronger roles. Over the longer term, a similar relationship has been reported for Parana State, Brazil, between 1900 and 1973, although few details are presented (Palo and Mery, 1986).

The implication is that agricultural expansion in response to population growth is a major factor influencing or determining forest clearance, and probably the predominant one. In addition, forest clearance means that supplies of fuelwood may decrease in some areas, leading to the increasing use of crop residues and dung as fuel and hence declining agricultural yields and thus further pressure for forest clearance (*e.g.* Pimentel *et al.*, 1986). This may be an over simplification, since some dung could be used for fuel during the non-growing season without undue effect on crop yields (Bajracharya, 1983a). On the other hand fuelwood collection may itself be an agent of deforestation if the population pressure is high and the forest growth rate low. Open tree formations such as those in savannah areas are usually the most likely to be affected by this process. The likelihood of deforestation from this cause is perhaps greatest around cities, especially in the Sahel and in India. In the case of the latter, forest cover within 100 kilometres of India's major cities decreased by 15 per cent or more in less than a decade (Bowonder *et al.*, 1987b). The effects of this pressure, however, are not confined to the vicinity of cities: the production f charcoal can affect forests far from the centres of consumption.

If there is a casual relationship between population growth and deforestation then there are two important implications. First, if population growth is the driving-force behind deforestation, then an increasing area of forest is likely to be cleared each year in response to it. Even if the absolute area cleared were constant, this would represent an increasing proportion of the dwindling forest area. At first a given area of deforestation might represent a very small percentage of the total area, but subsequently that percentage would rapidly increase. An accelerating decrease in the forest area may therefore occur. Since the later stages of deforestation may be so rapid, it is thus perhaps not surprising that some countries can effectively be almost completely denuded of forest before remedial action begins. Even so, deforestation

Global Warming and Forest

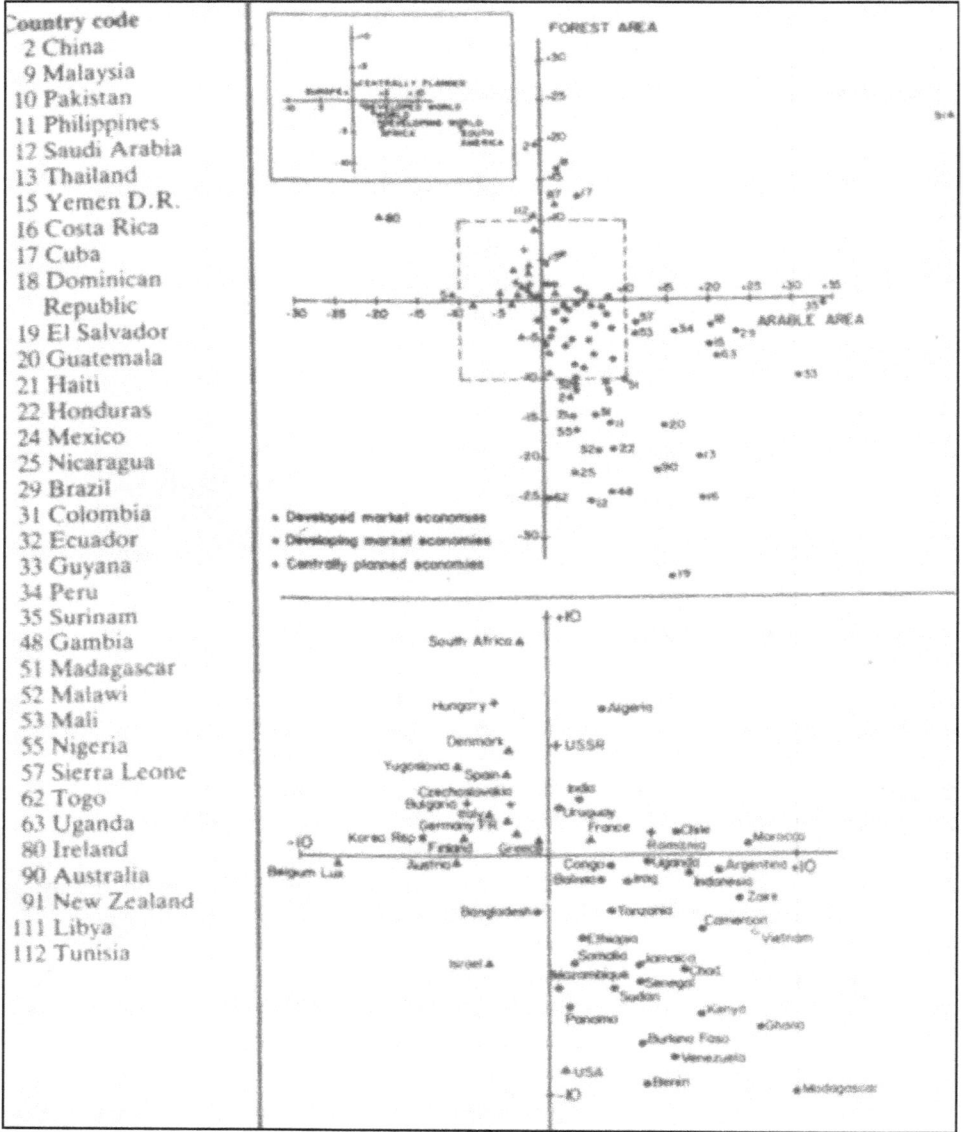

**Figure 12.5: Distribution of countries by percentages change in arable and forest area, 1979-85. The section of the upper diagram indicated by the dashed lines enlarged in the lower section.**

appears unlikely ever to be complete: small residual areas are likely to be perceived to have high conservation value and in any case are likely to be located in inaccessible areas where there is a measure of insulation against both agricultural and commercial forces.

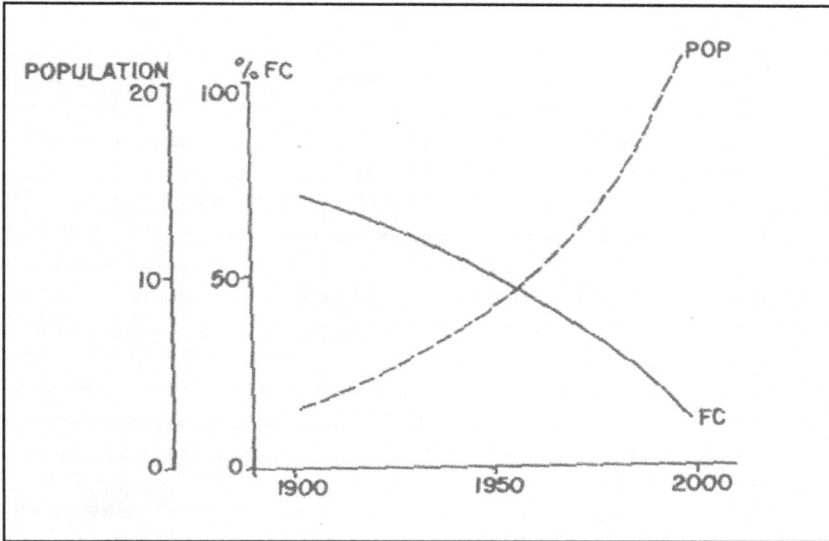

**Figure 12.6: Trends in population (POP) and forest cover (FC) in Sri Lanka, Population is in millions: forest cover is expressed as a percentage of land area.**

Second, the patterns depicted on Figures 12.4 and 12.5 combined with the concept of interrelationship between population and forest tends and the historical background of the use of the forest resource. During periods of rapid population growth, the forest area is likely to contract, while during periods of relatively stable population it is likely to stabilize or even to expand. Forest clearance in nineteenth-century Europe has now given way to forest expansion, and the high rates of forest destruction in the second half of the twentieth century. While precise, quantitative evidence to refine and evaluate the model is scarcely available at the national level, far less at the global scale, it is tempting to speculate that the model may also be valid globally, and that at present we are in a phase corresponding to that of parts of Europe and North America a century or more ago. The problem lies in predicting when and how the downward curve of the forest area will 'bottom out', and whether it will dos o while significant areas of forest still survive (Figure 12.6).

Third, it should perhaps be emphasized that the relationship between populations tends and deforestation is neither especially close nor deterministic. An association between the variables can be demonstrated, but this association may itself vary through time, as in the case of the United Sates, and other factors such as the upheavals created by war, revolution and political reorganization as well as more gradual changes in both popular and governmental attitudes may be followed by dramatic changes in forest trends. To attempt to portray the forest area 'transition' as a mirror image of the population transition is therefore an oversimplification, although this is not to say that it is devoid validity.

## Causes of Deforestation

The dominant trend in forest area at present is that of deforestation, which is largely concentrated in the tropical zone. Trends in tropical forest areas are discussed further in Chapter 8: here the major causes of deforestation are outlined briefly. As Figures 12.4 and 12.5 imply, an inverse relationship appears to exist between trends in population and the forest area. This relationship, however, is far from perfect. Several factors account for the imperfect nature of the statistical relationship between population trends and deforestation. One is that the pressure of population growth on the forest may be expressed in the form of shifting cultivation or of sedentary agriculture, and that different effects may result. The former is believed to account for around 45 per cen6 of the annual rate of loss of tropical forest (Lanly, 1982). Shifting cultivation is identified as the principal cause of deforestation in all three tropical regions, accounting for 70, 50 and 35 per cent respectively in Africa, Asia and tropical America. About 500 million people and 240 million hectares of closed forest are involved in shifting cultivation, which is increasing at an average annual rate of 1.25 per cent. More than 5 million hectares of new for4et fallow are being created annually (Lanly, 1985). As population increases, fallow periods grow shorter and forest recovery between periods of cultivation becomes impossible.

Another factor is that forest clearance occurs for reasons that are related to internal fiscal or development policies or to export markets in the developed world, rather than to population growth in the developing world. This may be one of the factors that accounts for the existence of a weaker relationship between population trends and deforestation in Latin American than exists in Africa and Asia. Large tracts of tropical forest in Latin America have been cleared for cattle ranching to produce cheap beef for North America (Myers, 1981), although in recent years beef exports to the United States have deceased as a result of a combination of factors including decreasing beef consumption as well as political tensions in Central America.

Deforestation for conversion to pasture, however, is not only a result of international commodity markets–the so-called 'hamburger connection'–but also occurs for domestic reasons, especially in Brazil (*e.g.* Hecht, 1989). Forest clearance for cattle-grazing has been carried out by small peasants as well as by large ranchers, not least in order to strengthen land claims and secure fiscal incentives and tax holidays. In addition, the market value of cleared land increases, and forest removal may be a hedge against inflation. Browder (1988) suggests that the rate of forest conversion is positively correlated with Brazilian inflation. Even although cattle-ranching per se may not be economic in Amazonia, the related economic circumstances are a powerful driving-force for defo5restation. Real-estate speculation in particular is a major factor (Fearnside, 1989).

Conversion to pasture has been identified as the foremost cause of deforestation in Brazil, accounting for 72 per cent of the forest clearance up to 1980 (Browder, 1988). This compares with a contribution of 10 per cent from mall farmers and shifting cultivators. It is believed that at least 325,000 square kilometers of Latin American tropical forests were cleared for pasture between 1962 and 1985, and at least in the case of Brazil almost none of the wood cleared was disposed of as commercial timber (Myers, 1983b). According to Myers, this represents a lost timber value of over $10 billion. Much of the Brazilian forest clearance for pasture development was encouraged indirectly by the development authority, the Super intendancy of Amazonia (SUDAM): merchantable timber was recovered only 18 per cent of SUDAM-assisted ranches, and the loss of timber on these ranches alone is estimated at over 200 million cubic metes, perhaps representing a social opportunity cost of $1 billion to $2 billion (Browder, 1988).

Forest clearance for pastures and for export-oriented commercial agriculture is not confined to modern Latin America: as Chapter 3 indicated, it occurred in areas such as Burma and the Philippines during the nineteenth century, and has continued in recent times in countries such as Ivory Coast (Bertrand, 1983) and Nigeria (Osemeobo, 1988).

Large-scale commercial clearance may occur alongside peasant colonization and government sponsored settlement schemes for small farmers, as in the case of Brazil (*e.g.* Fearnside and Salati, 1985). Deforestation for subsistence production is not a major cause of deforestation in Brazil at present (Fearnside, 1989a), but land cleared by small-scale farmers may in turn be amalgamated into large farms or ranches, and the process of small-scale clearing repeated. The government-sponsored 'transmigration' of millions of Javanese to outlying parts of Indonesia, including Sumatra, Kalimantan and Sulawesi, has resulted in extensive forest removal, perhaps amounting to 3 million hectares, in these areas (*e.g.* Ross, 1985; Repetto, 1988). Deforestation may also be carried out in some areas as evidence that a government has taken steps to establish an interest, and hence sovereignty, in a frontier area. Such action is reported, for example, from southern Venezuela (Buschbacher, 1987).

Timber production in general and commercial logging in particular may also be contributory factors. In some areas this has a yet been a minor factor. In Brazilian Amazonia, for example, loggers remove around 4 million cubic metres annually– around 0.01 per cent of the area's 50,000 million cubic metres of timber (Westoby, 1989). Commercial loggings is rapidly becoming major disturbance even in Amazonia, however, as the forests in other parts of the tropics are depleted (Fearnside, 1989a). In parts of south-east Asia timber production has been a much more significant factor.

Allen and Barnes (1985) found that per capita wood production was not significantly related to change of forest area in their 'short-run' model, but when they allowed for a time lag effect by relating wood production in 1968 to change in the forest area over the following decade, their conclusion was that wood-harvesting does result in deforestation, albeit with a delayed effect. In their analysis, harvesting for both fuel and export was considered. Commercial logging, often geared to export markets, may be an important cause of deforestation at the local or national scale if reforestation is not enforced, even although it is not a major factor at the global scale (Lundgren, 1985). And even if logging is not significant as a direct cause of deforestation, it plays a vital part in a two-step process. Commercial loggers construct roads in order to extract timber, and thereby greatly improve accessibility. Settles or forest farmers can then move in and proceed either to clear parts of the forest or to prevent its regeneration (Walker, 1987). In more general terms, the construction of roads and the resulting change in forest accessibility is a major factor in deforestation:

for example Malingreau and Tucker (1988) link recent acceleration of deforestation in part of Amazonia to the paving of highway BR-364 and the consequent surge of migrants.

Various other factors may contribute to deforestation. Overgrazing has been damaging in some parts of the world, especially in mountainous areas such as north Pakistan (Allan, 1986; 1987) and in open or dry forests such as those of Saudi Arabia (Abo-Hassan, 1983). Fire is also a major hazard, especially in parts of southern Europe: in Greece, for example, 25,000–120,000 hectares are damaged or lost each year through fire, urban growth or agricultural expansion, while reforestation averages only 3,000–4,000 hectares (Modinos and Tsekouras, 1987). Political instability or strife may also be a contributory factor. In Pakistan, for example, the resettlement of Afghan refugees resulted in the destruction of forest, and Pakistani nationals took advantage of the confusion to participate in illegal logging (Allan, 1987). In southern Africa, incendiarism arising from disgruntlement amongst the rural population has damaged the Zambezi teak forest, as have accidental fires started by cultivators, honey gatherers and hunters (Piearce, 1986).

During the Vietnam conflict, an area of around 1.25 million hectares was sprayed with herbicides and defoliants, and more than 4 million hectares were damaged by shells (Lanly, 1982). In the south; 5./6 million hectares of upland forest were damaged, but more forest–around 200,000 hectares per year–has been lost in Vietnam since the war than during it (Kemf, 1988). There is a long history of destruction of forests during wars. This dates back to classical t5imes (Chapter 3) and probably earlier. More recently, vast areas of forest in Pomerania were cut down by the Swedes during the Thirty Years War, with the result that many districts were invaded by sand dunes (Braudel, 1979). In the Soviet Union, forest depredation by the Japanese was suffered on Sakhalin between 1905 and 1945, and 20 million hectares of forest were felled or destroyed in areas occupied by the Nazis or subject to military activities (Barr, 1988).

Elsewhere, urban expansion may encroach on forest land. In the United State, the rate of conversion of forest land to urban uses increased dramatically during the 1970s, and amounted to almost 250,000 hectares 1981 (Miller and Rose, 1985). At the global scale the area of forest affected in this ways may be small, but the forest land affected may be of prime quality (as for example in the case of British Columbia (Redpath *et al.*, 1986) and the loss of timber production disproportionately high.

It is apparent, therefore, that several factors contribute to deforestation. Myers (1980a) has attempted to estimate their relative importance in terms of are. Of his estimated conversion (as opposed to deforestation) rate of 245,000 square kilometres per year, he estimates that commercial logging combined with follow-on cultivation accounts for 200,000 square kilometers, commercial logging alone between 19,000 and 29,000, fuelwood-cutting 25,000 and ranching 20,000 square kilometres.

The relative importance of these factors varies from place to place, and such estimates may have little validity at the national level. The nature of deforestation at all levels from the global to the local level is very complex. This is shown by the case of Jamaica, which is illustrated in Table 12.4. As the table indicates, a variety of causes including commercial and peasant cultivation, logging and the creation of coniferous plantations have all contributed. The compiler of data is careful to set them in context by referring to the possibility of cyclical tends related to climatic variability and to economic conditions (Eyre, 1987). An illustration of the possible significance of climate change or fluctuation in relation to woodland area comes from Mauritania, from where it is reported that drought during the 1970s seriously depleted the riverine woodlands, with the result that almost 7,000 out of a total of 16,000 hectares were lost (Pellek, 1983). Economic fluctuations were also associated with fluctuations in the forest area in the past: for example Eyre indicates that Jamaica had a smaller forest area in the plantation era than it has at present: during the second half of the nineteenth century regrowth occurred over substantial area. Changes in the forest area may therefore be complex in terms of cause and amplitude: although the long-term trend may be downwards there may also be periods of regrowth, at least t the local or national scale.

**Table 12.4: Causes of deforestation in Jamaica, 1986.**

| | Percentage of Deforested Land |
|---|---|
| Peasant agriculture | 52.2 |
| Pasture | 11.0 |
| Coffee | 9.3 |
| Residential etc. | 8.8 |
| Horticulture | 6.5 |
| Logging and fuelwood | 4.5 |
| Bananas | 2.9 |
| Cannabis (marijuana) | 2.7 |
| Other commercial agriculture | 2.2 |
| Conifer plantations | 0.2 |

Source: Eyre (1987).

## Causes of Afforestation

At the global scale, expanding forest area is correlated with low rates of population growth, in the same way as deforestation is associated with rapid growth. In many parts of the developed world, areas that previously suffered rapid deforestation are now undergoing reforestation. In the French Alps, for example, extensive reforestation ha occurred both by natural regeneration and by planting. If followed a decrease in population, the abandonment of agricultural land, and a reduction in numbers of grazing animals (Douguedroit, 1981). In the Alps and in many other parts of the developed world, population growth in the eighteenth and nineteenth centuries led to an expansion of the agricultural areas and increased agricultural pressures in previously wooded areas. More recent times have seen an abandonment o many of these agricultural areas and regrowth of the forest. Furthermore, afforestation has been carried out, or encouraged, by governments as a mean of reducing the agricultural area at times of overproduction, of providing protection against erosion, or of building up reserves and supplies of timber. Until recently, most afforestation was carried on poor or marginal agricultural areas, but there are now indications that it is extending upto better qualities of land in counties such as New Zealand, Britain, and other parts of the European Community.

## Long-term forest Tends

If there is great uncertainty over the present dynamics of the forest are, our knowledge about historical rates of defor4estation is even more deficient. Matthews (1983) has attempted to compare the present extent of forest and woodland with the 'pre-agricultural' extent, numerous atlases complemented by satellite imagery. Most of the atlases from which she worked wee published during the 1960s, and probably rely heavily on considerably earlier data. The results of such an exercise are, of course, no more robust than the original data, but they are nevertheless interesting and are summarized in Table 12.5.

**Table 12.5: Pre-agricultural and recent forest cover.**

| | Pre-Agricultural Area | Present Area* | Percentage Converted to Cultivation (million km²) |
|---|---|---|---|
| Forest: total | 46.8 | 39.3 | 15.2 |
| Tropical evergreen rain forest | 12.8 | 12.3 | 3.8 |
| Tropical/subtropical evergreen seasonal broad leaved forest | 4.0 | 3,.3 | 19.0 |
| Tropical/subtropical drought deciduous forest | 4.0 | 3.0 | 25.0 |
| Temperate evergreen seasonal broad leaved forest, summer rain | 1.2 | 0.8 | 32.0 |
| Temperate/sub-polar evergreen needle-leaved forest | 9.6 | 9.3 | 4.0 |
| Cold deciduous forest, with evergreens | 7.8 | 5.2 | 33.0 |
| Cold deciduous forest, without evergreens | 5.5 | 4.0 | 28.0 |
| Xeromorphic forest/woodland | 3.1 | 2.7 | 13.0 |
| Woodland | 15.2 | 13.1 | 13.8 |
| Shrubland | 13.0 | 12.1 | 6.7 |

*: Present' relates to data extracted from sources mostly published in 1960s and 1970s.

NB: Only forest types extending to over 1 million km² are indicated; the total figure therefore exceeds the sum of those indicated for specific types.

*Source*: Adapted from data in Matthews (1983).

The estimated reduction in the area of forest and woodland is approximately 900 million hectares, or approximately 15 per cent. Of this reduction, around 775 million hectares came from forest, and the remainder from woodland. A feature of her figures is the small reduction reported for the tropical rain forest. This amounted to only 50 million nectars or around 4 per cent of the 'pre-agricultural' extent: a similar reduction is indicated for the boreal coniferous forest. The clearing of mainly deciduous forests in Europe and the eastern United States, and of temperature evergreen forests in Asia, accounts for most of the reduction. Nevertheless, none of the conversion figures listed in the table exceeds one-third. These results seem low in comparison with current rates of change, and Sommer (1976) has suggested that the reduction in the tropical rainforest area alone may have amounted to 40 per cent. They nevertheless appear to be in broad agreement with the estimates of Viliams

(1989b), who concluded after assembling historical data on deforestation that a total of between approximately 750 and 800 million hectares had been cleared.

Such estimates can be interpreted in various ways. On the one hand, the apparent fact that around 85 per cent of the original forest areas has survived to the second half of the twentieth century (at least in extent, if not necessarily in terms of composition) is a useful counterbalance to current fears about the fate of the world forest. On the other hand, an area of around 400 million hectares of forest and woodland may have been cleared between 1860 and 1978 (Williams, 1989b; Richards, 1986). The area of forest and woodland may thus have contracted as much during the last century as it did throughout earlier times. Furthermore, a net contraction f around 5 million hectares pr year (as indicated by statistics in FAO Yearbooks) would indicate an accelerating rate of loss, even within the modern period. And it should be borne in mind that estimates of the annual gross loss of forest may be several times greater. The theme of accelerating loss is highlighted by comparing Matthews' estimated reduction of the area of the tropical rain forest (4 per cent) with recent estimates of annual loss of around 0.6 per cent, or even more strikingly by comparing her 'historical' figure of around 50 million hectares with estimates of recent annual losses of at least 7-11 million hectares.

Whilst it is impossible of construct a reliable set of global time-series data for the changing forest area (even with a time-scale of only a few decades), it is probable that rates of deforestation have closely paralleled rates of population growth at the medium and long terms as well as in the shorter tern. If the historical relationship between population and forest area persists, it seems unlikely that the latter will stabilize before the former does so, perhaps well over a century from now. Any projection of historical trends to that time is completely speculative: such an exercise, however, would suggest that a further 15-20 per cent of the 'original' area might disappear. Optimists may prefer to think in terms that such a projection would still indicate the survival of around 70 per cent of the 'original;' area. At a net rate of contraction of 5 million hectares per year, the estimated 2,800 million hectares of closed forest would disappear in 560 years (assuming rather unrealistically that all the contraction would be concentrated in the closed forest). More strikingly, the straight-line projections of Lanly's (1982) figures indicate the complete disappearance of the tropical rainforest by 2057 (Guppy, 1984). While there is, of course, no logical reason why such extrapolation should have any predictive value, it is rather disconcerting that the projected end of the forest should be so imminent, when human pressures throughout history have as yet caused the removal of perhaps only 15-20 per cent of the 'original' forest.

## Plantations

Plantations constitute a significant but inadequately known part of the world forest area. Many countries have adopted forestry policies which have incorporated elements of plantation establishment, especially during the twentieth century. Such policies have frequently been motivated by concern about forest tends (nationally or globally) and their perceived implications for timber supplies. Fears of timber famine or shortage have underlain policies of this nature in settings as diverse as the United

Sates during the 1920, in New Zealand in the same decade (Roche, 1986) and in Britain around 1980. A related goal is to achieve complete or increased self-sufficiency. In Venezuela, for example, around 200,000 hectares of plantations ha been established by 1986, with the goal of self-sufficiency of raw material for pulp and paper (Fahnestock *et al.*, 1987). On the other hand the export potential for forest products may also be a motivating factor, as in the cases of New Zealand in the 1970s (Le Heron, 1986) and Chile (Postel and Heise, 1988). Environmental protection has also been a motive for planting in some instances, such as, for example, in the French Alps (*e.g.* Buttond, 1986). Whatever the driving-force may be, plantations may be created by government itself through state forestry services, or encouragement or afforestation may be offered in the form of planting grants and tax incentives. There may also be less direct incentives for landowners to afforest in some countries: in Ecuador, for example, forest land is likely to be safeguarded against land reform and some landowners may carry out afforestation for this reason (*e.g.* Gondard, 1988).

No convenient source of data such as the FAO Production Yearbooks exists for plantation areas, and most of the available compilations are based on diverse sources that relate to a variety of dates and are based on uncertain or differing. Sometimes planting and regeneration are not clearly distinguished. Plantations may include both reforestation immediately after felling and the creation of new forests on land that at least in recent times (sometimes defined as within fifty years) has not supported forests, and it is not surprising that different definitions are adopted in different countries.

Despite these problems of data sources, there is some agreement that the total area of plantations world-wide in the early to mid-1980s was in excess of 100 million hectares, or approximately 3 per cent of the closed forest area. Evans (1986) estimates the area as between 120 and 140 million hectares, while in the early 1970s Persson (1974) assembled data, which he emphasized varied greatly in quality, indicating that the total area of man-made forest then was around 95 million hectares. That figure compares with an estimate of 81 million hectares in 1965 (Logan, 1967).

Postel and Heise (1988) have compiled estimates for the world-wide area of industrial forest plantations around 1985: these indicate that the total amounts to approximately 92 million hectares (Table 12.6). To the latter should be added areas of plantations established for fuelwood, environmental protection and other non-industrial purposes: for example 40,000 hectares/year or just under half of the total afforestation effort in Africa in recent years has been for these purposes (Evans, 1986): world wide, 54 per cent of tropical plantations in the late 1970s were intended for industrial purposes and the remainder for fuelwood, environmental protection, and a variety of minor purposes (Evans, 1982). Lanly (1982) quotes broadly similar proportions of 60 per cent industrial and 40 per cent non-industrial.

The total annual rate of planting world-wide is uncertain, not least because of problems of definition and in particular of distinguishing between different forms of regeneration and between reforestation and afforestation. It was estimated by the World Resources Institute (WRI, 1986) that by 1980, 14.5 million hectares of forest land had been reforested or renewed annually. This is likely to represent an overestimate of the area actually planted. Table 12.6 out rates for a variety of countries, but it should be emphasized that no definitions are given in many of the individual sources from which the table is compiled.

As Table 12.6 indicates plantations are very unevenly distributed. While some uncertainty may surround the extent to which the establishment of plantations in the Soviet Union and China has been successful, together they account for over 40 per cent of the plantation area. Europe, Japan and the United States are the other major areas, each having around 10 million hectares or more of industrial plantations. In Europe, Sweden alone has in excess of 5 million hectares of plantations (Savill and Evans, 1986).

It is notable that he four largest individual countries in terms of plantation area lie wholly or mainly in the temperate zone, an indeed do the majority of the countries listed in the table. Probably less than one-quarter of the plantation area lies within the tropics. Tropical regions account for around 17 million hectares if China is excluded

**Table 12.6: Estimated area of industrial forest plantation, c. 1985%**

| | (Million ha) |
|---|---|
| USSR | 21.9 |
| China | 17.5-28.0 |
| USA | 12.0 |
| Japan | 10.0 |
| Brazil | 5.0 |
| South Korea | 2.9 |
| India | 2.0 |
| Indonesia | 1.8 |
| Canada* | 1.5 |
| Chile | 1.3 |
| South Africa | 1.1 |
| New Zealand | 1.1 |
| Australia | 0.8 |
| Argentina | 0.8 |

*Plantations established since 1975

NB: Data for European countries are omitted. Most of the forest area in countries such as Britain and Hungary is in the form of plantations: in each of these two countries alone, the forest area is in excess of 2 million ha.

*Source*: Compiled from data in Postal and Heise (1987), and Savill and Evans (1986).

(Lanly, 1982) but that figure may rise to nearer 25 million hectares if southern China and other countries just outside the true tropics are included (Evans, 1986). Much uncertainty surrounds plantation area and planting rates in China. One estimate of the open-land area planted between 1950 and 1983 is as high as 124 million hectare, with a further 8.8 million hectares having been planted following logging in natural forest (Zhu *et al.*, 1987). The same source estimates recent planting rates averaging 4-5 million hectares per year. In addition, 'four sides' planting along rivers, roads, houses and streams amounted to 9.5 billion trees in 1981-2 (Hsiung, 1983). In 1981 it was resolved that all Chinese citizens above the age of eleven years should be obliged to plant three to five trees each year (Yuan, 1986). Although the reported planting rates are very impressive, survival rates have been low. They averaged only 31 per cent over the period from 1950 to 1983 (Zhu *et al*, 1987). It is officially claimed that 65 per cent of trees planted survive, but some estimates are as low as 10 per cent (Forestier, 1989).

In some north-west European countries and Japan, there is a long tradition o forest-planting extending back for several centuries. Most of the plantation area, however, has been established during the twentieth century. In Sweden, for example, a rapid increase in the replanting rate has occurred over the last fifteen years (hanger, 1986). Even in countries such as Japan with long traditions of forest expansion (Chapter 3), many of the plantations are comparatively young: a programme of reafforesttion involving approximately 10 million hectares began after the Second World Was (Guilland, 1983; Sedjo, 1987). Peak planting rates of over 400,000 hectares were achieved in the 1950s and early 1960s, as compared with post-war averages of around 200,000 hectares (Matsui, 1980; Tsay, 1987. Planting rates fell by half during the ten years prior to 1983. This decrease was related of an economic depression in Japanese forestry resulting from the effects of timber imports (Ohba, 1983).

Much of the 12 million hectares area of plantations in the United States was planed in the south-east of the country during the 1930s (see, for example, Clark, 1984). The annual area planted or direct-seeded in the United States increased from 200,000 hectares in 1950 to 800,000 hectares in 1978 (USDA, 1982). The agricultural Conservation Reserve Program introduced in the mid-1980s may lead to the planting o millions of hectares of farmland (*e.g.* Cubbage and Gunter, 1987), and is potentially one of the biggest afforestation exercises in American history. In countries such as the Soviet Union and Canada the replanting of forests is of recent origin. In the former, annul rates of forest establishment by seeding and planting increased from 582.000 hectares in 1955 to 1.33 million hectares in 1981 (Barr, 1984). In Canada, little attention was paid until recently to regeneration of cut-over areas, and the remaining forest stands became increasingly remote from mills (Fox, 1988). This has led to a rapidly expanding programme of replanting or regeneration. The total area planted per year increased from 128,000 hectares in 1975 to 192,000 in 1982, of which nearly half was in British Columbia (Canadian Forestry Service, 1988). In British Columbia, the annual planting rate doubled between 1973 and 1983. Even so, little more than half of the area logged in 1983-4 was replanted (Pearse *et al.,* 1986). Despite recent increases in replanting rates, over 450,000 hectares of forest land in Canada go out of production annually as not being satisfactorily restocked with commercial species (Honer, 1986). Only about 20 per cent of the area cut each year in the late 1970s and early 1980s was replanted (Weetman, 1986).

Substantial expansion of plantations in the form of new or replacement forests has also occurred in Europe. In Sweden and Finland, for example, over 170,000 hectares and over 100,000 hectares per year respectively were being planted by the early 1980s (Savill and vans, 1986; Tilastokestus, 1985-6). The forest area in Hungary increased by nearly 50 per cent between 1950 and 1980 (based on Keresztesi, 1984), and the planting rate in Bulgaria has exceeded 40,000 hectares per year (Grouev, 1984).

One of the most spectacular increases in the plantation area is in Brazil, where the plantation area increased from 500,000 to 3.7 million hectares between 1966 and 1979, and which has subsequently grown to over 5 million hectares (Postel and Heise, 1988). During the 1970s, Brazil established over 250,000 hectares of plantations per year (Sedjo, 1987). Most of this planting was carried out by private companies benefiting from the tax-saving incentives of forestry investment. An incidental effect of this type of incentive has been the dominance of large-scale investments and large plantations (Victor *et al.*, 1986). Another South American country in which the plantation area has rapidly expanded in Chile. Over 300,000 hectares of pine plantations were established by 1974, largely through the government-owned National Forestry Corporation. Planting accelerated to 77,500 hectares per year during the second half of the 1970s (Husch, 1982) and by 1985 the area had expanded to 1.1 million hectares, largely because of the incentive of a 75 per cent reimbursement of costs incurred by the private sector in establishing n maintaining plantations (Postel and Heise, 1988). Over 100,000 hectares were reforested in 1976, but by the end of the decade the rate had fallen to about 50,000 hectares as the National Forestry Corporation was phased out (Solbrig, 1984). There has also been rapid expansion in South Korea and in countries such as New Zealand, where planting targets increased from 9-12,000 hectares per year in 1958-9 to 45,000 in 1981 (Abbiss, 1986). Annual planting rates averaged 44,000 hectares between the mid-1970s and mid-1980s (White, 1987). In New Zealand afforestation grants met 45 per cent of establishment costs, but wer3e withdrawn in 1986. In Tasmania, planting rates increased from under 20 hectares per year prior to the early 1970s to over 1,500 hectares per year by 1982-4. This increase is attributable to the advent of a ready market for eucalyptus pulpwood with the growth of export wood chipping, and the introduction of planting assistance schemes (Tibbits, 1986).

Numerous other countries are also engage in planting programmes which although modest in absolute terms are very significant in relation to the national forest area. For example, afforestation is proceeding in Israel at a rate of about 2,500 hectares per year (Cohen, 1985), and over 60 per cent of the national forest area now consists of plantations (Gottfried, 1982).

Although 1,0-1.2 million hectares o new plantations are now established annually in the tropics, subtropical and non-tropical areas account for most of the plantation area and also for much of the current planning (Tables 12.6 and 12.7). Nevertheless, the tropical plantation area has been rapidly expanding in recent years. The establishment of plantations in the tropics began in the Indian subcontinent in the mid-nineteenth century, but until recently the area of tropical plantations was very small. More than 90 per cent of the area after 1975 (Lanly, 1982). Between 1965 and 1980 the forest plantation area in the tropics treble from under 7 to almost 18 million hectares (Evans, 1982) (Evans defined the tropics a lying between 27 degrees north and south of the equator). The tropics' share of the world plantation area increased from 8 per cent in 1965 to 13 per cent in 1979. This trend is likely to continue as more countries embark on planting programmes. For example, it was decided in 1981 that 188,000 hectares o forest plantations should be established in peninsular Malaysia, to offset a projected wood deficit and to reduce pressure on national forests. By 1988, 29,000 hectares had been planted (Mead, 1989). In Indonesia, it is projected that 4.4 million hectares of timber estates, consisting partly of fast-growing pines, will be established between 1985 and 2000 (Meulenhoff, 1986). In East Kalimantan, a programme aimed at establishing fifteen estates each of 50,000 hectares of industrial plantations has been agreed, together with the rehabilitation of 60,000-70,000 hectares of logged areas by enrichment planting each year (Priasukmana, 1986).

**Table 12.7: Annual rates of forest plantation establishment, selected countries, 1980s.**

| | (ha) | |
|---|---|---|
| China | 4.8 million | (Zhu *et al.*, 1987) |
| USA | 1.8 million | (WRI, 1986) |
| USSR | 1.3 million | (Barr, 1984) |
| India | 370,000 | (Evans, 1982) (Bowonder *et al.*, 1987a) report that government envisages 3-5 million ha of afforestation per year 1987-90) |
| Brazil | 250 000 | (Sedjo, 1987) |
| Indonesia | 200 000 | (Evans, 1982) |
| Japan | 200 000 | (Tsay, 1987) |
| Vietnam | 200 000 | (Kem, 1989) |
| Canada | 192 000 | (Canadian Forestry Service, 1988) |
| Turkey | 150 000 | (WRI, 1986) |
| Finland | 114 000 | (Tilstokeskus, 1985) |
| Chile | 77 500 | (Husch, 1982) |
| Sudan | 60 000 | (Evans, 1982) |
| Philippines | 58 000 | (Evans, 1982) |
| Nigeria | 45 000 | (Evans, 1982) |
| New Zealand | 44 000 | White, 1987) |
| Bulgaria | 40 000 | (Grouev, 1984) |
| Australia | 32 000 | (Bureau of agricultural Economics, 1985) |
| Britain | 20 000 | (Annual Reports, Forestry Commission) |

*Note*: See also Grainger (1986) for areas of plantation established in Africa 1976-80.

Within the tropics, some forest areas have been cleared for industrial plantations but most of the plantations are on savannah, *cerrado* or other forms of grassland that has not been forested in recent times. A complex pattern of transfer may occur whereby forest is cleared for agriculture, and eventually, after the abandonment of cultivation and reversion to grassland, afforestation takes place. An illustration of this process comes from Ecuador, for instance: high-altitude natural forest is eventually replaced, after clearance, abandonment and reversion to grassland, by plantations (Gondard, 1988). Huge areas of tropical grassland are physically suitable for planting, some of which were at one time wooded and at another cultivated, and it is in this zone that much of the expansion of the forest area is likely to occur in the foreseeable future.

While the plantation area is growing at an accelerating rate, it is apparent hat it matches the rate of deforestation in neither aggregate nor spatial terms. As has already been indicated, the net deforestation rate has in recent years been running at an annual rate of at least 5 million nectars, and the areas of deforestation and reforestation coincide at neither the global nor the national scales. Most of the deforestation is concentrated in low latitude, while much of the reforestation is in the temperate or subtropical zones. Some tropical countries such as the Philippines claim that reforestation now exceeds deforestation (Durst, 1981), but the experience of Nigeria, where deforestation exceeds reforestation by a factor of ten, is more typical (Osemeobo, 1988). A similar ratio is reported for the tropical world as a whole, while in individual continents the ratio ranges from 1.29 in the case of Africa to 1:4.5 in Asia (Lanly, 1982). In countries such as Brazil, most of the deforestation is in Amazonia, while most of the new plantations are in the south o the country. (There are, of course, some exceptions to this generalization, such as the huge plantations at Jari (*e.g.* Palmer, 1986).) Deforestation and reforestation are typically separate in both time and space, and it is apparent that the creation of plantations is not a substitute or the loss of natural forests.

This later point applies in terms of species composition as well as area. Modern plantations are typically simple in composition, and indeed are frequently composed of a single species. Species of *Eucalyptus, Pinus* and *Tectona* account for 85 per cent of all plantations in the tropics (Evans, 1982), and in countries such as Brazil almost all the new forests are composed of either eucalyptus or pine, which, according to Sedjo (1980), comprised plantations amounting to 3.8 million hectares by 1979. It is estimated that a total of nearly 1.5 million hectares of southern pine species alone were planted world wide in 1985 (McDonald and Krugman, 1986). A few species also dominate plantations in temperate and sub-tropical zones: for example plantations in Chile and New Zealand are composed mainly o *Pinus radiara*, while in the higher latitudes of Britain and Ireland Tice sitchensis and *Pinus contorta* play dominant roles. The latter species, of North American origin, is also widely used in Swedish plantations (Hagner, 1983; Gamlin, 1988).

While plantations cannot replace natural forests either in terms of area or of composition, they can play crucial roles in the overall pattern of forest resource use. They are characterized by high productivities compared with natural forests in similar environment, and they are playing an increasingly important part in the supply of industrial wood in particular. In Latin America, for example, industrial plantations make up less than 1 per cent of the forest are but account for 30 per cent of industrial wood production: this proportion will rise to at least 50 per cent by the end of trhe century (Evans, 1987). In Australia, production from plantations will begin to exceed that from native forests sometime between 1990 and 2000 (Booth, 1984). In the United States, where at present around 14 per cent of the commercial forest are is in the form of plantations, it is expected that plantations will account for half of the wood-fibre production by 2000 (Sedjo, 1987).

Especially in relation to the production of industrial wood, therefore, the significance of plantations is disproportionate to their area. By offering the possibility of supplying large quantities of timber from relatively small areas of land, they may help to reduce at least one of the pressures on the natural forest and hence help to slow down its rate of contraction.

# Chapter 13

# Forest Resource Use

The use of the forest resource is as varied as it is controversial. It varies with the type and location of the forest; ownership and status, and it varies through time time. In turn the character and pattern of use have a strong influence on the nature and condition of the forest, and through time on its extent. They also have implication for the wider environment at scales ranging from the local to the global.

Forests world-wide have two main classes of functions: production and protection. In the former, timber and a variety of other commodities are produced. In the latter, the emphasis in on the provision of services such as watershed protection and nature conservation, rather than on material commodities. In practice the distinction between the production of commodities and the provision of services is not always clear or rigid. For example, the forest may be perceived or managed to 'produce' an equable flow of water. Nevertheless, the distinction between production and protection forests is manifested in the form of zoning systems. And in addition to this spatial dimension, a time dimension may also be recognized. This dimension is more apparent in terms of broad sequences than absolute dates, but it nevertheless has validity and utility as an integrating framework.

Three major stages of development of the forest resource, in terms of its use, may be recognized as the 'pre-industrial' forest is typically characterized by common-property ownership and by the production of a wide range of products, of which timber for construction and fuel is only one. The 'industrial' forest is usually subject to use by private individuals or companies (although it may remain under public ownership). In contrast to the diversity of products of the 'pre-industrial' forest, the product range is narrow and simple. Priority is usually given to timber production, sometimes to the exclusion of other considerations. In the 'post-industrial' forest, the provision of services such as conservation and recreation is accommodated alongside (or even to the exclusion) of timber production.

This sequential model is no more rigid than that sketched for forest ownership in the previous chapter. Some forests, for example, undergo a direct transition from the 'pre-industrial' to the 'post-industrial' stages on classification as protection forests within a national system. And the model cannot at the global scale be calibrated in terms of years or dates. The forests of much of Europe and countries such s the United State and Japan may be regarded as collectively entering the post-industrial stage at present, although within these areas separate tracts of forest may be classified for production or for other purposes (in Europe more than 70 per cent of the forest area is managed primarily for wood production (Prins, 1987). On the other hand, large areas of forest in the developing world are making the (sometimes difficult) transition from the pre-industrial to the industrial stage, as timber production and state/private control replace multipurpose use and common-property ownership. Perhaps the key variable in the model is the primacy or extent of dominance of timber production. In many industrial forests the primary objective of management is timber production, and other goods and services, if acknowledged at all, are relegated to subsidiary positions. Value and utility are sometimes perceived by the managers to reside in wood alone, giving rise to the jibe that they 'cannot see the trees for the wood'.

'Timber primacy' is not to the same degree a characteristic of reindustrialize forests, nor indeed of post-industrial ones. Its relationship to a specific stage in the evolution of forest management is illustrated by the case of Austria, where it found expression in the Forest Act of 1852 and persisted until the Forest Act of 1975 was passed (Gluck, 1987). While the details of this example may be peculiar to Austria, general parallels may be seen in many other developed countries, where in recent times multiple use or the provision of conservation or recreation services has been incorporated in forest management.

The use of the forest resource will be discussed against the background of this model, and within the framework of Table 13.1. First, 'traditional' use and use for 'minor' products are considered briefly, in relation to both indigenous consumption and to the introduction of market economies. The pattern of production of timber for industrial purposes and for energy (fuel wood) is then considered. Trends in this use are considered in the light of the resource potential. Finally, the role of the forest in the provision of services such as recreation and conservation is briefly outlined. It should be emphasized that there is no perfect or rigid correlation between model stage and nature of production and control. The relationship is an imperfect one and is complicated by the transitions that are implicit in dynamic model such as this. The production of fuelwood, in particular, may span at least the first two states of the model, although the nature of is organization and the problems to which it gives rise change with 'progress' towards the second stage. For this reason, it is considered in a separate section.

## Traditional Uses and 'Minor' Products

Forests have traditionally yielded a great variety of useful products. The production of fodder, food and fiber typical of many African forests in recent times (Poulsen, 1982) also characterized forests in many other parts of the world in the past. Under common-property ownership, most products were consumed locally and never entered the market. In some parts of other world, however, the harvesting of 'minor' (*i.e.* non-timber) products has survived the transition to market economies and to state or private forest ownership.

**Table 13.1: Uses of the forest resource.**

| | |
|---|---|
| Traditional use and 'minor' products | Fodder, grazing, shifting cultivation |
| | Food-fruit, nuts, honey, game |
| | Medicines |
| | Fibres |
| | Latex gums, resins |
| | Building materials |
| | Wood for utensils and furnishing |
| | Fuelwood |
| Industrial use | Sawlogs |
| | Pulpwood |
| | Veneer logs |
| | Fuelwood and charcoal |
| | (Minor industrial products *e.g.* cork, turpentine |
| Non-consumptive uses | Soil conservation |
| | Water conservation |
| | Nature conservation |
| | Amenity |
| | Recreation |

In additional to providing the resource base for shifting agriculture, forests have provided food through the hunting of animals and the gathering of fruits, nuts and honey. They have also yielded products (or derivatives of products) that were or are perceived as useful as traditional or modern medicines, raw materials for domestic utensils and tools, and building materials and fuel. This traditional folk use of the forest characterized much of Europe until medieval times. In Scotland, for example, trees and woods even supplied the raw material for alcoholic beverages as well as for numerous more functional purposes. In Russia, the importance of the forest and its many functions was reflected in the language, which contained as many as 103 different words used to denote forest types and vegetation. The forest yielded wood for building, dead wood for fuel, fruit and berries, and honey and game (French, 1983).

Such use has largely died out in the industrial age in the developed world, but continues in many parts of the world. Forest- dwelling shifting cultivators many parts of the world. Forest dwelling shifting cultivators may number as many as 500 million, and are believed to use around one-fifth of the tropical forest are (240 million and 170 million hectares respectively of closed and open forest) (Lanly, 1985). In addition to providing land for cultivation, the forest offers for these cultivators and other forest dwellers grazing and fodder, as well as fuelwood and direct sources of food such as nuts, betties and fruits. There may be a considerable indirect use through domestic animals that graze and browse in the forest. This use involves leaves and other green (as opposed to woody) material, but nevertheless is both important in the functioning of the local economy and significant in terms of amount of forest biomass consumed. In Nepal, for example, domestic animals my annually consume twice as much forest biomass as is used for fuelwood (Agarwal, 1986).

These uses may be continued in perpetuity if the intensity is modest, but the resource may be threatened in terms of extent or productivity or both- if population pressures build up or when traditional systems of control break down. A shortening fallow rotation, perhaps combined with increased grazing pressures, may prevent the full recovery of the forest. Various modes of production may exists.

Forest dwellers may harvest products such as fruit and berries directly from the forest by simple gathering, or forms of management may be developed whereby plants perceived as useful are concentrated in special areas of the forest by human activity. Transplanting and selection amount to a semi-domestication of some plant species, and animal species of birds, fish, bees and mammals are also manipulated for use as food and game in areas such as Zmazonia (Posey, 1985).

In both temperate and tropical forests, numerous plants and animals are utilized at present, or were utilized in recent times. In eastern Canada, at least 175 food plants and fifty two beverage plants were gathered by native peoples, and over 400 plants were used in native medicine (Arnason *et al.*, 1981). As in Russia and earlier in much of Europe, the forest was indeed the resource base for the needs and wants of its human inhabitants, and provided a huge range of useful materials as well as a living environment. Much of the tropical forest fulfils a similar role for its inhabitants today. In southern Venezuela, for example, the Yekuana Indians regularly use nine species of terrestrial mammals, nine species of monkeys, and eighteen species of birds (Linares, 1976). In eastern Ecuador, as many as 224 plant species are utilized, mostly for foods, but also for construction, tools and medicines, while in northern Bolivia 80 per cent of the forest's trees, shrubs, vines and herbs are used (Myers, 1986a). Elsewhere in the Bolivian Amazon, one hectare contains ninety one species of which local Indians used severity five (85 per cent) in some way, while of 649 individual trees as many as 619 (95 per cent) were used (Prance, 1986). Around Iquitos in the Peruvian Amazon, one hectare of forest contains 275 species and 842 individual trees of 10 centimeters or more in diameter. Of these, 72 per cent of the species and 42 per cent of the individual trees yield products with a local market value (Peters *et al.*, 1989). Diversity of products is a characteristic especially of the tropical moist forest, but it also applies to other tropical forests. For example over seventy tree and other species are listed by Person (1986) as being used by village people in a savannah woodland area of southern Sudan. The uses of forest products in this area range from medicines and fish poisons through fruit, soap and cosmetic oils to ropes and constructional materials.

Although many non-timber products are consumed directly by forest dwellers, others do enter the market. Perhaps one of the most obvious is rubber. Prior to the establishing of large rubber plantations, the tropical forest was the main source of this commodity, and Amazonia experienced a rubber boom at the beginning of this century, and indeed the production of 'wild' rubber continues to the present. This boom was based on industrial demand from the developed world, and for the most part involved non-indigenous people but is usually characterized by a small scale of operation. The episode is a salutary reminder that the use of natural products of the forests is not necessarily geared to direct consumption, nor is it necessarily carried out by indigenous peoples; to this extent the 'pre-industrial' model is clearly an oversimplification.

Other products with commercial value include bamboos and rattans. (These, although woody materials, are not normally classed as timber). Rattan exports from Indonesia have been reported to be worth $90 million per year (Myers, 1988). In East Kalimantan, a range of minor products including rattan and resins and even birds' nests and reptile skins have been an important source of income for the rural population. Some still are, and are continuing to expand; others have decreased in production with the rise of commercial logging (Priasukmana, 1986). In China, the area of natural and planted bamboo has increased rapidly in recent decades, in response to government encouragement of more intensive management of this widely used product, and it now accounts for nearly 3 per cent of the total forest area (Hsiung, 1987). In Tanzanian forests, wild bees provide large amounts of honey for export, representing a value many times that of timber (Westoby, 1989).

In many forest areas in the tropics, 'wild' meat constitutes a high proportion of the animal protein consumed by local people. In Nigeria, the proportion is around one-fifth and in Zaire one quarter; in Cameroon, Ivory Coast and Liberia it may be as high as 70 per cent and in the Ecuadorian Amazon as much as 85 per cent (Myers, 1986a, 1988). When this meat value is computed, and added to the potential harvesting of cayman hides and of primates for biomedical research, the potential value amounts to over $ 200 per hectare per year, in comparison with a return of a little over $ 150 per year from commercial logging (Myers, 1988).

The commercial value of each 'minor' individual product may be modest, but when aggregated the total value may be impressive and may indeed be not insignificant alongside logging values. The Indonesian tropical forest for example, provides a range of products such as rattan, resin, sandalwood, natural silk and materials useful in the pharmaceutical and cosmetics industries. Exports of non-wood forest products from Indonesia in the early 1980s were worth around US$125 million annually, and had increased significantly over the previous decade, both in absolute terms and relative to total forest product export value. In 1973, for example, the share of non-wood products amounted to only 2.9 per cent of the latter, but it had increased to 11.2 per cent in 1981 and to 13.3 per cent by 1982 (Repettq and Gillis, 1988). Some estimates are even higher; Myers (1988), for example, puts the 1982 value at $ 200 million, compared with $ 28 million in 1973.

Compared with those from timber production, the benefits Arising from 'minor' forest products are often relatively widely distributed. In Indonesia, for example, huge numbers of smallholders earn around $ 200 per year from non-timber products such as rattan and orchids (Myers, 1986b). In India, up to 30 million people depend on minor forest produce for some part of their livelihood (Agarwal, 1986). The production of local cigarettes, rolled in the leaves of the native tree *Dionspyros melanoxylon* contributes £ 200 million to the local economy and provides at least some income for 3 million part time workers (Westoby, 1989). The total value of minor forest products in 1979-80 amounted to around 23 per cent of that of wood (Muthiah, 1987). It is reported that non-wood forest products accounted in the late 1970s for 40 per cent of the total net revenues accruing to the government from the forestry sector, and for 63 per cent of the exports. These figures exclude the estimated 60 per cent of such products that are consumed locally and do not enter the cash economy. Furthermore, the rate of growth of revenues from such products (including pharmaceuticals, gums and resins, bamboos and essential oils) was far higher than that from commercial timber, and non-wood products generated more than 70 per cent of the employment in the forestry sector as a whole (Myers, 1988).

The diversity and potential value of 'minor' products are perhaps greatest in the case of the tropical forest, a but they are not confined to low latitudes. Further polewares, forests have traditionally supplied a variety of products for local economies, and in some cases there is still a considerable commercial value. In the Soviet Union, for example, huge areas of forests are (or can be) harvested for 'minor' products (Table 13.2). Non-timber products can contribute around 50 per cent of the economic value of the forest (Cherkasov, 1988) and in some areas even more; revenue from 'minor' products from Belorussian forest could reach 80 per cent of that from timber production (Sankovich,1984). The full potential value is rarely realized, however, because commercial gathering of products such as berries and edible fungi is often laborious, and processing and marketing are often poorly developed. The harvesting of such products may depend on the pattern of ownership, amongst other factors. Where small, privately owned woodlots are owned by absentees, for example, there is less likelihood that mushrooms and betties will be gathered than if they are owned by resident farmers.

The significance and value of 'minor' products are often overlooked, for a variety of reasons. One is statistical coverage. It is almost impossible to achieve an adequate coverage of the reduction of a diverse range of products, many of which are consumed at or near the point of production and which do not enter domestic or overseas trade. FAO, for example, does not attempt to include coverage of such products in its *Yearbooks of Forest Products*. In the absence of such data, the value of the products can easily be overlooked, especially in comparison with that of timber, which can be relatively easily quantified. Timber enters international markets and brings in foreign exchange, whereas non-wood products are often sold locally, and are difficult to monitor and easy to ignore (Peters *et al.*, 1989). For various reasons, therefore, timber production may be perceived to be a paramount importance, and this perception in turn may be reflected in national forestry policies which give priority to it and which conversely tend to neglect non-wood products.

**Table 13.2: Minor forest products: Soviet forests.**

| a) | Fruit Bearing Area (ha) | Utilisation of Yield (Percentage collected) |
|---|---|---|
| Cedar nuts | 30000 | 5 |
| Japanese stone-pine nuts | 18000 | 3 |
| Edible chestnuts | 72 | 25 |
| Walnuts | 46 | 50 |
| Hazel nuts | 1700 | 50 |
| Raspberries | 10000 | 3.5 |
| Mushrooms | 50000 | 3.5 |
| *(b) Production (tons, 1986)* | | |
| All fruits and berries | 167.3 | |
| Honey | 22.9 | |
| Birch juice | 42.7 | |
| Mushrooms | 31.7 | |
| Nuts | 10.5 | |
| Medicinal herbs, etc. | 15.7 | |
| Hay | 338.3 | |

*Source*: Compiled from data in (a) Tseplyaev (1965) (b) Barr and Braden (1988).

Nevertheless, many of these products can be produced continually, and it has been concluded that 'It is not at all clear that the discounted present value of annual income (in perpetuity) per hectare from non-wood forest products must be less than the discounted present value of log extraction per hectare (Repetto and Gillis, 1988, p 65). On the basis of work on the Amazonian forest, Peters *et al.* (1989) have suggested that the annual collectionof fruit and latex in perpetuity could have a greater value than sustainable timber harvest: the former could account for as much as 90 per cent of the total value, and the latter 10 per cent.

The harvesting of non-wood products is often non-destructive, especially when carried out in traditional ways, and may have minimal impact on the forest ecosystem. It does not, however, necessarily mean that the resource is always conserved. For example, the exploitation of fauna in the Peruvian Amazon in the 1970s proved to be exhaustive, and led to a ban on the commercial hunting of game. As recently as 1973 it generated an internal gross product comparable to that from timber. More recently, it has been suggested that active management could restore the wild fauna to former levels of economic and social importance, and at the same time eliminate the risks of extinction brought about by indiscriminate hunting (Dourojeani, 1985). On the other hand, the development of logging, perhaps following the granting of logging concessions, can seriously affect the use of the forest for these traditional products, and obviously the removal of the forest effectively precludes it. Furthermore, external threats to the forest such as changing climate or acid rain can pose serious threats to the harvesting of minor products. For example a sudden increase in dieback in sugar maple in Quebec since 1982 has threatened a $ 40 million industry and the livelihoods of 10,000 syrup producers (Hendershot and Jones, 1989). Both traditional use and the commercial use of minor products, however, are more usually jeopardized by the advert of commercial logging and the transition to the industrial forests.

Many examples of conflicts between the traditional use of the forest and logging and clearing have been reported. One of the most celebrated is that involving indigenous groups and small-scale rubber producers in the face of large scale loggers and ranchers in Amazonia (*e.g.* Hecht and Cockburn, 1989; Schwartz, 1989), but there are many others that are less known. For example, the clearing of reverie frosts in southern Somalia has resulted in the loss of an important resource for bee-keeping and honey production, as well as for timber for building and browse for livestock (Douthwaite, 1987). Much of the forest clearance in this case has resulted from donor assisted refuge resettlement schemes, which have been carried out with scant regard for the interests of the local inhabitants. More generally, 'traditional' producers have frequently been poorly organized and poorly represented in conflicts with logging interests or with those who would clear the forests for other purposes. Furthermore, the harvesting of animal products is not always compatible with logging: in Sarawak, for example, heavy commercial logging is alleged to reduce the sustained yield harvest of wild meat from 54 kg per local resident per year to about 2 kg (Myers, 1989). In 1987 several tribal groups in Sarawak set up barricades across logging roads to protest against the damage caused by timber companies (Apin, 1987). 'Traditional' use may be compatible with the common property ownership which characterizes many pre-industrial forests, at leas unless and until rapid population growth occurs, but with the rise of industrial logging and private control it may suffer severe stress. Indeed the traditional users maybe in effect dispossessed of both their land and of their traditional forest activities.

On the other hand, more intensive management of the forest for timber production is not always detrimental to other products. For example, while other yields of some berry species suffer from the lowering of water tables following the ditching of peat land forests in Finland, other species benefit from such treatment, and the application of fertilizers improves the yields of mushrooms and berries as well as the growth rates of trees (Veijalainen, 1976). Whether the increased yields are actually harvested is another matter.

The transition from the 'pre-industrial' forest, characterized by a diversity of products, to the industrial stage, distinguished by primacy of timber production, has occurred at different times in different areas. For example it was largely achieved in countries such as Sweden by the end of the nineteenth century (*e.g.* Stridsberg, 1984). More recently it is well illustrated by the case of Indonesia. As late as 1938, the value of trade in minor products amounted to as much as 13 million Dutch guilders, as compared with 16 million for timber. By the 1980s, the ratio had widened to 5:95 (Jacobs, 1988). There may be some signs that it is narrowing once again (page 127), but there is no doubt that the transition has in general terms been difficult and painful. Part of the problem lies in the differing perceptions of traditional and small-scale producers on the one hand, and of logging interests on the other. It is all too easy for 'industrial' foresters to allege neglect of adequate forest management when traditional perceptions of the forest and its ownership and use encounter the modern economy. For example Kumar (1987) attributes damage to forests in Indian reserves in Alberta to the lack of management. Traditionally, such forests were used for hunting and for the small scale use of various products. Common property ownership suited such use, but could not cope with larger scale logging for commercial purposes.

Superimposed on these contrasting perceptions are differences in influence and power. Traditional and small scale users have usually lost out to commercial loggers, who are often backed by government. Various political factors contribute to the shift from the small-scale, multi product use of the 'pre-industrial' forest to the large scale timber production that characterizes the 'industrial' forest. Several are listed by Jacobs (1988). They include the fact that logging may permit the extension of government power over remote lands, whilst economic development may be perceived as synonymous with large-scale exploitation. It may be difficult for governments to collect revenues arising from small scale operations focusing on minor products, and these operations may not be attractive to aid agencies and their large-scale investments. Finally, small scale collectors and other forest users have no significant power base from which to protect their source of income.

Formidable political and economic problems therefore underline the shift from the 'pre-industrial' to the 'industrial' phase. In recent years there has been a growing awareness of the long-term benefits that may accrue from the use of the forest for minor products, and in at least some areas the value of output of these products is rising relative to that of timber. Nevertheless, major problems still confront the commercial exploitation of these products. The transition from local indigenous use of commercial harvesting has rarely been smooth. More often than not it has involved the growth of large-scale logging, and has been associated with both environmental and social disruption. Whether a smoother and less destructive transition can be made from the pre-commercial to commercial use in the remaining areas of pre-industrial forest remains to be seen.

## Wood Production

A diversity of products, including timber for construction and for fuel, is characteristic of the 'pre-industrial' forests, but in the modern age timber has assumed primacy, and indeed in many instances an overwhelming predominance, in forest production. Many forests are now managed and used solely or primarily for timber production for industrial purposes, and in other cases the production of fuelwood has emerged as the dominant use.

The measurement of wood production, like that of the forest area and other forest attributes is fraught with difficulty. Much production does not enter the market and cannot be precisely measured. In assembling statistics, therefore, bodies such as the FAO have to make assumptions about the level of use of fuelwood per head of population, for example, and even population totals may not be known with certainty or precision. Data for timber production should therefore be viewed in the light of these difficulties.

The *exploitable or operable* forest area is smaller than the total area of forest and woodlands since production in some areas is precluded by inaccessibility or management objectives of conservation or protection. The definition and measurement of the exploitable forest area are surrounded by uncertainty, not least because different types and orders of constraints may limit exploitation. Some are economic: the harvesting of some forests may not be economically viable at prevailing timber prices. This constraint may be linked to that imposed by transport or by costs of logging: new transport routes and new logging methods may radically alter perceptions of exploitability. And designation of forests for protection, conservation or recreation may also be an effective constraint.

Conflicting trends are probably operating in relation to the exploitable forest area. Some new areas are being opened up by new transport lines, whilst other are being designated for non-consumptive purposes. One of the potentially most spectacular examples of the former is the case of the forests along the Baikal-Amur railway in the eastern part of the Soviet Union. It is estimated by the FAO that an additional 40-50 million hectares of exploitable forest area will be added by this development (FAO, 1982a), and there are plans to develop wood-using industries along it (*e.g.* Eronen, 1983). (As yet, however, there has been little noticeable impact (Barr and Braden, 1988). On the other hand large areas have been designated as non-production frosts in countries such as the United States. Extent of exploitable area therefore changes through time, and also depends on definition. Unfortunately, estimates of this extent rarely specify date and definition.

Current estimates of the exploitale area are around 2,000 million hectares: a figure of 1,950 million hectares is quoted for exploitable closed forest by Kuusela (1987), while data assembled by Binkley and Dykstra (1987) indicate that the present exploitable area is around 2,150 million hectares. They estimate a total growing stock and net annual increment of approximately 300,000 and 5,200 million cubic metres respectively. (This estimate of net annual increment is considerably higher than some previous estimates such as that of Kind (1975, Chapter 4).)

Total production or removals of wood from world forests has been around 3,250 million cubic metres in recent years (FAO, annually). Overall, therefore, recent levels of production may amount to little more than 60 per cent of the estimated net annual increment in exploitable closed forests. On the other hand they may exceed the net annual increment of the forest area, depending on which estimate is accepted. Recent volumes of production are equivalent to an average of around 0.65 cubic metres per person, and correspond to around 0.8 cubic meters per hectare of the (FAO) forest and woodland area and to approximately 1.6 cubic metres per hectare of exploitable closed forest.

Of the total production, around 40 per cent is softwood (coniferous) and the remainder hardwood (non-coniferous). These proportions approximate to the relative extents of these forests types, but the coniferous share has been decreasing slightly in recent years, having fallen from 42 per cent in 1975 to 40 per cent in 1986 (Figure 13.1). This trend has been evident for many years: in 1955, for example, production was divided almost equally between the two types, as indeed it was earlier in the

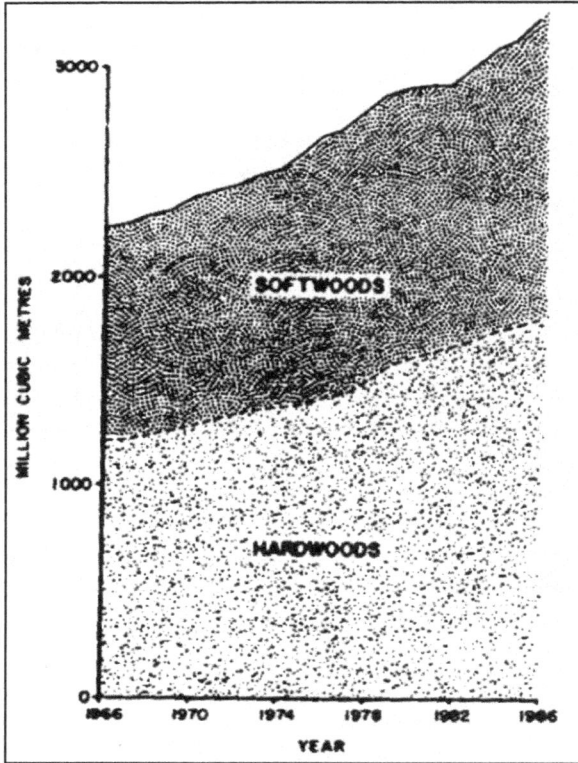

**Figure 13.1: Trends in wood production (removals): softwoods (coniferous) and hardwoods (non-coniferous).**

*Source*: **Based on data in FAO Yearbooks of Forest Products.**

century (Zon and Sparhawk, 1923). Furthermore, it seems set to continue, not least because hardwoods tend to be in surplus in the United States and many other developed countries (Bethel and Tseng, 1986), while softwoods are in shorter supply.

As will be discussed more fully later, around one half of the total production is used for industrial purposes (including construction and pulping) and the remainder is consumed as fuelwood or charcoal. These two sectors show marked contrasts in trends and in spatial patterns of production.

## Trends in Wood Production

Wood production has increased rapidly in recent decades, especially in the developing world (Figure 13.2 Table 13.3). It rose from 1,823 to 3,252 million cubic metres between 1956 and 1986, representing an increase of around 75 per cent. On a longer timescale, the rate of expansion appears to be considerably slower: for example Zone and Sparhawk (1923) estimated total production at 56 billion cubic feet, corresponding to approximately 1585 million cubic metres. They emphasized that their figure was an estimate: for individual countries they considered that it was

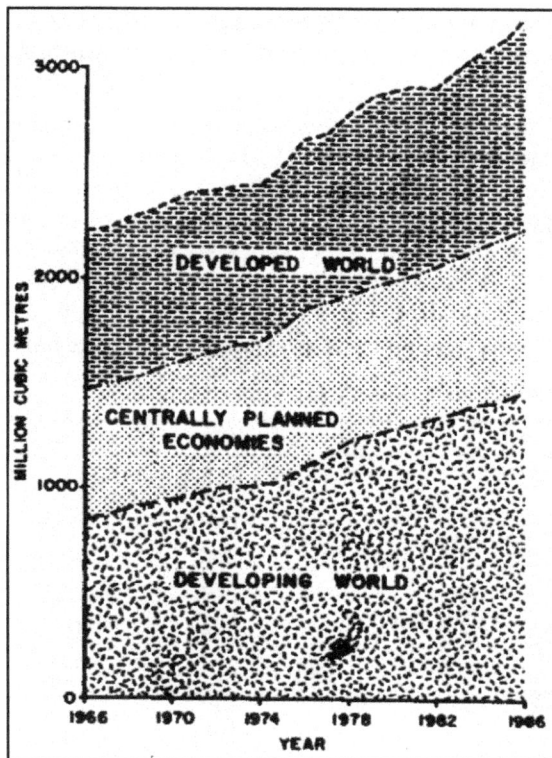

**Figure 13.2: Trends in wood production (removals) by world division.**
*Source*: **Based on data in FAO Yearbooks of Forest Products.**

within 15-25 per cent of the actual production of saw timber but that it could be more than 50 per cent in error for fuelwood. Their figure, however, may well be an overestimate, and according to FAO (1946) the volume cut in 1937 was 1,500 million cubic metres, compared with 1,674 million cubic metres around 1950 (FAO, 1966). Whatever the precise levels of production in the first half of the century may have been, it is highly probable that they were less than half of present day figures.

Fuller statistical coverage and better time-services data become available after the Second World War, although this improvement may bring its own problems. It is possible that part of the apparent trend in levels of production is statistical rather than real, and results from more comprehensive coverage and changes in procedures rather than from actual changes in levels of production. Statistics contained in FAO *Yerbooks of Forest Products* are sometimes revised (usually upwards) in successive volumes, posing problems for the assemblage of time-series data (and their graphical representation). For example, total production for 1975 was given as 2,452.7 million cubic metres and 2,579.2 million cubic metres respectively in the FAO *1976 Yearbook of Forest Products 1966-76* and *1986 Yearbook of Forest Products 1977-86*. The identification of the precise nature of long-term trends is therefore difficult, and there is a danger that real trends may be exaggerated by the more comprehensive coverage of recent years.

Nevertheless, the general trends are clear, and are dominated by a rapid rate of increase, especially since the 1960s (Table 13.3). During the ten year period from 1977 to 1986 for example, production increase by 20 per cent, compared with an increase of 11 per cent during the previous ten year period and one of 10 per cent during the previous ten year period and one of 10 per cent between 1954 and 1963. During the 1970s and 1980s, annual rates of increase often exceeded 2 per cent and sometimes approached 3 per cent, although almost no growth occurred from 1980 to 1982. In the 1950s and 1960s annual rates of increase averaged little over 1 per cent.

**Table 13.3: Wood production (million O³).**

|                     | 1950* | 1960      | 1970      | 1980      | 1985      |
|---------------------|-------|-----------|-----------|-----------|-----------|
| Total roundwood     | 1674  | 1901(14)  | 2365(24)  | 2927(19)  | 3164(8)   |
| Developing          |       |           | 1118      | 1527(37)  | 1743(14)  |
| Developed           |       |           | 1247      | 1348(8)   | 1422(6)   |
| Fuelwood            | 866   | 872       | 1091(25)  | 1476(35)  | 1646(12)  |
| Developing          |       |           | 914       | 1241(36)  | 1384(12)  |
| Developed           |       |           | 177       | 235(33)   | 263(12)   |
| Industrial roundwood| 8208  | 1028      | 1274(24)  | 1440(13)  | 1518(5)   |
| Developing          |       |           | 204       | 327(60)   | 358(9)    |
| Developed           |       |           | 1070      | 1113(4)   | 1160(4)   |

* Average 1950-2.

Figures in brackets indicate percentage increase from previous data.

*Source*: FAO *The stae of food and agriculture* (annual): FAO *Yearbooks of forest products* (annual); FAO World forest products statistics 1954-63: FAO Wood: world trends and prospects Unasylva 20 (1966) 1-135.

Until recently, the rate of increase in wood production has been less than that of human population, but is the last few years the relationship has been reversed and growth in wood production now exceeds that of population. Between 1975 and 1985, population increased by 18.7 per cent, but total wood production rose 22.7 per cent. The increase in production of industrial roundwood (17.2 per cent) was slightly less than population growth, but the rate of increase in fuelwood production, at 28.3 per cent, was substantially greater. Over the longer terms, however, there has been little change in per capita levels of production (and consumption): the average figures in cubic meters per person were 0.7 in 1937 (FAO, 1946) and 0.65 in 1985. Stagnation of levels of wood production in the early 1980s gave rise to a belief that the steady growth of consumption that had taken place since the Second World War was now slowing down (*e.g.* Kuusela, 1987), and that the flattened S-shape (logistic) curve that charactrerised growth in consumption of some other natural resource commodities such as oil and some minerals applied also to wood. The recent resurgence of rates of increase of production in the second half of the 1980s, however, may call such beliefs into question, and it is too early to reach definite conclusions. In many resource sectors, unusually high rates of growth in production and consumption during the third quarter of the present century fuelled fears of resource shortages, and prompted upward revisions of levels of production. In many instances, however, such trends have not been sustained, and the question arises as to whether the downturn in growth rates during the late 1970s and early 1980s was a temporary blip, reflecting the effects of the recession following oil price rises, or the beginning of a new phase. A major complication in considering wood production is that different trends apply in different sectors, as is discussed subsequently. Projections of overall trends are therefore fraught with difficulty.

In the forest sector, as in many other resource sectors, the rate of growth during the late 1960s and 1970s was unexpected, and many forecasts and projections made in the 1950s and 1960s were exceeded by actual levels of production and consumption. Conversely, many more recent forecasts and projections have turned out to be overestimates, and in the last few years expected rates of increase have tended to be revised downwards. The significance of such forecasts and projections is profound, as they influence decisions on issues such as afforestation and investment in the forest products industries.

## Sectoral Trends

Overall trends in wood production, however, conceal important contrasts between different sectors (Figures 13.1 and 13.3). The rate of increase in production of hardwoods or non-coniferous species has been greater than that of conifers. Between 1975 and 1985, for example, non-coniferous removals increased by 27.7 per cent, compared with 15.7 per cent for coniferous removals. Between 1960 and 1980, the estimated annual growth in softwood consumption (as industrial roundwood) averaged 2.4 per cent per annum, whilst the corresponding figure for hardwoods was 3.2 per cent (FAO, 1982a). These differential trends reflect a shift southwards in production, towards the tropical forests in particular.

In recent years the rate of growth in production of industrial roundwood has slowed down, while that of fuelwood has accelerated. Some uncertainty arises since the growth in fuelwood is relatively steady, whilst trends in industrial roundwood fluctuate considerably from year to year in accordance with the state of the world economy. The choice of study period therefore strongly influences the apparent rate. Between 1962 and 1974, annual growth rates averaged 3.0 per cent, while from 1974 to 1984 they fell to 0.6 per cent, prompting Kuusela (1987) to conclude that the steady growth in consumption since World War II was slowing down, and that a turning point has been reached. The picture is complicated, however, by the stagnation that characterized demand for industrial roundwood during the recessions of the mid-1970s and early 1980s. Almost no growth occurred from 1974 to 1976 and from 1980 to 1982. Over the period from 1970 to 1982, annual growth rates averaged 1.0 per cent (FAO, 1986a). Since 1982, however, more rapid growth has been resumed, and during the ten-year period from 1977 to 1986, growth in production of industrial roundwood amounted to 14.0 per cent, compared with 13.9 per cent in the previous ten years. The

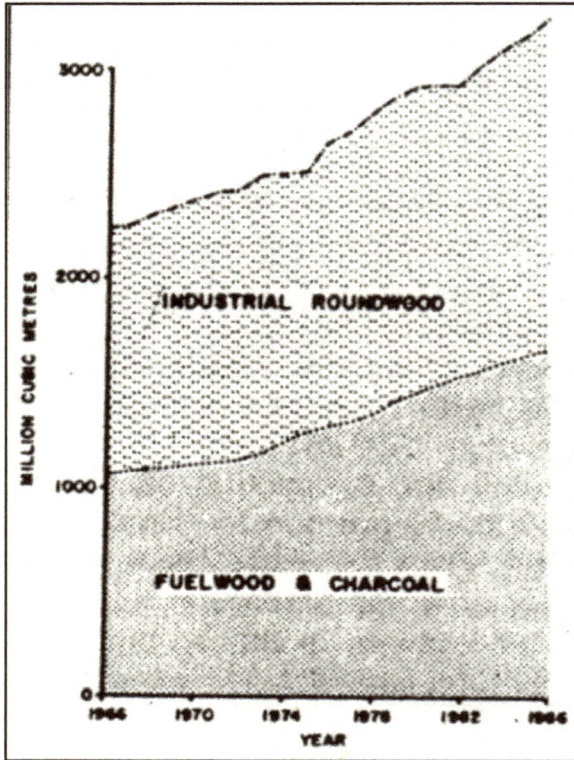

**Figure 13.3: Trends in production of industrial roundwood and fuelwood and charcoal.**
*Source*: **Based on data in FAO Yearbooks of Forest Products.**

slowing of growth that seemed apparent in the early 1980s is therefore called into question, and its significance in relation to long-term demand supply relationships is debatable. Nevertheless, projected rates of increase in consumption for the period between 1980 and 2000 are substantially lower than those for 1960-80. Annual rates for softwoods are expected to average 1.8 per cent (compared with 2.4 per cent), while those for hardwoods are 2.3 as compared with 3.2 per cent (FAO, 1982a).

Whatever the long term significance of apparently decelerating rates of increase in the industrial roundwood sector may be, however, it is clear that growth rates for the production of industrial roundwood have been slower than those for fuelwood. Annual rates of increase for industrial roundwood between 1963 and 1983 averaged 1.46 per cent, compared with 1.66 per cent for fuelwood, For fuelwood, there is a clear pattern of accelerating production, with increases of 9 per cent and 26 per cent respectively for 1967-76 and 1977-86. Fuelwood production has increased rapidly and steadily, while the production of industrial roundwood has grown more slowly and less continuously. These trends have been completely reversed since the 1950s and early 1960s. In the second half of the 1950s, for example, almost all the increase in total roundwood production was accounted for by an increase in the production of industrial roundwood. The net increase in fuelwood production of less than half a million cubic metres against a total of nearly 900 million cubic metres conceals the fact that the growth of fuelwood production in the developing world was counterbalanced by a long term decrease in its use in Europe, the Soviet Union and North America (FAO, 1966).

Much of the recent growth in total production, therefore, is accounted for by the production of fuelwood,. External factors such as the major increases in oil prices during the 1970s have had a major influence on trends in fuelwood production, not only in the developing world but to some extent also in the developed world. In Europe, for example, the downward trend in fuelwood consumption which began in the 1950s was reversed in the late 1970s. In the United States, the production of fuelwood increased fivefold between 1976 and 1981. On the global scale, a growing proportion of total production in the post war period has been accounted for by fuelwood. For example in the mid-1980s (1984-86) its percentage was 51.7, compared with 49.7 per cent in 1975 and an average of 49.6 per cent in the early 1960s. In the mid-1950s, the proportion was 48.5 per cent. On a longer time-scale, however, the trend may have been different: Zone and Sparhawk 91923) estimated that 53.6 per cent of the removals at their time of writing were for fuelwood.

## The Spatial Pattern of Wood Production

The spatial patterns of wood production is complex and varied. Disparate factors underlie it, including the extent and nature of the resource, environmental conditions in relation to tree growth, intensity of management and relative proportions of 'natural and 'man made' forests, forest classification (whether for protection or production) and accessibility. Production may also depend on the status and condition of the resource, and in particular whether sustained-yield management is being operated or whether the resource is diminishing as a result of over-exploitation. These factors operate on varying scales, and give rise to contrasting patterns of wood production at levels ranging from the continental to the local.

The pattern on the scale of continents and other major divisions is shown in Table 13.4. Just under half of the world area of forest and woodland lies in the developed world, and an almost identical proportion of wood production comes from that area. In general terms, tree growth rats in the non-tropical forests that make up most of the forest and woodland area in the developed world are slower than those in the developing world, but on the other hand the intensity of management is often higher. In general terms, however, a slow shift towards the developing world is apparent: in 1975, for example, its share of production amounted to 53.3 per cent, and it now has a large share of a larger total. In other words, the rate of increase has been considerably greater in the developing world than in the developed world.

In contrast to the accordance of shares of production and of the forest and woodland area at the scale of two fold division into developing and developed world, marked imbalances occur at the continental level. Relative shares of production and area rarely match at this level. The relatively intensively managed forests of Europe, for example, make up less than 4 per cent of the world's forests, but account for nearly 11 per cent of the production. Conversely, South America has more than one fifth of the forest, but less that 10 per cent of the production. When total production is broken down into industrial roundwood and fuelwood, however the distribution becomes very different, as Table 13.3 indicates. The developing world accounts for more than four fifths of fuelwood production, but less than one quarter of that of industrial roundwood. Nevertheless, its share of industrial roundwood production is increasing, having risen from 19.8 to 23.0 per cent between 1975 and 1985. Similar contrasts occur at the continental level, with Europe and North America in particular being characterized by contributions to industrial roundwood production that are far greater than their share of the world forests area. These two areas produce well over half of the world's industrial roundwood, from little more than one fifth of the world's forest area.

**Table 13.4: Area of forest and woodland and wood production 1986.**

|  | Percentage of World Forest and Woodland | Wood Production | Percentage of World Industrial Roundwood | Fuelwood |
|---|---|---|---|---|
| Developing World | 54.4 | 54.6 | 23.0 | 84.3 |
| Developed World | 45.6 | 45.4 | 77.0 | 15.7 |
| Africa | 16.9 | 13.8 | 3.4 | 23.6 |
| North and Central America | 16.8 | 22.2 | 36.1 | 9.3 |
| South America | 22.4 | 9.7 | 5.9 | 13.2 |
| Asia | 13.2 | 30.7 | 15.6 | 44.8 |
| Europe | 3.8 | 10.8 | 18.7 | 3.4 |
| Oceania | 3.8 | 1.2 | 1.9 | 0.5 |
| USSR | 23.1 | 11.6 | 18.5 | 5.2 |

*Source*: Based on FAO Yearbook of Forest Products 1975-86 and FAO Production Yearbook 1987.

The pattern of production at the national level is indicated in Figure 13.4. As the map suggests, production is strongly concentrated. Six countries accounted for 50 per cent of production in 1980, while eighty other countries collectively produced less than 5 per cent of the total (Styrman and Wibe, 1986). A small number of large, extensively forested countries such as the United States, the Soviet Union, Canada and Brazil stand out as major producers. In general terms, production is roughly proportional to the extent of the forest resource, but there are exceptions to this generalization, and various qualifications need to be made. The productive potential of a national forests resource depends not only on its extent, but on environmental factors and intensity of management. Production levels also depend on accessibility, logging costs and trade: relatively cheap imports may discourage home production in high cost countries such as Japan. Furthermore, in some countries little use of the forest has as yet been made for wood production, while in others the forest resource has been over harvested (in relation to growth). Removals exceeded the potential cut in Finland in the 1960s and Sweden in the 1970s, prompting attempts to stimulate long term wood supply (FAO, 1982a; Gamlin, 1988). More seriously, over cutting may reduce the extent and productivity of the forest resource on the long term. In

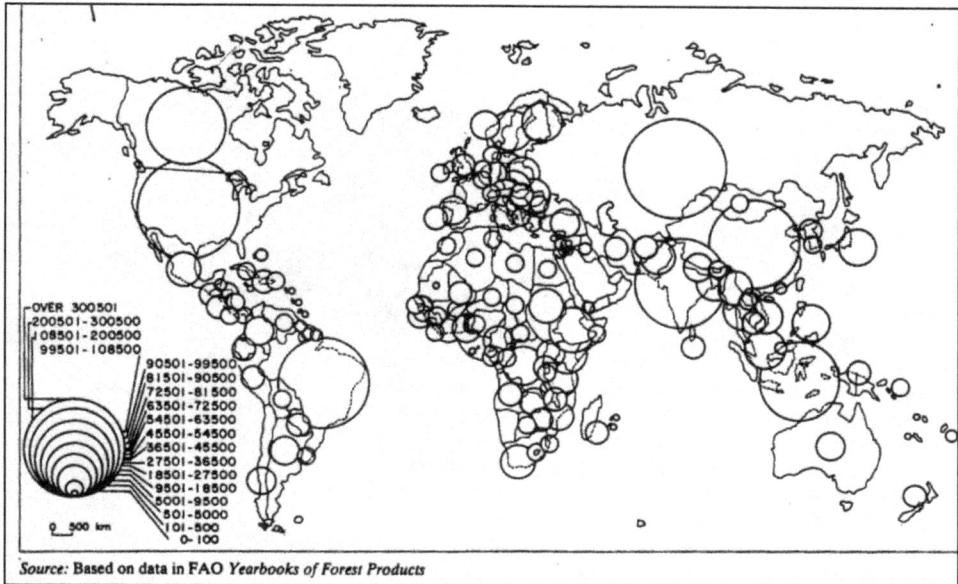

Figure 13.4: Roundwood production (thousand m³, 1986).

P:akistan, for example, annual growth accounts for only 62 per cent of the annual wood harvest, and the forest resource is being eroded (Biswas, 1987).

## Intensity of Production

Harvesting intensity varies in terms both of growing stock and area. On the basis of an estimated total of 300,000 million cubic metres of growing stock, the overall harvesting intensity in the early 1980s was 0.8 per cent (Peck, 1984). This intensity was lightly higher in the developed world, 0.9 per cent, where generally slower natural growth rates were offset by higher intensities of management, than in the developing world for which the average was 0.7 per cent. There much of the forest was completely unmanaged. The highest harvesting intensities were recorded in Europe, where they averaged 2.2 per cent. Even there, however, removals were less than the net increment, and the standing volume is increasing. Indeed, it is expected that it will increase by around 8.5 per cent between 1970 and 2000 (FAO, 1982a). Growing stock is also increasing in the United States: between 1952 and 1977 the softwood growing stock inventory increased by 7 per cent and that for hardwood by 43 per cent (USDA, 1982). In much of the world, therefore, the intensity of production is such that the wood resource is expanding, despite record levels of production. The pattern of intensity of production per unit forest area is shown in Figure 12.5. A small number of mainly low latitude countries are shown as having very high levels of production in excess of 5 cubic metres per hectare of forest and woodland. This group contains a number of small countries such as Haiti, Costa Rica and Nepal, where the rate of deforestation is high, and also a number of countries such as Egypt and Libya where the forest area is very small. In contrast with the scattered pattern of very high production countries, that of countries within the 2-5 cubic metres per hectare category

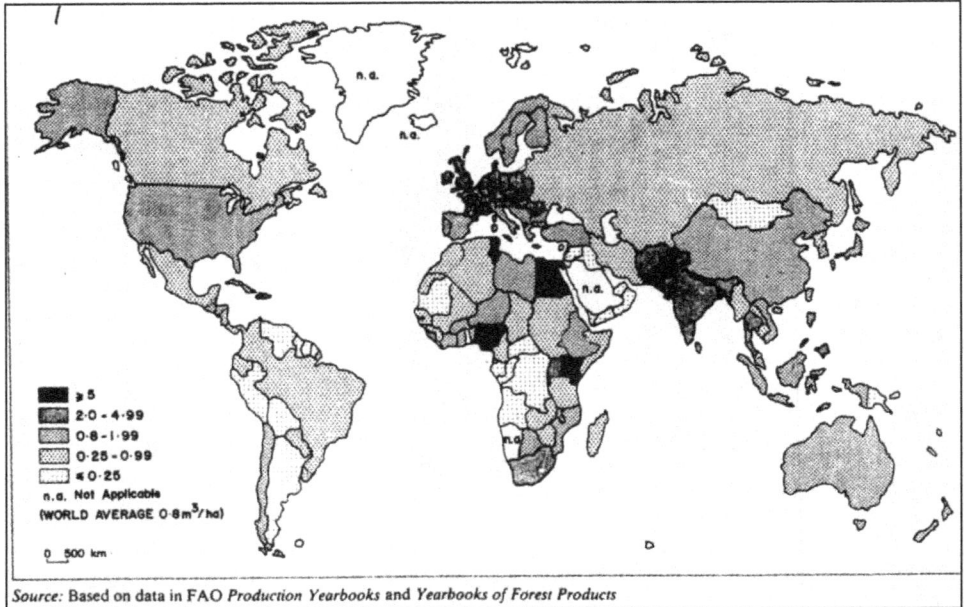

Source: Based on data in FAO *Production Yearbooks* and *Yearbooks of Forest Products*

**Figure 13.5: Averae roundwood production per hectare of forest and woodland 1986 (m³).**

is more compact. Most European countries lie in this group, and her there is a combination of moderate 'natural' growth rates and relatively intensive management.

Most of the large countries with large forest areas have production ratio below the world average. Both Canada and the Soviet Union are characterized by relatively low levels of production, resulting in large part from a combination of hostile climatic conditions which limit growth rates and inaccessibility which precludes the exploitation of much of the forest. Much of the tropical zone, however, also is in this group, and it is noticeable that many tropical countries in Africa and Latin America have low or very low levels of wood production per unit area. While climatic conditions are of course well suited to high ecological productivity in such areas, again the problem of inaccessibility has constrained the level of production to date.

If, the forest is to be harvested on a long term sustained yield basis, the ultimate limit on removals or levels of production is net annual increment (NAI). Various natural and human factors determine NAI, including type and intensity of management. Climate is one of the main natural variables, and climatic gradients are clearly reflected in the special pattern of net annual increments per hectare, as reported by UNECE/FAO (1985). On the maritime margins of north west Europe for example, NAI values for exploitable closed forests are 8.51 and 7.29 cubic metres per hectare respectively for Denmark and Ireland. In the colder and more continental climates of Sweden and Finland, the values fall to 3.01 and 3.18, while in the Soviet Union and Canada they are 1.40 and 1.66 cubic metres per hectare respectively. For the Soviet Union, the average per hectare of forest covered land (as opposed to exploitable closed forest) is quoted by Blandon (1983) as 1.15 cubic metres. This average conceals a marked contrast between the European part, at 1.58, and the Asian section with an average of 1.01. NAIs also decrease towards the drier areas of southern Europe and the Mediterraneana, with values of 2.05 for Greece and 1.01 for Cyprus being reported. In the more arid areas, for example of Soviet Central Asia, average increments may be less than 0.1 cubic metres per hectare (Blandon,1983).

The ratio of fellings to net annual increment depends not only on ownership, control, accessibility and forestry policy, but also on factors such as the age and maturity of forests, since very young forests are unlikely to be productive in this sense, irrespective of their growth rates. Fellings in most European countries usually amount to around 70 to 80 per cent of the net annual increment (UNECE/FAO, 1985). In the continent as a whole, removals have recently amounted to just under 70 per cent (Prins, 1987). In the United States, they amount to around 63 per cent (UNECE/FAO, 1985), whereas in 1920 wood consumption was more than four times the level of wood growth (Cox *et al.*, 1985).

These national average ratios vary both spatially, as will be shown subsequently, and with ownership. Much of the surplus of growth over removals in the United States, for example, is concentrated in non-industrial private forests owned by farmers and other small scale owners (USDA, 1982). Comprehensive figures for other parts of the world are not available, but strong contrasts are likely to exist between the 'high' and 'low' producing countries. Many of the former are likely to be characterized with ratios well in excess of unity, and with rapid deforestation, while very low ratios obtain in the cases of lightly peopled countries, such as Surinam, which are still extensively forested and which have experienced relatively few and light pressures on their forest resources. In other instances such as Japan, production levels are relatively low because of the high costs of logging and extraction, competition form imports, and low quality timber from some of the forests.

Within individual countries, strong variations may exist in both levels of production and in ratios of actual to allowable cuts. These variations reflect not only environmental conditions, but also forest classification and accessibility. Production maybe inhibited by the classification of the forest for purposes of protection, conservation or recreation, and also by the need in some instances for harvesting to be incorporated within multiple use frameworks. Remoteness and inaccessibility are also powerful influences, which are perhaps most clearly demonstrated in the case of large countries such as the Soviet Union. In 1975, for example, 70 per cent of its total wood production came from the European Uralian section of the country, which contained only 18 per cent of Soviet mature timber. Conversely, only 30 per cent of the production came from Siberia and the Far East, which contained 82 per cent of the timber. In the former area, the cut represented 3 per cent of the growing stock: in the latter it amounted to only 0.3 per cent (Barr, 1984), while production as a percentage of defined allowable cut ranged from 91 per cent in the European part of the country to 29 per cent in Western Siberia and 32 per cent in the Far East (Blandon, 1983). The annual allowable cut is itself alleged to exceed the net increment by a factor of 2.5 for the country as a whole, and by ratios of 4 and 5 for the Ukraine and Lithuania respectively (Barr, 1988). In countries such as the Soviet Union and Canada, apparently low ratios for the country as a whole may therefore conceal the fact that some of the more accessible forests are over-utilised, while the more remote ones are almost unused.

The spatial pattern of production is characterized by some elements of continuity on the timescale of the present century but more especially by change during recent decades. Throughout the century, production has been dominated by the United States and the Soviet Union, but their degree of dominance has been weakening. In the early part of the century, these two countries accounted for 57 per cent of the total wood production (Zon and Sparhawk, 1923). In 1965 their share had fallen to 35 per cent, and by 1985 to just over one–quarter. In the 1920s the United States accounted for over 40 per cent of world production, its production being there times greater than that of the Soviety Union and nearly item times that of the next largest producer, Canada. In the second half of the twentieth century, American production has actually been lower than it was in the 1920s. In the Soviet Union, on the other hand, production increased more than six fold between 1922 and 1975 (Blandon, 1983): by then it was the leading producer although more recently it has been overtaken once more by the United States.

The weakening degree of dominance of the United States and the Soviet Union is accompanied by the rapid expansion of production in the developing world and especially in tropical countries. In the Philippines, for example, it has expanded from Zon and Sparhawk's estimate of just under 1 million cubic metres in the early 1920s to 35 million cubic metres in 1985: in Nigeria the increase was from 2.5 to 98 million cubic metres, while in Indonesia (Dutch East Indies) it was from 5 to 154 million cubic metres. Much of this growth has taken place within the last 20 years: in Brazil for example, production increased from 148 to 238 million cubic metres between 1966 and 1986, while the corresponding rise in Indonesia was from 93 to 158 million cubic metres. In comparison, production in northern countries has been relatively static. In Sweden, for example, total production amounted to 52 million cubic metres in 1986, compared with 51 million in 1966. In some tropical countries, much of the rapid increase in recent decades is accounted for by commercial logging. For example, log production in Indonesia is reported to have increased from 2 million cubic metres in 1967 to 26 million in 1973 (Walker and Hoesada,1986). In others, the production of industrial roundwood has actually decreased, but that of fuelwood has increased rapidly. In Ivory Coast, for example, industrial roundwood production fell from 5.5 to 3.6 million cubic metres between 1976 and 1986 (after increasing rapidly during the previous decade),while fuelwood production increased from 5.5 to 8.25 million cubic metres (Figure 13.6).

The detailed patterns of changes in total production at the national level is known in Figure 13.7. Over the period from 1976 to 1986, large increases in production occurred in most of Africa and Asia, and in much of Latin America. Outside the tropics, countries showing an above-average increase include a group in north-west Europe (including Britain, Ireland and Spain) where twentieth century afforestation is now being reflected in rapidly increasing production. Another groups, including Finland and Canada, is characterized by a more stable forest area combined with more intensive management. Several European countries, on the other hand, have below average increases, which in some cases may conceal the fact that both intensities

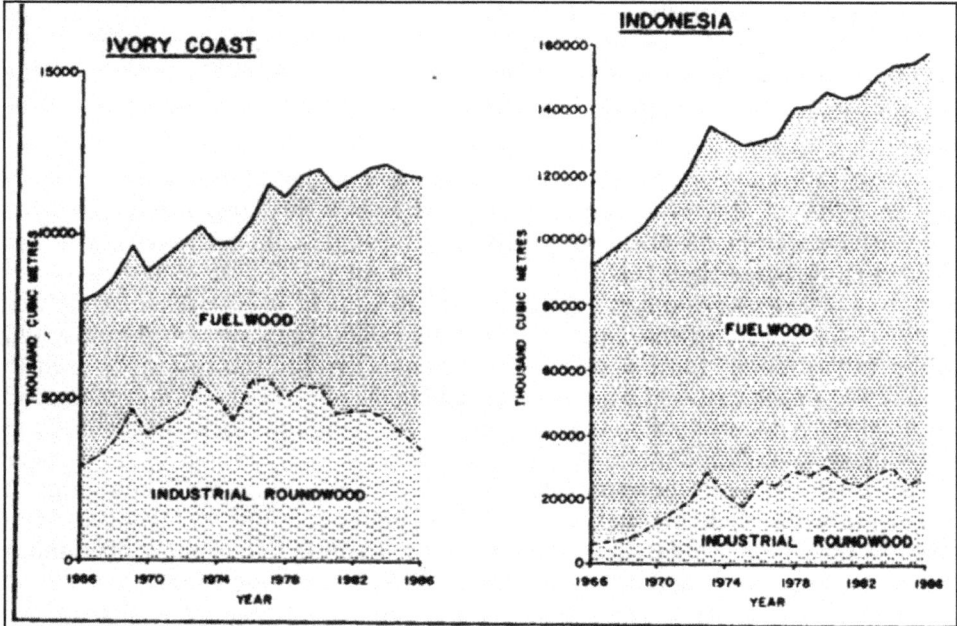

Figure 13.6: Wood production in Ivory Coast and Indonesia.

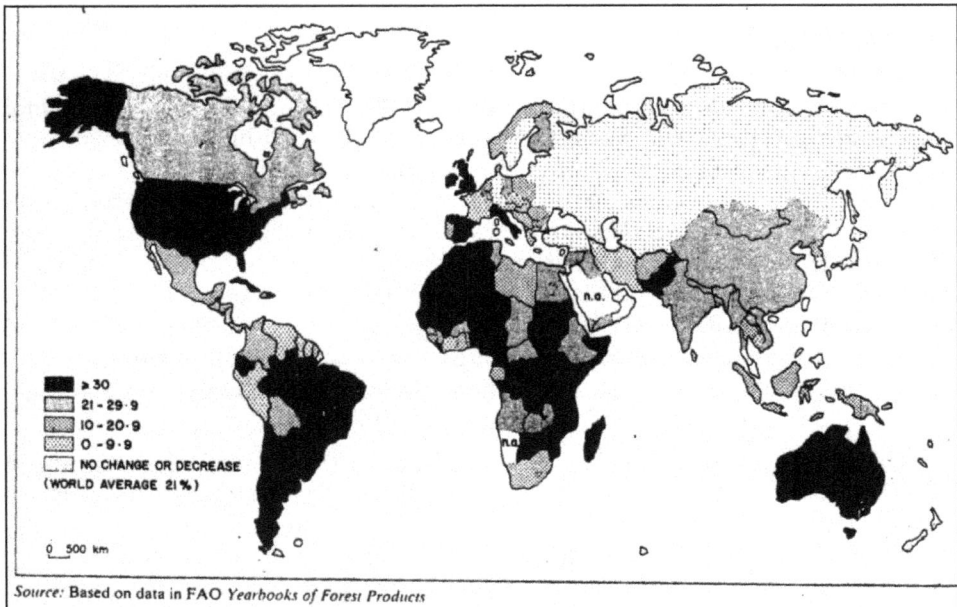

Figure 13.7: Percentage change in wood production 1976-86

of management and production levels were relatively high before the baseline of 1976. A number of neo-tropical countries, including Venezuela, Guyana and Surinam, also have production increases well below the world average: in these cases the tropical forest has been exploited much less than in neighboring countries such as Brazil and Ecuador.

While the detailed spatial pattern of roundwood production at the national level is complex, the overall pattern on the global scale is very clear. Production from low-latitude areas in general and from the tropics in particular is rising relative to that from Europe, the Soviet Union and North America. On the basis of the production estimates of Zon and Sparhawk (1923) for the early part of this century, the three continents of Africa. Asia and South America accounted for 20 per cent of total world production. By 1965 that percentage had doubled, and by 1985 had increased to around 54 per cent. A clear shift towards lower latitudes is evident in roundwood production. With production levels in many northern countries already high in relation to growth rates (although still capable of some increase with more intensive management and with the sue of exotic species). With considerations of environment and amenity playing increasing roles in forestry policy and forestry management in these countries, this trend is likely to continue, especially if high yielding plantations are extensively established in tropical latitudes. Within Europe, a less obvious shift is likely as the century draws to an end. Maritime countries such as Britain and Ireland are likely to expand their production rapidly, as postwar plantations become productive, while rates of increase in central and northern Europe are likely to be slower. On both the global and European scales, therefore, there are likely to be signs of a shift towards the areas of high potential productivity, while the dominance of the northern coniferous forests zone weakens in relative terms.

## Industrial Roundwood

Major contrasts exist between the industrial roundwood and fuelwood sectors in terms of composition, pattern and trends. While around 60 per cent of total roundwood production is from non-coniferous species, almost 70 per cent of industrial roundwood is coniferous. The extent of coniferous domination in this sector is slowly decreasing, as higher rats of annual increases in production characterize the hardwood sector of industrial roundwood as well as in fuelwood. Little change occurred during the first half of the century. According to Zon and Sparhawk (1923) three-quarters of the supply of industrial wood came from coniferous forests, compared with 76 per cent in 1955. By 1985, however, the parentage had fallen to 69, reflecting in particular the increasing exploitation of the tropical forest for this purpose. Nevertheless, the production of industrial roundwood is still strongly concentrated in the developed world (Table 13.4 and Figure 13.8).

The domination of industrial roundwood production by the United States and the Soviet Union is clearly defined and long established. Together, these two countries accounted for over 44 per cent of production in 1985, and if the contribution of Canada, the third major producer, is added, then the share amounts to 56 per cent. The same three countries were the largest producers in the early part of the century, accounting

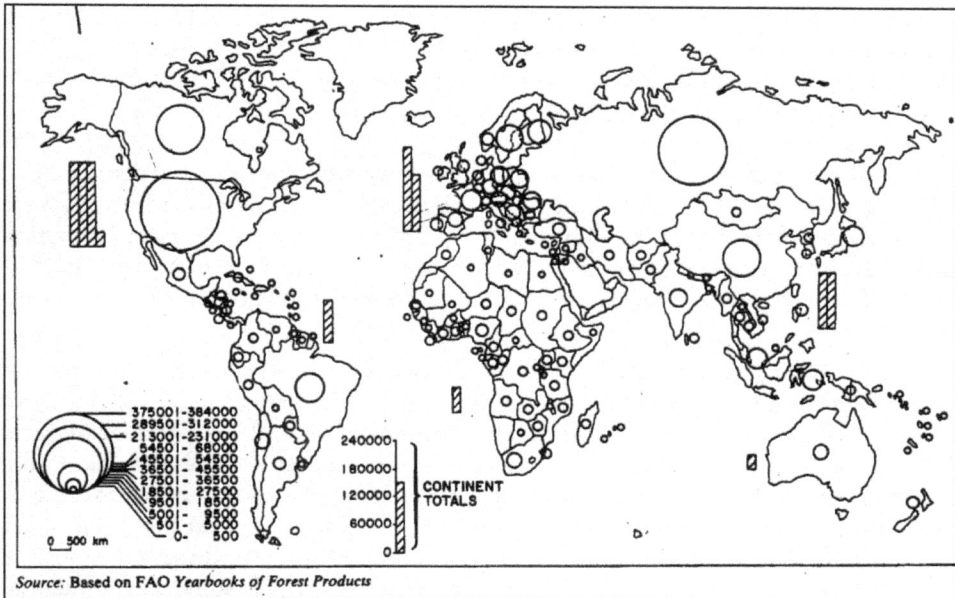

**Figure 13.8: Production of industrial roundwood (thousand m³) 1986.**

for around 74 per cent of the production of industrial timber (Zon and Sparhawk, 1923). The United States along produced 53 per cent of the world's industrial timber at that time. Its share has decreased subsequently, to around one-quarter, as have those of the two other leading producers.

Although the United States and the Soviet Union remain dominant, their share of production is likely to continue to decrease, as production in tropical and other low latitude countries increases. North America's share of softwood production, for example, is projected to decrease from 39 per cent in 1980 to 34 per cent in 2000, and Western Europe's share will also fall while those of Japan and Latin America are likely to increase (FAO, 1982a).

Brazil is now the fifth largest producer, whereas it was the seventh largest in 1975 and the eleventh in 1965. A number of other tropical counties are now major producers, including in particular Malaysia and Indonesia. In contrast to old-established counties such as Sweden where production levels have been relatively stable in recent decades, outpu6t from most tropical countries has expanded very rapidly. Commercial logging on a large scale began in many tropical countries in the 1960s and expanded in the 1970s. In Malaysia production rose by 49 per cent between 1975 and 1985, while in Indonesia it increased by 45 per cent. East Kalimantan (Indonesia) exemplifies the rapid expansion. Mechanical logging began in a joint venture with a Japanese company in 1960. In 1962 there were two concession-holders on an area of 400,000 hectares; by 1983 the number of concession-holders had grown to 106, working over 12.35 million hectares (Priasukmana, 1986). In some countries production rapidly grew to levels that could not be sustained, and soon dropped. For example, production fell by 46 per cent in the Philippines during the same period, and in Ivory Coast the decrease was 10 per cent. A similar tend had previously been experienced in some other African countries such as Nigeria and Ghana. In the latter, timber exports decreased spectacularly from 124 million cubic metes in 1973 to 11 million cubic metres in 1982 (WRI, 1985). In these countries the resource was over-exploited rather than husbanded, and the result was its erosion in the same way in which very different forest resources of counties such as the United States were damaged earlier in the century. Cycles of expansion and contraction reminiscent of the 'boom and bust' cycles of the Lake Sates and other parts of the United States a hundred years ago have been experienced in successive tropical countries over recent decades.

In the long term, the prospects of tropical countries as major producers of industrial round wood are likely to depend largely on the successful establishment of a major plantation element in their forest resources. Table 13.5 illustrates estimates of potential returns to industrial plantations in representative schemes around the world. Sedjo (1984, 1986) concludes that these returns are large in a number of regions, including parts of the tropics and Southern Hemisphere, and on the basis of his figures such areas clearly outshire areas such as Scandinavia and the Northwest of the United States. On the other hand start-up costs may be high in some non-traditional locations, and both political and silvicultural risks may be high.

**Table 13.5: Representative plantations: internal rates of return (1979 constant prices in perpetuity).**

| | Regime | |
| --- | --- | --- |
| | *Pulpwood* | *Integrated** |
| South USA (*Pinus taeda*) | 12.02 | 12.45 |
| North-West USA (*Pseudotsuga menziesti*) | 7.11 | 7.07 |
| Brazil Amazoni (*Pinus caribaea*) | 17.89 | 20.44 |
| (*Gmelina* spp.) | 27.53 | 23.54 |
| Central (*Eucalyptus* spp.) | 20.16 | 15.54 |
| Chile (*Pinus radiate*) | 23.39 | 17.50 |
| New Zealand (*Pinus radiate*) | 11.90 | 13.11 |
| Australia (*Pinus radiate*) | 10.68 | 10.06 |
| Gambia/Senegal (*Gmelina* spp.) | 18.42 | 17.52 |
| Nordic counties (*Picae abies*) | 4.61 | 5.57 |

* With sawtimber

*Source:* Based on Sedjo (1984).

A number of countries, notably in subtropical latitudes, have been successful in establishing plantations to meet domestic requirements of industrial wood. For example, Chile, Kenya and Zambia succeeded in this respect over a relatively short time-scale of round twenty years (Spears, 1983). In Brazil, 4 million hectares of plantations now supply 60 per cent of domestic needs, compared with the 10 per cent from the 280 million hectares of Amazonian forest. At present industrial plantations comprises only 1 per cent of the forest area in Latin America, for example, but they already supply one-third of the continent's industrial wood (Sedjo and Clawson, 1984). By the year 2000, production from industrial plantations will be almost four times higher than in the mid-1980s, and they will account for about half of the continent's production (Sedjo, 1987). By 2000, perhaps as much as one-third of all industrial roundwood removals in tropical countries will come from plantations, compared with 7 per cent in 1975 (USDA, 1982). In absolute terms, supply may increase tenfold between 1980 and 2000, reaching 100 million cubic metres per year (FAO, 1982a).

## Composition of Demand

The composition of demand for industrial wood has changed radically during the twentieth century. Demand for constructional timber and for pitprops has fallen, while that for wood for processing into pulp or board has increased. The trend is most apparent in old-established industrial countries such as the United States, for which the changing pattern of utilization of wood is shown in Figure 13.3. In addition to the underlying tr4end of decreasing use of fuelwood, there is a clear shift from lumber towards manufactured wood products. Lumber consumption per capita has fallen to around one-third of its peak value at the beginning of the century, although the rate o decrease has slowed since around 1940. AT the same time, both total and per capita consumption of pulp and plywood have risen steadily throughout the century. Worldwide, the ratio of population of sawlogs to that of pulpwood has decreased from around 4:1 in the 1940s (FAO, 1946) to little more than 2:1 in the 1980s. This trend reflects a steady decrease in the ratio of consumption of sawn wood to that of pulp and other highly processed products. Pulp and reconstituted panels accounted for around one-third of all industrial consumption in 1960, but by 1980 was almost 50 per cent and by 2000 may approach 60 per cent (FAO, 1982a). This log-established trend reflects differential growth rates for different wood products. Consumption of fibre-based products has grown much faster than that of 'solid wood': the comparative annual rates of increase between 1960 and 1980 averaged 4.6 and 1.3 per cent respectively (FAO, 1982a). For the period from 1970 and 1982, annual growth in demand for industrial roundwood has averaged 1.0 per cent, compared with percentage rates of 2.0 for pulp, 3.2 for wood-based panels and 4.2 for paper for writing and printing (FAO, 1986a). The corre3sponding figure for sawn wood is around 0.5 per cent (Kuusela, 1987). In all these sectors, growth rates are much higher in the developing world than in the world as a whole or in the developed world.

In the early days of the wood pulp industry, spruce was the most suitable raw material, and early pulping plants were located mainly in areas such as northern Europe and the Northeast of the United States where spruce was abundant. Subsequently processes using pine were developed, initially for packaging materials and later also for fine paper. Thereafter, forests containing little or no spruce, which could previously be used only for lumber, became potential sources of paper-making fibre. New areas were therefore opened up, especially in the South American, where large areas of pine plantations were established on worn-out agricultural land (Clark, 1984), and more recently in Latin America and Oceania. This process of expansion has continued with the development of processes for pulping hardwood. The combination of technical developments and availability (and hence price) of hardwoods has led to a rapid growth in the use of broad-leaved species in pulping. This trend has been in operation even in traditional softwood areas such as the Nordic counties, where the proportion of hardwoods in pulpwood removals increased from 3 per cent in 1950 to 16 per cent by 1972 (Pringle, 1977). The tend, however, has been most spectacular in the case of Japan, where the contribution of hardwoods to pulpwood supply increased from 11 o 60 per cent between 1955 and 1970 (Shimokawa, 1977). This increase was accompanied by a rapid growth in imports of hardwood chips from the mid-1960s. These 3 chips came initially from eucalyptus forests in Australia and from rubber and mangrove trees in Malaysia. During the 1970s, trade in chips of mixed topical hardwoods ('jungle wood') developed between Papua New Guinea and Japan. While tropical rain forests still have limited industrial value (Ryti, 1986), in Papua New Guinea mixed forest containing 120 tree species has been used for the production of chips for export to Japan (Fenton, 1986). Such a change could revolutionize the use of the mixed tropical forest, where for long only a handful of tree species were considered commercially attractive (Whitmore, 1984). On a more modest scale, the same trend towards a widening of the range of species perceived as useful is evident in Indonesia. There the change in emphasis from log production for export to use for local plywood and sawmill industries has been accompanied by a widening of the range of harvested species.

The rise of the manufacturing sector has also been accompanied by an increas4e in the level of utilization. While wasteful exploitation characterized the American forest industry around the turn of the century, efficiency has since increased markedly. For example, in the United States the Weyerhaeuser Corporation reported that the level of timber utilization per unit area increased from 21 per cent in 1950 to 79 per cent in 1975 (Cox *et al.*, 1985). In Indonesia, almost one-third of the timber cut is regarded as waste because of rot or shattering (Schreuder and Vlosky, 1986), and logging waste, including damaged trees, amounts to 30-40 per cent of the timber harvested (Priasukmana, 1986). There is still plenty of scope for further improvement in efficiency, but in the United States, for example, the overall efficiency of wood utilization has increased steadily at an annual rate averaging 0.8 per cent between 1950 and 1979 (USDA, 1982a).

World-wide, the use of mill residues has increased sharply in recent decades. Their contribution to total wood consumption rose from 5.7 per cent to 11.4 per cent in 1980, and may reach 12.7 per cent in 1990 (FAO, 1982a). It may amount to the equipment of 300 million cubic metes of roundwood by 2000. Furthermore, waste papers can be recycled as inputs into some manufacturing processes, thereby reducing the requirement for roundwood. In the United States, the waste-paper recovery rate is around 25 per cent (US Statistical Yearbook 1987), while in Japan it has reached 50 per cent (Riethmuller and Fenelon, 1988), thereby achieving a major saving in primary raw material and in energy requirements in paper manufacture. In short, recycling an increased use of residues are likely to mean that net demand on the forest will increase less rapidly than consumption of wood products. In this respect, trends in the use of the forest/timber resemble those in other resource sectors such as mineral ores. With changing technology, perceptions of useful resources are revised, and increasing use is made of non-primary inputs (such as scrap metal). Both these trends tend to reduce pressures on the natural resource.

## Trade in Wood and Wood Products

A complex pattern of international trade in wood products has evolved, although less than 10 per cent of the total world production of wood leaves its country of origin (Gammie, 1981). Wood itself is a low grade and poorly transportable material, and only a small and decreasing volume of roundwood enters international trade. This volume amounted to around 3.4 per cent of total roundwood production in 1985, compared with 3.9 per cent in 1975. Almost all the roundwood entering international trade is intended for industrial purposes: trade in fuelwood is negligible. The proportion of production entering international trade increases with degree of processing, as Table 13.6 indicates. A marked differential exists between the shares of coniferous and non-coniferous saw logs entering international trade, reflecting he export of tropical hardwoods to the developed world. Nevertheless, the overall proportion of saw logs traded in this way is well under 10 per cent, while the corresponding proportions for several processed or manufactured wood products are around one-fifth.

**Table 13.6: Forest product: proportions entering international trade 1986.**

| Product | Percentage Entering Trade (Export) |
|---|:---:|
| Fuelwood and charcoal | 0.1 |
| Industrial roundwood | 6.8 |
| Saw logs and veneer logs | 6.4 |
| (coniferous) | 5.0 |
| (non-coniferous) | 11.7 |
| Pulpwood | 10.2 |
| Sawnwood | 18.5 |
| Wood-based panels | 17.0 |
| Wood pulp | 16.1 |
| Paper and board | 21.2 |

*Source*: Based on statistics in FAO Yearbook of Forest Products 1986.

The declining share of roundwood entering the international market is explained by the widespread wish to add value in the producing country, expressed in some cases in attempts to ban log exports. Furthermore, the trend in recent years has been for trade in more highly processed products such as paper, plywood and wood-based panels to grow faster than that in more basic products such as pulpwood or pulp, as Table 13.7 indicates. Between 1960 and 1974, for example, the proportion of paper production entering trade rose from 12 to 19 per cent, while that of pulp increased only slightly from 15 to 16 per cent (Pringle, 1977).

Table 13.7: Forest products exports: composition and tends 1975-85.

| | Composition by Value (Percentage by total) | | Percentage in Change in Value 1975-85 |
| --- | --- | --- | --- |
| | *1975* | *1985* | |
| Industrial roundwood | 14.2 | 10.9 | 47.2 |
| Sawlogs and veneer logs | 10.2 | 8.3 | 55.6 |
| Pulpwood | 3.1 | 2.3 | 41.9 |
| Other in, roundwood | 1.0 | 0.4 | -28.3 |
| Sawnwood | 19.3 | 20.4 | 86.9 |
| Wood-based panels | 8.8 | 9.1 | 98.2 |
| Pulp | 20.7 | 15.8 | 46.7 |
| Paper and board | 36.7 | 43.6 | 127.3 |
| Charcoal, | 0.1 | 0.1 | 102.6 |
| Fuelwood | 0.1 | 0.1 | -46.4 |
| Forest products | | | 91.8 |

Source: Based on statistics in FAO Yearbook of Forest Products 1986.

The overall pattern of imports and exports of forest products is very complex, as Figure 13.9 suggests. Many 'northern' countries are both importers and exporters, tending to import relatively basic products and to add value to these in industries developed initially on the basis of domestic wood production: Despite the complexity of the pattern, however, a number of general points stand out. First, a large proportion of trade in forest products comes from developed countries. These counties accounted for 90 per cent of the trade in 1963 and 86 per cent in 1985 (by value). North America and northern Europe have been the major exporters of manufactured forest products, most of which have been based on softwoods, and account for over half the exports. Second, much of the trade, is concentrate in a few major flows. Around half of world trade in forest products in recent decades has occurred in three main spheres: within North America, within Western Europe, and between northern and western Europe (Kornai, 1987). In addition, Japan alone accounts for around 10 per cent of all imports, and is the focus of a rapidly expanding exports trade from south-east Asia and the Pacific rim countries. On the basis of the gross value of imports of forest products, Japan is outranked by the United States, but it is the leading country in terms of net imports. In the United States, the huge value of imports in offset by exports. Japanese trade in forest products differs from that of the second largest net importer, the United Kingdom, in several respects. While the latter is poorly endowed with forest resources, Japan is one of the most extensively forested countries in the world. Paradoxically, it is also the largest importer. Japan also has a very different trading sphere centred in the western Pacific, while British supplies have traditionally come from North America and northern Europe. Furthermore, Japanese imports of saw logs and pulpwood are several orders of magnitude greater than those of the United Kingdom. Japan has favoured the import of primary products, while most of the forest-product

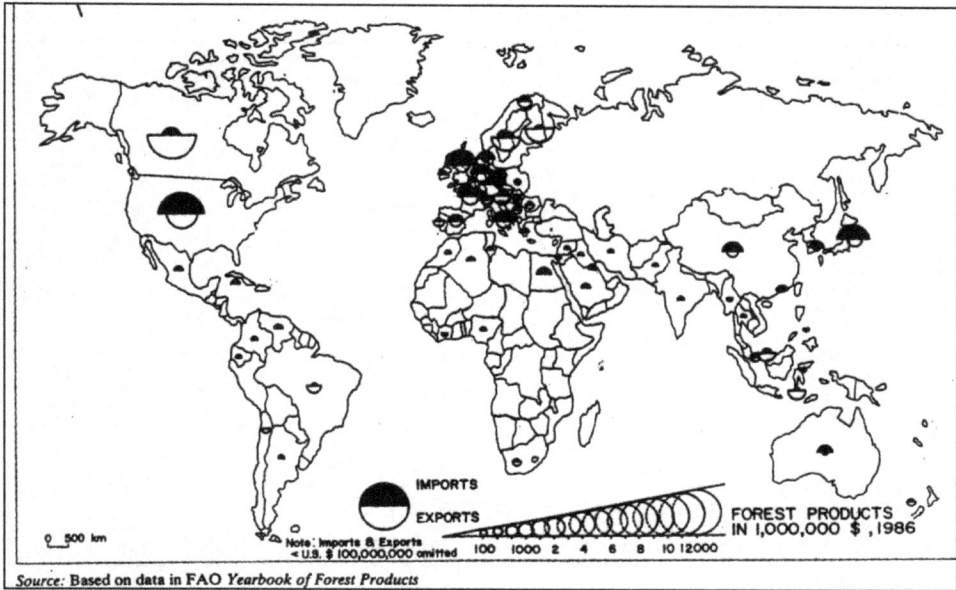

Figure 13.9: Trade in forest products.

imports into the United Kingdom are processed or manufactured. This policy of importing raw material rather than processed products has been supported by differential tariffs which discriminate against value-added products and which protect domestic industries. Such differentials are not unique to Japan, as Table 13.8 shows, although they are pronounced there. Tariffs on wood products generally increase with the level of processing, and the effect has been to favour the maintenance of processing industries in the 'traditional' forest-industry countries of the developed world, whilst discouraging the growth of manufacturing capacity in the developing world. At the same time, however, tariffs in developing countries are often even higher than in their developed counterparts, and also escalate with the level of processing.

For various reasons, this pattern of concentration of forest-product industries in the developed world in general and in Japan in particular is now changing. One reason is the general liberalizing of trade and reduction or dismantling of tariff barriers. Another is the growing production of industrial roundwood from plantations in non-traditional locations such as Latin America, south-east Asia and the Pacific rim countries, at a time when there is little short-tem scope for further expansion of production in traditional forest-product countries such as those of northern Europe. For example, Brazil by the mid-1990s is likely to produce quantities of pulp that are large in relation to world production, and is therefore likely to have a significant effect on world markets (Sedjo, 1980). National bans on log exports imposed by some countries in an effort to encourage the growth of domestic wood-using industries also encourage the trend. Also exports of logs from North America gave rise to much controversy from the 1960s and led to bans on exports of logs from federal forests

**Table 13.8: Tariff rates for forest products.**

| (a) Japan 1986 | Product | Percentage rate |
|---|---|---|
| | Saw logs | 0 |
| | Sawn timber* | 0 |
| | Woodchips | 0 |
| | Wood pulp | 0 |
| | Mews print | 3.1-4.6 |
| | Printing and writing paper | 3.1-4.6 |
| | Finisher timber | 10 |
| | Laminated timber | 20 |
| | Plywood (hardwood) | 20 |
| | (softwood) | 15 |

*rate for pine under 160 mm is 7 per cent

**(b) EEC and Japan (Post Tokyo Round, concluded 1979)**

| | Product Class | Percentage Rate for Imports From | | |
|---|---|---|---|---|
| | | Developing countries | Developed Market ecs. | Socialist countries |
| Japan | Wood in rough | 0 | 0 | 0 |
| | Primary wood products | 7.4 | 0.2 | 1.9 |
| | Secondary wood products | 4.8 | 4.3 | 4.6 |
| EEC | Wood in rough | 0 | 0 | 0 |
| | Primary wood products | 1.9 | 0.8 | 0.8 |
| | Secondary wood products | 1.5 | 1.7 | 3.2 |

*Source*: (a) based on Riethmuller and Fenelon(1988); (b) based on Olechowski (1987).

(Cox, 1988). Furthermore, some forest products industries may be perceived as polluting, and their expansion may be less than welcome by the environmentally conscious citizens of developed countries. In the case of Japan, for example, a preference is now being shown for expanding imports of pulp rather than woodchips, partly because of environmental concerns, partly because of instability in the supply of woodchips, and partly because of decreasing Japanese competitiveness in pulping. Much of the country's pulping an paper-making capacity is based on outdated machinery more than twenty years old. Rather than modernize such plant, several companies are actively seeking to develop overseas operations, including joint ventures in locations such as Tasmania (Riethmuller and Fenelon, 1988). Considerable advantages in terms of costs of raw materials, as well as in fuel, labour and management may also be enjoyed in countries such as Australia, New Zealand, Brazil and Chile, compared with Japan and other countries more traditionally associated with forest-products industries. Some examples of these advantages in terms of timber production and other costs are illustrated in Table 13.9. In short, the global structure of the wood-processing industry is being reshaped. 'Traditional' forest-product countries such as Finland have responded by becoming increasingly specialized in higher-value products in which wood costs are relatively less important. A recently as the early 1960s it was the leading exporter of roundwood. This role has since passed to less developed countries, and the emphasis has switched to wood-based panels and then paper (Kiljunen, 1986). Some developing countries have undergone similar changes, and have switched to panels as new countries have emeged on the scene as log exporters.

**Table 13.9: Relative costs: pulpwood and paper production.**

**(a) Growing and harvesting pulpwoods 1960s**

| Area | Species and Growth (m³/ha) | | Growing US$/m³ | Logging US$/m³ |
|---|---|---|---|---|
| Chile | *Pinus radiate* | 20 | 1.27 | 1.15 |
| East Africa | *Pinus* spp. | 17 | 1.90 | 2.00 |
| SE USA | Pine | 6.8-18.6 | 0.57-1.55 | 3.00 |
| Widen | Pine, Spruce | 4.5 | 15.65 | 5.25 |

**(b) Relative costs of major inputs in paper production 1984-5**

| | Pulpwood[1] | Fuel[2] | Labour[3] | |
|---|---|---|---|---|
| | | | Operator | Staff |
| Australia | 30 | 130 | 27 | 38 |
| New Zealand | 20 | 170 | 20 | 30 |
| Brazil | 27 | 210 | 17 | 27 |
| Japan | 75 | 310 | 28 | 35 |
| Canada (west) | 37 | 175 | 50 | 61 |
| United States (south) | 47 | 180 | 52 | 69 |
| Sweden | 57 | 295 | 28 | 36 |

[1] Per cubic metre (under bark) at mill door, A$. [2] Per tonne oil equivalent, A$. [3] Per year, AS000.

*Sources*: Based on Streyferrt (1968) and Riethmuller and Fenelon (1988).

Nevertheless, several factors retard or inhibit this global restructuring, and in particular the growth of woodpulp industries in developing countries. These include requirements of very large capital investments and a high level of technical skill, as well as substantial and dependable supplies of timber suitable for pulping (Bethel and Tseng, 1986). On balance, however, the pattern of pulp production as shown in Figure 13.10 is likely to continue to shift away from its traditional concentration in the northern coniferous forest zone.

While the pattern of trade varies from product to product, it is in general terms characterized by a high degree of involvement of the developed world. Collectively, the developing countries have accounted for no more than 16 per cent of total exports of wood products (Laarman, 1988). Large proportions of the international trade of most forest products are between developed countries, and the main foci, in North

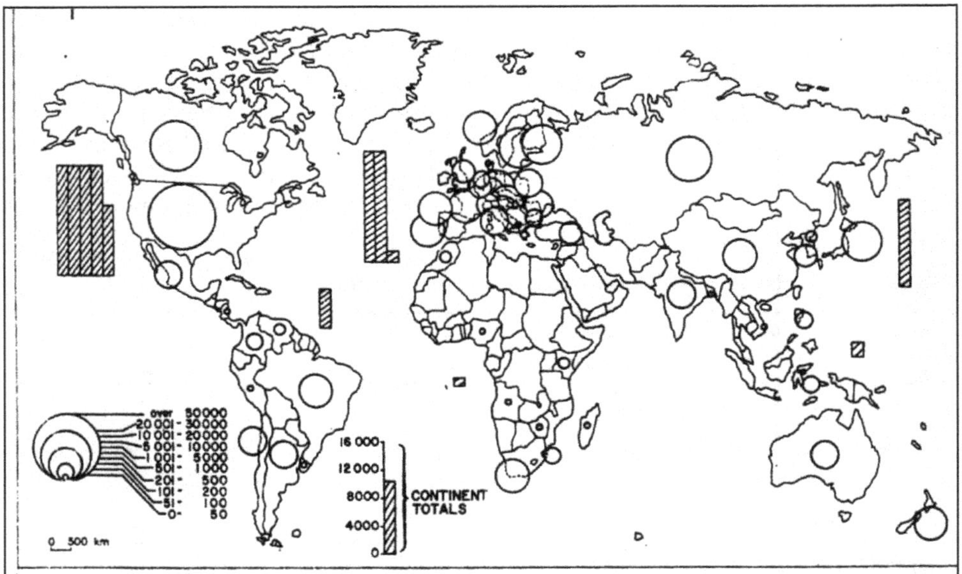

Source: Based on data in FAO *Yearbook of Forest Products*

**Figure 13.10: Production of wood pulp (thousand tones) 1986.**

America, northern and western Europe and Japan, all lie in the North. Examples of the pattern of flow of one product–saw logs–are shown in Figures 13.11 and 13.12 Japan and to a lesser extent (South) Korea and China stand out as by far the main centre of imports of coniferous sawlogs, most of which. Come from the Pacific Northwest of the United States and western Canada. The major areas with exportable surpluses of softwood are North America and the Soviet Union: smaller flows occur from Chile and New Zealand. The export of logs and saw timber from New Zealand to Japan began in 1958 and 1967 respectively, and by 1972 they represented some 25 per cent of roundwood removals (Fenton, 1985). While the Japanese economy was growing rapidly during the 1960s and early 1970s, it seemed that export growth could continue indefinitely. Such expectations led to the rapid expansion of plantations, but both Japanese consumption and New Zealand exports have 'subsequently declined (Fenton, 1986).

In contrast with Pacific theatre, which includes significant flows from the Soviet Far East to China and Japan, the European system (consisting of flows between Western Europe countries and between the (western) Soviet Union and its neighbours) is a relatively minor one, as is also trade between the United States and Canada.

International trade in hardwood (non-coniferous) logs is different in scale and pattern (Figures 13.12). While Japan remains a major focus, more than half of the trade flows between the developing and developed world. By far the major direction of trade is between south-east Asia and Japan, but there are also distinct systems of interaction amongst western European countries (mainly from France to neighbouring counties) and, more clearly and simply, from West Africa to Europe.

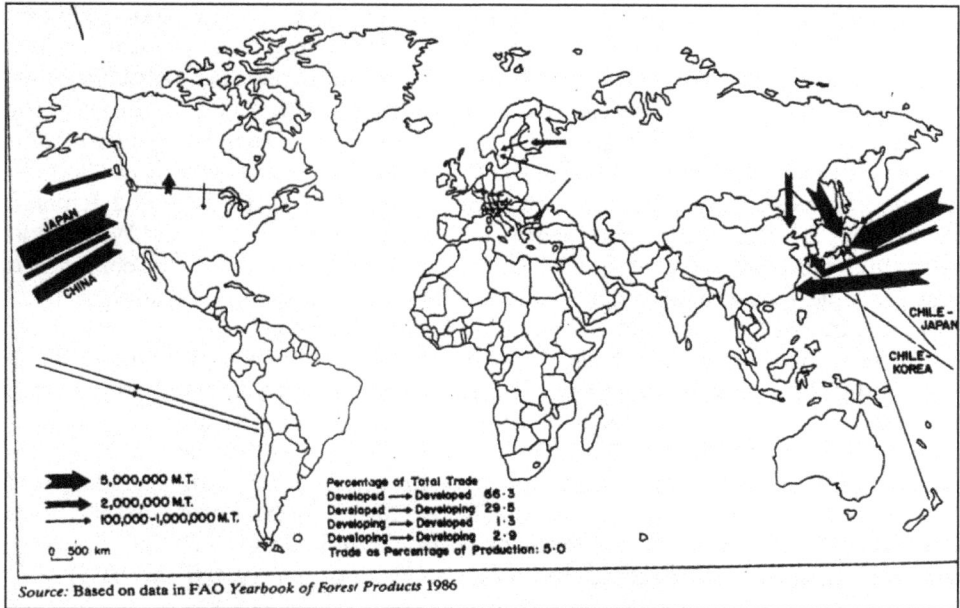

**Figure 13.11: Directions of trade (EXPORTS): CONIFEROUS SAWLOGS, 1985.**

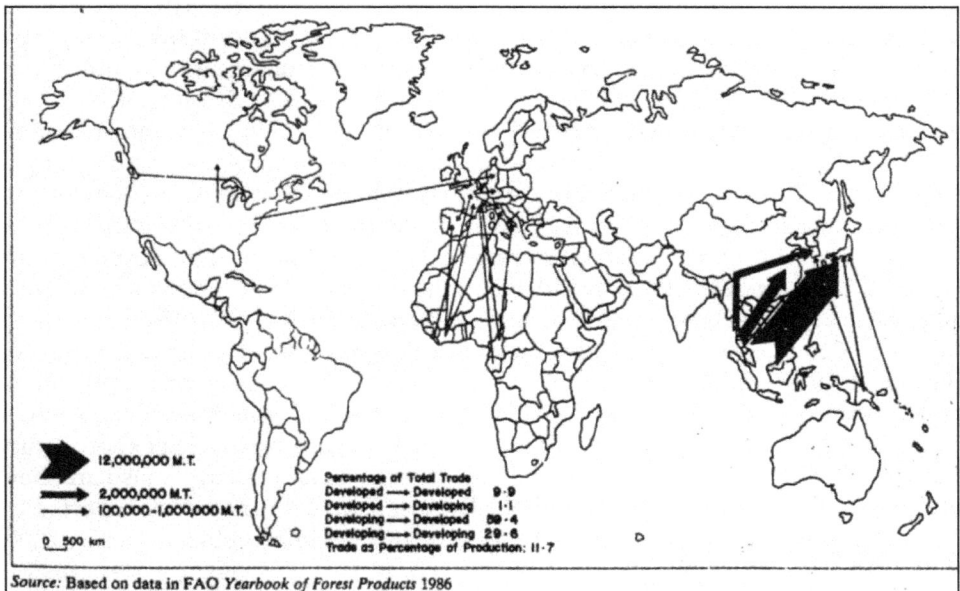

**Figure 13.12: Directions of trade (exports): non-coniferous sawlogs and veneer logs, 1985.**

Perhaps the most striking feature of international trade in wood is the remarkable growth in exports of tropical hardwoods over the last few decades. In the early part of the century, much of the trade consisted of dyewoods and tannin woods, and Latin America was the leading exporter. Pre-war, most of the exports from south-east Asia were of teak. Between 1946 and 1980, the volume of trade increased by twenty-four times, and the cumulative volume of trade during the 1970s exceeded that of the previous seven decades. This great increase reflected changes in both demand and supply. Growth in income and population, combined with a dwindling availability of certain species an grades of temperate hardwoods, led to greatly increased demand. On the supply side, mechanized logging and transport after the Second World War permitted the harvesting of species and areas that were previously considered unexploitable. Furthermore, improved technology for producing veneer and plywood from tropical hardwoods such as lauan and meranti opened up new markets for products such as doors and panels, whilst there have also been improvements in the technology of pulping. Although tropical hardwood logs comprise only 4 per cent of all trees harvested world-wide, production has been growing far more rapidly than that of other forms of timber. During the period from 1961 to 1980, annual growth in production averaged 4.8 per cent, compared with 1.2 per cent for softwood logs and 0.7 per cent for temperate hardwoods (Takeuchi, 1983). Much of the growth is attributable to Japan, which accounts for approximately half the world volume of traded tropical hardwood, compared with only 4 per cent in 1950 (Laarmn, 1988).

The growth of exports of tropical hardwoods was accompanied by shifts in the pattern of production. In the early post-war period, th4 main source was Africa, but by the mid-1950s the Far East had become dominant and maintains that position to the present. Latin America accounts for only a very small proportion of exports. Within both Africa and the Far East, shifts have occurred from country to country, as production reached unsustainable levels and as bans have been imposed on log exports. The pattern of growth and decline experienced in the various forest regions of the United States in the late nineteenth century has been repeated, on a larger spatial scale if in a more subdued form, over the last thirty years. In Africa, log exports from Ghana and Nigeria peaked around 1960 and then declined, while those from Ivory Coast, Gabon and Cameroon increased. There have been similar trends in the Far East. The emphasis has moved from the Philippines to Indonesia and then to Malaysia, and in particular to Sabah and Sarawak. In the late 1970s, Indonesia and Malaysia accounted for nearly half of the production and three-quarters of the exports (Pringle, 1979): the latte alone was the source of 36 per cent of hardwood logs exported and 25 per cent of the sawn wood (Kumar, 1986). Thailand, which was once a major exporter of hardwood, is now a net importer. In the same way that log exports from individual countries within south-east Asia have risen and fallen, so also are they likely to do so from the region as a whole. Exports from the Far East are expected to decrease from 18.6 million cubic metes in 1980 to 9.2 million cubic metres by 2000 (FAO, 1982a).

Concurrent with the growth in production and trade in tropical hardwoods there have been changes in the composition of exports. Whilst most exports from individual countries are initially in the form of logs, the emphasis then moves towards processed products such as plywood. In the case of Indonesia, for example, hardwood sawlog exports dropped to negligible levels in the early 1980s in response to a log export ban imposed from 1 January, 1985, while production of plywood took off. In 1973 Indonesia and two plywood mills with an estimated output of 9,000 cubic metres but by 1986 the industry had grown to 108 mills producing more than 4.5 million cubic metres (Sinduredjo, 1986). Exports of plywood rose dramatically from 1,000 cubic metes in 1975 to more than 3.75 million in 1985. By 1982 Indonesia was the world's third largest producer of plywood, after the United States and Japan but ahead of Canada and the Soviet Union (Figure 13.13) (Schreude and Vlosky, 1986). Sawmilling also expanded dramatically, as Table 13.10 indicates: the clear tend in Indonesia, as in many other countries, was towards the production and export of manufactured wood products rather than logs.

**Table 13.10: Log production, plywood production and sawmilling in East Kalimantan, Indonesia.**

| | No. of Concessions | Area (1000 ha) | Log Production (1000 m³) | Workers |
|---|---|---|---|---|
| 1969-70 | 13 | 1866 | 2527 | 2878 |
| 1971-72 | 34 | 4951 | 5537 | 9495 |
| 1973-74 | 65 | 7025 | 8411 | 12103 |
| 1975-76 | 70 | 7355 | 7260 | 19246 |
| 1977-78 | 86 | 8311 | 9883 | 23079 |
| 1978-79 | 93 | 8659 | 10159 | 26378 |
| 1979-80 | 99 | 9353 | 8809 | 26460 |
| 1980-81 | 101 | 9465 | 5634 | 15556 |
| 1981-82 | 104 | 9985 | 3150 | 10251 |
| 1982-83 | 104 | 9985 | 3418 | 11269 |
| 1983-84 | 104 | 9985 | 4174 | 9822 |

| Year | Plywood | | | Sawmill industry | | |
|---|---|---|---|---|---|---|
| | Mills | Production (1000 m³) | Workers | Mills | Production (1000 m³) | Workers |
| 1978-79 | | | | 61 | 87.3 | 1588 |
| 1979-80 | 8 | 18.7 | 3368 | 63 | 115.7 | 3685 |
| 1980-81 | 5 | 52.4 | 6034 | 287 | 277.0 | 6962 |
| 1981-82 | 7 | 172.6 | 7460 | 315 | 361.2 | 6678 |
| 1982-83 | 13 | 266.0 | 10816 | 318 | 122.5 | 6703 |
| 1983-84 | 22 | 602.7 | 19018 | 339 | 365.0 | 7499 |

*Source*: Compiled from data in Priasukmana (1986).

Transit processors such as Singapore, which built up large-scale exports of plywood on the basis of log imports, are very sensitive to such changes. In he long term, it seems probable that the tend will continue to be towards processing in the country of production, rather than in importing or transit countries such as Japan or

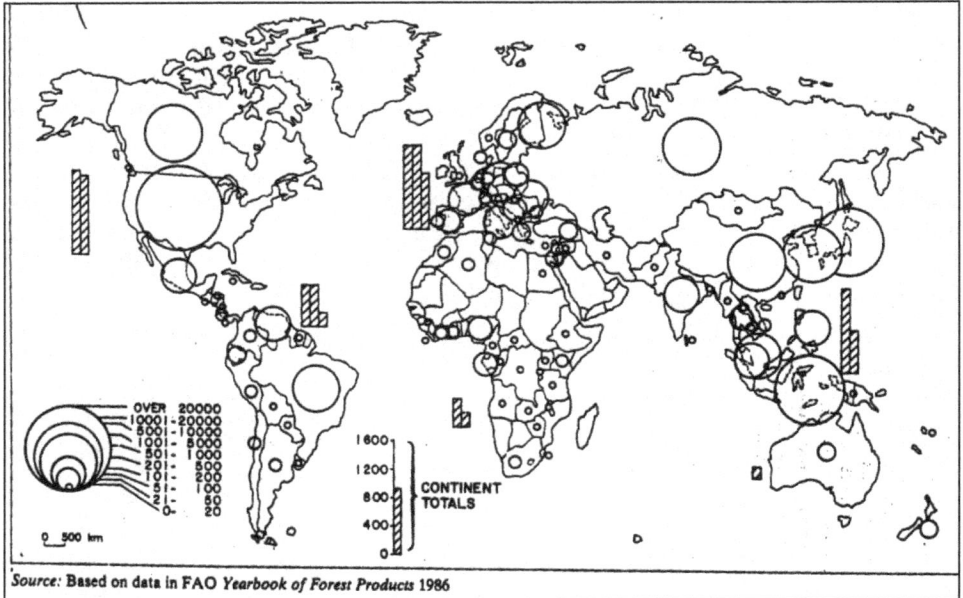

Figure 13.13: Plywood production (thousand m³) 1985.

Singapore. Transit processors in particular face serious problems as log export bans take effect.

In general terms, a slow and gradual shift is taking place in the forest-products industry. This shift is following that in the production of industrial roundwood, namely towards the parts of the world with high productivities. Different sectors of the industry are displaying this tendency t different rats and with different degrees of clarity. One important variable is the cost of wood relative to other inputs: another is the relative costs of transport of raw materials and of finished products. In some respects the shift is almost imperceptible: large degrees of inertia result from he capital investments in existing plant and skilled labour forces, and radical shifts are unlikely to occur in the short term. As constraints of wood supply and of environmental controls become increasingly significant in some of the traditional forest-products areas such as the Nordic countries and eastern North America, then much of the industry's expansion is likely to be set in areas such as the American South and tropical counties such as Brazil and Indonesia. These areas have already experienced major developments in forest-products industries: further expansion in these and similar areas can confidently be expected.

## Prospects for Industrial Roundwood

For decades shortages of industrial roundwood have been predicted. At the beginning of the century, for example, the specter of a timber famine haunted the United States: In mid-century, FAO asserted that a 'wood-wide wood shortage exists and threatens to become critical' (FAO, 1946, p. 68). Most of these dire predictions have failed to materialize, as supply has greatly increased and as demand has changed or grown less rapidly than predicted. Growth in demand for industrial roundwood in particular has frequently been overestimated. For example in 1966 FAO estimated that the world would require 1,500 million cubic metres of industrial wood per year by 1975 (FAO, 1966) while actual production in that year was under 1,300 million cubic metres (incidentally forecast fuelwood requirements and actual production were very close (1,200 and 1,180 million cubic metres respectively). Similarly, many national forecasts have exceeded actual consumption, for example in countries such as Japan the United Kingdom and the United States. Part of the reason or such overestimates is that the rate of increase in demand for saw timber for each unit of increase in per capita income has decreased. This has been especially true in Japan (*e.g.* Nomura, 1986). As a result many projections for national and global consumption have been revised downwards. On the global scale, demand for wood fiber is now expected to rise at an average rate of 1.8 per cent annually between 1980 and 2000, compared with an average of 2.4 per cent for the period from 1960 to 1980 (FAO, 1982a). At the same time as forecasts of growth in demand have been revised downwards, the prospects for supply have at worst not deteriorated. Inventories have continued 5o increase3 in a number of countries such as the United States, and in many countries removals have been less than net annual increments.

The changed perception of prospects for industrial wood is exemplified by that of FAO. In contrast to its earlier views, FAO concluded in 1982 that "The world's supply of industrial roundwood is considered adequate to meet its growing fiber needs throughout the balance of this century' FAO, 1982, p. 197). This does not imply that no problems of supply will be encountered. On the contrary, it is expected that a shortage of high-quality hardwood logs will be encountered as supplies from the Far East forests diminish (and the small supplies expected from plantations will fall far short of compensating (Grainger, 1988). There may also be a strain on supplies of softwood logs and pulpwood. Furthermore, all types of wood may be scarce in some regions, such as Western Europe and Japan (FAO, 1982a). Nevertheless, regional or sectoral scarcity is quite different from global scarcity. Furthermore, some of the expected shortages are offset by relative abundance in other sectors, or by continued expansion of plantations. For example mixed tropical hardwoods could physically contribute more, in terms of quantities, to industrial wood requirements, and the potential supply of hardwood fiber logs is abundant, especially in countries such as the United States and France where they are currently under-utilized. Much of the projected growth in demand for hardwood supplies is for pulp and reconstituted panels, rather than or 'solid wood' products. And the annual afforestation of the 1-2 million hectares of land that would be required to match an annual increment of up to 20 million cubic metres of softwood supplies is feasible (FAO, 1982a). Wood volumes equivalent to projected industrial needs for he year 2000 could in theory be met by production from 100-200 million hectares of plantations, or 3.7-7 per cent of the closed forest area (Sedjo and Clawson, 1984).

In short, the prospects for the adequacy of supplies of industrial wood, are largely if not invariably bright. Those for the fuelwood sector are another matter.

## Fuelwood

Fundamental contrasts exist between and industrial wood in terms of patterns and tends of production, ownership and management of the resource, and adequacy of supply. Much f the fuelwood supply comes from woodland or farm trees rather than closed forest. The greater part of it comes from hardwood trees, and its production is centred in low latitudes. Supplies of industrial woo are likely to be adequate to meet demand to the end of the century and beyond, while severe shortages of fuelwood already exist. The so-called fuelwood crisis has emerged as a major issue, and fuelwood is one of the most problematic sectors of natural resources in terms of adequacy of supply and resource destruction.

Like industrial wood, fuelwood is used in a variety of ways. Direct consumption as firewood for domestic cooking and heating is usually by far the largest use. Some fuelwood is converted to charcoal, which my then be used for domestic or for industrial purposes. Substantial quantities are used in industry, and although domestic use represents the lager sector, industrial consumption remains significant in many parts of the world.

Unlike industrial wood, fuelwood is usually obtained by self-collection by households, and only relatively small quantities enter the market. This mode of procurement prevails in the developed and developing worlds alike. In the United States, only one-quarter is purchased (Skog and Watterson, 1984), while in India the proportion is under 13 per cent (Agrawal, 1986). It does not follow, of course, that no costs are incurred in acquiring supplies, but these costs are not necessarily of a monetary nature.

The analysis of the fuelwood sector encounters especially severe data problems, since much of the production does not enter the market. In compiling statistics on fuelwood, FAO relies heavily on assumed per capita rates of consumption, which themselves vary with a number of factors including the relative availability of the material. Some imprecision and uncertainty are therefore likely to exist, especially in relation to trends in production. Nevertheless, it is clear that the production of fuelwood is increasing more rapidly than that of industrial wood (Table 13.1 sand Figure 13.3). It is also clear that the pattern o production is very different from that of industrial wood (compare Figures 13.8 and 13.14).

## Pattern of Production: The Global Scale

The production of fuelwood is overwhelmingly concentrated in the developing world. Around 85 per cent of annual production in the mid-1980s was located there. In this respect, production of fuelwood is the converse of industrial wood. The composition of supply also strongly contrasts with that for industrial wood, with over 80 per cent coming from hardwood sources. Although significant quantities of fuelwood are produced in many parts of the developed world, the northern coniferous forest is much less significant in this sector than it is or industrial roundwood. A striking feature of Figure 13.14 is the degree of dominance of low latitude countries in the pattern of production. As might be expected, large counties such as Brazil, China and India are the leading producers, accounting between them for just over one-third of total production. The production of fuelwood is less concentrated than that of industrial roundwood, for which top two producers account for around half of the world output.

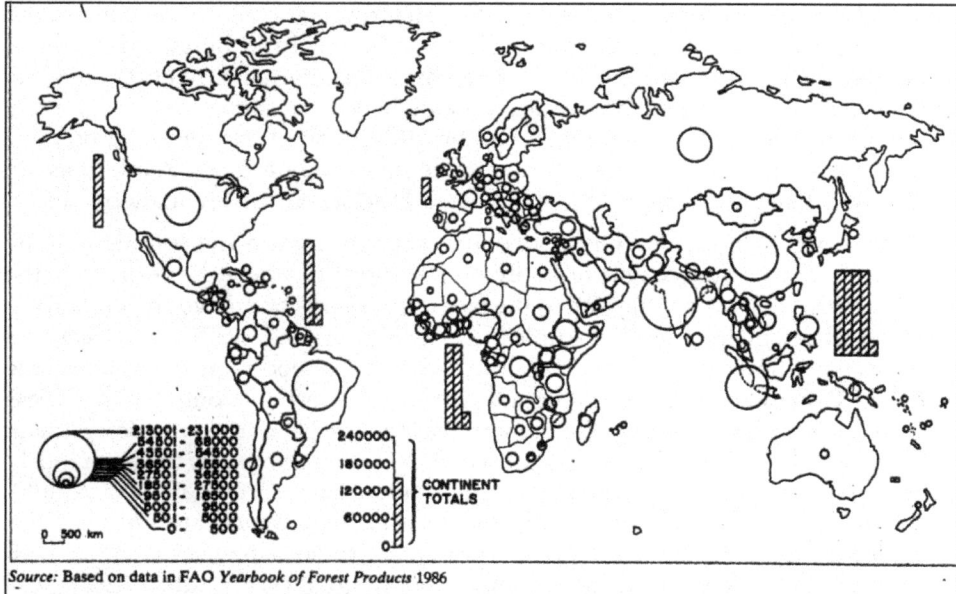

**Figure 13.14: Production of fuelwood and charcoal (thousand m³), 1985.**

Less immediately obvious from Figure 13.14, but equally or more significant, is the scale of production from countries with modest forest resources. Large volumes of fuelwood are produced from the savannah or *miombo* woodland (Chapter 2) of African countries. In many of these countries both the volume of growing stock and of net annual increment per unit area are low, but the pressures of demand for fuelwood are very high. The consequence of such pressures on a limited resource is all too frequently resource depletion and fuelwood scarcity.

At the same time, however, fuelwood production cannot be considered solely in relation to forest resources, and indeed much of it comes from non-forest sources. For example, in Thailand during the 1970s some 57 per cent of fuelwood came from outside the forest or from wood residues. In Sri Lanka over half came from coconut and rubber plantations. While in Tunisia four-fifths came from shrubs and tree crops (Arnold and Jongma, 1977). Much of the fuelwood produced in many countries comes from farms, and 2-5 per cent of the farm area can usually grow trees for fuel (and for other purposes) without loss in agricultural production. In eastern Java, for example, 63 per cent of the fuelwood consumed by farmers comes from farm trees: as few as six coconut palms can supply the fuelwood requirements of a family of five (Ben Salem and van Nao, 1981). In Kenya and neighbouring areas of east Africa, almost all fuelwood requirements can be supplied by a combination of sources including dispersed farm trees, hedgerows around homesteads and field boundaries, and small woodlots on soils unsuitable for cropping (Winterbottom and Hazlewood, 1987). Furthermore, fuelwood is regularly available in areas where fallow systems of agriculture are practiced. The clearing o fallow areas provide considerable amounts of woody material (Foley, 1987). In Mali, for example, a cropping cycle o four to five years is followed by eight to ten years of fallow, from which 30-50 per cent of the fuelwood consumption of small-scale farmers can be met. In addition, an average of seven trees per hectare, representing he remnant of the original vegetation, survives on field and fallow and also contributes to the fuelwood requirements (Ohler, 1985). It is concluded by Bailly *et al.* (1982) that the integration of trees into agricultural systems is the most worthwhile solution to the fuelwood problem. Plantations are expensive, while the improvement of the natural forest is cheap but inadequate.

The supply of fuelwood therefore depends on the nature and condition of the agricultural land resource and not only on the forest resource. Nevertheless, the forest resource is itself contracting in the face of pressures of fuelwood demands in many countries, including for example Pakistan (*e.g.* Biswas, 1987). Dead trees and branches have traditionally supplied much of the fuelwood requirement, and the cutting of livewood usually becomes a threat to the forest area only when a stress facto comes into play. This stress factor may take the form of rapid population growth, urbanization, or restriction of access for wood gathering as parts of the (previously common-property), forest become privatized or nationalized (Goodman, 1987; Arnold, 1987a). In addition, the clearance of forests may bring new pressures to bear. For example in Kenya, the removal of forests for military reasons in the 1950s an for cash crops had a profound effect on local communities (Ngugi, 1988), and one result was the fuelwood 'crisis' that developed as pressures were focused on the remaining resources of woods and trees. A similar process operated over a longer time and on a larger scale in India. Following the passing of the India Forest Act in 1965, large areas of forest were reserved for commercial purposes. Other areas were set aside to meet the needs of the local rural population for fuelwood and fodder. Many of these forests have suffered disproportionately in the face of growing population, and most of them have either been converted to agricultural land or have been totally degraded (Shyamsunder and Parameswarappa, 1987). Rural development projects may themselves lead to problems. The development of an irrigation scheme in the Tana basin of Kenya, for example, resulted in the trebling of the local population and severe impacts on the floodplain forest: Inadequate consideration had been given to the fuelwood requirements of the increased population (Hughes, 1987).

The relative importance of fuelwood production, compared with that of industrial roundwood, is illustrated in Figure 13.15. This map is a reminder that few individual countries conform to the overall pattern of production on the global scale, which is approximately evenly divided between fuelwood and industrial wood. Many developing countries use 80 per cent or more of their wood production as fuelwood, and in some case the proportion is as high as 90 per cent. In some countries, such as those of the Sahel, fuelwood provides around 90 per cent of total energy consumption, as Table 13.11 shows.

Most of the biggest producers in absolute terms, including Brazil, China and India, are characterized by a much lower fuelwood share in energy supply, although fuelwood share in total wood production remains high. In much of the developed world, fuelwood accounts for under 10 per cent of wood production, and supplies a very minor or even negligible share of energy requirements.

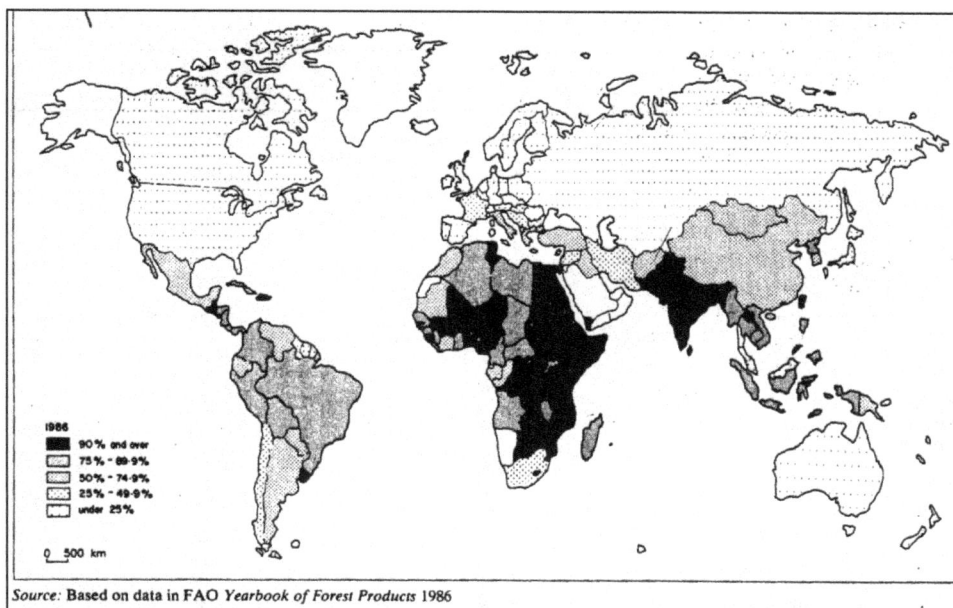

Source: Based on data in FAO *Yearbook of Forest Products* 1986

**Figure 13.15: Production of fuelwood and charcoal as a percentage of total roundwood production, 1985.**

The relative importance of fuelwood production in this respect is greatest in a number of African countries. Here the forest resource has not attracted development geared to large-scale production of industrial wood, but it has been intensively exploited for fuelwood. A notable feature of Figure 13.15 is the fact that the fuelwood sector remains very large or even dominant even in countries such as Indonesia and Ivory Coast, where the forest resource has been used for commercial logging for industrial purposes in recent decades.

**Pattern of Production: The Local Scale**

The growing scarcity of fuelwood in parts of the developing world mans that supplies have to be drawn from increasing distances. This tendency has to some extent characterized the use of the forest resource for the production of industrial wood as the world economy has expanded in recent centuries, but in the fuelwood sector it is especially apparent. Distances of several kilometers have frequently to be traveled to obtain firewood even in rural areas, while supplies for urban centres may have to be transported over distances of several tens of kilometres. The depletion of forests around cities is snot new. In south-east Brazil, for example, removals for wood and charcoal constituted a demanding and extractive use of the forest around towns and cities by the nineteenth century (Dean, 1983), while concentric circles of deforestation were becoming evident around the cities of the Sahel by 1935 (Thomson, 1988). The extent and scale of depletion around cities have, however, greatly increased in recent decades.

*Global Warming and Forest*

## Table 13.11: Wood fuel consumption.

| Country | Consumption per Capita (cubic metres) | Fuelwood as Per cent of Total Energy Consumption |
|---|---|---|
| Angola | 0.96 | 71.7 |
| Benin | 1.05 | 89.8 |
| Central African Rep. | 1.06 | 89.4 |
| Chad | 1.62 | 96.2 |
| Ethiopia | 0.83 | 89.9 |
| Ghana | 0.60 | 61.9 |
| Ivory Coast | 0.78 | 47.6 |
| Kenya | 1.48 | 82.8 |
| Madagascar | 0.59 | 74.7 |
| Malawi | 0.90 | 85.8 |
| Mali | 0.58 | 86.8 |
| Niger | 0.58 | 80.2 |
| Nigeria | 0.84 | 60.0 |
| Senegal | 0.55 | 47.8 |
| Sudan | 1.75 | 87.4 |
| Uganda | 1.77 | 95.9 |
| Zaire | 0.91 | 81.5 |
| Zimbabwe | 0.96 | 33.6 |
| Brazil | 1.25 | 37.2 |
| Chile | 0.50 | 15.6 |
| Colombia | 0.50 | 15.8 |
| Honduras | 1.03 | 59.5 |
| Bangladesh | 0.32 | 68.5 |
| China | 0.16 | 8.2 |
| India | 0.29 | 32.5 |
| Indonesia | 0.74 | 51.1 |
| Malaysia | 0.50 | 14.4 |
| Pakistan | 0.21 | 23.3 |
| Philippines | 0.55 | 35.6 |

*Source*: Compiled from data in Agarwal (1986) and FAO Yearbooks o Forest Products.

In rural areas substantial and increasing pats of the day are now occupied for many people in the developing world by firewood collecting. In parts of Sudan, for example, fuelwood collectors by the 1970s had to walk for one to two hours from their villages, whereas ten years previously adequate supplies were available within ranges of fifteen to twenty minutes (Digerness, 1979). In some areas of the developing world, women and children and spend up to 40 per cent of their daylight hours in scavenging for fuelwood (Goodman, 1987). In central Tanzania, it may take 300 man days (or more usually woman days) of work per year to provide the fuelwood for an average household (Grainger, 1982). In the Kigoma region of that country, villagers now have to walk 10 Kilometres for firewood whereas twenty years ago they had to go only 1 kilometre (Furgus, 1983). In parts of India, colleting distances in some localities increased from 1.5-2 kilometres in the 1970s to 8-10 kilometres in the 1980s (Agarwal, 1986). In parts of Nepal, the collecting of fuelwood, which might have taken an hour or two per day in the previous generation, by he 1970s was taking a whole day (Eckholm, 1975).

The problem of diminishing resources and increasing distances is especially severe around some of the Sahelian cities. By the 1970s, fuelwood for Ougadougou mostly came from a belt lying from 50 to 100 kilometres from the city (Li-Zebo, 1981), while almost all woody growth had disappeared within a 70-kilometre radius of Niamey (Agarwal, 1986). In the case of Dakar, fuelwood and charcoal are supplied from as far as 400 kilometres (UN, 1977). 'Islands' of depleted fuelwood resources therefore occur at various scales around both villages and cities. In many instances the resource has been impoverished at the very time when rapid growth in population has occurred, giving rise to numerous local shortages of fuelwood. Increasing scarcity around the urban centres especially is directly reflected in rising costs of fuelwood. In the Sahel and also in parts of east Africa, up to 40 per cent of salaries may be taken up by the purchase of fuelwood (Ki-Zerbo, 1981; Mnzava, 1981). In parts of Tanzania, almost hall of the income of families on minimum wages may go on fuelwood (Fergus, 1983). In Addis Ababa or Maputo, a family can spend up to half a week's wages in buying enough fuelwood or charcoal for survival (Goodman, 1987). Even it, fuelwood does not have o be purchased, there is an indirect or opportunity cost as a result of the time taken to procure supplies. Substantial parts of the day may be given over to obtaining fuelwood, and this cost falls especially on women, whose responsibility fuelwood supply has traditionally been in many parts of the developing world.

Significant differences exist between patterns of production for rural and urban consumption. For rural consumption, production is largely from dead tees, bushes, farm trees and other non-forest sources. Much of it is in the form of twigs, small branches or roots. The scale of production is usually small, and it is carried out for and by the household. Firewood collection of this type is unlikely to cause serious deforestation. Production for urban or industrial use, however, may be quite different. Its scale may be far greater, it may b organized commercially, and it is far more likely to result in deforestation. If a market exists, there may be a strong incentive to cut live trees, and organized gangs of woodcutters, perhaps operating illegally or in collusion with forest guards, may plunder even nominally protected forests. This can happen not only around cities, but also at greater distances, if the wood is converted to the more easily transportable charcoal (Eckholm *et al.*, 1984). The switch from fuelwood to charcoal which often accompanies the migration of rural dwellers to the city is potentially destructive of the source. The loss ratio on conversion of wood to charcoal is around 2.5:1. A poor family moving to an urban area may therefore require two and a half times as much wood (assuming that their fuel requirements remained constant) and may therefore exert two and a half times as much pressure on the resource (O'Keefe and Kristoferson, 1984).

The use of fuelwood in cottage industries and for rural processing activities such as tobacco-curing and tea-drying can have similar results in terms of forest removal. These activities are estimated to account for between 11 and 25 per cent of the fuelwood used in the developing world (Agarwal, 1986). In Tanzania salt production through brine evaporation in Kigoma region over a thirty-year period consumed a quantity of fuelwood equivalent to the clear felling of 2200 squares kilometers: the surviving area of forest in the region amounts to only 9,000 square kilometers (Fergus, 1983). In the same country, the wood from one hectare of savannah is required to cure the tobacco crop from a similar area (Mnzava, 1981). During the early 1970s tobacco-curing consumed more than 1 million cubic metres per year (Arnold and Jongma, 1977). With rapid expansion o each crops such as tobacco, there has been rapid contraction of parts of the savannah woodlands. The production of 'industrial' fuelwood in dry areas in particular may have long-term detrimental effects. For example, in Australia the environs of some nineteenth-century mining towns are surrounded by a zone up to 50 kilometres wide from which fuelwood was collected and within which regeneration has not occurred (UN, 1977).

## Consumption Levels and Resource Adequacy

A feature of fuelwood use is the widely varying level of per capita domestic consumption both between countries and within countries. Consumption levels vary greatly even between countries where wood constitutes the main source of energy, as Table 13.11 indicates. In many counties, per capita consumption is around one cubic metre per year, a figure which, incidentally, is similar to that for the consumption of industrial wood in many developed countries. Considerably higher levels characterize some African countries such as Sudan and Uganda, while much lower rates are typical of many Asian countries. As a very broad generalization, per capita consumption levels correlate with the extent of forest and woodland and with the availability o the resource. In parts of the developing world where fuelwood is plentiful, 2000 kilograms or more may be used per person per year, while in areas of scarcity only one-quarter of that amount may be used (Eckholm *et al.*, 1984). Economic factors are also, of course, a major influence on levels of consumption. They tend to reach a maximum in developing countries in the middle range of income (Laarman, 1987). As income rises, more fuelwood is initially consumed, but beyond a certain level its use decreases as other fuels such as oil are substituted. According to Foley (1985), price influences the amount of fuel that is consumed, but does not have a great influence on choice between fuels. As the acknowledges, however, this conclusion leaves unanswered the question of what determines the transition from wood to other fuels as a country becomes richer. 'Real income and an index of commercial energy prices are significant factors in relation to demand, but adjustments in consumption in response to these factors tend to be very slow and slight (Laaman and Wohlgenant, 1984). These workers also found that a contracting forest area was a significant variable in terms of supply, but that it ha a stronger breaking effect on fuelwood consumption in middle-income than in low-income developing countries.

Since levels of consumption appear to depend on the availability of the resource and on income, it is not surprising that consumption levels also vary greatly within count5ries. In Tanzania, for example, household consumption in villages near wooded areas is there times higher than in villages with little or no woodland (Agarwal, 1986). Similarly in Nepal, people moving to the well-wooded plains, where firewood is relatively abundant, consume twice as much as those remaining in the forest-depleted hills (Earl, 1975). In essence, people use more wood when it is readily available than when it is scarce.

This truism may partly explain the enormous range of estimates of per capita consumption reported for countries such as Nepal. A factor of 67 separates the highest and lowest estimates (6.67 and 0.1 cubic metres respectively) reported by Thompson and Warburton (1988). Even when the extreme highest figures were excluded, a range factor of 26 remained Amongst the possible reasons cited for such variation are confusion (not least over units of measurement) between local people and visiting consultants or researchers, and misstating of actual levels of consumption as a result of suspicion or apprehension of tax collection or forest regulations. Levels of use of fuelwood also, of course, depend on the availability and price of alternative fuels, and therefore, vary through time. Much uncertainly therefore exists over the concept of fuelwood 'needs', and this uncertainly is matched in some areas by that surrounding the availability of fuelwood. In Nepal, for example, estimates of forest productivity (the ultimate determinant of fuelwood supply) range from 0.2 to 15 or even 30 cubic metres per hectare per year (Thompson and Warburton, 1988). When both consumption and potential supply are so variable or uncertain, therefore, it is extremely difficult to quantify the dimensions of the fuelwood crisis.

Perhaps the best-known of the assessments of these dimensions is that produced by FAO (1982b), It is summarized in Table 13.12, 'Acute scarcity' is defined in terms of the depletion of fuelwood resources to the point where sufficient fuelwood cannot be obtained even by over-cutting, and consumption is below minimal needs. Approximately 100 million people lived in such areas in 1980, approximately half of them in Africa (Figure 13.16). A further 1,000 million lived in 'deficit situations', where minimal needs could be satisfied only by over-cutting and depletion of the resource. In 1980 almost 1,400 million people lived in areas of acute scarcity or deficit, and by 2000 that number may increas4e to close on 3,000 million. It does not necessarily follow, of course, that each of these persons will directly suffer from a shortage of fuelwood, as higher income groups can substitute, oil. Nevertheless, around 57.7 per cent of the 2,000 million persons dependent on fuelwood in the developing countries in 1980 did not have access to sufficient supplies, and the proportion may rise of 89 per cent by 2000 (Montalembert and Clement, 1983). The problem is by any standard a major one, and as Figure 16 indicates, a striking feature of its spatial pattern is its widespread occurrence. While little scarcity occurs or is expected to occur in the equatorial zone (except for parts of Andean Ecuador and of east-central Africa), several separate areas of acute scarcity occur in locations ranging from the Andes and Caribbean through Sahelian Africa to the flanks of the Himalayas. Huge tracts of eastern Brazil, the savannah lands of Africa, and India also face actual or prospective deficits. The crisis indeed has international if not global dimensions, and as such represents one of the most notable resource problems of history.

**Table 13.12a: Populations involved in fuelwood deficit situations (million).**

|  | Acute Scarcity | Deficit | 1980 Prospective Deficit | 2000 Acute Scarcity or Deficit |
|---|---|---|---|---|
| Africa | 55 | 146 | 112 | 535 |
| Near East and North Africa |  | 104 |  | 268 |
| Asia Pacific | 31 | 832 | 161 | 1671 |
| Latin America | 26 | 201 | 50 | 512 |
| Total | 112 | 1283 | 323 | 2986 |

**Table 13.12b: Fuelwood situation in Africa south of Sahara, 1980.**

|  | Population Depending Mainly on Fuelwood (millions) | Fuelwood Needs | Availability m³/ inhabitant/ year | Balance | Total Balance million m³ |
|---|---|---|---|---|---|
| Acute scarcity Arid and sub-arid areas | 13.1 | 0.5 | 0.05-0.1 | -0.45 | -6 |
| Mountainous area and islands | 35.7 | 1.4-1.9 | 0.5-0.7 | -1.1 | -40 |
| Deficit Savannah areas | 131.4 | 1.0-1.5 | 0.8-0.9 | -0.5 | -66 |
| Prospective deficit Savannah areas | 65 | 1.0-1.5 | 1.8-2.1 | +0.7 | +45 |
| High forest areas | 36.5 | 1.2-1.7 | 1.8-2.0 | +0.5 | +20 |
| Satisfactory High forest areas | 6.2 | 1.2-1.5 | 5-10 | Over 4 | +50 |

*Source*: FAO (1982b).

# The Fuelwood Shortage

Fuelwood has become increasingly scarce over the last twenty years. Since 1970 fuelwood and charcoal have increased in price relative to other goods at an annual rate of 1.5-2 per cent. Prior to 1970 there was a similar rate of decrease (Wardle and Palmieri, 1981). The resource has been deplete as a result of the pressures exerted on it. Various interrelated factors have contributed to the degradation: urban growth, the spread of charcoal production and the transition of fuelwood from free good available to all into a commercial commodity (Montalembert and Clement, 1983).

As major resource issue, the fuelwood shortage has attracted much attention and has generated some controversy. The effects of the shortage have proved to be almost as controversial as possible solutions. The woodland resource has undoubtedly been severely6 depleted in many parts of the world, and in addition to the ensuing economic and social problems, various environmental problems such saw accelerated soil erosion have been reported. Furthermore, it has frequently been suggested that a fuelwood shortage initiates a vicious downward spiral in the welfare

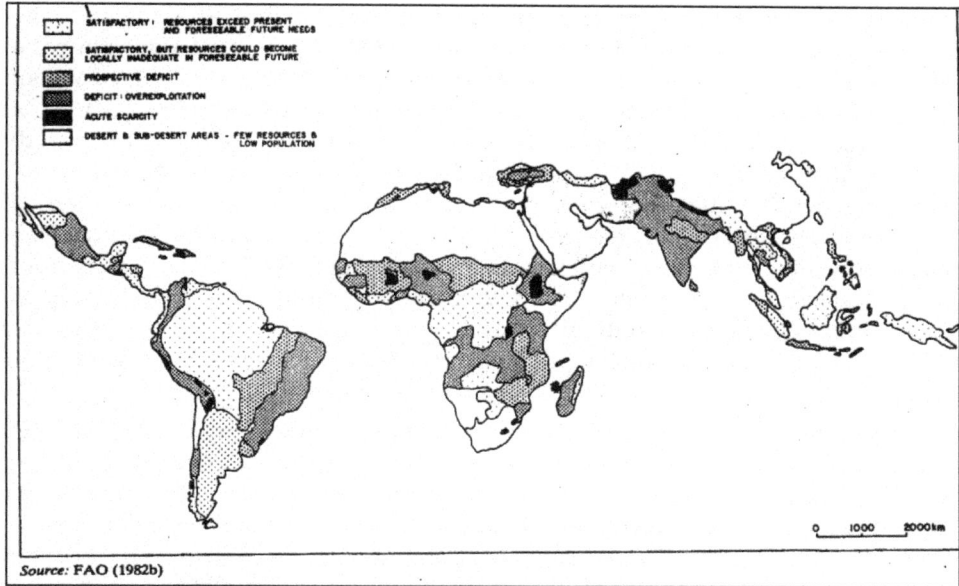

Source: FAO (1982b)

**Figure 13.16: Fuelwood supplied.**

of some rural communities, whereby increased walking distances to supplies lead to a greater use of dung as an alternative fuel, and hence less use of it as a fertilizer for crops. The loss o dung as manure has been estimated to amount to a loss of as much as 20 million tons of grain production annually (Agarwal, 1986). In short, depletion of the woodland resources leads to a deterioration of farmland and hence to an increasing shortage of food as well as fuel (e.g. Eckholm et al., 1984). As indicated in Chapter 4, the general truth of this relationship has been questioned by commentators such as Bajracharya (1983a), who report that in Nepal at least, dung is used as an alternative to fuelwood in winter, when it is not required as a fertilizer. Nevertheless, there is little doubt that many communities have suffered greatly because of shortages of fuelwood.

Numerous problems are encountered in the search for solutions. To many governments of developing countries, investment in fuelwood resources has been perceived as a retrograde step in the pursuit of progress. Furthermore, state forestry services have traditionally been oriented towards the protection of forest reserves and the commercial production of industrial timber, rather than towards fuelwood (Noronha, 1981). These perceptual and institutional factors have been major handicaps.

It is apparent that simple single solutions are not available. The use of more efficient wood-burning stoves may help to alleviate the problem in some areas, whilst in others a variety of measures, including agro-forestry and more intensive management of the existing resource, fuelwood plantations and the use of charcoal, may help Agro-forestry systems my help of increase household production of fuelwood in some rural areas, but even here rapid population growth may be a serious constraint. In Kenya, for example, it has been established that the intensity of management of farm woodlands increases with population density: planted and managed woody biomass increases as a percentage of total on-farm woody biomass as population density increases. And the amount of land devoted to the production of woody biomass also increases with population density (Bradley, 1988). Nevertheless, population densities may be so great in some areas that large deficits are experienced, despite these trends. Despite these local problems, however, agro-forestry probably remains the most promising solution to th4e problem of fuelwood shortage. Around 300 million hectares of agro-forestry could (in theory) supply the fuelwood needs of 2,000 million people on a continuing basis, assuming a yield of 5 cubic metes per hectare per year and an annual requirement of 0.75 cubic metres per person per year (Goodman, 1987).

Population pressures may also militate against more intensive n effective management of existing resources. Such management becomes more difficult, as well as more necessary, as the traditional institutional systems and structures suffer increasing stress. Common-property land constitutes a par4ticulr problem in some areas. The rural poor are heavily dependent on fuelwood from such land: in pats of India, for example, 90 per cent of the landless labourers and small farmers rely on it (Agarwal, 1986). The extent of common land contracts as private ownership becomes established and over-harvesting and rapid depletion become almost inevitable as population pressures increase. The poor, as well as the women, therefore tend to suffer disproportionately.

Fuelwood plantations and the use o charcoal imply the monetization of fuelwood production. In many areas fuelwood has been regarded as a free good, and farmers therefore are not readily persuaded to plant it, and are even less inclined to view it s a cash crop. The perceptions of both consumers and potential producers may therefore have to be altered if a successful transition is to be achieved. Furthermore, production from some plantations, for example in the Sahel, has proved to be disappointing and far below the projections employed in investment appraisal (*e.g.* Eckholm *et al.*, 1984). In the Philippines yields from plantations established under a 'dendrothermal power' programme have been far less than expected: annual growth rates of 75-100 metres per hectare were predicted but some yields have been less than 100 cubic metres over entire four-year cycles of growth (Durst, 1986). Enormous sums of money have been devoted to forestry products in the Sahel in particular, but there has been a decline in the availability of fuelwood. Most of the projects have involved plantations: few have been directed at managing and utilizing existing forests, although these would also contribute food, fodder and a range of 'minor' products (Gritzner, 1988).

Nevertheless, the combination of fuelwood plantation and charcoal production offers many attractions. For example, this mode of production could be freed from proximity of the market, and located with greater consideration for climate and soil (*e.g.* Kelta, 1987). Use of high-value, relatively transportable charcoal, as opposed to wood, could mean that plantations could be located in areas with optimal growth potential or alternatively on poor land with minimal impact on agriculture, and not necessarily in optimal location with respect to the market. While 30-50 per cent of the heat value of wood is lost on its conversion to charcoal (Arnold and Jongma, 1977), the use of charcoal rather than fuelwood has clear advantages in some instances. In addition to permitting the development of 'remote' fuelwood plantations, it can be produced from wood residues and from non commercial species available, respectively, from wood-using industries and from land clearing.

Fuelwood plantations have been established for many years in a few areas. For example, farmers on coastal saline soils around Madras grow casuarinas trees on short rotations, maintaining a tradition dating back to the nineteenth century (Foley, 195). And energy plantations consisting mainly of eucalyptus were established around Addis Ababa in the 1890s in response to a severe wood shortage (Eckholm *et al.*, 1984). More recently small commercial fuelwood plantations designed or supplying local urban markets have been expanding rapidly in parts of India (Arnold, 1987b). In Gujarat the rate of private planting increased fourfold between 1975 and 1979, doubled again by 1981 and yet again by 1983. By then the equivalent of more than 150,000 hectares had been planted. This rapid expansion was stimulated by very attractive rates of financial return (Arnold, 1983, WRI, 1985).

In India some attempts have also been made to integrate fuelwood production in industrial plantations. Degraded land is offered to industry for planting, on condition that up to 30 per cent of the biomass, in the form of top, lop and bark can be removed free of charge by local people, and also that a proportion of fodder species be grown (Shyamsunder and Parameswarappa, 1987). As yet only a very small area is committed to such schemes.

These examples of fuelwood and multi-purpose plantations, however, are on small and localized scales. South Korea offers an example of a much more spectacular scale of reforestation oriented to fuelwood. Under a village fuelwood programme launched in 1973, more than 1 million hectares of land were planted. In addition, 600,000 hectares of existing fuelwood plantations and more than 3 million hectares of other forest land were brought under more intensive management (WI, 1985). This scale is most unusual, and its community basis is, with the possible exception of China, unique. Extensive establishment of community plantations elsewhere has proved very difficult, especially where there is no surviving tradition of community forestry.

An enormous expansion of plantations or other supplies will be required if significant inroads are to be made into fuel shortages. It is estimated that around 48 million hectares of fuelwood plantations would have been required to meet the tropical fuelwood deficit in 1980, and 105 million hectares that are expected in 2000. In Africa the requirement to supply the expected shortfall in 2000 would correspond to a 6,000-kilometres belt extending from Senegal to Ethiopia, and averaging 34 kilometres wide (Grainger, 1986). Against such requirements achievements to date are modest almost to the point of insignificance. Attempts to create community planta6tions for fuelwood on the degraded Jos Plateau of Nigeria, for example, began just after the Second World War, but by 1987 covered only 0.5 per cent of the area (Buckley, 1987). In Africa as a whole, the annual rate of establishment of non-industrial plantations averaged around 40,000 hectares per annum around 1980; a forty-fold increase would be required to meet the expected shortfall in 2000 (Grainger, 1986).

## Fuelwood in the Developed World and Large-Scale 'Industrial' Use

The scale of the domestic fuelwood shortage in the developing world has overshadowed the whole of the fuelwood sector. As a result, relatively little attention has been focused on the use of wood as an energy source in other parts of the world or even for industrial purposes within the developing world. In particular, the fact that the use of fuelwood has increased far more rapidly in the developed world than in the developing world in recent years has been largely overlooked. Nevertheless, between 1979 and 1986 it rose by 57 per cent in the former compared with 23 per cent in the latter. Wood now provides 2-3 per cent of energy requirements in the United States, and is thus of the same order of impotence as nuclear and hydro sources (*e.g.* Schreuder and Vlosky, 1986). In Europe it accounts for around 2 per cent of energy consumption (Prins, 1987).

Most developed countries at some time in their history have been as dependent on wood as an energy source as many countries in the developing world are today, and indeed many have experienced similar wood shortages or famines. A recently as mid-century, the developed world accounted for over half of world consumption of fuelwood, while at the beginning o the century the proportion was two-thirds (Tillman, 1978). In the United States in the latter part of the nineteenth century, wood was the primary energy source, and most of the wood produced annually was used as firewood. In earlier centuries, the same was true of countries such as England, where fuelwood was used for both domestic heating and cooking and for industrial purposes such as iron-making. By the seventeenth century, severe problems of supply were being encountered, and this early fuelwood crisis stimulated the transition to coal (Nef, 1977). A similar transition to coal and oil only duly occurred in the United States at the turn of the present century even although an acute shortage of firewood was not encountered. During the nineteenth century, firewood was cheap and abundant, especially where the forest was being cleared for agriculture, and its use was profligate. By the 1960s, the use of firewood in the Unite States had fallen to a negligible level, and consumption had also fallen rapidly in many European countries and in the Soviet Union.

Since then, however, there has been a partial revival, not least because of rises in oil prices in the 1970s. In Europe, fuelwood removals by 1986 had increased by around one-fifth compared with those of the mid-1970s. The revival has been spectacular in the United States, with a fourfold increase between 1975 and 1980. It remains to be seen whether this renewed interest is permanent, and it is noticeable that many countries have experienced temporary increases in the use of firewood when other fuels were scarce, for example during and after wars. Nevertheless, it seems that the higher levels of use of the 1970s have been at least maintained during the 1980s in many developed countries.

While the amounts used remain modest compared with industrial use, the production of firewood is a significant element in the use of the forest resource in many developed countries. In the United States, for example, around half of the 7.8

million private forest owners cut firewood from their own forest land (Birch *et al.*, 1982). Firewood production my therefore be at least one of the objectives of numerous small owners, not only in the United States, but also in European countries and Japan. It may be a management objective in a far greater area of forest than its apparently small-scale use would suggest. Rising oil prices may therefore stimulate increased levels of intensity of management of farm woodlands in the developed world, as well as imposing additional stress on fuelwood supplies in the developing world.

Increasing demand for firewood in the developed world has had little effect on timber markets. Firewood prices are usually very low compared with those offered by industrial consumers and much of the firewood comes from trees that are unsuitable for industrial purposes. In the United States, for example, only a quarter of the firewood cut by households comes from trees suitable for pulpwood or saw logs (Skog and Watterson, 1984). In any case, much of the firewood comes from small private (non-industrial) forests, rather than from larger industrially owned or managed ones. These small forests are often viewed as providing various service as well as goods. The colleting of firewood may be perceived by households as an enjoyable activity in itself, and in at least some North American, European and Australasian countries may be at least partially seen as a form of forest recreation and not just as a chore (*e.g.* Trotmann and Thomson, 1988).

While it is unlikely that forests will be greatly expanded in the developed world to produce domestic firewood (except perhaps on farms), there is a greater likelihood that sizeable plantations will be established for commercial energy production. Rises in oil prices during the 1970s led to renewed interest in alternative energy sources, including short-rotation forests geared to producing fuel for power stations. The feasibility of producing woody biomass for energy has been investigated especially within the European Economic Community (*e.g.* Hummel, 1988), and it has been shown that systems based on coppiced hardwoods with short cutting cycles may have some potential. For example, it has been shown that energy plantations of sycamore on cut-over peat land can have attractive returns in pats of Ireland (Lyons and Vasievich, 1986). Such systems may have some attraction as alternatives to conventional agriculture in times of food surpluses, but their viability will ultimately depend on the prices of other energy sources.

More conventional use of wood for energy for transport and industry is, of course, long established in some parts of the world. In the United States in the nineteenth century, for example, large tracts of forest were harvested on a rotational basis in order to produce charcoal for iron-making, and a number of metal smelters and refineries were still fuelled by wood in the 1970s (Tillman, 1978). Perhaps the largest-scale industrial user of fuelwood, however, is Brazil, where charcoal was used in the production of around 30 per cent of the annual output of pig-iron during the 1970s (Tillman, 1978). The projected expansion f iron production in Brazil has major implications for the forest resource (Fearnside, 1989b). To feed the proposed expansion of smelting capacity at Carajas in eastern Amazonia, the charcoal from either 700,000 hectares of eucalyptus plantations or from the clear-cutting of some 50-70,000 hectares of dense native forest annually would be required. On the other hand, the use of wood as a fuel in some other industrial applications is much less significant in relation to the forest resource. Much of this use involves wood residues or wastes, rather than material removed from the forest as fuelwood. In Europe, for example, fuelwood represents less than half of the total volume of wood used for energy (Table 13.13). Around 40 per cent of volume of wood removed annually is eventually used as a source of energy (Prins, 1987).

**Table 13.13: Wood as an energy source in Europe around 1980.**

| | |
|---|---|
| Total removals (million cubic metres) | 342 |
| Fuelwood | 72 |
| Wood and bark residues | 40 |
| Forest products after original use | 11 |
| Estimated wood equivalent of pulping | 44 |
| Liquors bunt in chemical pulping | |
| Total | 167 |

*Source*: Based on Prins (1987).

The forest-products industry is the biggest single industrial user of fuelwood in most countries. In the United States, the forest-product sector accounts for 50 per cent of all the wood used for energy (Schreuder and Vlosky, 1986). Up to half of the energy consumed in the course of manufacture of plywood or pulp and paper may be supplied by wood, and much of that comes from mill residues and other waste material. The growth of wood-using industry in the developing world may offer some potential for increasing production of charcoal from waste, and hence facilitate the use of wood fuel in cities remote from forests. In Malaysia and Indonesia, for example, charcoal production increased by an estimated 26 and 21 per cent respectively between 1976 and 1986, and charcoal exports from the former increased almost fourfold in the same period. World-wide, exports almost doubled between 1975, but much of them were destined for the developed world or for oil-rich but forest poor counties in the Middle East, rather than for the areas of greatest need of domestic fuel.

## Non-consumptive Functions

The forest has traditionally been regarded as fulfilling two major types of functions. On the one hand, it is a source of wood and in some instances also a variety of other products. On the other hand, it may provide a variety of services such as oil and wildlife conversation, general environmental production, and recreation. This function is sometimes described as non-productive, in the sense that physical products (such as wood) are not involved. Here the term 'non-consumptive' is preferred to 'non-productive', as it may have fewer negative connotations.

In practice the distinction between the production of goods and the supply of services is not always sharp, and in particular the frequently used classification into 'productive' and 'non-consumptive' or 'protective' functions is sometimes inaccurate or misleading. Recreation, for example, may be as valuable a 'product' of the forest as wood, while the economic value of watershed protection may in some instances be as real as that of wood production, although it my be far more difficult o estimate.

The forest is of major environmental significance. The existence and nature of forest cover has a strong influence on the hydrological characteristics of an area, and this influence may extend far beyond the forest. Similarly, the forest has traditionally been perceived as protecting land from soil erosion. Environmental protection may therefore be a major function of the forest, and indeed in some instances the primary or sole one. The interaction between the forest resource and the environment issue is section the focus is on the relative extent of the function (in so far as it can be established) and on the managements issues involved.

Since this protective function does not yield revenue in any direct sense, it has rarely been associated with private ownership. Similarly, the use of forests for informal public recreation rarely involves payment at the forest gate, and is more typical of publicly owned forests than of those in private ownership. In short, non-consumptive uses are often associated with public of state forests, while management for wood production is equally associated with private ownership. This correlation, however, is by no means perfect, and in the long term it is probably becoming weaker. This trend perhaps characterizes the transition from the 'industrial' to the post-industrial' forest. In the former, wood production is the primary or indeed sole objective. In the latter, wood production, wildlife conservation, recreation and various other functions are combined as multiple objectives, to which varying priorities are ascribed. The trend is apparent in both private and public forests in some pats of the world, and indeed one of the features of the 'post-industrial' forest is perhaps the luring of the distinction between public and private ownership as government influence and regulation increase. This is not to say that all forests are now subject to multiple use, or that management under private and public control is identical. Huge areas of single-purpose forests remain, as of major contrast in management objectives. In some pats of the world, however, and especially in Europe, there has been both diversification of use and convergence of management. Private forest management has been increasingly influenced by a combination of regulation and incentives, while in at least some countries the management of state forests has become more commercial in outlook and orientation. At first sight forest privatization and other developments in some counties raise questions about the strength of government influence and the continuation of the long-term trend of growing government involvement. On closer examination these measures may turn out of be less a reversal of the long-term trend than a tuning-point in how government influence is exerted.

## The Extents of Non-Consumptive Uses

Non-consumptive uses of the forest are almost impossible o quantify meaningfully in terms of areas and extents. Recreation, for example, may range from high-intensity, primary use to low-intensity occasional use for which no management is provided. To attempt to quantify such use in terms of hectares is not necessarily helpful, and even when estimates are bed on official classifications of forests and management objectives their value may be limited. In some instances recreation may be the sole or primary use, whilst in others it is a subsidiary activity. Estimates of the extent of non-consumptive uses should therefore be viewed with the greatest of caution. The problem is further exacerbated by differing international definitions and procedures, which mean that the simple addition of national statistics (assuming that they are available) is unlikely to produce a meaningful global estimate. A particular problem arises over the terms 'protection' and 'protected'. Some forests are managed primarily for forms of environmental 'protection' such as soil or water conservation. Some forests are 'protected' as national parks or other reserves. Some forests may be both protective and protected, while some may be one or the other. The semantic confusion is sometimes increased by the use of statistical categories such as 'protection etc'. Furthermore, forests may be classified for protection against very different hazards. In Bavaria, for example, woods and forests may be separately classified for protection against soil erosion, avalanches, air pollution and noise (especially along motorways) (Woodruffe, 1989).

Around half of the world forest area is considered to be 'exploitable' (which term, incidentally, reflects the perceived primacy of wood production). 'Unexploitable' forest is characterized by low physical productivity, by inaccessibility and high transport costs, or by classification for special (non-wood or non-consumptive) purposes such as conservation. These characteristics may overlap in many areas, and it is not therefore always possible to identify the relative importance of forest designations in rendering a forest 'unexploitable'. In any case, classifications are not nec4ssarily permanent, and although increasing areas in many parts of the world are being designated as reserves of one kind or another, the designations are not always immutable. Nor does official designation of a forest as a nature reserve or other protected area always mean that wood production does not take place.

Despite these problems and provisos, some indication of the relative extents of different functions can be obtained for at least some parts of the world, and variations in the relative significance of different functions can be identified at the national level. Table 13.14 shows estimates of the relative extents of 'recognized major functions' for Europe, the Soviet Union and North America, representing most (in area terms) of the developed world. In this area, wood production is the 'recognized major function' in approximately three-quarters of the area of forest and wood-land, while recreation has that role in around 1 per cent of the area. In the 'unexploitable closed forest', of course, he relative proportions are very different. Protection and recreation are classed as recognized major functions' in 52 per cent and 6 per cent respectively. Together, these two functions are therefore likely to constitute the main reason for 'unexploitability'. In terms of the total area of closed forest, however, they represent only 15 per cent although they amount to 25 per cent of the total area of 'forest and other wooded land'. Wood production is clearly the primary function, in terms of area, in both the closed forest and in the category of 'forest and other wooded land'. Protection-related functions are next in importance in terms of area, while recreation is a 'recognized major function' on around 1 per cent of total area of 'forest and other wooded land' and 0.1 per cent o 'exploitable closed forest'.

**Table 13.14: Forest proportions by recognized main functions (percentages of respective areas).**

| | Total Forest and Other Wooded Land | | Exploitable Closed Forest | | | Unexploitable Closed Forest | | |
|---|---|---|---|---|---|---|---|---|
| | Wood Production | Protection etc. | Recreation | Wood Production | Protection etc. | Recreation | |
| Nordic Countries | 90.5 | 7.5 | 2.0 | 96.6 | (-) | 3.3 | (45.1) | 0.1 | (54.9) |
| EEC | 89.0 | 10.6 | 0.4 | 98.8 | (44.1) | 1.1 | (53.4) | 0.1 | (2.5) |
| Eastern Europe | 80.3 | 12.4 | 7.3 | 85.3 | (16.1) | 10.7 | (49.7) | 4.0 | (34.2) |
| Southern Europe | 55.2 | 43.7 | 1.1 | 98.4 | (6.5) | 1.6 | (83.1) | <0.1 | (10.4) |
| USSR | 65.1 | 32.9 | 2.0 | 100.0 | (27.3) | - | (65.4) | - | (7.3) |
| USA | 89.5 | 10.5 | - | 100.0 | (100.0) | - | (-) | - | (-) |
| Canada | 95.7 | 4.3 | - | 100.0 | (-0) | - | (100.0) | - | (-) |

Source: based on data in UNECE/FAO (1985).

In Europe, wood production is the major function on 78 per cent of 'total forest and wooded land', while protection and recreation account for 19 and 2 per cent of he area respectively. These continental percentages, however, conceal a distinct gradient of increasing importance of the protective function from north to south, and a corresponding reduction in the relative importance of wood production. In some countries such as the United Kingdom and Ireland, protective functions are of negligible significance in terms of area, while in others they are as important as wood production. In Turkey, for example, protection is a 'major recognized function' in two-thirds of the total area of forest and woodland, and in Spain in almost half of that area. In the case of Spain, much afforestation has been carried out over the last half-century in areas affected by or prone to soil erosion. In general terms, the relative areal extent of the 'protective' function increases towards the south and east of Europe, and is generally low in the more humid areas of the north and west.

Outside Europe, the relative extent of protective forest is very variable. As Table 13.14 indicates, around one-third of the forest area in the Soviet Union has a protective function, but the inevitable problems of definition need to be borne in mind in interpreting this figure alongside that of just over 10 per cent (31 million hectares) shown for the United States, where a complex variety of sometimes overlapping designations apply to forest land. In the Soviet Union around 100 million hectares of forest fulfill a 'predominantly protective' function. In addition, more than 5.5 million hectares of protective forest stands have been established, mainly on previously agricultural land (Pavloskii, 1986). In Japan, around 8 million hectares of the total forest area of 25 million hectares are defined as protection forest (Hebbert, 1989). Three-quarters of the former area is protected or the conservation of headwaters and the remaining one-quarter for soil erosion. In Chile, over 60 per cent of the forest area is either protective or protected (WRI, 1985). In Malaysia, around one-third of the intended permanent forest area is classed as protective (Tang, 1987), while the corresponding proportion in Indonesia is 27 per cent (Haeruman Js, 1988).

**Protected Forests**

While large areas of forest have environmental protection as a 'recognized major function', the relative extent of protected forest, lying within designated national parks and nature reserves, is much smaller. Around 4 per cent of the closed forest and 17 per cent of the open forest are thus designated: overall approximately 8 per cent of the area of forest and woodland is protected (Table 13.15).

Proportions of protected closed forest area do not differ greatly between the developed and developing worlds, but a much higher percentage of open forest is protected in the former than in the latter. This means in turn that the protected extent of the overall forest and woodland area is nearly three times greater in the developed world than in the developing world. In the United States, around two-third of the forest area is classed as commercial timberland (capable of producing at least 20ft$^3$ (0.566 cubic metes) of industrial wood per acre per year), and approximately 5 per cent of the area thus capable has been withdrawn by statute or regulation as national park or wilderness area (USDA, 1982). Overall, in addition to 32 million hectares of national parks (WRI, 1986), over 30 million hectares land, mot of which is forested, are officially defined as wilderness (Daniels *et al.*, 1989), in which lumbering is not permitted. Most of this area is remote and inaccessible, and indeed two-third is in Alaska.

**Table 13.15: Protected forests.**

| | Percentage of Forest Area in Protected Forest | | |
|---|---|---|---|
| | Closed Forest | Open Forest | Total |
| Tropical Africa | 4.3 | 8.6 | 7.3 |
| Tropical America | 2.1 | 1.0 | 1.8 |
| Tropical Asia | 5.8 | 1.9 | 5.4 |
| Tropical countries | 3.4 | 6.1 | 4.4 |
| Europe | 1.3 | 58.3 | 12.3 |
| USSR | 2.5 | 100.0 | 17.0 |
| North America | 7.9 | Na | 4.9 |
| ECE Region | 4.2 | 35.4 | 11.7 |
| Total | 3.8 | 17.2 | 8.0 |

*Source*: Based on WRI (1986).

This case illustrates the general tendency in many countries for highest environmental and timber values to be non-coincident in spatial terms. In much of Alaska the costs o harvesting timber would exceed its returns, and hence there may be no opportunity cost involved in designating the forests for conservation purposes (Hyde and Krutilla, 1979). Similarly in New Zealand, where around 10 per cent of the land surface lies in various reserves where the prime concern is the preservation of flora and fauna, most of the protected area lies in upland forests or alpine areas where the threat from agriculture or exotic afforestation is unlikely to be great (Halkett, 1983). Perhaps it is not surprising that protected forests often tend to be those of lowest commercial value. This tendency raises the questions o the purpose of designation if few threats exist, and also the problems of establishing protected status in the commercially (and some-times ecologically) more valuable areas.

On the other land, in the Soviet Union much of the 'Protection etc' area indicted in Table 13.14 consists of highly accessible forest where thinning and sanitary and regeneration cuts are permitted (Holowacz, 1985). On the approximately 300 million hectares of protected forest indicated, around 20 million hectares are in national parks (WRI, 1986). In Japan, over 3 million hectares of national parks, most in forested country, overlap with the protective forest area, an thereby exemplify the problems of functional definitions. An Australia, national parks and similar areas from which log production is precluded amount to around 8 million hectares, compared with a native forest area of 42 million hectares (South, 1981). The relative extent of protected forests in the developing world tends to be smaller (but not invariably so), although the definitional problems are as great if not greater than in the developed world. In tropical Africa, America and Asia, less than 5 per cent of the combined open and closed-forest area was 'protected' around 1980 (WR, 1986). Around 3 per cent of the closed tropical forest is safeguarded, at least on paper (FAO, 1986a).The relative extent of protected closed forest is not dissimilar to that in the developed world, and indeed each of the thee major tropical continents has a higher proportion of their closed forests in protected areas than has Europe (Table 13.15). On the other hand the relative extent of open forest with protected status is very much smaller, at 6 per cent compared with 35 per cent. Protected forests are most extensive in relative terms in Africa, where some game and forest reserves date from the early par of this century, and least extensive in Latin America.

A characteristic of the tropical continents is the degree of concentration of protected forests in a few countries. For example, Zaire accounts for 60 per cent of the reported area of protected closed forest in Africa, Indonesia has over 50 per cent of the corresponding area in Asia, and Brazil and Venezuela together account for over 50 per cen6t of that in South America (based on data in WRI, 1989). The apparent degree of concentration partly reflects the different rates and degrees of progress made in classifying forests as a basis for management policy. In the case of Indonesia large absolute and relative areas have been assigned functions of protection or conservation (Table 13.16). Concentration applies also at the intra-national scale as well as at the international level. In Indonesia, for example, more than thee quarters of the forest area in Bali has been zoned for protection and nature conservation, while only around 10 per cent is designated as production forest (Mc Taggart, 1983).

**Table 13.16: Forest classification examples of Indonesia and Finland.**

|  | *(million ha)* |
|---|---|
| Indonesia |  |
| Permanent production forest | 33.6 |
| Limited production forest | 30.4 |
| Protected forest | 30.3 |
| Nature conservation forest | 18.7 |
| Total | 113.0 |
| Finland |  |
| Timber production forests | 18.2 |
| Nature reserves | 0.4 |
| Preserved high latitude/altitude forests | 0.6 |
| Recreational areas | 0.1 |
| Private forests used for recreation | 0.5 |
| Other | 0.3 |
| Total | 20.1 |

*Sources*: Indonesia—Schreuder and Vlosky (1986); Finland—OECD (1988).

Zero or negligible areas (less than 500 hectares) are indicated for around 5 per cent of counties for which protected forest areas are listed by WRI (1989). Most of these cases are in Europe, and include countr4ies such as Belgium, The Netherlands, Ireland and the United Kingdom, where surviving remnant of native forest are of negligible extent. The fact that the elative extents of protected forests in some developed countries are so small is to some extent understandable in terms of the limited surviving areas of 'natural', unmodified forest. It does little, however, to strengthen their case when trying to persuade some developing counties to adopt more conservation-minded policies towards their forests. On the other hand the stark breakdown of forest classes in cases such as Finland may be misleading: nature conservation and recreation are also given some priority over one-quarter of the area of 'timber production forests' (Table 13.16). Just as 'protected' status may have different meaning in different settings, so also classification as production forest may have very different significance for nature conservation (and recreation) in different countries.

Small though the relative extent o protected forest areas may be, there is a clear trend towards expansion. Globally, the extent of all kinds of protected areas has grown rapidly, with an approximately fourfold increase between the mid-1960s and mid-1980s. In Central America, for example, the number of national parks and other protected areas increased from 25 to 149 between 1969 and 1981, with a corresponding expansion of the protected area from under 200,000 to over 600,000 square kilometers (Neumann and Machlis, 1989). Financial assistance from American conservation bodies has aided this expansion.

Numerous other instances of increases in the extent of protected forest areas could be quoted. Two examples from contrasting countries will suffice. In Australia, the native forest area from which log production is precluded following designation as national parks or similar areas increased from 1.8 to 8 million hectares between 1971 and 1981 (South, 1981). In the very different setting of Bulgaria, protection forests, national parks and other special-purpose forests increased from 400,000 to 1 million hectares between 1965 and 1985, when they comprised some 29 per cent of the total forest area (Nedelin and Gulev, 1987). Almost world-wide, increasing areas of native forest have been designated for some form of conservation, but formidable obstacles have been encountered in some areas. In Papua New Guinea, for example, the nature of forest landownership (Chapter 5) has meant that it has proved very difficult for the government to acquire land for parks and similar purposes, and the need to select areas representative of all altitudinal bands is an additional problem (Diamond, 1986).

The expansion of designated areas reflects the emergence of environmental issues and environmental pressure groups. In countries such as Australia and New Zealand in particular, the role and future of native forests have emerged ad a major issue, and this is reflected in the designation of increasing extents of the surviving forest under some form of protection. In Australia, for example, opposition on the pat of environmentalists to woodchip developments in native forests led to the setting aside of forest reserves (Conacher, 1977). Similarly, in New Zealand, increasing extents of lowland forest on the west coast of South Island have been reserved in the face of environmentalist pressures (Tilling, 1988). Previously, in the 1970s, similar pressures led to the adoption of sustained-yield use of such publicly owned forests in place of clear filling. While the issues may be sharply defined and the controversies especially heated in these countries. They are present in some form in almost every developed country and are increasingly emerging in many developing countries.

While a standard response to environmentalist pressures in many countries is the designation of protected forest, the process of designation is itself likely to have implications or non-designated forest areas. On the west coast of New Zealand's South Island, for example, the earlier failure to manage the native forests on a sustained-yield basis has aliena6ted environmentalists. In response, more and more of these forests have been protected, and the reduced area available for timber production has curtailed efforts a promoting sustained-yield management in the remaining production forests (Tilling, 1988). In shot, protected forests may be subject to single 'use' in the same way as timber production may be the single use on non-protected, exploitable forests, and an increase in the protected area can simply lead of an increase in the intensity of management of other, non-protected forests for timber production.

Furthermore, the granting of protected status does not necessarily guarantee the effective conservation of the protected area. Full legal support for the status of strict natural reserves is rare, and even when it exists it does not necessarily guarantee full protection (Hall, 1983). Illegal felling is often a major threat, whilst the collection of forest produce may also conflict with the objectives of reserve management (*e.g.* McKinnon *et al*, 1986). Local people may be as alienated or disaffected by their exclusion from the for4est for reasons of conservation as they are for reasons of commercial exploitation of timber. In Thailand, for example, nearly half of the virgin forest area is now formally protected, but the loss of the forest through activities such as poaching and opium growing ha not been halted (Ewins and Bazely, 1989). In Brazil, a fairly comprehensive system of protected areas has been established, extending over 3 per cent of Brazilian Amazonia, but protection remains inadequate in practical terms (Johns, 1989).

Despite the disparity that may exist between designation and practical protection, protected status is of fundamental importance in relation to the conservation of habitats and speci4es, and therefore it is important that an adequate coverage of forest types is achieved. Table 13.17 shows the distribution of protected areas by biome and biogeographical realm, but as the note to the table indicates, percentage coverage cannot yet be determined and hence adequacy of coverage cannot be assessed.

**Recreation**

If increasing area of protected forest is a feature of recent decades, so also is increasing recreational use. While recreation in the form of hunting has been a forest use since time immemorial, the use of the forest for informal recreation such as walking, hiking and camping is a characteristic of the twentieth century, and more especially of the period since the Second World War. In much of the developed world, levels of recreational use have generally grown rapidly in recent decades, usually at rates far surpassing those of timber production. In the United States national forests, for example, there was a thirty-fold increase over the forty-year period from the late

1920s (Clawson, 1976). In the state of Victoria in Australia, the number of visitor-days at the mainly forested national parks was increasing at a rate of 11 per cent per annum in the early 1980s (Algar, 1981).

**Table 13.17: Protected forest areas by biomes and realms 1985.**

| Biome/Realm | No. of Areas | Total Area (million ha) |
|---|---|---|
| Tropical humid forests | 280 | 39.1 |
| Afrotropical | 44 | 8.9 |
| Indomalayan | 122 | 5.0 |
| Australian | 53 | 7.8 |
| Neotropical | 61 | 17.3 |
| Subtropical/temperate rainforests/woodlands | 275 | 18.5 |
| Nearctic | 18 | 4.3 |
| Palaearctic | 48 | 1.7 |
| Australian | 26 | 0.9 |
| Antarctic | 145 | 2.8 |
| Neotropical | 38 | 8.8 |
| Tropical dry forests/woodlands | 581 | 65.5 |
| Afrotropical | 240 | 48.7 |
| Indomalayan | 238 | 10.4 |
| Australian | 10 | 0.9 |
| Neotropical | 93 | 5.5 |
| Temperate broadleaf forests | 483 | 11.5 |
| Nearctic | 82 | 1.9 |
| Palaearctic | 400 | 9.6 |
| Temperate needle-leaf forests-woodlands | 175 | 38.8 |
| Nearctic | 53 | 30.3 |
| Palaearctic | 122 | 8.5 |
| Evergreen sclerophyllous forests | 475 | 12.0 |
| Palaearctic | 122 | 3.4 |
| Afrotropical | 41 | 1.6 |
| Australian | 301 | 6.9 |

(Realm areas of less than 0.5 million ha omitted)

*Source*: Based on IUCN (1985)

*Note*: Biome type is not synonymous with habitat type: a protected area within a tropical humid forest biome, for example, may not necessarily contain tropical humid forest. Furthermore, the total area of each biome in each realm has not yet been determined with sufficient precision to assess percentage coverage. Biogeogrphical realms are as defined by Udvardy (1975).

Rapidly growing recreational use is a characteristic of almost all accessible and state-owned forests in the developed world, an increasingly it is also taking place in less accessible and privately owned forests. In the United States, for example, recreational use in national forests amounted to around 188 million visitor-days per year by 1970 (Clawson, 1976). In the state forests of he United Kingdom, the annual number of recreational visits is estimated at around 24 million (Willis and Benson, 1989). In Ireland, recreational access to state forests was actively discouraged until as lat a 1968, but by the early 1980s around 1.5 million visits were being made annually (UNECE/FAO, 1985). In West Germany, visits to forests for informal recreation my number as many as 1.2 billion per annum (UNECE/FAO, 1985). In Denmark, 90 per cent of the adult population visits the forest at least once a year, compared with 79-85 per cent in Sweden and 92-96 per cent in Norway (Koch, 1984).

While recreational use is often strongly associated with publicly-owned forests, it is by no means confined to that sector. In the United States, although only 4 per cent and 1 per cent respectively of non-corporate and corporate private owners have commercial recreation as a primary management objective, 29 per cent and 54 per cent of the non-corporate and corporate forest land is open to the public for some form of recreation. In addition, large areas that are not open to the public are available to family members, friends, or employees (USDA, 1981). Over much of the private area recreation is permitted because it would be too difficult to prohibit, but on the other hand more positive attitudes such as 'good neighbour' policies or profit-seeking prevail. Furthermore, as private ownership changes and non-farmer owners become increasingly common in some countr4ies, so a shift in the perceived value of the forest holding takes place, with timber production being accorded less significance and recreation value more importance. Such shifts may have implications for levels of timber production, especially in countries such as Finland (OECD, 1988).

While recreational use and protected status are not necessarily mutually exclusive (for example in national parks and wilderness areas), protected forests and recreational forests often have different locational characteristics. The former are often in remove areas, where relative inaccessibility has afforded at least partial protection from inroads by lumbering and agriculture. The most heavily used recreational forests, on the other hand are usually to be found in more accessible locations, especially around large cities. In small, densely populated countries such as The Netherlands and England, recreation may be the principal function of some forests around urban areas, and indeed forests have been and are being created primarily for recreation in such settings. In the Netherlands, for example, the 1984 Forestry Plan aimed at expanding the forest area by around 10 per cent by the end of the century, and one-third of the expansion was to be close to the main cities (van den Berg, 1989). In general terms, highly accessible forests may have high values in terms of both timber and recreation, and hence conflicts in management objectives are sometimes encountered.

Intensive recreational use may also be found in some remoter areas, but there it is usually highly localized in a few prime landscapes such as national parks. In such settings the revenue generated by forest camping, for example, may far exceed that from timber production. In some of these areas timber values are low in both absolute and relative terms. Around the northern timber line in Finland, for example, timber-cutting is restricted for environmental reasons, and the value of tourism and recreation is twice that of timber production (Saastamoinen, (1982).

Recreational use of the forest may conflict with timber production in a number of ways, such as impairment of tree growth through soil changes resulting from trampling, or as a result management decisions involving choice of species, silvicultural systems and rotation lengths. It may also conflict with nature conservation, through, for example, disturbance. And there may be internal conflicts within the recreational sector, especially between hunting and informal recreation. In West Germany, for example, this conflict arises from the disturbance of game by other recreational activities, while conflicts between hunting and timber production result from damage caused by game and disputes as to the choice of tree species that are optimal for timber production or game (*e.g.* Lang, 1986).

Forest recreation is not usually associated with topical areas, but there are signs that he perception of tropical forests as a resource for recreation and tourism is growing. The recreational use of Malaysian forests, for example, is growing as the population becomes more urbanized (Hamzah *et al.*, 1983). In Indonesia, around 0.5 million hectares have been designated as recreational forests (Haeruman Js, 1988). In Bali some areas have been earmarked as forest reserves accessible to tourists, in order to assist with the development of tourism (Mc Taggart, 1983). In Costa Rica, forest national parks have begun to attract foreign exchange through tourism (Green and Barborak, 1987). In the past, tropical rain forests were perceived o have little tourist or recreational valued, and this is one reason why few national parks were established in lowland forest areas in Africa (Pullan, 1988). The growing perception of their value for tourism and recreation is a welcome step that may help to halt their loss. Perhaps it is significant that one of the factors contributing to the forest 'turn-round' in the United States at the end of last century was a growing awareness of recreational values and wilderness qualities .

Increasing recreational use and an increasing absolute and relative area of protected forest is a feature of recent decades, but a more distinctive characteristic of the 'post-industrial' forest in the incorporation of several or multiple objectives–including protection of conservation–alongside timber production. This trend is not easily quantified, and differences in definitions, procedures and management frameworks make international comparison difficult. Nevertheless, UNECE/FAO (1985) have attempted to assemble data on relative extents and relative importance from most of the developed world, and some examples are indicated in Table 13.18.

While qualitative terms such as 'high', 'medium' and 'low' mean that detailed interpretation requires great caution, some general points stand out clearly. Timber production is defined as a function in almost the whole of the forest area of the specimen countries listed in the table, but its relative importance varies greatly. In the mall, densely populated Netherlands, for example, wood production is ascribed 'high' importance in less than half of the forest area, while recreation has 'high' status in 90 per cent of that area. In more sparsely populated Finland, on the other hand, timber production has 'high' importance on 90 per cent of the area, compared

**Table 13.18: Forest and woodland areas by importance of function (1000 ha).**

| Country/ Category | Importance of Function | Function | | | | | | Area | |
|---|---|---|---|---|---|---|---|---|---|
| | | Wood Production | Recreation | Hunting | Protection | Nature Conservation | Range | Total | Closed Forest |
| Sweden (closed forest) | High | 11800 | 5000 | 100 | 500 | 200 | 0 | 24400 | 24400 |
| | Medium | 11000 | 12400 | 24000 | 1500 | 1000 | 2600 | | |
| | Low | 2600 | 7000 | 300 | 22400 | 21800 | 21800 | | |
| France (forest & other wooded land) | High | 7140 | 1000 | 11600 | 738 | 95 | 0 | 15075 | 13875 |
| | Medium | 5200 | 121000 | 2520 | 4500 | 12580 | 620 | | |
| | Low | 2735 | 1955 | 955 | 9837 | 2400 | 14455 | | |
| W. Germany (forest & other wooded land) | High | 6395 | 293 | - | 4917 | 118 | 0 | 7207 | 6989 |
| | Medium | 202 | 1133 | - | 0 | 1898 | 0 | | |
| | Low | 610 | 5781 | - | 2290 | 5191 | 7207 | | |
| Netherlands (closed forest) | High | 120 | 270 | 92 | 5 | 40 | 0 | 300 | 300 |
| | Medium | 130 | 30 | 104 | 15 | - | 0 | | |
| | Low | 50 | 0 | 104 | 180 | 260 | 300 | | |
| Spain (forest & other wooded land) | High | 1800 | 100 | 6973 | 5433 | 60 | 750 | 12511 | 6906 |
| | Medium | 2900 | 1000 | 2693 | 3770 | 40 | 350 | | |
| | Low | 7811 | 11411 | 2845 | 3308 | 12411 | 11411 | | |
| Soviet Union (Forest & other wooded land) | High | 548400 | 18900 | 754600 | 176300 | - | 138700 | 1185900 | |
| | Medium | 442900 | 304900 | 0 | 73600 | - | 110600 | | |
| | Low | 194600 | 862100 | 431300 | 936600 | - | 936500 | | |
| United States (forest & other wooded land) | High | 141288 | 80898 | 56634 | 87409 | 17300 | 30583 | 298076 | 195256 |
| | Medium | 56933 | 103134 | 155000 | 44711 | 15030 | 143911 | | |
| | Low | 99855 | 114044 | 86442 | 165956 | 265746 | 123582 | | |

Source: Compiled from UNECE/FAO (1985).

with a recreation ranking of only 3 per cent. In some Mediterranean countries such as Spain and Turkey, larger areas of forest are of 'high' importance for protection than for timber production, but in most of these cases nature conservation has low importance.

As Table 13.18 suggests, various combinations and priorities of functions of objectives exist, and large areas of forest are in practice subject to multiple use. In many if not most cases, wood production was the original function, to which others have been added incrementally and chronologically. In state forests in New Zealand, for example, management for soil and water conservation was added in he mid-1950s, recreation in the early 1960s, nature conservation in the early 1970s, landscape conservation in the late 1970s and provision of educational opportunities in the early 1980s (Tilling, 1988). Non-consumptive objectives are now firmly incorporated into national policies, and indeed the concept of multiple use (at least for state for state forests) is explicitly adopted in countries such as the United States (under the Multiple Use-Sustained Yield Act of 1960) and New Zealand (under the Forest Amendment Act of 1976). Multiple use may also be encourage in various indirect ways in private forests, especially in European countries, through guidance and encouragement if no by strict enforcement.

While concepts of multiple or balanced use are widely espoused, they bring with them various problems. As Zivnuska (1961) observed, multiple use means multiple problems. If a forest is managed for purposes such as recreation, hunting or nature conservation, there is likely to be a cost in terms of wood production foregone. This cost may be expressed in a variety o ways ranging from straightforward damage resulting from recreation (for example through fires) or from game to the choice of species, techniques or rotation lengths that are sub-optimal from he standpoint of timber production. Some examples of such costs are illustrated in Table 13.19. These costs can be expressed in timber volumes or monetary values. In Sweden, for example, the present species composition in 100,000 hectares of broad-leaved forest is to be preserved, with a potential annual loss of around 0.5 million cubic metres of wood, while in Finland cutting restrictions will lead to a reduction of the total allowable cut by about 2 per cent. In France, restrictions on cutting and on choice of species are estimated to 'cost' FF 200 million: in Denmark to 100,000 cubic meters of wood worth around $US 4 million annually. In the United Kingdom it is claimed that recreation and conservation functions cause losses amounting to an estimated £1.8 million per year, as a result of sub-optimal rotation periods and species (UNECE/FAO, 1985). On the other hand, clear-felling causes recreational use to decrease (*e.g.* Kadell, 1985), and recreation itself has a value whether or not it produces revenue at the forest gate. In short, in multiple use a trade-off occurs between wood and other goods or services offered by the forest.

Some of these other goods or services may generate revenue directly-for example, forest campsites or hunting licenses–but most of them have values that are very difficult to quantify in monetary terms. Numerous attempts have been made to estimate the value of informal recreation, but that of species conservation is much more difficult to quantify, although perhaps no less real. One obvious value is in the role o the native forest as a source of species or provenances suitable for planting elsewhere.

One indirect indication of this value is the fact that 4 million hectares of eucalyptus plantation have been established outside Australia (Algar, 1981). The fact that benefits may accrue far beyond the immediate area of forest grea6tly adds to the problems of evaluation in monetary terms. Using a travel-cost method of evaluation and grossing-up from a small number of accessible forests. Villis and Benson (1989) have estimated that the recreation 'value' of state forests in Britain amounts to between £14 and £45 million per annum, compared with timber sales of around £60 million (these will rapidly increase as the young forests mature). Impressive as such figures may be, however, they are difficult to interpret. Since access to the forest or informal recreation is not normally charged, the values are notional or at best indirect. Furthermore, since no charge is usually made for recreational access to state forests, private forests owners are rarely in a position to charge for recreational access to their forests. Nevertheless, the relatively easily quantified 'losses' of wood production resulting from management for recreation or conservation need to be viewed against the 'gains' from these other functions.

**Table 13.19: Losses in wood production or increased costs due to recreation, protection or nature conservation.**

| | Species | Rotation Periods | Non-Optimal Techniques | Silvicultural Measures | Cutting Restrictions |
|---|---|---|---|---|---|
| Canada | | | | | + |
| Denmark | + | + | + | | + |
| Finland | | | | | + |
| France | + | + | + | | (+) |
| Hungary | + | + | | | |
| Ireland | | | | | + |
| Italy | | | | (+) | (+) |
| Netherlands | + | + | | | + |
| Poland | | + | + | + | + |
| Sweden | + | | | | + |
| Turkey | | | | | + |
| United Kingdom | + | + | | | |

+ indicates that losses occur: (+) losses occur in some areas only

*Source*: Compiled from UNECE/FAO (1985).

## Problems of Multiple-use Policies

The formal adoption of multiple-use policies for the management of state forests is a recognition that forests have values that extend beyond wood production, and is a reflection of growing pressures from environmentalists who perceive high value in the continued existence of native forests in particular. While an official endorsement of environmental and recreational values is widely welcomed, multiple-use policies are not without their problems. Since different units or currencies apply to the various uses, it is difficult to optimize total 'output', and the relative weights or priorities have still to be ascribed to the various functions. Most multiple-use policies are rather vague about priorities. In New Zealand, for example, clear guidelines about the roles and places of different uses in forest management were not given (Tilling, 1988), and in the United States, the Multiple Use-Sustained Yield Act evaded the issue of priorities entirely (Bonnicksen, 1982), Multiple use 'has all to often meant a little of everything everywhere' (Clawson, 1979).

This vagueness does not mean that the adoption of such policies has been without effect or meaning. On the contrary, it has provided an arena where competing single-use groups come together and a means whereby goals other than timber production can be promoted by interest groups or pressure groups. In the United States, for example, the clear-felling of forest stands (with its association with the destructive forest exploitation around the turn of the century) was strongly opposed by environmentalists during the 1960s and 1970s. This led to the passing of the Forest Management Act of 1976 which confirmed that all management decisions had to be in accord with multiple-use policy. Selective cutting was encouraged, and while clear-cutting was not prohibited it was made subject to guidelines (Cox *et al.*, 1985: see also Chapter 9). The very vagueness and lack of specificity of the concept of multiple use may have made such policies politically acceptable to governments and to conflicting parties and pressure groups such a timber producers and environmentalists, but at the cost of presenting almost insurmountable problems for the forest managers.

For example, since the adoption of 'balanced use' policies for state forests in New Zealand in 1976, the 'highest-attainable goal for managers (has been) a state of moderate dissatisfaction among all client groups hardly inspirational or motivating and virtually guaranteed to reinforce and latent tendency to fortress mentality (Kirkland, 1988). Under such conditions, priorities amongst objectives are likely to be ordered according to the effectiveness of client or pressure groups, and their relationship to the prevailing political philosophy and government of the day, rather than or more rational or popularly based criteria. It is therefore perhaps not altogether surprising that the decision was taken in New Zealand in 1985 to separate commercial and non-commercial forest management, with the Forest Service having responsibility for the former while most of the native forests were transferred to the Department of Conservation.

It remains to be seen whether this separation in New Zealand will set a trend and whether it marks the beginning of a reversal of the trend towards multiple use and non-consumptive functions that characterize the 'post industrial' forest. Paradoxically, it comes at a time when increasing regulation is being imposed on the management of private forests in much of the developed world, to the extent that the distinction between private and public forests is being blurred. This trend is more apparent in some countries than in others, and indeed in the case of the United States, for example, in some states than in others. In is manifested especially in terms of constraints of management practices, especially in relation to environmental protection, and in some cases also in public recreational use of private forests. In West Germany, for example public entrance to private forests is permitted under the Federal Forest Law of 1975. Forest owners must allow the forest authority to install trail-marking signs, and visitors have the right to pick flowers and berries (for non-commercial purposes) (Lundmark, 1986). In Britain, grants towards the costs of establishing private forests are now given only where certain environmental criteria are fulfilled. In short, government influence is exerted on the design and use of private forests, and is certainly not restricted solely to state owned forests. As the means of exerting such influences increase, the perceived need for continued state ownership may decrease. Under such circumstances, the privatization of publicly owned forests may accelerate.

If a blurring of the distinction between the management objectives of state and private forests is in prospect, perhaps a similar trend may become apparent between native and exotic or man-made forests. Whilst nature conservation values will remain highest in native forests, man-made forests may increasingly be designed, under the influence of a combination of government 'carrots and sticks', to provide for nature conservation as well as timber production. Similarly, man-made forests may be designed to cater for recreation, and for some types of recreation may do so very effectively, as for example in countries such as Britain and The Netherlands. Even in countries such as New Zealand where sizeable proportions of native forests remain, exotic forests accommodate a wide range of informal recreational activities, and indeed these activities are there practiced in private and state forests alike (Trotmann and Thomson, 1988).

# Chapter 14

# Forest Resource and Environment

Environmental issue are of fundamental importance in the of the forest resource at the present day. The fate of the tropical rain forest, for example, has emerged as a major issue on the global scale, and attracts the attention and concern of both citizens and governments far removed from tropical latitudes. Similarly, symptoms of forest decline in central Europe and parts of North America give rise to much concern, not least because of the symbolic significance of dying forests as indicators of environmental malaise. This significance is heightened by the image of the forest as a clean environment, yielding equable flows of pure water, protecting against soil erosion, and providing wildlife habitats and recreational setting. Such an image has for long been promoted by forest services and proponents of forestry: it is an image that suffers badly when the reality is seen to be marred by unhealthy trees, muddy stream and the loss or impoverishment of wildlife.

During much of the present century the forest has been perceived in very positive terms, and any attempt at afforestation or reforestation has usually been welcomed as an environment groups have begun to oppose the expansion of forest in countries such as Britain and Denmark, because of the perceived environmental effects of the type of afforestation and the way in which it is carried out. Both the use of the forest resource and changes in its extent attract widespread attention on environmental grounds, and environ-mental issues are now taking their place alongside economics and politics as major influences on forestry policies.

The relationship between the forest resource and the environment is of the utmost complexity, and some aspects of it are characterised by uncertainty. This short review can outline only a few of its main features. First, the nature of the effect of some general environmental trends on the forest resource will be considered. Second, the environmental effects of the use and management of the resource will be reviewed. Finally and very briefly, some of the environmental effects of afforetation will be outlined.

## Environmental Trends and Forests

While timber harvesting and other forms of resource use often have significant effects on the forest environment, the forest itself is affected by changes in climate or other element of the wider environment. As we near the end of the twentieth century, two issues in particular are attracting much attention. The first of these I the increasing content of carbon dioxide in the atmosphere, and the resulting increase in temperature through the 'greenhouse effect'. Forest growth and distribution may be affected by this warming effect, while tree growth-rates may be directly modified by changing concentrations of carbon dioxide. Second, acid precipitation and other forms of atmospheric pollution have been suspect-ed of being responsible (wholly or partly) for the symptoms of forest decline reported from the forests of part of Europe and North America. Both the issues have important implications for the future of the forest resource—the former on the long-term and the latter on the medium-to-short time-scale.

### Climatic Warming

The distribution and extent of forests are not static, as might be implied by maps such as Figure 14.1, but have undergone major changes in response to climatic fluctuations in recent (geological) times. During glacial periods, the forest belts have migrated equator wards, only to move pole wards again during inter-glacials. Such shifts may seem irrelevant on the human time-scale and in the context of resource use, but they are reminder of the dynamic nature of forest patterns and distribution. It is possible that current climatic trends may give rise to comparable changes in forest distributions over the next few decades and centuries.

### The Role of Deforestation

The much-publicised 'greenhouse effect' results from the role of carbon dioxide and other 'greenhouse' gases in absorbing long-wave terrestrial radiation, with consequent warming of the atmosphere. Carbon dioxide ($CO_2$) is only one of several 'greenhouse' gases, but it is the most important and accounts for at least half of the 'effect'. As is well known, concentrations of $co_2$ in the atmosphere have been rising steadily, and are likely to continue to do so in the foreseeable future. Most of the increase in $co_2$ comes from the combustion of fossil fuels, but substantial amounts also result from the clearing and burning of forests. Estimates of the relative importance of forest clearance and other biotic sources compared to fossil fuels vary greatly. For example, a 'deforestation' output of between 0.4 and 2.5 billion tons of carbon is reported by Houghton and Woodwell (1989), as compared with an output of approximately 5.6 billion tons form the combustion of fossil fuels. Using FAO/UNEP data on deforestation rate in 1980, Houghton *et al.* (1985) concluded that the net flux from non-fallow forests was between 0.7 and $1.4 \times 10^{15}$ g of carbon, compared with a

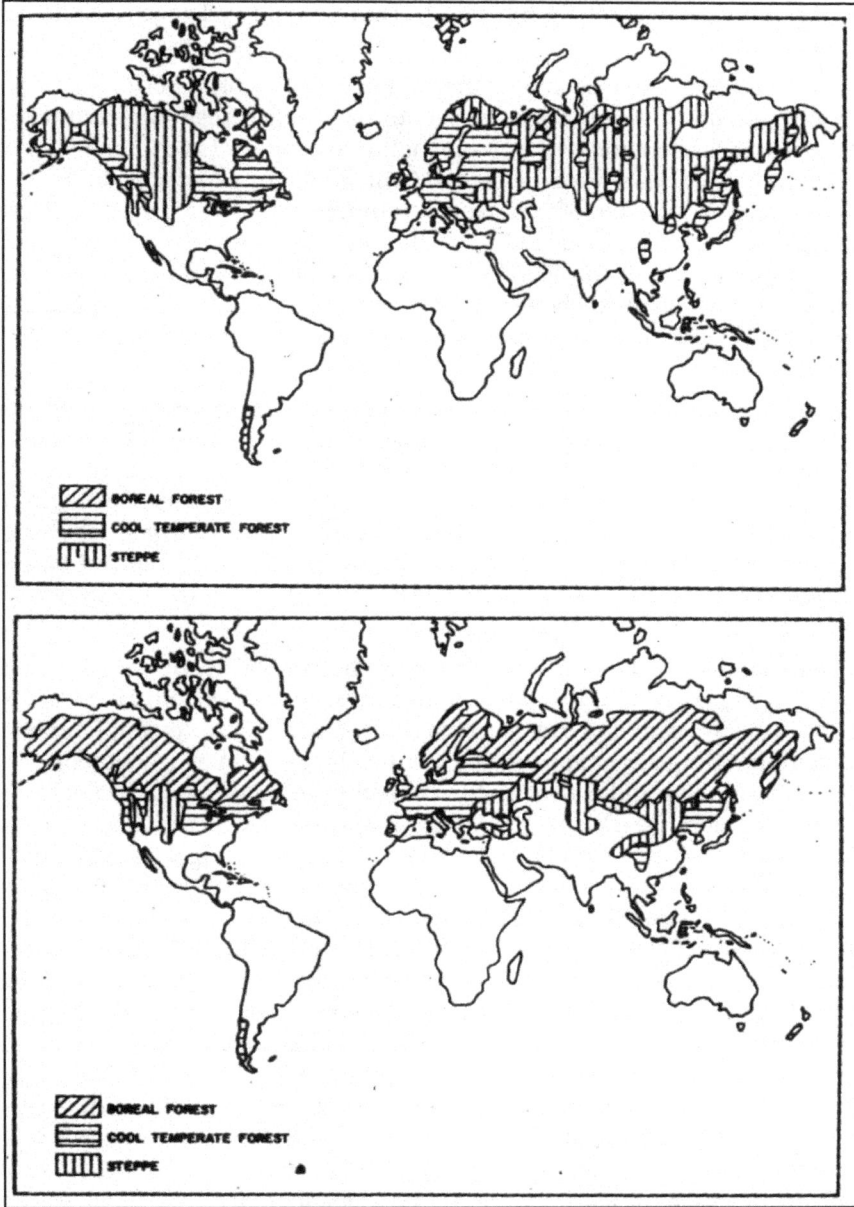

**Figure 14.1: (a) and (b) Ecological zones (Holdridge classification) (a) base case (b) elevated $CO_2$.**

*Source*: Modified after Emanuel *et al.* (1985).

release of $5.2 \times 10^{15}$ form the combustion of fossil fuels. Lower estimates of at least 0.4 but not more than $1.6 \times 10^{15}$ of carbon in 1980 are suggested by Detwiler and Half (1988). Of their estimates, between one-third and one-quarter comes from decreases in soil organic matter, while the burning and decay of cleared vegetation account for the remainder.

Some disagreement exists about the nature of the flux of carbon from the clearing of forest vegetation. Combustion is frequently incomplete, resulting in conversion of biomass into charcoal and soil organic matter and hence functioning as a carbon sink rather than carbon source (Seiler and Crutzen, 1980). More generally, the fate of wood cut from tropical and other forests determines how rapidly that organic matter is returned, to the atmosphere as $CO_2$. While fuelwood is quickly returned, industrial wood product are usually broken down much less rapidly. Paradoxically, even some deforestation may create carbon sinks, if it is followed by a vigorous regrowth with rapid uptake of $CO_2$, and if a proportion of the initial biomass remains on site (Lugo and Brown, 1980).

Considerable uncertainty therefore surrounds the contribution of forest removal to rising levels of $CO_2$, but deforestation may account for around 20 per cent of the total increase in $CO_2$ (Brunig, 1937). It is a minor, but nevertheless significant, contributor to the total output, and hence to the 'greenhouse effect'. Furthermore, the use of woodfuel is an additional source of $CO_2$ over and above that resulting from forest fires associated with forest clearing or arising from natural causes. Its contribution is estimated to amount to around one-tenth of the fossil fuels (Rotty, 1986).

Deforestation (and other biotic sources such as the cultivation of grassland soils and the drainage of wetland) and fossil fuel combustion have both increased rapidly over the last 150 years. For area such as south-east Asia, for example, it has been suggested that the current flux of $CO_2$ is much larger than at any time in the past, because of the growth and density of the population, reflected in increasingly extensive and intensive land use (Palm et al., 1986). Conversely, however, a change in forest trends can lead to a reversal of function in terms of carbon budgets. Temperate forests now function as $CO_2$ sinks, whereas in the past they have been sources, like the tropical forests of today (Lugo and Brown, 1980). The Southeast of the United States, for example, was a net source of carbon from around 1750 to 1950, as the forests were cleared for agriculture. Since 1950, it has become a sink (albeit a minor one) as reforestation has proceeded (Delcourt and Harris, 1980). This North American example may be repeated in other areas where the forest area has been increasing, and perhaps even for non-tropical forests as a whole. Afforestation has been suggested as a possible tool for moderation global warming (e.g. Sedjo, 1989). Furthermore, the recent case of the American power company which agreed to plant 500 square kilometers of forest in Guatemala gives rise to some hope for the future. This forest will absorb at least as much $CO_2$ as it new power station in Connecticut, using fossil fuels, will emit (Tyler, 1980). On the other hand, differences in biomass between natural and planted forests mean that reforestation can never completely offset the carbon loses to the atmosphere resulting from initial deforestation (Delcourt and Harris, 1980).

Whatever the detailed role of forest trends may be, the role of tropical forests is crucial. The rapid rates of carbon recycling in tropical rain forests in particular mean that their conservation, together with the upgrading of already degraded tropical habitats, is potentially the most effective demand-side means of dealing with the global $CO_2$ problem (Goreau and de Mello, 1988).

The greenhouse effect results not only from increasing concentrations of $CO_2$ but also from changes in concentration of other gases such as methane. The burning of tropical rain forests may indirectly lead to increase in such gases, as well as in $CO_2$. It has recently been suggested that the burning of tropical forests and savannahs generates at least as much carbon monoxide (CO) as does the burning of fossil fuels, and that this in turn leads to the accumulation of methane as well as ozone, which can be toxic to plants (Newell *et al.*, 1989).

## The Effect on Forest Patterns

Atmospheric $CO_2$ concentrations have increased by almost 25 per cent over the last hundred years. They are rising (at a rate of around 10 parts per million during the 1980s, against a level of around 350 ppm at the end of the decade) and are likely to continue to do so in the foreseeable future. By one hundred year from now, they may have risen to between 500 and 600 ppm. By then global average temperatures could be several degrees higher. Such increases in temperature are unlikely to be evenly distributed around the globe, but are likely to be relatively small in the tropics and much higher (of the order of several degrees) in high latitudes.

On the basis of temperature changes resulting from projected $CO_2$ concentrations, major changes in forest extent and distribution could result. One well-know model, based on a doubling of $CO_2$ and on Holdridge Life Zones (indicating the type of vegetation expected from climatic parameters) was developed by Emanuel *et al.* (1985). Its results are spectacular: forest zones would shrink from 58 to 47 per cent of the land surface, and there would be some expansion of the tropical forest zone and a dramatic contraction of the boreal coniferous zone (Table 14.1; Figure 14.1). Much of the present boreal zone would be replaced by cool temperate forest or steppe, and its poleward migration would be impeded by the relative paucity of land in polar latitudes.

**Table 14.1: Climatic change and possible changes in forest ecosystems.**

|  | ($10^6$ km²) | |
| --- | --- | --- |
|  | Base Case | Elevated $CO_2$ ($X^2$) |
| Tropical | 19.004 | 24.965 |
| Sub-tropical | 11.998 | 8.689 |
| Warm temperate | 15.951 | 15.540 |
| Cool temperate | 11.226 | 12.230 |
| Boreal | 17.375 | 0.835 |
|  | 75.554 | 62.259 |

*Source*: Emanuel *et al.* (1985).

The main value of such a model probably lies in its indication of the possible scale of impact rather than in the detailed pattern of forest changes. Possible changes in precipitation, which would be of major importance at the forest-steppe boundary, were not incorporated, nor were the effects of altitude or of soils. Furthermore, changes in mean annual temperature were employed, and it has been suggested that the seasonal distribution of temperatures would be more relevant (*e.g.* Harrington, 1973). More recent work has drawn attention to the roles of other factors such as oil moisture in determining the detailed pattern of response, and also factors such as the changing availability of soil nitrogen , which may be both a cause and an effect of changing vegetation (*e.g.*Pastor and Post, 1988). Even with these qualifications, however, the result of the work of Emanuel *et al.,* are starting, and there is agreement that large reductions in the extent of boreal forests, together with a poleward shift in their boundaries, is a distinct possibility (*e.g.* Shugart *et al.* (1986).

## The Effect on Forest Growth

In addition to these shifts in the forest belts, dramatic change in forest productivity could occur, especially in the boreal forest zone. In absolute terms, the greatest increase in growth would b expected in warmer, southern parts of the zone, but in relative terms the increases would increases northwards (Kauppi and Posen, 1985). Figure 14.2 shows an estimated increase in potential productivity resulting from the $CO_2$ levels expected to exist in the middle of next century, compared with present conditions (Kauppi, 1987). Production would more than double in some areas. Potentially significant increases are not confined to the boreal zone. In theory, sizeable increases might be expected over much of Australia, for example, but in practice other limiting factors, such as availability of water or nutnents, may come into play (Pittock, 1987).

In practice there would b a time-lag between climatic change and biological response, and the response indicated in Figure 14.2 would certainly not be immediate. Such a time–lag would also, of course, characterize distributional changes in forest area.

On the other hand there is some indication that changes in growth rate and in forest boundaries are already occurring. Under laboratory conditions, an increase in the $CO_2$ level I usually found to increase the rate of photosynthesis, and changes in growth rates may therefore result directly from this 'fertilization', as well as from climatic warming. In the western United States sub-alpine conifers have been found of have had increased growth since the middle of the nineteenth century (La Marche *et al.,* 1984).

This increase exceeds that which would be expected from climatic trends, but is consistent with global trends in $CO_2$. In high-level forests in the Cascade Mountains of Washington, net primary productivity at a number of widely separated sites has increased by up to 60 per cent during the present century, but it has been concluded that changes in summer temperatures, rather than direct $CO_2$ fertilization, are responsible (Graumlich *et al.,* 1989). Elsewhere in North America, increased tree-ring widths and/or advancing forest margins are reported from Alaska, sub-Arctic Canada and Quebee (*e.g.* Garfunkel and Brubaker (1980) and Payette *et al.* (1985). It has also been suggested that similar changes resulting from climatic warming (with or without $CO_2$ fertilization) have occurred in Europe. For example, in southern Finland an

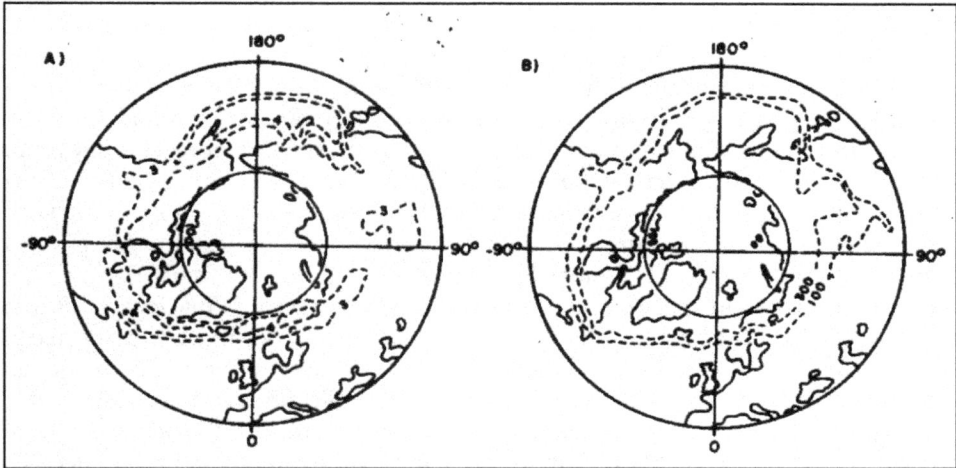

**Figure 14.2: A scenario of the possible forest effects of warming. Estimated increase in the potential productivity of boreal forests resulting from mid-twenty first century CO$_2$ levels (a) in m³/ha/hr and (b) in percentage terms. Note that there is likely to be a time lag between climatic change and forest response.**

*Source*: **Based on Kauppi (1987).**

increase in volume increment has been observed between successive inventories and its magnitude could not b explained in terms of climate conditions or improved silviculture. The tentative conclusion is that the increase is due to rising CO$_2$ levels and/or nitrogen deposition (Arovaara *et al.*, 1984).

While changes in growth rates and forest boundaries in high-altitude or tundra-margin forests may be of little significance in practical terms of tundra-margin forests may be of little significance in practical terms of resource use, those in areas such as Finland may be more important. Also the possible ramifications of increasing CO$_2$ levels extend well beyond growth rates and forest boundaries. Changing climate could mean changing frequencies of climatic events such as gales or severs frosts, and could thus have implications for the choice of species, as well as possible effects on growth and yield (Cannel *et al.*, 1989). In Canada, for example, fast-growing species, including some hardwood and weed species, may be favored, and a mismatch may develop between the present forest types and the climatic regions that they occupy (Pollard, 1985). Under such condition, pest outbreaks and fire damage may intensify, and in addition operational problem, for example in winter logging, may be encountered. In short, rising levels of CO$_2$ may well have both positive and negative effects: in the words of Hoffman (1984, p.166) '(it)' presents neither an unmitigated blessing nor a disaster for forestry'. Nevertheless, the changes in high-latitude forests such as those of Canada are expected to be considerable, and commentators such as Harrington (1987) have urged the responsible agencies to devise appropriate strategies to deal with them, especially since the lifetime of current plantings will extend well into the period of expected climatic change.

## Forest Decline

While growth rates in some forests may already be increasing as a result of changes $CO_2$ concentrations, other forests are showing sings of decline and even of death. One of the major environmental issues of the 1980s was the forest decline reported from much of Europe and from parts of North America. Forest decline is not a completely new phenomenon. Local-scale damage to forests form air pollution has occurred widely, both in the past and at present, and is associated especially with metal-smelting industries. Furthermore, periodic declines have been reported from a number of areas and for a number of species in the past. The forest decline that became so apparent in the 1980s, however, is of a different scale. While local pollution effects such as those around Sudbury, Ontario, extended to an area of nearly 20,000 hectares (Smith, 1985), the scale of the forest decline of the 1980s is regional or even continental, extending to millions of hectares. And while the pollutants responsible for local declines in the past have usually been readily identified, the causes of recent forest decline remain uncertain. Symptoms include discoloration and loss of needles and leaves, decried growth rates and abnormal forms of growth, and in some instances death of trees. The effects are in due course transmitted throughout the forest ecosystem. In the Netherlands, for example, a deteriorating quality of eggshells of some hole-nesting birds, attributed to acid precipitation and hence insufficiency of calcium, is reported (Drent and Woldendorp, 1989). In Hungary, a decline in oakwoods, believed to be related to soil acidification which in turn is related to air pollution, has been accompanied by a disappearance of mushrooms (Jakucs, 1988).

During the 1970s, signs of decline were observed in white/silver fir (*Abies alba*) and then in Norway spruce (*Piece abies*) in West Germany. By the early 1980s, the symptoms were observable also in pine, larch and a number of broad-leaved species. By 1983, up to 75 per cent of the area of fir in West Germany was affected, while over 40 per cent of pine and spruce, 26 per cent of beech and almost 15 per cent of oak were also identified as damaged in 1983 (CEC, 1987). In 1982, 8 per cent of West German forests showed signs of damage: by 1985 the proportion had increased to 55 per cent (WRI, 1986). Very rapid increases in the extent of damage were reported for some for some species between 1982 and 1984 in particular, as Table 14.2 indicates: in the Land of Baden Wurttemberg the percentage of healthy fir trees fell from 62 to under 5 between 1980 and 1982, whilst that of spruce damage increased from 6 per cent in 1981 to 94 per cent in 1983 (CEC, 1987). The extent of damage appeared to stabilize in the second half of the 1980s (Figure 14.3). While forest decline or Waldsterben is usually associated with West Germany, other countries in western, central and eastern Europe have also suffered serious damage. The extent and progress of this damage in one instance–part of Czecho-slovakia–is shown in Figure 14.4.

Various species in North America have also displayed symptoms of decline. These include sugar maple in the Northeast of the United States and the south-east of Canada, white pine and high-altitude red spruce in the eastern United States, and Ponderosa and Jeffrey pines in California (*e.g.* Chevone and Linzon, 1988). A sudden increase in dieback of sugar maple was observed in Quebec in 1982, and since then symptoms of defoliation, dieback and mortality have worsened and spread to other species. The heaviest mortality corresponds with localities suffering severe insect attack, seasonal climatic extremes, low soil nutrient status and highest fallouts of wet acid sulphate and nitrate (Henershot and Jones, 1989). Annual growth of red spruce in New England increased consistently from 1910-20 to 1960, and then fluctuated around a generally declining trend. By the early 1980s, it was 13 to 40 per cent below its peak levels (Hornbeck and Smith, 1985). Uncertainty exists as to whether the observed declines in growth rates are natural or anthropogenic (Zedakaer *et al.*, 1987). Defoliation by spruce budworm, climatic change, increasing maturity Of the forest and acid deposition have all been identified as possible explanations.

**Table 14.2: West Germany forest damage surveys (1982-A).**

| Tree Species | Per cent of Forests Damaged | | |
|---|---|---|---|
| | 1982 | 1983 | 1984 |
| Spruce | 9 | 41 | 51 |
| Pine | 5 | 44 | 59 |
| Fir | 60 | 75 | 87 |
| Beech | 4 | 26 | 50 |
| Oak | 4 | 15 | 43 |
| Others | 4 | 17 | 31 |
| Total | 8 | 34 | 50 |

*Source*: Based on WRI (1986).

Various hypotheses have also been advanced to explain forest decline in Europe. These include oil acidification, accelerated by the deposition of acid substances from precipitation or directly from the atmosphere, leading to aluminium toxicity and possibly magnesium deficiency; gaseous pollutants such as ozone, sulphur dioxide and ammonia; excess nitrogen from the atmospheric deposition of nitrogen compounds (especially from automobile engines); organic compounds; and general stress, perhaps from multiple pollutants, which expose the trees to increased risk of damage from drought or pests (see, for example, Hinrichsen, 1986). Different causes may be responsible in different areas: for example, local effects result from high levels of ammonium sulphate around intensive livestock units in the Netherlands and around industrial plants and power stations in countries such as Czechoslovakia and Poland (Pitelka and Raynal, 1989), but other, larger-scale factors must be responsible in areas such as the German and Swiss mountains. Ozone, produced by photochemical reactions involving emissions of nitrogen compounds and hydrocarbons from power stations and motor vehicles, is believed to be at least a contributory factor in West Germany (Ashmore *et al.*, 1985) and North America (Chevone and Linzon, 1988). Rigorous proof of cause has been established for ozone damage in eastern white pine and in southern Californian forests, while there is circumstantial evidence of causality in the sugar maple forests of New England and neighbouring areas and in low-altitude coniferous forests in the eastern United States from New England to Florida (Cowling, 1989).

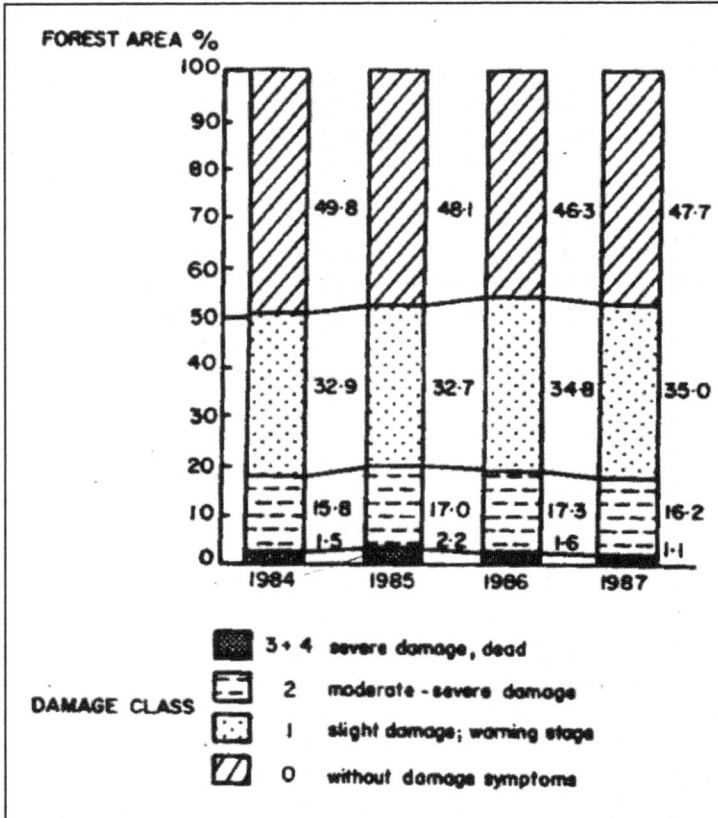

Figure 14.3: Forest decline in West Germany: results of forest-damage surveys 1983-87.

*Source*: Blanka *et al.*: reproduced from Nature, 336: 27-30, MacMillan Magazines Ltd. (1989).

While ozone is the only regionally dispersed pollutant that has been rigorously proven to cause detrimental effects on forests, others such as acid deposition (especially in rainfall and from cloud and fog) are strongly suspected of being damaging. In Canada, the opinions of surveyed scientific experts are that acid deposition has already significantly reduced forest productivity in the east, and may soon do so in west (Fraser, 1985). The estimated decline in productivity may be of the order of 5 per cent (Crocker and Forster, 1986).

Soil acidification is widespread, especially under conifer crops, and while it may be natural process, there is also evidence that it has accelerated. The strong and deep-reaching acidification which is widespread on all except limestone soils in Germany is traced back to acid deposition (Ulrich, 1986). In Sweden it has been established that the rate of acidification between the 1920s and was higher in the south than in the north, and the most likely explanations is the larger load of acid deposition in the south (Tamm and Hallbacken, 1988). In southern Sweden soil

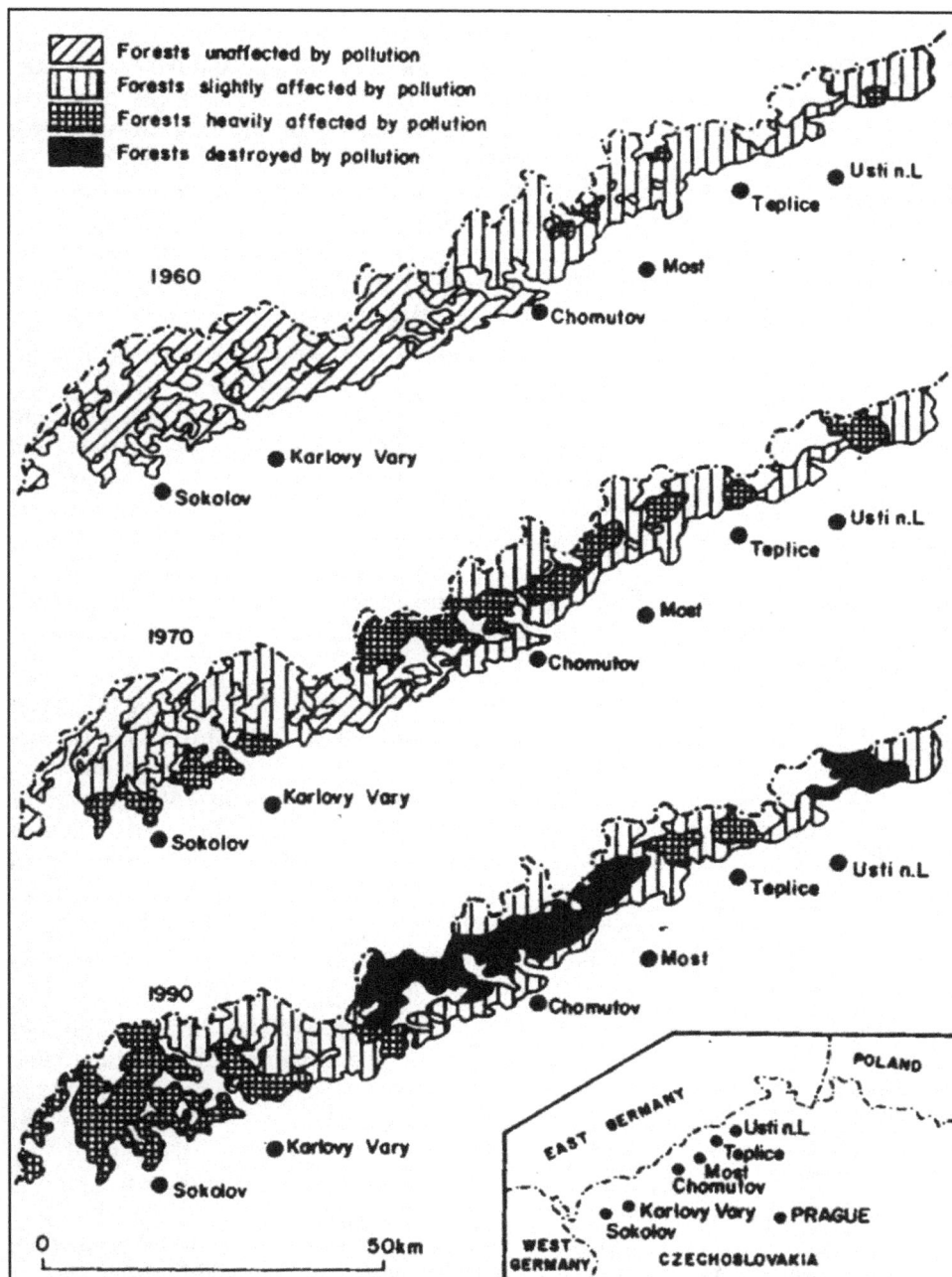

Figure 14.4: Pollution of forests in north-west Czechoslovakia (Predicted extent in 1990).

*Source*: Based on Carter (1985).

acidification of up to one $p^H$ unit has occurred (Anderson, 1986). On the other hand, however, the geographical pattern of symptoms of forest decline in Scandinavia is Scandinavia is not correlated with levels of gaseous pollutants (Pitelka and Raynal, 1989). Nor is there a close correlation between acid deposition and forest decline in West Germany (Rehfuess, 1985). Despite the synchronous appearance of symptoms of decline across different regions and species in Europe, it seems that no single cause is responsible. In the case of Norway spruce, for example, several different diseases and causes may be responsible, and the synchronicity of onset of symptoms is perhaps due to climatic tress. Climatic conditions are also considered to be the key factor underlying changing growth rates in the vosges Mountains of France, although regional air pollution may have had and exacerbating effect (Becker, 1989).

While it has been concluded that, at least in some areas such as parts of west Germany, 'it has been proven through complementary circumstantial evidence that air pollution is the dererminative factor causing the forest decline' (Schopfer and Hradetzky, 1984, p. 284), the occurrence and severity of the decline are not correlated with any specific pollutant. Probably no single casual factor is responsible (Klein and Perkins, 1987). In this state of uncertainty, the prognosis is obviously unclear. Climatic warming may itself impose stress on some species, and predispose them to attack by pets or pests or disease, and various other factors may also contribute. One may be the structure of the affected forest. Uniformly structured European plantation forests are compared by Mueller-Dombosis (1987) with similarly structured natural forests in New Zealand and Hawaii.

Extensive die-back occurs in the latter when they enter the senescing life-stage, even in the absence of air pollution. The underlying basis of the process in Europe may therefore be natural, and factors such as air pollution may simply accelerate it. Another factor may be land-use history, especially in relation to observed reductions in growth rates in pine stands in the south-eastern United States. Many of the stands planted between 1945 and 1956 were on abandoned farmland, while many of the more recent plantings have been on cut-over timber land, giving lower growth rates (Sheffield and Cost, 1987). This trend, combined with other social and institutional factors, may lie behind the recent downward turn of net annual growth of softwood in the region, following a long climb (Knight, 1987).

Whatever the causes of forest decline may be, the potential practical significance in terms of lost timber production is enormous. In North America, if it is assumed that the effect of acid deposition is to reduce forest productivity by 5 per cent, then the annual losses to the commercial timber industry would be around $197 million (1981 dollar) in Canada and $600 million in the United States (Crocker and Forster, 1986). Estimates of the annual costs of *Waldsterben* to the West German timber and related industries amount to hundreds of millions of dollars per annum and may eventually amount to a billion or more dollars (WRI, 1986). One study suggested that the total direct and indirect monetary cost, based on loss to the forest industry, reduced recreational value and additional costs arising from the increased need for protection against avalanches could amount to a staggering $200 billion between 1984 and 2060 (Blank *et al.*, 1988). In Bavaria, where the loss of annual increment was estimated at 17 per cent in 1984, the cost was believed to amount to over DM 200 million per annum (Kroth, 1985). Behind thee aggregate figures lie serious consequences for numerous owners. For example, it has been estimated that a model farm- forest enterprise in Bavaria, consisting of 80 hectares of forest and 7 hectares of grassland together with tourism and hunting, could witness a decrease in gross annual income from DM 114,400 under damage-free conditions to DM 48,400 under a scenario of constant-damage intensity and to DM 8,400 with a moderate increase of damage (Netsch, 1985). Any increase of damage would seriously jeopardize the viability of the farm forest. On a wider scale, fears have been expressed that a general collapse of the forest-products industries in West Germany could occur unless strict controls are imposed on air pollution (Jager, 1985). In addition to the obvious economic costs, indirect losses will be incurred n terms of reduction of intensity or quality of recreation in damaged forests. Further-more, protected forests will be affected as much as production forests in the regions in question.

Perhaps the biggest cost, however, is the alarm and concern generated in the minds of the population: if the supposedly pure forest environment displays symptoms of decline, what does this indicate about the health of the wider environment? The fact that the causes of the decline have not been fully identified makes it seem even more ominous and sinister. On the other hand the rapid increase in symptoms of decline observed during the first half of the 1980s has not continued through the second half of the decade (Figure 14.3). While damage has increased in some area, others have shown some signs of recover. And the predicted financial disruption to the timber market resulting from unplanned salvage felling has not materialized (Blank *et al.*, 1988).

External environmental changes are of major significance in relation to the use of the forest resource. On the one hand, climatic warming may lead to a reduction in the forest area and a shift in the forest belts. On the other hand it, in combination with $CO_2$ fertilization, may enhance growth rates and necessitate changes in cultivation practices. Little can be done to halt these trends, as the output of $CO_2$ is so huge and widespread, but uncertainty remains as to their strength and pattern. Their effect will probably be increasingly felt in the future, while the symptoms of forest decline are already all too apparent in parts of Europe and North America. While regional-scale air pollution is suspected as a major cause or predisposing factor, the details of the mechanisms involved are imperfectly understood. Whilst in theory emission levels of ozone and acid deposition could be reduced more easily than that of $CO_2$, in practice it will be many years before regional controls become effective. The outlook for the boreal coniferous and cool temperate forests, therefore, is uncertain.

## Forest Resource Use and the Environment

While the growth, extent and health of forests may be affected by air pollution and changing atmospheric composition, the use of the forest and changes in its area can also affect the wider environment. Some of the impacts arising from resource use are well known, but in other cases there is greater confusion or uncertainty. One of the main problems is that resource use can take various forms, ranging from complete removal of the forest to more modest and benign operations such as the harvesting of 'minor' forest products. Deforestation itself may take various forms, ranging from sudden and drastic clearance to a slow and gradual attenuation of the forest. Furthermore, the same operation carried out in different ways can have very different effects. For example 'recovery' from forest clearance in areas such as Amazonia may take only a few years if the clearing is carried out manually, but thousands of years if it is done mechanically.

## Hydrology

The forest has been widely perceived as supplying equable flows of pure water and linkages between forest cover and rainfall have been widely believed to exist. For example, in the United States in the nineteenth century it was thought that afforestatio on the Great Plains would lead to increased rainfall (*e.g.* Williams, 1989a). Myth and fact are difficult to separate in this area, and relationships that are well established in some areas cannot necessarily be extrapolated to other situations.

## Forest and Rainfall

On the local scale, it has usually been assumed that no definite connection exists between forest cover and rainfall, and allegations of declining rainfall following forest removal have rarely been supported by convincing evidence. More recently, however, some evidence has accumulated from places as different as India, Peninsular Malaysia, the Philippines, Ivory Coast and the Panama Canal area to support the hypothesis that deforestation may be followed by lower and/or less reliable rainfall (Myers, 1989), or at least a reduced availability of water. The actual mechanisms by which rainfall might be reduced are uncertain, although it has been suggested that changed surface roughness may be involved. One special case is that of high-altitude cloud forests and forests in coastal fog belts. Occult precipitation (from fog and cloud) may be of very significant amounts on the large surface areas of leaves, needles, stems and branches. In Hawaii, for example, occult precipitation in such forests contributed an extra 760 millimetres compared with 2,600 millimetres in a similar non-forest area (Hamilton, 1988). It follows that removal or attenuation of cloud forests will be followed by a reduction in rainfall.

On the larger scale, there is also some uncertainty about the relationship between rainfall and forest cover. It has been speculated that extensive deforestation in areas such as Amazonian forest plays an important role in recalculating water to the atmosphere: about half of the incoming precipitation is returned to the atmosphere as evaporation, and the rainfall intercepted by the forest canopy is a significant component of this evaporation (Shuttlewrth, 1988). A reduction in this recirculation could therefore lead to a reduction in rainfall. Salati and Vose (1984) conclude that continued large-scale deforestation could lead to reduced precipitation, which in turn could adversely affect climate and hence agriculture in south-central Brazil. Furthermore, it has been asserted that Henderson-seller and

> Total annual rainfall will decrease considerably when a certain percentage of Amazon forest has been destroyed and the seasonality of rainfall will become more pronounced. This will probably have a disastrous effect on the survival of pared forest areas which are intended as 'nature reserves' or the like (Soili, 1985 a and b).

Gornitz (1984) have attempted to simulate the climatic effects arising in changes in albedo resulting from tropical deforestation. They simulated the effects of removing almost 5 million square kilometers of forest (the current global rate continued for thirty-five to fifty years) from Amazonian, and replacing it with a grass/cover. This extreme scenario yielded a decrease in rainfall of only 0.5-0.7 millimetres per day (annual rainfall is 3,000 millimetres or more in much of the basin). Earlier work by Potter *et al.* (1975) in modeling the climatic effects resulting from albedo change following removal of the entire tropical rain forest indicated slight decreases in precipitation in tropical and high latitudes and slight increases in the subtropical zone. (They also suggested that the albedo effect would result in slight global cooling.)

While the general direction the changes in rainfall indicated by such modeling is the same as all alleged in much popular and environmentalist writing, the modest dimensions of change contrast with many of the more extreme prediction of withering droughts following tropical deforestation. Furthermore, the climatic effects of deforestation in a continental area such as Amazonia will not necessarily be replicated in other parts of the would with different situations and climatic regimes.

## Water Yield and River Flow

Similar uncertainty and disagreement characterize the role of the forest in relation to water yield and river flows. The forest and its soil have frequently been claimed to function like a sponge, absorbing rainfall and releasing it gradually in equable flows (*e.g.* Myers, 1988). The corollary is that the removal of the forest will result in decreased and less equable flows, and hence tend to lead to drought and floods. Both of these components of this belief are open to question, qualification or refinement.

Whatever it effects on rainfall may be, the removal of trees and forests usually reduces losses through evaporation and transpiration, and hence increases water yield. Could forests may represent a special and exceptional case, but there is widespread evidence to suggest that an increase in water yield is the normal response to a decrease in forest cover. For example the conversion of tropical forests in the high rainfall belt of Zambia to agricultural use had resulted in an increase in streamflow (Mumeka, 1986). On the basis of a review of ninety-four catchment experiments around the world, Bosch and Hewlett (1982) found that the increase in yield was greatest in areas of high rainfall, that it was proportional to the reduction in the forest canopy, and that for every 0 per cent reduction in forest cover there was an increase in water yield equivalent to about 40 millimeters of rainfall in the case of coniferous and eucalyptus forests and around 25 millimeters for deciduous types. In other words, deforestation is more likely to increase water yield than to decrease it. Conversely, afforestation s likely to reduce streamflow. In a study of ten large river basins in Alabama, Georgia and South Carolina, where the forest area increased from 10 to 28 per cent between 1919 and 1967, annual stream discharge were found to decrease by amounts varying between 4 and 21 per cent (Trimble and Weirich, 1987). A change in the type of forest, and hence in its function of rainfall interception and evapotranspiration, may also result in changes in streamflow. For example, ion the southern Appalachians, annual streamflow decreased by around 20 per cent following the conversion of a mature deciduous hardwood forest to white pine (Swank and Douglas, 1974). Young, rapidly growing eucalyptus plantations in the humid tropics have been found to consume more water than natural forests (FAO, 1986b), and decrease in water yield of as much as 28 per cent have been reported from cases in India (Mathur et al., 1976).

Some evidence exists to support the popular belief that forest cover helps to regulate river flows, especially in comparison with crops. In the Ivory Coast, for example, the dry-season flows of rivers are much higher from primary forest than from coffee plantation: the ratio of flow varies from approximately 2:1 during mid-season to 3-5:1 at the end of the dry season (Dosso et al., 1981). In was found that low flows of rivers from primary rain forest were roughly double those from rubber and oil-palm plantations (Daniel and Kulasingam, 1974). To this extent, then, the 'sponge' theory is supported, and has been invoked to explain the problems of water availability for the Panama Canal which first became evident in the late 1970s. Destruction of the forest by peasant cultivators was believed to be responsible (Simons, 1989). On the other hand, the relationship between deforestation and flooding is complex. There is a long-held belief that forests prevent or reduce floods: large areas of land were planted for this reason in the Tennessee Valley and elsewhere in the United States during the 1930s, for example. Rates of run-off may be much higher from cropland than from natural forest, but it dose not necessarily follow that allegations that major floods in rivers such as the Ganges and Brahmaputra are 'caused' by deforestation or the harvesting of fuelwood in the Himalayas (see, example, Guppy, 1984; WRI, 1985; Myers, 1983c; Jacobs, 1988). According to Myers (1983c, p. 66), for example, over the last thirty years 'forest cover in the upper catchment territories (of the Ganges) has been reduced by 40 per cent with the result that monsoonal flooding now causes appreciable damage'. Other commentators, including Hamilton (1987) and Hamilton and Pearce (1988), however, conclude that it is simply not defensible to attribute major floods in the rivers of the lower parts of the Indian subcontinent to fuelwood harvesting in the Himalayas. Indeed, Ives (1988, 1989) asserts that deforestation in the Himalayas has caused serious impacts on the plains and deltas of the Ganges and Brahmaputra must be challenged at all levels.

While deforestation may intensify downstream flooding by increasing the frequency of small floods, it is unlikely to have much effect in terms of major floods. The storage capacity of the forest 'sponge' is limited, and once exceeded, especially in high-rainfall, mountainous areas such as the Himalayas and New Zealand it will have little effect in preventing major floods or even in reducing their magnitude. Nevertheless, even if upstream deforestation is not responsible for disastrous floods in areas such as Bangladesh, there may be a signification effect in terms of the heights and frequency of small floods. For example, it has been suggested that agricultural colonization in Peruvian Amazonia has increased flooding downstream, to the detriment of human settlements and crops along the rivers. The height of the annual flood crest of the Amazon at Iquitos is reported to have increased markedly during the 1970s, when the population of Peruvian Amazonia doubled but rainfall patterns apparently did not change (Gentry and Lopez-Parodi, 1980). This claim, however, has been disputed by Nordin and Meade (1982) and by Sternberg (1987). The latter found no clear statistical evidence to support the allegations of an increasing height of annual flood, and even if such a change had occurred, it was not necessarily due to deforestation.

What it is appropriate that 'sponge' theory has been subjected to critical review in recent year and that allegations of a direct relationship between deforestation and major floods be treated with some scepticism, there is perhaps a danger of over-reaction. In theory grassland and scrub may provide as much storage and protection as forest, but in practice that protective role may be greatly reduced by overgrazing or burning (smite, 1987). Quite apart form any other benefits it may yield, the maintenance of the protective forest may well be justified by hydrological reasons, although the effectiveness of the protective function depends on the type of forest. Eucalyptus plantations, for example, have been found to regulate flow less well than natural forests (FAO, 1986b).

The macro-scale climatic and hydrological effects of extensive deforestation in areas such as Amazonian have attracted much popular attention, but other forms of impact also occur, as do impacts arising from smaller-scale operations carried out as normal parts of forest use and management. An example of the former is a rising water table following clear-felling of indigenous forest. With forest removal there is a reduction in evapo-transpiration, and in addition to the increases in river flow that have already been discussed, there may also be a rise in the water table. Resulting problems of salinity have been encountered in areas such as south-west Australia, where native sclerophyll forests have been felled for wood chips (Conacher, 1983).

Clear-felling of small area usually results in an increase in streamflow (*e.g.* Likens *et al.*, 1978). Following the clear-felling of two small catch-ment in central Sweden, run-off increased by 119 and 75 per cent respectively, but these results are more dramatic than those reported from other Scandinavian instigations (Rosen, 1984). Over larger river catch-ments, the nature of the effect may be both small and complex, depending on, amongst other factors, the size and location of the coupes. For example, in central Sweden the total effects on peak flow of a 10 per cent clearcut in a large basin is small compared to the effects of the effects of extreme weather conditions. The locational pattern of cuts, pattern of cuts, however, has a differential effect which is greatest in spring because its influence on the pattern of snowmelt (Brandt *et al.*, 1988).

On the scale of small clear fells or coupes, these hydrological effects are likely to be exceeded in significance by those of sediment transport, and numerous reports from around the world confirm that both forest removal and timber harvesting are likely to laid to increased rates of sedimentation.

## Soils

The relationship between forest vegetation and soils is close, and any change in forest vegetation–for example through clearance or selective felling–may be followed by changes in the physical or chemical characteristics of the soil. Harvesting and other mechanical operation may also lead to changes in soil condition, and theses changes may have off-site consequences–for example, in increased sediment loads downstream- or on-site consequences such as reduced growth rates in subsequent rotations or cycles.

## Forest Clearance, Soils and Water Budgets

When natural forest is cleared, the nutrient and water budgets of the soil are fundamentally altered. With the removal of trees and shrubs, evapo-tranpriration is reduced, and hence both water tables and water yields may rise. Circulation of nutrients from soil through trees and shrubs and back through plant litter to the soil is disrupted. The practical significance of these consequences varies with climate and ecological situation. In cool temperate regions such as the maritime fringes of north-west Europe, for example, increased waterlogging following the initial forest clearance thousands of years ago led to the development of peat bogs which have persisted ever since. In addition to the direct hydrological effect of reduced losses from evapotranspiration, the removal of deep-rooting trees meant that the soil surface was no longer enriched by nutrients brought up from the deeper layers of the soil, while increased leaching of the surface layers meant impoverishment of their nutrient status. In both physical and chemical terms, therefore, the soil deteriorated, and might become incapable of supporting forest growth, or at least growth of the kind typical of the original forest. Such changes are not confined to cool, maritime areas. In the Mediterranean zone, for example, a comparison of soils under cultivation and under oak forest in the south of Spain has shown that removal of the forests has resulted in a decrease in organic matter, in cation exchange capacity, and in available water. Deforested soils have undergone physical and biological degradation, as well as water erosion (Delgado-Calvo Flores *et al.*, 1985).

While such deterioration has undoubtedly occurred in some parts of countries such as Scotland and Ireland as well as in the Mediterranean world, the soils of most of the long-cleared forests of the temperate zone have been transformed by cultivation, drainage, and enrichment by fertilizers as they have been converted to agricultural use. In the tropics, however, the effects of forest clearance may be more dramatic and more serious. In some instances hard crusts of laterite may form on the soil surface, following the transformation of the micro-climate and drying out of the soil. Crusts of indurated material of around 2 metres in thickness are reported to have developed in area such as the Cameroons in less than a century (Goudie, 1981). These crusts are by no means inevitable consequences of forest clearance in the tropics, and their frequency of occurrence has perhaps sometime been exaggerated. Nevertheless, serious deterioration of the soil is a widespread consequence of forest removal.

In addition to the transforming of the micro-climate, clearance of the tropical forest disrupts nutrient cycling. A characteristic of the tropical rainforest is the fact that the biomass contains a very high proportion of the total nutrient content of the ecosystem. With the removal of the forest, much of the nutrient content will be lost. Despite the luxuriance of the forest, the nutrient status of soil may be modest, and down through the years numerous forests have been cleared in the hope that their apparently fertile soils could be equally productive under agriculture or plantation crops. Initially the nutrients contained in the ashes of the burned biomass may help to maintain the illusion, but all too often a marked deterioration in productivity has set in within a very few years. The 'export' of nutrients in products such as timber or beef may simply accentuate the trend, and represents a major change from the tight and closed nutrient cycles of the original forest. In one example from Amazonia, it is

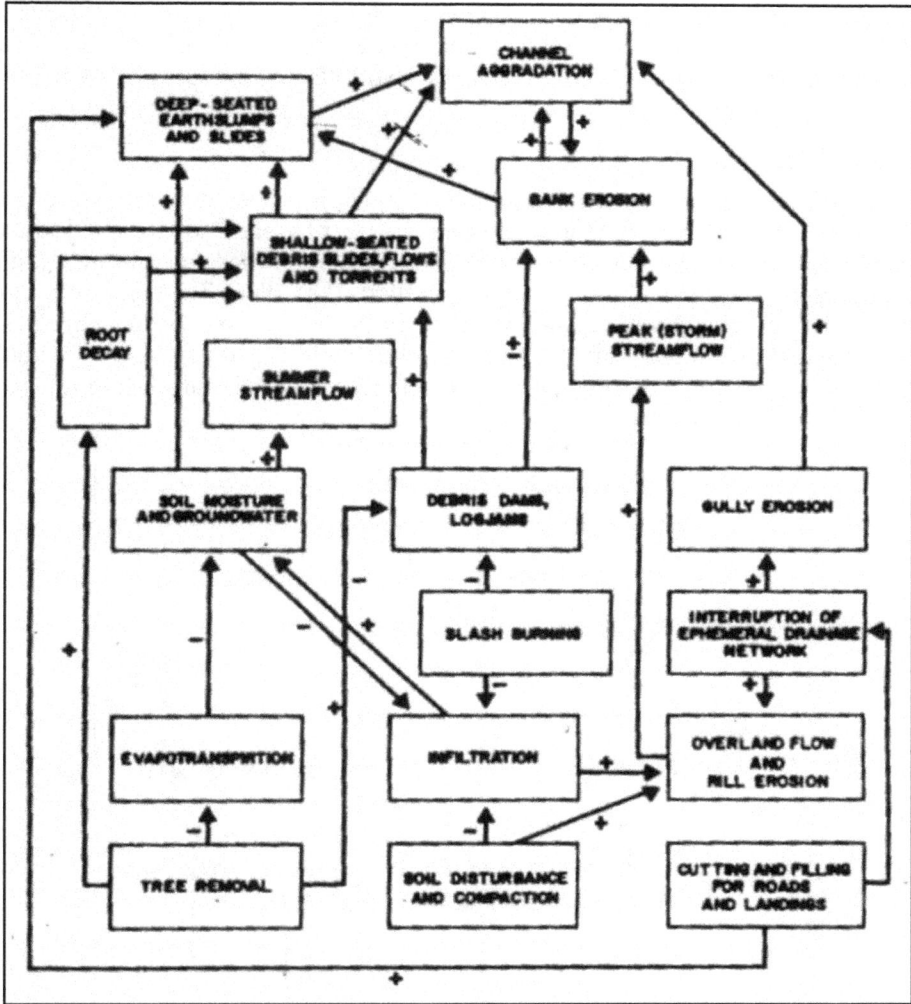

**Figure 14.5: The influence of timber harvest activities on hydrological and erosional processes in a watershed.**

*Source:* **Based on Coasts and Miller (1981).**

reported that the carrying capacity of pastures created on former forest land fell from 0.9-1 head per unit area to 0.3 head within six years (Siloli, 1985 a and b). Such land is then likely to be abandoned, and forest regeneration may begin. Where abandonment has been almost immediate there is little depletion of total nutrient tocks and a 25 per cent recovery of mature rain forest biomass may occur within eight years. The regrowth forest, however, will not necessarily have the same species composition as the original one.

The nature and significance of these effects of forest clearance depend on how the land is used after conversion as well as on the original ecological characteristics.

They may also depend on how the clearance is achieved (Figure 14.6). In Amazonia, bulldozing, herbicide application and chronic fire disturbance may effectively prevent forest recovery on about 10 per cent of the land cleared for pastures and then abandoned (Uhl *et al.*, 1988).

## Physical Effects

A body of evidence is building up to show that large-scale mechanized clearance is likely to have greater effects than small-scale clearance by 'slash and burn'. Clearance by tractor in Africa has been reported to result in a more than fivefold increase in erosion compared with traditional manual methods (Lal, 1986). Table 14.3 compares some of the physical characteristics resulting from different modes of clearance of virgin tropical forest. Soil bulk density, reflecting soil compaction, is considerably higher where the forest has been cleared mechanically, while infiltration rates are

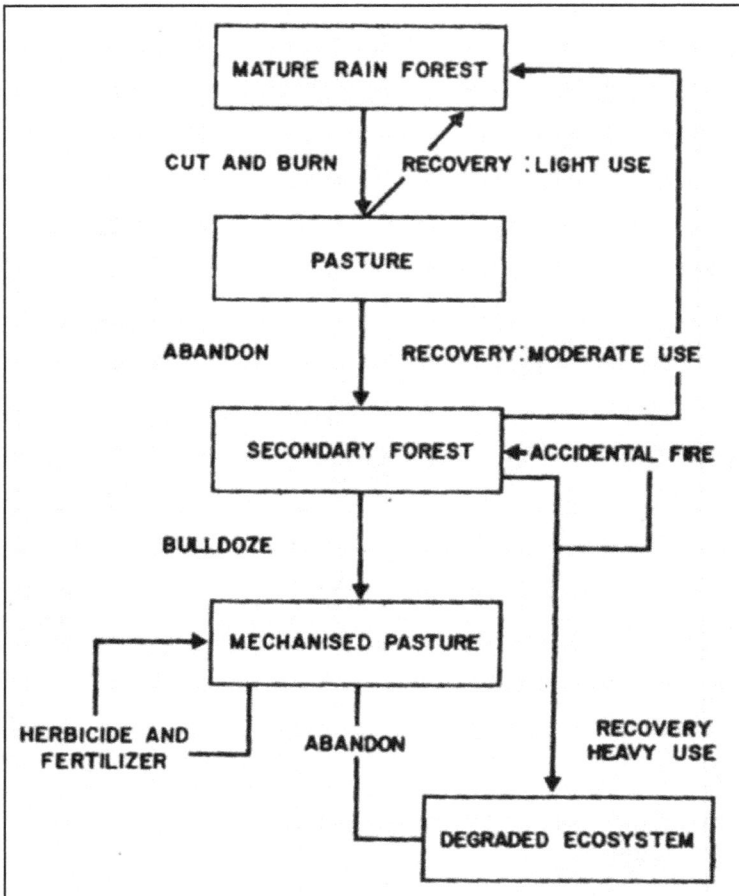

Figure 14.6: Suggested pathways of ecosystem degradation and recovery following deforestation.

*Source:* Based on Uhl *et al.* (1988).

much lower, indicating higher rates of run-off and higher probabilities of soil erosion (Nortcliff and Dias, 1988). At the same time, the protective canopy of the forest has been removed, exposing the soil surface to the kinetic energy of raindrops and hence to the initiation of soil erosion by impact splash.

**Table 14.3: Forest clearance and physical effect on soil.**

| Dry Bulk Densities Soil Depth (cm) | Virgin Forest | 'Slash and Burn' clearance | Bulldozer Clearance |
|---|---|---|---|
| 0-5 | 0.79 | 0.84 | 1.31 |
| 5-10 | 1.04 | 1.06 | 1.25 |
| 10-15 | 1.12 | 1.13 | 1.17 |
| 15-20 | 1.11 | 1.12 | 1.18 |

*Source*: Based on Nortcliff and Dias (1988).

The removal of the forest may therefore have a double effect in terms of soil erosion, and any modification of the forest canopy can lead to altered rates of soil loss. Under primary rain forests, sediment yields are usually very low not only because of canopy protection, but also because of the protective litter layer and dense root mat. If the forest is converted from its original form to one of high tree cover with bare ground underneath, then bare soil splash may increase by up to 6.6 times its value under the primary forest (Brandt, 1988). Thus while annual erosion rates under undisturbed natural tropical forest are usually very low, they may increase dramatically under plantations, as Table 14.4 indicate. Different types of forests mean different rates of erosion, and some types may have little beneficial effect in reducing erosion rates. For example, eucalypts are not effective means of erosion control under dry conditions, since the ground vegetation cover is suppressed by root competition (FAO, 1986b). In plantations of Eucalyptus globules in Portugal, 66 per cent of the ground surface was found to be exposed soil without vegetation or litter, compared with percentages of 1.9 and 27.4 respectively in plantations of pine (*Pinus pinaster*) and oak (*Quercus suber*) (Kardell *et al.* (1986).

**Table 14.4: Forest soil erosion: annual rates (tons/ha/year).**

| Undistrubed Natural Forest | Teak Plantation Widely Spaced Mixed Understorey | Teak Plantation Closely Spaced No Sndestorey |
|---|---|---|
| 0.2-10 | 2-10 | 20-160 |

*Source*: Based on Brunig *et al.* (1975).

Modification of forest vegetation, whether in the form of complete removal and replacement by pastures or crops or of conversion into plantations or managed forests, may therefore be accompanied by changes in soil characteristics. Processes of forest management, such as timber harvesting, may also have significant effect. Adverse effects resulting from logging operations have been reported from almost all types of exploited forests, from the tropics to the sub-Arctic.

## The Effects of Logging

Low-technology logging in the tropical forest, using human and animal power combined with river transport, may have little lasting environmental impact (*e.g.* white, 1978). In contrast, mechanised logging in the tropical forest causes extensive baring and compaction of the soil, even although only a few trees are extracted per hectare. The extraction of only eleven trees per hectare in East Kalimantan, for example, was found to leave around 30 per cent of the ground surface bare and compacted, with a reduced rate of water infiltration and hence an increased rate of run-off (Rochadi Abdulhadi *et al.*, 1981). In the same area, selective logging at a rate of twenty trees per cent of the logged area. On these tracks, water infiltration rates were found to be very much lower than in undisturbed primary forest (0.63 centimeters/minute compared with 4.62 centimetres/minute) (Kartawinata *et al.*, 1981). In the tropical forests of Ecuador, the haphazard development of skid rails, combined with skidding during wet weather, was found to result in very high levels of ground disturbance, with up to 75 per cent of the surface under trails (DeBonis, 1986).

The magnitudes of such effects depend to a large degree on the way in which the operation is carried out. For example, the avoidance of skidding during wet weather in the Ecuadorian example I advocated as an effective means of reducing damage. In the Queensland rain forest of northern Australia, a marked reduction in sediment yields occurred after a few simple conditions were imposed on the loggers (Gilmour, 1971). These included a prohibition on hauling through stream, and a requirement that the logger should construct and maintain drains along their roads. A major problem, however, is that of communication between the concession-holder, feller and skidder operator (*e.g.* Boxman *et al.*, 1987). General instructions from the former may be modified in course of transmission to the latter, and much unnecessary damage may be done by the skidder in searching for the logs to extract. In short, the environmental consequences of exposed soil and increased rates of erosion depend to a large extent on the way in which management operation are conducted, rather than on the type of use of the forest resource.

These effects of soil compaction and erosion are not confined to the tropical zone, but have been reported from numerous types of forests around the world. The complex physical consequences following tree removal in temperate latitudes are illustrated in Figure 14.5, while Table 14.5 illustrates the dimensions of increases of sediment production related to forest operations in one major forest area, the Pacific Northwest. Growing environmental awareness during the late 1960s and led to greatly increased attention to logging effects in this area in particular, and has subsequently spread to some other parts of the world, including countries such as New Zealand where forests in Steep County are being logged (*e.g.* Carson, 1983).

Extraction roads and stream crossing are especially susceptible to high rates of sediment yield. In a Quebec broad-leaved forest, for example, a two hundredfold rise in sediment concentrations was found following the skidding of logs across streams (Plamoudou, 1982), while the construction of extraction roads in an Idaho forest increased sediment production by a factor of around 750 compared with the 'natural' rate (Megahan and Kidd, 1972). It follows that the avoidance of stream skidding, the maintenance of buffer strips along streams and the design of roads can greatly reduce sediment yield in temperate forests as in the tropics. For example a 24 per cent reduction in sediment yield could have been achieved if conventional engineering methods had been employed in the construction of roads in a Californian forest (McCashion and Rice, 1981). In the Pacific Northwest of the United States, which has been one of the main settings for research on sediment yield related to forest operation, a paved road surface was found to yield less than 1 per cent of the amount of sediment yielded by a heavily used road with a gravel surface (Reid and Dunne, 1984). Intensity of use and road design have a major influence on sediment yields. The problem is one of implementation: methods of erosion control on forest roads are well known but frequently are not employed (Partic, 1980).

**Table 14.5: Sediment production from forests: representative rates.**

| Forested slopes | 4-10m³/km² road per year |
|---|---|
| Logged slopes | 16 m³/km² road per year |
| Roads during use | 10000-15000 m³/km² road per year |
| Roads, first year | 1000-2000 m³/km² road per year |
| Abandoned roads | 100-500 m³/km² road per year |

*Source*: Compiled from data in Roberts and Church (1986).

Although the effects may be greatest on roads and skid trails, they are not confined to such extraction routes but extend to some degree over much of the harvested area. While erosion rates, in the form of debris slide, were a hundred times higher on roads and landing than in undisturbed areas in Oregon forests, they were seven times higher in harvested areas (Amaranthus *et al.*, 1985). In New South Wales, a hundredfold decrease of water infiltrability was found on roaded areas, while a fivefold decrease was recorded on clear-felled areas (Riley, 1984). The more modest changes in logged areas, compared with roads and skid trails, are usually also more short-lived. In Oregon, overall values of infiltration capacity and erodibility measured there to six years after logging did not differ significantly from those on unlogged areas (Johnson and Beschta, 1980). Similarly, differences in sediment yield between harvested and control areas faded to in signification by the fourth year in a study in the Ouachita Mountains reported by Miller (1984). On the other hand extraction tracks on sandy soils under Pinus radiate forests in Australia were found to be still compacted fifty years after use (Greacen and Sands, 1980).

Therefore, while some physical effects of timber-harvesting on the soil (for example surface erosion) may be short-lived, others can be delayed or semi-permanent. The decay of roots after timber-cutting may lead to slope failure after a number of years, and slope stability may thus be adversely affected (Ziemer, 1981). In the case of Pinus radiate forests in New Zealand, for example, the most susceptible period ranges from about two years after harvesting. When the root systems are in an advanced state of decay, to about seven or eight tears when the root systems of the replacement crop begin to provide substantial reinforcement to the soil (O'Loughlin and Owens, 1987). The potential significance of this effect is obviously greatest on steplands, into which forest explotation may begin to move after the more accessible and easily worked lowland forests have been fully utilized. Slope failure in such situations may add to the problems of soil disturbance at the time of logging, and the net effect may be a sizeable reduction in soil depth. In a comparative study in virgin beech-podocarp forest and logged forest in South Island, New Zealand, Laffan (1979) found that soil depth were 20 centimetres or less in 33 per cent of observations in the former, but in 56 per cent in the latter.

Sediment yield from timber-harvestion depends on the techniques used as well as on the nature of the site. The method of logging is of primary importance. Soil disturbance, which in turn is correlated with sediment yield, is directly related to logging method, with ground-skidding systems usually giving rise to much higher percentage rates of ground disturbance than cable-based systems. In British Columbia, the extent of disturbance former was found in one study to be 40-45 per cent, compared with 22-30 per cent from the latter (Krage *et al.*, 1986). In the Pacific Northwest, Swanston and Dyrness (1873) found that with the use of tractors for extraction, 35 per cent of the ground surface was bare soil, while the figure fell to12 per cent when cable extraction was employed. Average values for percentage soil disturbance reported by Megahan and King (1985) for the United States and Canada range from 35 per cent for tractor systems to 23 per cent for ground cable systems, and to 9 and 5 per cent respectively for and aerial (balloon/helicopter) systems. Unit erosion rates may be as much as 3.7 times greater from tractor-logging than from cable-logging systems.

Different types of ground extraction machine, as well as different numbers of loaded passes, have been found to give rise to different severities of damage (Murphy, 1982). Even apparently minor details of the extraction method can have an influence on the extent of disturbance. Power requirements are much higher for whole-tree skidding than for the skidding of tree lengths, and also vary depending on whether the logs are skidded butt-end or small-end first (Wingate-Hill and Jakobsen, 1982).

Logging has various effects on the soil, including 'scalping' of the soil surface and shallow roots, the formation of rut, and beating and compaction (Rotaru, 1985). Its consequences, and hence its significance, may be manifested in various ways both in the locality in which it is undertaken and sometimes also downstream. The consequence of greatest direct concern to the forest manager is a reduction in tree growth rates. Both accelerated erosion and soil compaction lead to reduced growth rates. The height growth of Douglass fir on landslide scars in Oregon was found to be reduced by 62 per cent compared with that on similar sites, and in addition substantial proportions of the scars were incapable of supporting tree growth because of instability or impenetrability of the surface (Miles *et al.*, 1984). Stand volumes of the same species thirty-two years after tractor logging were found to average 34.1 cubic metres/hectare. On skid roads compared with 128.9 cubic metres/hectare on undisturbed areas, and the corresponding densities were 693 and 1,180 stems per hectare respectively (Wert and Thomas, 1981). Growth of Ponderosa pine on compacted soils along skid trails was found to be reduced by 6 per cent on moderately impacted areas and by 12 per cent where the impaction was heavy (Froehlich, 1979). More generally, Greacen and Sands (1980) conclude that compaction by harvesting machines may reduce current and future growth of tree, but that it is difficult to predict the extent of such reduction because of the complex interactions involved.

In addition to possible reductions in tree growth and timber production, disturbance from logging may have other adverse effects. If the forest has a primarily protective function any increased in sediment yield resulting from logging, forest conversion or other operations may be undesirable, and the functions of production and protection may be difficult to combine. Furthermore, increased sediment yield means increased turbidity of forest streams, and in addition to potential problems of downstream river pollution there may be effects on fisheries or marine life. Such effects have been reported from New Zealand by Johnston *et al.* (1981), and contamination of coastal fishing grounds as a result of logging- induced erosion is alleged to be a serious problem for the Philippines and many other tropical countries (Hodgson and Dixon, 1989). Coarse sediment may also be delivered to stream channels and remain there semi-permanently, giving persistently poor aquatic habitats (Roberts and Church, 1986). Again, however, the magnitude of the effect of sedimentation following logging can be greatly reduced by precautions such as the maintenance of buffer strips along the streams. For example, in New Zealand Graynoth (1979) found that suspended sediment concentrations were respectively 8, 188 and 27 milligrams/litre in a control, in an area logged with no protection, and in an area logged with riparian protection.

These off-site effects may focus attention on forest operations and in some instances lead to controls being imposed. In the United States, for example, federal water-pollution legislation is concerned with non-point sources of pollution (such as forest and farm land) as well as with point sources. In addition, some states such as California have introduced forest-practice act. In the case of California, a forest owner wishing to harvest timber must first lodge a timber-harvest plan prepared by a registered professional forester. Logging must be carried out in accordance with regulation and standard established by the State Forestry Board, and violations are subject to criminal penalties.

Both practical and theoretical problems, however, are encountered in attempting to achieve regulation and control. In addition to problems of policing, there are more basic difficulties of design of regulatory frameworks. One problem is the nature of cumulative effects in catchment: harvesting practices on individual sites may give rise to only limited effects off-site, but cumulatively the effects could be more serious. The duration of effects is also problematic. Short-term impacts (such as surface erosion) may be mitigated by dispersing timber harvest in space and time, but long-term effects are less easily alleviated by such approaches (Coats and Miller, 1981).

Although most research work on the effects of forest use and management on soil erosion and compaction has been concentrated in only a few parts of the world (such as the Pacific Northwest of the United States and to a lesser extent in New Zealand), there is convincing evidence that significant effects can arise both on-site and off-site. While it may be asserted that forest land can be managed so that there is little or no increase in soil erosion (Patric, 1976), it is clear that increases have occurred in numerous management regimes around the world. Some situation are more susceptible than others, depending on factor such as slope and climate, while management practices vary greatly in their damage potential. On-site effects may be of direct concern to the forest user, if reductions in tree growth rate result. Off-site effects are attracting increasing attention from the wider public and their representatives in central and local government, resulting in at least in the imposition of controls and regulatory frameworks relating to logging and other management practices.

While timber-harvesting has been the focus of concern about forest use and its effects on the soil, it should perhaps be borne in mind that other uses may also have similar effects. Recreational use may, through compaction resulting from trampling along trails and around campsites, also cause reductions in tree growth rates. For example growth reductions of 20-40 per cent are reported from around campsites in 60-110-year-old Scots pine stands in southern Finland (Nylund et al., 1980. finally, it should perhaps also be borne in mind that alternative land uses such as agriculture may have erosion rates as high or higher than those characteristic of the forest. Typical sediment yields of around 0.25 tons per acre per year from minimally disturbed forest land in the eastern and western parts of the United States (excluding the Pacific Northwest) are contrasted by Patric et al. (1984) with yields of 2-5 tons that are considered tolerable for agricultural land. The natural forest usually has very low sediment yields. These are likely to increase overall to managed forest, and to increase dramatically on conversion to agricultural land and following harvesting. The magnitude of the increase, however, can to a large degree be controlled by the means of conversion and the technique of harvesting.

## Forestry and Nutrient Budgets

Forestry practices may have important chemical effects on the soil as well the extraction of plant nutrients in timber harvest has been suspected of improving the soil under some management regimes, and hence of leading to declining growth rates in subsequent rotations. By its very nature, this long-term process is less obvious than accelerated sediment yield, and is characterized by a greater degree of uncertainty. This uncertainty stems partly from conflicting results from studies that have been undertaken, and partly from the relatively sparse research effort that has been devoted to the subject.

In Australia, second-rotation planted stands of *Pinus radiate* have been found to have lower productivities than first-rotation crops on the same sites (Keeves, 1966). Declines in productivity in second and subsequent rotation have been tentatively linked to losses of soil organic matter and of certain nutrients such as phosphorous (Blyth and McCallum, 1987). Declines in productivity in teak have also been observed in India after the first rotation, and teak productivity is also likely to decline in Venezuela, especially on highly productive sites (Hase and Foelster, 1983). On the other hand, no evidence of any decline in site productivity has been found after several rotations of pine, eucalyptus and water mono9cultures in South Africa (Schutz, 1986), while both better and poorer rates of growth have been reported from some forests, for example in Swaziland (Evans, 1982). A second-rotation decline in productivity in eucalyptus plantations has been reported from India, and conflicting reports exist from Portugal. One account suggests that the third successive coppice of *Eucalyptus globulus* yielded only 80 per cent as much wood as the second (wadsworth, 1983), while another states that a decline of this type has not been observed (Kardell *et al.*, 1986). Nevertheless, the upper mineral soil horizons under E.*globulus* plantations show a clear depletion of calcium at end of each rotation, and the soils show an increase in compaction and decrease in aggregate stability compared with those under native vegetation (Madeira, 1989). The cropping of eucalyptus on eucalyptus on short rotations has been found to lead to rapid depletion of nutrients in the soil, and other species grown in similar regimes will probably have a similar effect (FAO, 1986b). In Britain a parliamentary inquiry concluded that the effects of plantation forestry in respect of nutrient reserves were uncertain (House of Lords, 1979-80).

In managed and man-made forests, nutrient cycles are usually much more open than those in natural forests where the uptake of nutrients in the biomass is largely compensated by the rel of nutrients on decom-position. In forests from which timber is produced, the 'leakage' of nutrients in timber and sometimes from the burning of slash is much greater. The significance of this leakage depends partly on the frequency with which it occurs. With logging at intervals of 70-120 years, the clear felling of New England hardwoods, for example, is not likely to have a significant adverse effect on site-nutrient capital (Hornbeck *et al.*, 1986). The short rotations typical plantations may give rise to problems of nutrient reserves under some conditions (*e.g.* Conrnforth, 1970). Also with shortened rotations geared to 'biomass' or wood energy' plantations, the risks of nutrient depletion increase, and deficiencies may appear after repeated harvest (Anderson *et al.*, 1983). Short rotations mean not only frequent removal of nutrients in biomass, but also the loss of nutrients as a result of accelerated erosion during the harvesting-replanting part of the cycle (Raison and Crane, 1986). Nutrient loss through soil disturbance on harvesting may be a greater problem than nutrient depletion *per se* (Van Hook *et al.*, 1982). Dyck and Beets (1988) conclude that the displacement of nutrients by the piling of logging slash and the disturbing of topsoil can results in a substantial decline of *Pinus radiata* plantations in New Zealand, in addition to any effect arising from nutrient export in harvested material. Whole-tree harvesting also increases the magnitude and significance of nutrients loss, and raises the spectre of soil impoverishment on many forest sites (Kimmins, 1977). Nutrient exports may be increased by a factor of two or more (Van Hook *et al.*, 1982). Whole- tree harvesting usually removes one and a half to four times more nutrients than bole-only harvesting (Borman and Gordon, 1989). Work on nutrients budgets in Nova Socotian forests, for example, suggests that one or several whole-tree harvests of hardwood stands on rotation of around fifty years would be unlikely to result in important depletions of site nutrient capital overall, but calcium removals would be a cause of concern as they would amount to a large percentage of total site capital (Freedman *et al.*, 1986). And some problems of nutrient availability have been encountered even in less intensive management regimes, involving manipulation of the forest from its natural from to even-aged stands of one or more preferred species, rather than clearance followed by replanting. One case where problems of unavailability or inadequate cycling of nutrients has been experienced is in jarrah (*Eucalyptus marginata*) forests in Western Australia (Hingston *et al.*, 1981). The mode of management appears to have an important bearing on the nature of effects on soil chemistry: soil acidity has been found to decrease following clear-felling on northern coniferous soils, but the removal of slash was associated with decreased pH values compared with site where slash was not removed (Nykvist and Rosen, 1985). In more general terms, various techniques may mitigate the adverse effects of short-rotation energy forest. For example, practices such as fertilization, dormant-season harvesting, cable logging, harvesting on frozen ground or snowpack, and buffering stream channels are advocated by van Hook *et al.*, 1982.

Increasing concern about the sustainability of modern resource processes, combined with changes in forestry practice involving shortened rotations and whole-tree narvesting, has led to heightened concern about nutrient budgets. While nutrients lost through soil erosion or biomass extraction can be replaced by artificial fertilizers, this practice may be undesirable on economic or environmental grounds. Fertiliser application may not be financially viable, and even where it is there may be objections because of its ecological effects, especially on aquatic ecosystems. Usually less than half of applied Fertiliser is utilised by trees (van Hook *et al.*, 1982). Some may be washed or leached into the drainage water, especially when applied to ploughed surfaces at the time of planting, and hence lead to nutrient enrichment downstream.

## Ecological Effects

The use and management of the forest resource have a variety of ecological effects in addition to physical effects. These extend from changes in the species composition of the forest to less direct effects on wildlife, and occur in range of intensities varying from a complete loss of habitat (when the forest is removed or destroyed) to minor disturbance. In general terms, the removal of forest cover is likely to result in a loss of species of plats, animals and invertebrates, and similar effects may result where forest management is directed solely or primarily at timber production. Much of the popular concern about the nature of forest resource use stems from these ecological effects and especially from the consequences for wildlife in general and for bird life in particular.

## Forest Management and Wildlife

In temperate latitudes the effects result mainly from the intensification of management and the drive for increased wood production, rather than from the complete destruction of the habitat. Most of the forests are managed primarily for timber production, and to that end the species composition has habitats modified by drainage or by treatments with fertilizers or chemicals. Such management has significant effects on the forest ecology. In southern Sweden, for example, the managed forests consisting mainly of spruce and pine contrast with the mixed natural forest of deciduous and coniferous species. Comparing natural and managed areas of similar age, Nilsson (1979) found that bird densities were three times higher in the natural forests, and that the number of species represented was also much higher than in the managed forests. Bird densities in young spruce plantations were only one-ninth of those in the natural forest. The width of this ratio is probably at least partly a function of the age of the plantation as well as of its design. In France Blondel (1976) found that most of the woodland bird communities around Mont Veloux were as rich and well balanced in the man-made forests as in the natural ones, but confirmed that the diversity of the forest structure and it richness in bird life were directly related.

The intensification of forest management is Sweden has included not just alterations in the structure of natural forests, but also the use of drainage, fertilizers and exotic tree species. Many species have been threatened by this drive for increased timber production. It is believed that forty species of vertebrates are now seriously endangered, while fifty species of fungi, lichens and flowering plants are on the verge of extinction and a further 220 are in some danger (Gamlin, 1988).

The expansion of plantations of exotic tree species in countries such as Australia and New Zealand, and in particular the replacement of native forests by exotic plantations has also generated much controversy. Although the coniferous plantations in Australia are not the 'wildlife deserts' alleged by some opponents, they lack some of the species of plants, birds and mammals found in native forests, and their proportion of introduced species tends to be higher (Friend, 1980;1982). On the other hand, exotic afforestation on degraded grasslands and other habitats may lead to an increase in species diversity (Zobel *et al.*, 1987). Similarly, short-ration energy forestry involving, for example, willow coppice, can lead to an increase in habitat diversity in regions dominated by coniferous forests or in intensive agricultural areas (Gustafsson, 1987). The nature of the ecological effect will be determined by the management system, and the use of herbicides.

The design and structure of the forest are important variable, with species numbers and population densities barying not only age of stand but also with location in relation to native forest. Declining wood production from forests in the European Community in recent years may mean that they become of increasing ecological significance, as most mixed and deciduous forests become ecologically richer as they age (CEC, 1987). Richness and density tend to be highest alongside native forests, and in inliers of native forest, in New Zealand, long-ration pine forests are claimed to provide useful habitats for some but not all native animals, and again faunal diversity has been found to increase with age. Nevertheless, it rarely if ever becomes as high as in unmodified native forests on similar sites (Bull, 1981).

## Tropical Deforestation and Extinctions

Similar trends to those in temperate latitude are now in evidence in much of the tropical zone, and in addition there I widespread loss of forest habitat as the tropical forest area shrinks. The nature and magnitude of the resulting ecological effects depends on the initial forest environment, the way in which it is used and managed, and the purposes for which it is cleared.

In Kenya, for example, the clearance of indigenous forest to make way for plantations of exotic conifers has severely impoverished the native bird fauna in terms of both and densities. On the other hand, the new pine plantations may extend the wintering range for some Paleactic migrant species of birds (Carlson, 1986). The clearance of the native forest for purpose other than plantations is likely to have a similar or greater effect on bird and animal species. In Puerto Rico this clearance has proceeded further than in most low-latitude area. During the development of human settlement from the early sixteenth century, the island was deforested at an increasing rate, until less than 1 per remained by 1990 (although a further 9 per cent consisted of coffee plantations shaded by trees remaining from the native forest). At the same time as deforestation was proceeding, the island's inhabitants selectively persecuted some of the larger bird species. The avian extinction rate resulting from these factors was at least 11.6 per cent: deforestation was an important ultimate factor behind this loss of species, although humans were the proximate cause (Brash, 1987).

The possibility of a similar or greater extinction, on the scale of the tropical world as a whole rather than that a small island, lies behind much of the current concern about the fate of the tropical moist forest. The ecological richness of these forests is well knows: they contain 40 to 50 per cent of all species but occupy only 7 per cent of the land surface (Myers, 1980b). It is asserted that one or more species become extinct every day as a result of the clearance, conversion, or disturbance of tropical forests (Myersm, 1983b). In the projection prepared for the *Global 2000* report, Dr. Thomas Lovejoy of the World Wildlife Fund (as it then was) estimated that between 15 and 20 per cent of species would become extinct as a result of deforestation by the end of the century (Barney, 1980). Assuming that the tropical rain forest is being cleared at a rate of 1 per cent per year, Wilson (1989) estimates that 0.2-0.3 per cent of all species in the forest are lost each year. On the basis of a further estimate that the forest contain a total of around 2 million species, he concludes that this rate corresponds to a loss in absolute terms of 4,000-6,000 species per year. During the last quarter of the century, accordingly to Myers (1979), an 'extinction spasm' accounting for one million species will be witnessed. This is roughly equivalent to one species per day.

Both the reliability of such estimates and their potential significance are difficult to evaluate. Much uncertainty surrounds the total number of species world-wide, which may amount to anything between 5 and 30 million (WRI, 1986). As yet fewer than 2 million have been identified, although the identification rate for mammals, birds, fishes and plants is far higher than that for invertebrates. Confusion is sometimes engendered because some estimates of numbers and percentage relate only to plants, birds and animals, while others refer to all species. Some focus on deforestation alone, whilst others include other forms of human impact. The confusion and flimsiness of basis for the estimates of Myers and Lovejoy have been attacked by commentators such s Simon and Widalsky (1984) and Simon (1986). They regard them as little more than exaggerated guesses, and also question their perception of the significance of extinctions. Such perceived significance depends on value judgments of the intrinsic worth of a species, as well as any material benefits that may flow from it, and it is perhaps not surprising that disagreement exists.

Some firmer evidence exists, however, for current rates of extinction. On the basis of a recent study of extinctions of birds and flowering plants in the neo-tropical moist forests, the continuation of current trends would mean a loss of 12 per cent of the bird species and 15 per cent of the plant species, 2000 (Simberloff, 1986). These rates would correspond to losses of around 15,000 plant species and 100 bird species, but numerically would not be comparable to the major mass extinctions of the geological past. If deforestation were to continue unabated through the next century, until the tropical forests of the New World were reduced to those currently projected as parks and reserves, extinction rates would reach 69 and 66 per cent respectively for birds and plants. Such rates would be similar to those experienced in episodes of mass extinctions in the geological past.

## Ecological Effects of Logging

While the prospects for such mass extinctions remain controversial, evidence has begun to accumulate in recent years about some of the immediate effects of the exploitation of the tropical forests, although much remains unknowns. In Peninsular Malaysia, for example, the forest cover contracted from 84 per cent of the land area in 1958 to 51 per cent in 1975. This was accompanied by reductions of between 23 and 56 per cent in populations of six primae species (Aiken and Leigh, 1985). The same authors concluded that selective logging did not necessarily result in a very depauperate fauna, and indeed some large mammals were found to be more common in logged over areas than in undisturbed forest. Much depends on the availability of undisturbed refuges to which animals may move during logging, and from which they can subsequently return. Many species with large territories move away while logging is in progress and return afterwards, but others do not adapt well to logged over forests. After logging, there is a major change in the species mix, and then a gradual return. There is disagreement, however, about the rate of this return and the degree to which the logged over forest biota can approach those of the virgin forest (*e.g.* Shelton, 1985). On the basis of work in East Kalimantan, Wilson and Johns (1982) concluded that logged forest could be successfully recolonised providing that adjacent areas of undisturbed forest remain to provide a population pool for re-colonisation, and that hunting levels are low. Species diversity was found to be similar in undisturbed forest and in forest selectively logged there to five years previously, although densities were lower in the latter. Selective logging may mean a drastic reduction in the over all availability of food sources because of the high levels of damage. The extraction of 3.3 per cent of the trees from a Malaysian forest, for example, meant that a total of 50.9 pr cent were destroyed (Johns, 1988). In Amazonia, the monkey *Chiropotes satanas* was found to be eradicated by the removal of two or three trees per ha and resulting damage to 50-60 per cent of the stand (Johns, 1989). In Malaysia, the extraction of 11 trees per ha was found to cause damage to 40 per cent of the residual trees, and to result in a decrease in both tree species and densities (Abdullah *et al.*, 1981).

Selective logging of this kind may have a significant effect on the usefulness of the resource, as well as consequences for wildlife. Its main impact is argued by Esteve (1983) to be on economic value, rather than in biological or ecological terms. Future timber producing value is reduced, even if the forest extent is not reduced.

In theory, selective logging on a rotational basis ins an environmentally benign form natural disturbance, and natural regeneration may be rapid. Furthermore, only the logs are removed. These usually have low nutrient contents, and the more nutrient-rich leaves are left behind. In practice, however, the way in which selective logging is carried out can mean that the resource deteriorates. Poorly controlled logging has meant that some species that are useful sources of timber or 'minor' products such as rattans, gum and latex have effectively disappeared over large areas in the tropical uplands, and in regions such as the Amazon (Siebert, 1987; Fox, 1976). The same process that affected North American forests in the nineteenth century is therefore in operation in many areas of tropical forest today. In addition, the selectively logged forest may become more prone to fire. While the undisturbed rain forest is usually too wet to burn, logged over areas may burn much more readily. Human disturbance dramatically increase the risk of fire. Combustible material in the form of tree debris is left on the ground, and the forest floor becomes drier as the canopy is opened out (Uhl *et al.*, 1989). In Amazonia, firs set to control weeds on degraded pastures commonly spread onto logged forest, but not to undisturbed forest (Uhl and Buschbacher, 1985). Thus the effects of selective logging and of pasture burning interact to produce more detrimental effects than either operation acting separately.

While selective logging is closely associated with the tropical forest today, it was carried out in huge areas of the temperate forest in the past, especially in the nineteenth century. Even in areas where the forest was not completely removed, the long term ecological effects may be considerable, and the net effect may be a reduction in the usefulness of the forest as a resource for timber and other products. Some of these effects in the Great Lakes forest of Michigan have been recorded by Whitney (1987). Prior to settlement, the Great Lakes (pine) forest was conditioned to fire at intervals of 130-260 years. Selective logging of white pine, and later of hemlock and hardwoods transformed the forest (Table 14.6). Waves of fires, following logging in quick succession, upset the equilibrium of the forest. The result was a poorly stocked forest of oak and aspen, which had previously played a subordinate role in the pre-settlement forest. A reduction in the frequency of fires after control measures were imposed from the 1920s allowed the oak and aspen to mature, and set stage for a new pulp-oriented industrial forest in the 1950s. More generally the very extensive removal or modification of the Midwest forest has apparently resulted in few tree species being eliminated completely, but the relative abundance of the more common species has shifted significantly (Whitney and Somerlot, 1985). A reduction is species diversity or general ecological impoverishment is a frequent result of the use of the forest. One recent example is reported from southern China (Young and Wang, 1989). Here the original broad leaf evergreen forest has been replaced by a secondary forest of pine, which is characterized by a lower bio-diversity and density.

Ecological changes of this type can have significant 'knock-on' effects. In the forests of eastern Canada, for example, until recently only the best timber was taken, leaving impoverished stands of less desirable species. Balsam fir, which readily colonises disturbed sites, is often a major component of the modified forest. With protection against fire provided by forest management, stands develop to maturity and become susceptible to attack by outbreaks of budworm. Whilst in theory it would be possible to harvest the most susceptible stands before or during an attack, the resulting accelerated harvest rate is feasible only if the market can absorb a sudden influx of timber. One alternative to this strategy is to use insecticide sprays as a control measure (Dunster, 1987) and these may in turn have their own 'knock-on' effects on the forest ecology, as well as giving rise to adverse popular perceptions of forest management. In short, any modification of the forest ecosystem such as the harvesting of the most valuable timber species may have far reaching and unforeseen effects, which may extend both to the nature of the forest resource itself and to the wider environment.

**Table 14.6: Pre and post-settlement extent of various forest types in Crawford County, Michigan.**

| | Percentages of Land Area (i.e. of 145630 ha) | | |
|---|---|---|---|
| | 1836-59 | 1927 | 1979 |
| Hemlock-white pine-northern hardwoods* | 16.4 | 3.7 | 4.5 |
| Aspen-birch | 0 | 12.0 | 18.9 |
| Pin cherry | 0 | 2.2 | 0 |
| Mixed pine | 27.4 | 0.3 | 7.8 |
| Pine-oak | 22.3 | 0 | 0 |
| Oak | 0 | 24.2 | 19.5 |
| Jack pine | 24.4 | 23.7 | 28.2 |
| Jack pine openings | 0.5 | - | - |
| Lowland conifers and alder swamps | 8.2 | 6.2 | 6.1 |
| Grass land and upland shrubs | <1.0 | 20.6 | 8.6 |

*: Predominantly sugar maple after 1920.

+: Probably included under poorly stocked jack pine forest after 1920.

*Source:* Whitney (1987).

# The Environmental Effects of Afforestation

Although the forest is widely perceived as environmentally benign and as a provider of protective services, paradoxically the creation of forests it increasingly seen in some quarters as undesirable for environmental reasons. This paradox may perhaps be partly explained by a negative reaction to dramatic landscape change of the kind represented by aforestation, and partly by the way in which afforestation is carried out. The use of exotic species, such as, for example, *Sitka spruce* in Britain, *Radiata pine* in New Zealand and eucalypts in countries as diverse as Portugal and India, attracts particular attention and opposition. Deeper seated reasons, however, also exist for the antipathy than exist in some quarters to continuing afforestation in countries such as Britain, Denmark and Ire land , ands which to some extent matches the opposition to the conversion of native forests to exotic plantations in Australia and New Zealand. This antipathy is based on various grounds, including social issues such as employment as well as on environmental concerns. These concerns, however, are often expressed vociferously and many of them have been shown by recent research to have a factual basis.

## Physical Effects

Ground preparation, including ploughing and ditching, in advance of planting leads to major increases in sediment yield. In Scotland, the analysis of lake sediments has shown an increase in sedimentation rates as a result of afforestation of some catchments (Battarbee *et al.*, 1985). Sediment loads in streams may increase by orders of magnitude (*e.g.* Robinson and Blyth, 1982), and although this scale of effect is usually short lived the effect itself is long lasting. In one study area in the north of England, for example, suspended sediment concentrations were found to stabilize at double the pre-planting level (Blackie and Newson, 1986).

Increase in sediment loads may do damage to aquatic life and to anglings, as well as posing problems of sedimentation in reservoirs and creating difficulties in filtration systems in public water supplies. Sediment loads can be greatly reduced by terminating furrows and fitches back from stream courses, and codes of practice have been introduced in Britain in order to achieve this objective. Field supervision however, does not always match the quality of design of the operation (in this sphere any more than in, for example the logging of tropical forests). The combination of this effect with the reduction in water yield which usually follows afforestation may mean that the water industry will oppose afforestation schemes. The effects-and opposition-are likely to be greatest in small catchments undergoing complete or extensive afforestation, but they may also extend to larger scales. For example in the 906 square kilometer Tarawere catchment in central North Island, New Zealand, 250 square kilometers were afforested between 1964 and 1981, and the mean river flow was reduced by 13 per cent as a result (Dons, 1986).

Water chemistry may be affected as well as water quantity. Streams draining afforested catchments in Scotland have been found to be more acid than those in open moorland, and to have lower fish stock (Harriman and Morrison, 1981). This effect may be partly due to the direct effect of the coniferous afforestation, and partly to the effect of the coniferous trees in filtering dry acid deposits from the atmosphere. These pollutants are then washed from the trees by subsequent rainfall, and the drainage system provided by the pre-planting ditching means that the water runs off before it can be buffered by the deeper and less acid horizons of the soil. Fertilizers may be applied at the time of planting, and large proportions of the application, especially where falling on furrows and bare ground, find their way into the streams during the first year, and losses continue for up to three years. Phosphate concentrations, for example, may be ten times higher than the original levels for several months after treatment (Binns, 1986). In some instances this effect may be perceived as beneficial in increasing stream productivity, but in other cases leaching into ponds or lakes could have ecologically undesirable effects (Harriman, 1978), especially if nature conservation sites were affected. Again, the method of application influences the strength of the effect. If the fertilizers is applied before ploughing or to established crops, less will be lost to the streams.

In addition to effects on water quantity and quality, afforestation may also result in acidification of the soil. Soil acidification is associated with afforestation with conifers such as the extensively used Sitka spruce in Britain, for example (Hornung,

1985). In Canada, the soils under coniferous plantations in abandoned farmland showed a significant increase in acidity over a period of forty six years (Brand *et al.,* 1986). Soil acidification maybe especially pronounced if an entire catchment carries the same age of fast-growing stand. The rate of acidification, which at first may be rapid, is likely to slow down later in the rotation (Nilsson *et al.,* 1982). On the other hand, some species such as birch may reduce the acidity of heathland or moorland soils (Miles and Young, 1980). And afforestation may induce physical changes such as reduced water logging and improved aeration in wet soils (Pyatt and Craven, 1978).

## Ecological Effects

Ecological effects may also result, in addition to these physical ones. Pre-afforestation vegetation is transformed as the forest canopy develops, and the heavy shade cast in many plantation means that vascular plants on the forest floor are largely eliminated. The extent of this elimination depends on the choice of species and design of the forest. In the new habitat, bird life is also transformed. Compared with open moorland, songbird densities are found to be around three times greater in young plantations, and four to six times higher in thinned plantations (Moss, 1978). But this increase in numbers is accompanied by a change in species, with birds of the open moorland being replaced by those of woodland. The former may be much more rare than the latter, and the perceived to have a much higher conservation value. This issue lay at the heart of a major conflict between forestry and nature conservation interests over the afforestation of moor-land in the far north of Scotland during the 1980s (Ratcliffe and Oswald, 1986), and in practical effect reduced the area available for afforestatin in that part of the country.

While the image of the forest may in a general sense be one of beneficence, it does not follow that afforestation will be welcomed on environmental grounds in all situations. Environmental interests are now seen as major obstacles to afforestation in Britain, and indeed commercial afforestation using conifers is now to all intents and purposes outlawed in England. The roots of opposition perhaps lie more in the way in which afforestation has been carried out and the types of forest that have been created than in antipathy to afforestation per se (Small scale broad leaved planting are still welcomed in England. Many of the forests created in recent decades have been designed primarily or exclusively as industrial or production forests, with scant regard for either the provision of other goods or services or for the environmental consequences of the mode of afforestation. The reactions have been strong, and now strongly influence the amount and type of afforestation that is permitted and the ways in which it is carried out. At present such opposition is largely restricted to a small number of countries: it remains to be seen whether it will become a major factor elsewhere and whether it may yet have a significant influence in a possible transition from contracting to expanding forests on the global scale.

The environmental effects of afforestation in countries such as Britain are usually perspective in terms of wildlife, landscape and amenity, but the new forests may themselves be at risk from biological hazards. Plantations of exotic species in particular can suffer attack from pests and pathogens, whilst native trees species in neighbouring areas remain unscathed. In Scotland, for example, plantations of Lodgepole pine have been affected by larvae of the Pine Beauty moth, which pose few problems in Scots pine forest. They have been treated by aerial application of a pesticide, which in turn has given rise to concern about wider ecological effects although no evidence of resulting birds mortality has been found (Spray *et al.*, 1987). Monterey (Radiata) pine has been attacked by pine needle blight in settings as far apart as Chile, New Zealand and southern Africa. In some instances spectacular outbreaks of pests have resulted as local insects have adapted to exotic tree species. This process of adaptation can take several decades, and it may still be under wav in parts of areas such as Africa and Latin America where plantations of exotic softwoods have been established in recent years (Perry and Maghembe, 1989). And unexpected consequences can result from the creation of plantations. Dieback in an area of tropical rain forest in Uganda has recently been reported from the vicinity of conifer plantations. While the cause of the dieback remains unclear, it has been suggested that any one or several factors such as fungal pathogens, toxins or hydrological effects may be involved (Struhsaker *et al.*, 1989).

## Forest and Environment

Environmental interests now have a major influence on deforestation and on forest utilization world-wide. The environmental consequences of forests use are the focus of unprecedented interest and concern. Whereas in the past the protection of the timber resource was usually the main motive for government intervention in forest management, today it is equally likely to be environmental protection. The forest is perceived not only as a source of timber, but also as a valued environment in its own night. This perception is held more strongly in some parts of the world than in others, but the emergence of environmental groups in countries such s Brazil and Malaysia suggests that it is widening in spatial terms as well as intensifying through time.

Much uncertainty still exists about the nature of physical and ecological effects arising from the use of conversion of the forest: myth and fact are not always clearly distinguished. Nevertheless, environmental issues and interests affect the use of the global forests resource as never before. Previous episodes of deforestation and destructive utilization, such as in medieval Europe and nineteenth century North America, failed to generate the volume of environmental interest and concern now associated not only with the tropical forests but also with more local forestry issues around the world. The significance and effectiveness of this state of affairs for forestry policy and management await evaluation.

Finally, the environmental effects of the use of the forest resource need to be seen within the context of cyclical phases and trends in the forest area. The effects of both deforestation and reforestation or afforestation tend to be concentrated in certain areas, rather than randomly distributed. For example, the environmental effects of deforestation are concentrated in the tropical world today, whereas in the nineteenth century they were mainly felt in higher latitudes such as those of North America, parts of Scandinavia and Russia, and in some of the more mountainous parts of western and central Europe. In the French Alpos, for example, episodes of severe erosion occurred as a result of deforestation which in turn resulted from rapidly increasing population. As population pressure lessened and reforestation took place, this episode of erosion was succeeded by one of greater stability (*e.g.* Combes, 1982). Cyclical patterns of forest removal and replacement, therefore, are likely to be matched by cyclical patterns of environmental effects extending over wide areas.

# Chapter 15

# People and Policies

The focus falls on the social and political background to the use of the forest resource; Initially the popular perception of forest issues is considered as a basis for reviewing forest policies and their formulation. A number of themes emerge, including the general lack of involvement of the population at large in decisions about forest management, the roles of forests in social and economic development, and the formulation and significance of forest policies.

## People and Forest Issues

Throughout history the forest has been a resource of primary importance o large numbers of people. Management and control of the forest, however, have usually been in the hands of a relatively few individuals, and the people of a country, for example, have rarely been consulted directly on how they would wish their forest resource to be used. As forestry has developed in technical terms, there has been a tendency for decisions about forest management to be seen as the prerogative of professionally trained foresters. This tendency has extended to decisions about what the forests should be used for as well as how they should be managed. Forestry issues have seldom occupied leading positions on the political agenda, and usually have been overshadowed by concerns such as economic policy, defence or agricultural matters. Furthermore, national policy-making has been prone to 'capture' by special interest groups, such as the forest-products industry or forest owners (*e.g.* Davis, 1984), and the views of members of such groups have often had a disproportionate influence compared with those of the rest of the population.

## Population Opinion and Forest Management

As Chapter 1 indicates, the fate of the world forest resource is a major population issue of the day, at least in the developed world. Around one-third of the populations in the European Community, for example, appear to be concerned about it. Much of this concern is directed at the tropical rain forest, which has been promoted as a major issue by a number of international environmental groups. It also extends, in some countries at least, to concern over domestic forest issues. This concern embraces, in varying proportions, environmental questions and issues such as control of the resource and the provision of employment and other social and economic benefits from it. Broadly-base surveys of population attitudes to the forest and its management are conspicuous mainly by their absence, but one large-scale investigation in a country with huge forest resources–Canada–has yielded interesting results, which are summarized in Table 15.1.

**Table 15.1: Forest issues and public opinion in Canada.**

|  | *'Companies should be Free to Harvest without government Regulation'* | *'Chemicals such as Pesticides are Necessary for Taking Care of Forests'* | *'In Recent Years more Trees were Cut Down Compared to the Bumber of Trees Planted'\** | *The Forests should not Exploited Economically at all* |
|---|---|---|---|---|
| (Percentage of respondents) |  |  |  |  |
| Agree strongly | 4 | 12 | 35 | 18 |
| Agree somewhat | 5 | 46 | 31 | 20 |
| Disagree somewhat | 17 | 24 | 16 | 39 |
| Disagree strongly | 73 | 15 | 6 | 20 |
| Don't known | 1 | 3 | 12 | 3 |

\* 22 per cent of cut was replanted or reseeded in period 1976-80.

Based on survey of sample of 1960 respondents, 1981.

*Source*: Statistics Canada (1986).

One of he features emerging from the table is the apparent strength of feeling that Canadian forests should not be exploited at all. The forest-products industry is a major sector of the Canadian economy, and Canadian forests have a social and economic importance far greater than in most other countries. Yet a very substantial minority, amounting to 38 per cent of the people questioned, agreed strongly or 'somewhat' with the proportion that "The forests should not be exploited economically at all'. Presumably this section of the population viewed the forest primarily as a non-mat4erial resource, of value for recreation, wildlife conservation and wilderness qualities rather than as a source of timber. The prominence of the 'non-exploitation' view as reported in this survey contrasts with the recent perception that 'within the general public [in Canada] there exists a view, undiminished by the years, that here are plenty of forestlands to be exploited and that technology will somehow keep on improving accessibility, growing methods and utilization' (Gillis and Roach, 1986, p. 259). Perhaps policy-makers perceive public attitudes on such issues inaccurately, and unless specific and particular issues emerge to provide foci for public attitudes to be expressed, will continue to base their policies on such misperceptions. It is encouraging, however, that *A national forest sector strategy for Canada* recommends that the forest sector should encourage public participation in developing the objectives for forest management (Canadian Council of Forest Ministers, 1987).

Another striking feature is the strength of opinion about the need for regulation of forest use. Of the persons questions in the survey, 90 per cent disagreed with the view that 'Companies should be free to harvest our forests without regulation'. White the resource quality of Canadian forests has not been maintained, at least the forest area has been conserved to a far higher degree than in most countries. It may therefore seem surprising that there should be such an overwhelming support for government regulation of forest harvesting.

Responses to the other two statements are perhaps less striking, but it is noteworthy that a substantial body of opinion (39 per cent) disagrees with the view that the use of chemicals is necessary in the forest, while a clear majority accurately perceive that replanting does not match harvesting. More than one-third of the respondents agreed strongly with the statement, and the pattern of response suggests that he population at large had at least a general awareness of the forest trends in this respect. The table does not indicate directly whether this awareness is matched by concern, but the high proportion agreeing 'strongly' may suggest that concern is indeed present.

Three important conclusions may be surmised, if not logically deduced, from Table 15.1. First and on the basis of the admittedly limited evidence, it seems that the population at large accurately perceives the nature of basic characteristics such as the ratio of cutting to planting. If this is indeed so, then the argument that the population in general has neither an awareness of nor interest in forest issues in untenable. In turn, the view that forest issues are matters to be reserve for professional and technical personnel is untenable. Second, there is strong support for public or government regulation of the use of the forest. Third, almost two-fifths of the population considers that the forests should not be exploited economically *at all*. Presumably a larger proportion place value or utility on non-consumptive uses of the forests. In short, popular demand for the 'post-industrial forest' appears to be running ahead of any actual shift in the nature of use an management of the forest, and professional forest manages and policy formulators may b lagging far behind that demand.

Whilst it would be wrong to place too great emphasis on the results of one survey in one country, the indications they provide are of considerable interest. They deserve further testing in other countries, both in the developed world where many forests may be in transition from the 'industrial' to 'post-industrial' stages, and in the developing world where the transition is more likely to be from 'pre-industrial' to 'industrial', and where a completely different set of public opinions may be encountered.

The fragmentary evidence that is available suggests that Canadian perceptions are not atypical, at least of the developed world. For example, the proportions of respondents indicating concern about the use of insecticides and pesticides in American forests in the early 1980s were respectively 37 and 36 per cent (Hendee,

1982)–figures comparable with those from Canada. Hendee concluded that the public distrusts many of the methods used by the forestry profession, and in particular considers that forestry managers fail to consider properly the non-commodity, environmental and human aspects of forestry.

## Specific Issues: Clear-Cutting and Native Forests

Clear-cutting, with its historical associations with destructive exploitation, is one particular manifestation of conflict over forest resource values. According to Hendee, it was identified by 43 per cent of respondents as subject to widespread abuse. Clear-cutting has been a major source of conflict in the United States, resulting in court action between conservation interests and the US Forest Service, which as perceive by the former as giving priority to timber production (*e.g.* Bonnuicksen, 1982; Culhane and Friesema, 1979). In Australia clear-cutting has attracted opposition even when used within ecologically based regimes for the management of fire-climax species, and followed by even-aged regeneration (Bartlett, 1988).

In the United States, the clear-cutting issue that came to a head in national forests in Wet Virgina and Montana in the late 1960s and early 1970s gave rise to the most widespread discussion of forest management since the beginning of the century (Dana and Fairfax, 1980). Whereas in the early 1900s there was widespread public support for government management of forest resources, the US Forest Service was now seen as being excessively influenced by the timber industry, and as neglecting its multiple-use mandate. Clear-cutting symbolized a perceived overemphasis on timber production and under emphasis on other products and services. Eventually limits were imposed on the area extent of individual clear-cuts, and these maximum sizes varied according to forest location.

Another specific issue, which perhaps emerges more widely around the world, is the fate of native forests (and their replacement in some instances with plantations). This issue is epitomized by the case of the California redwoods. Since the late nineteenth century, the preservation of some of these woods was vigorously advocated, in the face of threats from logging. The preservation movement continued through the twentieth century, culminating in the 1970s with extensive purchases of surviving redwoods stands for preservation. By then, however, most of the old redwoods had disappeared, except for the few per cent protected in public parks (Schrepfer, 11983).

This issue is not confined to North America. In Australia, for example, a survey of 5,000 members of the Australian Conservation Foundation showed native forests to be seen as the top priority for action, amongst a list of eighteen environment issues (Florence, 1983). In the same country, intense controversy has been generated by a policy of expanding wood production from native forest areas, with the consequence of severely modifying the native forests or replacing them completely with plantations. In the course of the sometimes bitter conflict between the holders of different resource values, the public were not consulted, and indeed it was alleged that conversion plans were kept secret (Routley and Routley, 1974). It was alleged by the same authors that the role of the forest service was primarily as a servant of the wood-using industry, and that non-commodity values are generally overlooked or discounted by the forestry 'establishment'. In support of this view, they calculated that over 80 per cent of the articles in the journal Australian Forestry between 1956 and 1971 had been solely concerned with wood production or associated matters, and that only 2 per cent were on aspects of forest ecology not directly affecting wood production.

Whilst the native-forest issue may have generated more heat in countries such as Australia and New Zealand than in many other parts of the world, the basic nature of the conflict, lack of consultation with the public and perceived orientation of the forestry profession towards wood production are common to many other parts of the developed world. It remains to be seen whether native-forest issues develop domestically (as opposed to internationally) to the same extent or intensity in developing countries ass they have done in much of the developed world.

## Population Opinion:
## The Transition to the 'Post-Industrial' Forest

Many native forests remain under public ownership, and therefore few obstacles might be expected to impede their transition towards 'post-industrial' status. Indeed concrete indicators of this transition exist, including for example the designation of areas as national parks or designated wilderness. In the United States, the net transfer (or 'loss') of timberland to non-timber purposes has averaged around 2 million hectares per decade in recent times (Bonnicksen, 1982).

Where forests are under private ownership, the transition might be expected to be less smooth, and the relevance of popular opinion might be though to be less direct. In practice, however, increasing state influence has been brought to bear by means of a combination of regulation and incentives such as grants (Chapter 5). The public is consulted as seldom over the nature of these influences as on the management of forests on public land, but environmental interest groups may have considerably effect on the policy objectives underlying these measures. For example, environmentalist concern about the dwindling broad-leaved forests in Britain in the early 1980s led to be emergence of a new broad-leaved policy encompassing favourable planting grants an presumptions against the 'coniferisation' of existing broad-leaved woodlands. In this and most other comparable cases, however, the outcome was the result of campaigning by interest groups, rather than the product of any systematic attempt to base policy on public opinion. It was concluded by Douglas (1983) that forest services in developing counties are highly conservative and traditional, lacking mechanisms for meaningful reference to the broader economic and political perspectives of society. Perhaps his conclusion also has some validity in the developed world.

In the classical 'industrial' forest, policy objectives are solely or primarily utilitarian. The classical view was well expressed by the first director of what was to become the US Forest Service: 'The main service, the principle object. Has nothing to do with beauty or pleasure. It is not, except incidentally, an object of aesthetics, but an object of economics' (Fernow, 1896, quoted in Kennedy, 1981). Such views are, of course, deplored by environmental interest groups and perhaps by the wider public in many countries today. Nevertheless, powerful inertial factors have retarded their modification. One of these was (and is) the difficulty in incorporating non-material and non-economic values into decision-making. This difficulty was related in turn to a cultural problem: professional foresters, reared in a system in which priority was accorded to timber production, tended to discount concern for outdoor recreation, landscape or (non-game) wildlife as peripheral or as 'weak', emotionally based and generally unprofessional (Kennedy, 1981). Attuned to an utilitarian tradition and lacking clear and direct channels of contact with changing popular attitudes and values, they tended to react defensively at first. In the United States, for example, the Forest Service and the Society of American Foresters either resisted or failed to support legislation such as the Wildness Act (1964) and the National Environmental Policy Act (1970), despite strong support from Congress. In Australia, most foresters were taken unawares by the vehemence of the criticism leveled at the first proposals to reduce native eucalyptus forests to woodships for export to Japan. For the foresters the discovery that certain eucalypts were suitable for chipping provided an opportunity for a long-standing dream to convert these 'unproductive' forests into productive ones (Carron, 1979). This opportunity was quite consistent with the traditional 'production' orientation of the profession, though it conflicted head-on with the changing resource values of the day.

According to Kennedy (1985), part of the problem has been that foresters, like engineers and physicians, do not welcome advice and criticism from persons not trained in their discipline. He also considers that many students entering the forestry profession are attracted by the perceived simplicity and tranquility of the production forest, and that they do not welcome a role as conflict manager adjudicating between different interests and values. Furthermore, many professional foresters have looked upon themselves as the custodians of the community interest in forest management, and as the appropriate judge of where that long-term interest lies (Husch, 1987) In short, the age of the industrial forest fostered an outlook in which objective, as well as the techniques of management, were assumed to be the prerogative of the professional forester, and in which timber production was paramount. It is understandable, therefore, that the challenge represented by the changing values, aspirations and attitudes of the public should encounter some resistance, and hence that obstacles should be encountered in the transition from the 'industrial' to the 'post-industrial' forest.

## Popular Opinion: The 'Pre-Industrial' and 'Industrial' Forests

Popular opinion may play a limited part in forest management in the developed world, but it ha nevertheless been successful in effecting a partial transition from the 'industrial' to the 'post-industrial' forests. By comparison, it has probably had less effect in the developing world. The transition from the 'pre-industrial' forest, with communal control and use of the forest for a variety of purposes such as production of fuel, food and fodder as well as timber, to the 'industrial' forest with external control and an emphasis mainly or wholly on timber production, is often abrupt. It is more often imposed on local people than sought by them. The opinions of local people are rarely taken into account, an in less sophisticated societies it is more difficult for interest groups to form and function effectively.

This transition has frequently met with opposition, and it (or aspects of it) have often had persistent, long-term effects. One of these is resistance to conservation measures imposed by colonial governments, often with little or no reference to traditional patterns of rights and use. Post-independence governments in Africa, for example, have found it difficult to overcome these negative perceptions (*e.g.* Anderson, 1987). Another example is the case of India where resistance has been manifested in the form of forest Satyagraha. The Chipko movement that evolved in the early 1970s was a reaction against the management and use of forests for wood production to supply non-local needs. Its demand for the Himalayan forests to be seen as protection rather than production forests was ultimately at least partly successful (*e.g.* Bandyopadhyaya and Shiva, 1987; Shiva and Bandyopahyay, 1988). Examples of successfully attempts o halt the transition to the 'industrial' forest have been rare, but the last decade has seen the emergence of social or community forests as alternatives. Here the emphasis of production is on fuel wood, poles and various other 'minor' products, rather than on industrial wood.

During the 1950s and 1960s, many national forest services concentrated on industrial forestry in accordance with theories and policies of industry-led development, with a corresponding neglect of other aspects of forest use and management. This orientation is unsurprising: many post-independence national forest services were descended from former colonial services, and in addition many of their personnel were trained and educated in Western institutions. The forestry schools in many of these institutions were probably strongly geared towards timber production, and indeed many of them were probably strongly influenced by timber companies through grants and other means. In short, the Western, 'industrial' view of the forest was widely inculcated. Furthermore, the attempted segregation of land uses, as commonly practiced in temperate-zone forestry, has often led to detrimental effects and confrontation in the tropics (von Maydell, 1985). Eventually in the 1970s widespread interest emerged in agro-forestry in reaction to this.

By the 1970s the growing realization of the importance of fuel wood and other non-industrial products led to growing interest in social and community forestry. In the view of Kenndy (1985), who is himself a forester, the fact that the concept of social forestry had to be invented in order to emphasise broad and varied social needs is an indictment of the narrow value focus of 'traditional' forestry and its close association with wood production and market prices. In addition, there is the problem of social fiction or conflict arising from the fact that the forester has had both an industrial orientation and an association with powerful industrial or landed interests, rather than with the local population. A widely held view is that 'down through the ages, the forester has usually been looked upon as the gendarme of landed property and rich forest owners' (Westoby, 1989, p. 80), and that forest administrators have often been insensitive to the needs of local communities (Husch, 1987). This view is largely shared by Filius (1986, p. 191): 'The needs of the local people have almost always been ignored in forest planning. This has mainly been oriented towards the prime or classical goal of forestry, which is the production of industrial wood.' He goes on to emphasis that the socio-economic and cultural systems of the local population must be studied if social forestry is to be successful and not only physical site characteristics and the wood market.

The outstanding and oft-quoted example of successful community forestry is the Republic of Korea. Here more than 1 million hectares of fruit, fuel and timber trees were planted in five years in the 1970s by a network of over 20,000 village forest co-operatives supported by the government, which provided legislation to make land available, as well as free planting stock and extension services (Arnold, 1987b). The resulting forests to some extent resemble typical 'pre-industrial' forests, with community involvement and multi-purpose use, but they are of course man-made rather than natural an their creation was to a considerable extent a 'top down' rather than 'bottom up' process. Whilst some success was subsequently achieve with this style of community forestry in a few other countries such as Nepal, collective approaches of this kind generally have enjoyed at the best only slow progress. Projects involving individual rather than group approaches have often been found to enjoy greater success. Perhaps the re-creation of the 'pre-industrial' forest in its typical, communally-controlled form, can b achieved only rarely, irrespective of whether the driving force is government or people. If the forest resource under communal control has almost disappeared in a locality it is not surprising that it is difficult to re-establish it in that original form.

The transition from the 'industrial' to the 'post-industrial' phase may be driven by public opinion and that from the 'pre-industrial' to the 'industrial' stage is usually imposed by government or external would be exploiters of the timber resource. The re-establishment of 'pre-industrial' forests seems to require strong external stimuli such as government or aid agency. It remains to be seen whether direct transitions from 'pre-industrial' to 'post-industrial' stages will occur widely. If they can and do, then perhaps far more of the tropical forest will survive than if the transition can be effected only through the intermediate stage of the 'industrial' forest. Perhaps an encouraging, if very tentative indicator is the nationwide logging ban imposed in Thailand in 1989, imposed in response to a combination of pressures from rural villagers, conservationists, and widespread public reaction against a series of damaging mudslides perceived to be related to deforestation (Lohmann, 1989).

## Forests and Employment

One of the main foci of public interest in forests, and in some national forest policies, is in the employment they offer either directly or in wood using industry. Forest employment is widely seen as a significant issue both in counties with extensive forest resources, and in those with continuing afforestation programmes. In comparison with the attention focused on it, however, forest employment is relatively limited in scale. It also suffers from a number of characteristics that limit its usefulness as a mans of encouraging rural development. Nevertheless, the creation of employment is identified by FAO (1986d) as one of the main contributions that forest industries can make to socio-economic development. It goes on to assert that small-scale forest industries provide the principal employment for between 20 and 30 per cent of the rural labour force in many developing counties.

On the national scale, the contribution of forests and forest-based industries to total employment is usually very much less. In many counties it amounts to only 2 per cent or less. It is under 5 per cent even in most countries with extensive forest resources and well-developed forest industries. In the United State, for example, the proportion of employment attributed to timber in the 1970s was 4 per cent of all civilian employment, and of this proportion only 10 per cent related to timber management and harvesting (USDA, 1982). The remainder was in primary and secondary manufacturing (13 and 27 per cent respectively), construction (24 per cent) and transport and marketing (26 per cent). In Chile, where there has been a rapid growth of a plantation-based forest industry in recent years, forest-based activities (including transport) employ 3 per cent of the active population (Solbrig, 1984). The overall proportion is similar in Malaysia, which has one of the most highly developed forest industries in the developing world. Here the forest sector absorbs only 3 per cent of the total labour force, and logging accounts for less than one-third of that proportion (Rauf, 1983). Small as the overall contribution is, however, it has grown markedly in Malaysia as the forest industry has developed, with an increase of 46 per cent occurring between 1972 and 1976 (Kumar, 1986). On the other hand the forest-dependent labour force has generally declined through time as forest operations have become increasingly mechanized. In Canada, for example, almost half of the adult male populations were involve in the timber and lumber industry in the late nineteenth century (Gillis and Roach, 1986), while now only 7 per cent of the labour force is forest-dependent (Environment Canada, 1989). The long-term trend is almost invariably and inevitably downwards, because mechanization and improvements in labour productivity mean loss of employment after the stage of maximum sustainable yield has been reached. In Sweden, for example, forest-sector employment reached a peak at the end of the 1930s. Since then, it has decreased at a rate of around 2 per cent per annum, although wood removals increased by about 1 per cent year until the early 1970s (Lonnstadt, 1984).

The mechanization of forest operations and decreasing labour requirements pose serious problems in areas where the economy is poorly diversified and opportunities for alternative employment are few. Forest-based employment is often of greatest relative importance in such areas. Ever-increasing areas of forest are required to maintain or provide one job, and conflicts arise between the wish to manage the forest on a sustained-yield basis and the desire to maintain stability of employment in the forest and in wood-using industries. Thee problems are of long standing in Scandinavia and parts of North America. In addition in countries such as Canada current forest management and harvesting practices threaten long-term sustainability of production and hence of employment, and in addition their environmental effects may adversely interact with employment in other resource sectors such as fisheries (Environment Canada, 1989). Similar problems are likely to become increasingly prominent in the developing world. Even if the worst excesses of destructive exploitation, and of resulting instability of communities, are avoided, it may still be difficult to reconcile social, commercial and sustained-yield objectives.

In addition to the secular trend of deceasing demand for labour, problems are also caused by cyclical fluctuations in demand for timber. In order to help to stabilize logging-dependent communities during periods of downturn in demand, timber sales from national forests have at times been maintained by the US Forest Service (*e.g.* Repetto, 1988). This practice may, in turn, merely destabilize logging activity on private forests.

There is a long tradition of using mobile or transient labour in the forest, rather than settled workers. In many pats of the develop world today, as in North America in the past, this labour is housed in camps rather than in permanent communities. Such a mode of labour is understandable in terms of cyclical exploitation of the forest resource, but it can mean that few social and economic benefits accrue locally. For example, one study in East Kalimantan in Indonesia revealed that only 12 per cent of the total jobs provided in logging camps were taken by local people, and that local foods accounted for only 5 per cent of camp expenditures on consumable goods (Kartawinata *et al.*, 1981). In short, few benefits from logging were retained locally. Furthermore, the overall intensity of employment creation is very limited: in Indonesia, for example, 53 hectares were logged for each job created, including jobs in industrial processing (Repetto, 1988).

Secular changes in labour requirements and fluctuating demand depending on the age or stage of development of the local forest are world-wide problems that impair the usefulness of the contribution that the forest can make to socio-economic development. They limit that contribution in the developed and developing worlds alike, as they do irrespective of whether the forest is a natural one, whether it has been managed for many decades, or whether it is a recently established plantation.

The initial stages in establishing plantations in countries where the forest resource is expanding through afforestation have a relative high labour demand for ground preparation and planting. Thereafter the need for labour drops to very low levels, and remains there until thinning begins or clear-felling occurs. Peatland afforestation in Ireland, for example, typically requires seven workers per thousand hectares during years 1-5, but only one during years 6-17 (Gallagher and Gillespie, 1984). Such a strongly, fluctuating requirement does not fit well with the characteristics of typical areas of afforestation, which are often remote and lightly populated. Nor does it make a substantial contribution to socio-economic development in such areas in the short term, although in the longer term dome benefits may be provided by wood-using industry.

Fluctuating labour requirements associated with afforestation are frequently met by mobile squads or migrant workers rather than local residents. In addition to providing limited or minimal local benefits, such patterns of provision sometimes appear to lead to social friction. A rise in petty crime, for example, is reported from Northland, New Zealand (Farnsworth, 1983). On the other hand, the same author indicates that forest development programmes there have helped to arrest and reverse trend of rural depopulation, and have led to greater diversity of community structure and higher proportions of young married couples in local communities. In contrast, afforestation has apparently not helped to stem rural depopulation in areas such as North Wales (Johnson and Price, 1987).

Although labour requirements may fluctuate, they are often at least as great as for previous or alternative land uses. In Otago in South Island, New Zealand, for example, direct employment in forestry and pastoral farming is similar, but when downstream employment is included, forestry employs more than four times as many workers per unit area as does agriculture (Aldwell and Whyte, 1984).

Whilst forestry may have employment densities similar to or higher than those in pastoral farming, the conversion of land to forest plantations by no means always results in increases in employment. In Portugal, for example, employment densities in eucalyptus plantations are lower than in previous land uses involving live groves and vineyard: the creation of 5,000 hectares of plantation means a loss of 2,000 man-years of employment (Kardell *et al.*, 1986). Similarly, while a typical small farm in Latin America can sustain several labourers, a typical 100-hectare eucalyptus plantation provides jobs for only two or three (Joyce, 1988).

Regardless of whether forests or alternative land uses provide more employment, there appears to be a widespread tendency to overestimate the number of jobs that forestry and forest-products industries can provide. Employment potential is frequently an important element in forest policies or in justifying individual forest projects, and exaggerated or over-optimistic estimates have often been made. For example, it was forecast in the early 1970s that employment in forestry and the wood industry in Australia would rise substantially: in fact it has declined (Dargavel, 1982). In Scotland the chief minister in promoting a progamme of forest expansion in the 1940s looked forward to a day when forestry would employ workers as many as agriculture and coal-mining (around 150,000) (Johnson, 1952). In fact it now employs little more than 10,000.

Advocates of forestry expansion have frequently used augments of social an environmental benefits in support of their case. Objectives incorporating these benefits have often been included in national forestry policies and programmes. Unfortunately the way in which the expansion has been implemented or carried out has not always meant that the benefits have materialized in the ways expected. This in turn has on occasion led to disappointment and disillusionment.

## Forests and Development

This disappointment and disillusionment have been felt at a variety of scales, ranging from local projects to the international level. Nowhere are they better reflected than in the changing views of J.C. Westoby, who was a forestry official in FAO. These changing views are chronicled in a series of papers and books published between the early 1960s and the 1980s, culminating in the publication of his collected papers under the title of the *Purpose of Forests: Policies of Development* (Westoby, 1987). Initially, both he and FAO believed that forest industries in the developing countries could look forward to very bright prospects. In terms of resource endowment and environmental potential they were placed to meet the growing timber needs of the developed world. Linkages between the forest sector and other industries and other branches of the economy would mean that the sector could make a special contribution to the overall development process. In short, the forest sector deserved special consideration in development strategy: 'Industrialization based on the forest can both contribute to and promote the general economic development process' (Westoby, 1962, p. 200).

By 1973, doubts were creeping in: 'progress has not been as rapid as we thought we had a right to expect' (Westoby, 1987, p. 207), and with them the realization that the world's industrial wood needs could be met from plantation which would amount to only a tiny percentage of the world forest cover. By 1978 the gloom was deepening: 'as yet, forest industries have made little or no contribution to socio-economic development in the underdeveloped world-certainly not the significant contribution that was envisaged from them a couple of decades ago' (p. 246) this time he had concluded that forest development projects, like parallel developments in food and agriculture, were geared primarily to the needs of the developed world, and that they thus served to promote socio-economic underdevelopment. According to Westoby, the forests in developing counties and in tropical latitudes were exploited primarily for the benefit of the developed world, and the 'development establishment' (including international agencies) had assisted in this process or colluded with it. In his view, international aid in forestry has helped to identify, for the benefit of foreign capital, forest resources suitable for exploitation, and it has helped some irresponsible governments to alienate and eliminate substantial parts of their forest-resource endowment. In particular, almost every country now had a forest service, but many forest services in developing countries were woefully understaffed and underpaid: 'because they exist, exploitation is facilitated; because they are weak, exploitation is not controlled' (p. 248). He went on to conclude: 'the basic forest products needs of the peoples of the underdeveloped world are further from being satisfied than ever …. The famous multiplier effects are missing. Few new poles of development have been created' (p. 248).

Linkages between the forestry sector and other parts of the economy proved harder to forge than had been expecte3d. Policy-makers have usually overestimated employment benefits from forest-based industries, as they have from afforestation programmes, and benefits in terms of regional development have also been overestimated (*e.g.* Repetto and Gillis, 1988). In short, the bright hops of the 1950s and 1960s that the forest resources of developing countries could offer a springboard to socio-economic development have not been fulfilled. The same phases of destructive exploitation that characterized the use of the forest resources in eighteenth-century Russia and nineteenth-century North America have all too often been repeated, and perhaps even facilitated or promoted under the guise of development aid. Local resource values have been submerged under those of the developed world: the 'pre-industrial' forest has given way to the 'industrial' forest in many areas, and to one that has been rapidly degraded.

In a few developing countries the forestry sector has made a significant contribution to socio-economic development. One example is Malaysia, where the forestry sector has made significant contribution in both overall development and in terms of distributing the benefits of this development (Douglas, 1983). In peninsular Malaysia, forest utilization has been relatively well integrated with rural development and the growth of a export-oriented processing sector. Value is added through sawmilling and plywood manufacture, and government has attempted to control rates of harvesting, to improve management and utilization, and to expand plantations. The scene there does not quite conform to the gloomy picture or model sketched by Westoby, nor does it wholly do so in the case of Sabah and Sarawak, Whilst the emphasis in these areas in still on logs (rather than on processed products), control of forest clearing and of land allocation for agriculture has been retained by the Malaysia government, although this has not prevented conflict with indigenous peoples (Chapter 5). The rate of growth of forest-products exports during the 1960s surprised even government economic planners, and by the mid-1980s the forestry sector contributed around 5 per cent of Gross Domestic Product compared with under 3 per cent in 1960 (Kumar, 1986),/Douglas (1983) concludes that the overall performance of the forestry sector can be interpreted in a favourable light because control over forest exploitation has been maintained, rural development has to some extent been achieved, and forestry-industry activities have been specialized in areas of comparative advantage.

Such example, however, are rare, and not invalidate the Westoby view. Until the mid to late 1970s, most of the development effort was focused on large-scale industrial forestry. With widespread disenchantment about the results of such effort, attention then began to swing towards a recognition of the importance of forestry in rural (as opposed to industrial) development (e.g. Guess, 1981). With this shift came a re-evaluation of the financial characteristics of forest projects geared to objectives of rural development rather than to wood using industry, and with it new problems of funding and of co-ordinating the various institutions that provide the infrastructure within which the projects can be carried out. This shift is reflected in the dramatic change in the nature of funding of forestry projects by the World Bank. Prior to 1977, around 85 per cent of funding from this source went to industrial aspects of wood production and utilization: only four out of seventeen projects funded between 1953 and 1976 were specifically intended to benefit rural people (World Bank, 1978). Since then the share of a greatly increased volume of funding has fallen to 35 per cent. The other 65 per cent is now allocated to projects concerned with fuelwood an timber for domestic (as opposed to industrial) use, environmental functions, rural development, and the management of natural forest ecosystems (e.g. Brunig, 1984). Over half of the forty projects in the Bank's forward lending programme at the end of the 1970s were regarded by I was 'people-oriented' as opposed to 'industry-oriented' (World Bank, 1978, p. 9).

While the significance of involvement of bodies such as the World Bank in relation to forest destruction and forest conservation is debatable, there is little doubt that some reorientation away from industrial forestry projects has occurred.

## The Role of Forest Policy

The form and content of forest policy are enormously variable. Forest laws and institutional arrangements have a bewildering complexity and diversity. In some counties, specific laws and codes ate back for centuries, relating for example to matters such as ownership and control. Many of them relate, literally, to the tees rather than to the forest, and in many countries they have been added sporadically and incrementally. They are often fragmentary and piecemeal, lacking integration into coherent policies. Few countries have coherent, consistent and comprehensive policies, and where such policies do exist they are almost invariably the product of the twentieth century. The nature an content of forest policies may reflect something of the character of the state. This character and role are of fundamental importance. Whilst ideally the state may function as the trustee of its peoples, and manage the forest and other resources on their behalf, in practice it has often functioned in a way favourable to sectional interests. In varying degrees it still does so. A popular (*sensu stricto*) basis to forest policy has generally been lacking.

Forest policy seems at first sight to be the key to effective management of the resource. It seems to be self-evident hat the objectives of the management of the forest resource need to be clearly defined, and effective means of achieving them require to be identified. Yet many countries, including large countries such as Canada and Australia, have found great difficulty in formulating and implementing forest policies. During the 1970s, the Seventh and Eighth World Forestry Congresses urge that all counties devise and declare national forest policies, and that existing policies be updated. By no means all countries yet have forest policies, and there is scarcely a country which has a formal, thought-out and declared forest policy' (Westoby, 1989).

If they exist at all, national forest policies are often characterized by vagueness, confusion, or uncertainty. They may attract much attention, but less agreement and even less commitment. In a paper entitled 'National Forest Policy–myth, manifesto, mandate or mandala?', Carron (1983) chronicles the sorry saga of national forest policy in Australia, which has extended over a period of more than seventy years. He concludes:

> Anyone reading the mass of material on and around the subject might well be pardoned for not being sure whether we have one: whether we had one but don't have it any longer; whether we think it would be nice to have one but it isn't practicable.

It is true that particular problems are encountered in federal counties such as Australia (and many others including Malaysia, Canada and the United States). Different provinces or states pursue different policies: for example, in Australia, New South Wales allows woodchipping for export, while Victoria does not, and Tasmania places tight controls on the management of private forests while the other states do not (Bartlett, 1988). And the Commonwealth government is keen to protect the Queensland rain forest, while the state government proposes to have it logged on a sustained-yield basis (MacDonald, 1989). Under such conditions, broad forest strategies, rather than precise national policies, may be the most that can be expected.

This Australian example reflects the peculiar problems faced by federal countries, as well as the common and basic problems of will, commitment and agreement. It illustrates something of the general problem that arises from the distribution of power between different levels of government. This problem is experienced in unitary as well as federal countries, and difficulties in formulating national policy are encountered under both types of political structure. These difficulties include, amongst many others, the position of forestry issues on the political agenda, and the simultaneous requirement for long-term consistency in defining policy objectives and for flexibility in the face of changing pressures and concerns. Policy statements are sometimes regarded as stereotyped and irrelevant, because of their rigidity and failure to keep pace with changing conditions. It is sometimes argued that policy-making should become a continual process in order to keep pace with changing conditions (*e.g.* Gane, 1983). If too flexible, on the other hand, they may be overwhelmed or overshadowed by emphemeral or transient issues, and may fail to convey the impression of long-term commitment that forest management requires.

Part of the problem is that forestry issues have rarely had a sufficient degree of immediacy or urgency to ensure them a high priority on the political agenda. Even today, when interest and concern are widespread, it does not necessarily follow that forestry issues are the leading concerns: matters such as defence or social policy are likely to occupy more prominent positions in the agenda. Forestry issues rarely emerge prominently at general elections, and when they do attract much attention it may be at local or regional rather than national levels, or from relatively small interest groups rather than the population at large. Major influences, disproportionate to the numerical strength of the membership involved, may be exerted by groups representing forest-industry and environmental interests in particular.

## The Evolution of Policy

It seems that forest policy rarely evolves gradually in response to popular views. Instead, it is sometimes devised rapidly by government in the face of a particular problem or crisis. The classic example is the case of British Forest Policy, which was largely non-existent until the crisis year of the First World War brought home the disadvantages of dependence on timber imports. Policies devised in this way are likely to have simple, single, well-defined objectives, and they are the direct product of government rather than the reflection of popular interest or concern. The urgency and immediacy of such a crisis may very effectively focus the attention of government on an otherwise neglected issue, and may lead to the general acceptance of the idea of a national policy.

Its origins will obviously determine the initial nature of a policy, but inertia may ensure that the original policy objectives and instruments persist while conditions and circumstances change. It has been suggested, for example, that the 'objectives of official forest management in post-Independence India have not [yet] been liberated from colonial legacies' (Shiva *et al.*, 1985).

Subsequent crises–whether financial, military or political–may lead to radical revision of the original policy, but unless such crises arise any change is likely to be minor and incremental. Pressure from interest groups or other branches of government may result in adjustments, and new objectives may be added or the priorities of existing ones revised. In some instances new objectives are simply added to the existing ones without any clear indication of priorities, and without the addition of new instruments. As a result, the policies become increasingly complex, and even those charged with implementing them may lack clear direction of priorities. The principle of providing at least one policy instrument for each policy objective is probably ignored more often than it is honoured.

The origin and content of forest policies depend on a variety of factors including political will and the extent of the forest resource. In countries where the forest resource has largely disappeared, the stimulus of war or economic crisis may give rise to a policy geared to expanding the forest estate. In well-wooded countries, on the other hand, the gradual shrinking of the forest resource may lack the immediacy or urgency of such a crisis. When a policy is eventually devised, it is likely to be concerned initially with the protection of the resource, especially against threats such as fire. The combination of different degrees of resource endowment, different political system and different levels of development mean that the search for common characteristics of forest policies, on the global scare, is at best likely to be only partly successful. Nevertheless, a number of common features can be identified.

Most forest policies refer to goals of timber production, as well as to environmental and social objectives. In many instances concern about the availability of timber led to the initial formulation or reformulation of forest policy, and in some cases to the setting-up of a state forestry service. During the seventeenth-century reign of Louis XIV, for example, it was reported that 'la forest de range set en mauvais etat' (ONF, 1966, p. 8), despite attempts dating back to at least the fourteenth century to protect it by means of a forest code and ordinances. Fears about the availability of timber for naval purposes led to the famous ordinance of Colbert in 1669, and the laying of the foundation of modern French forest legislation. Another classic example is the case of the Britain, which embarked on a programme of forest expansion after the First World War: other countries have followed similar courses on the basis of fears that imports will become difficult to obtain in the future, and will become increasingly expensive as timber shortages develop. Increasing self-sufficiency, for reasons of physical or economic security, has been a primary objective in many national policies, and has in particular been the basis of which forest expansion programmes have been advocated.

Timber is one o the very 'few natural resource commodities to have increased in real price over prolonged periods of several decades, and thus arguments of increasing scarcity cannot be discounted. Furthermore, real wood shortages have already been encountered in many countries, especially in the more arid parts of the world. Nevertheless, the shortages predicted at various times in countries as diverse as the United States, Britain, Australia and New Zealand have not materialized. In the case of Britain, arguments of increasing scarcity, and hence rising prices, were used by the state forest service as recently as the 1970s in support of continued forest expansion (Forestry Commission, 1977). In the case of Australia, Carron (1980) notes that the 'scarcity' argument has been used for seventy years without the prediction coming to pass. Most 'scarcity' arguments have been directed at industrial wood: few (at least until recently) have been directed at fuelwood, where paradoxically real and acute shortages have been experienced in various parts of the world, including much of Britain by the seventeenth century and many pats of the tropics today. Perhaps the emphasis on industrial wood and the relative neglect of fuelwood simply reflects an 'industrial' bias in the forestry profession and in forestry policies.

Self-sufficiency or adequacy of timber reserves for emergency use are impermanent goals. In the case o Australia, for example, forest expansion during the

1960s and early 1970s was geared to self-sufficiency, but with a lowering of population estimates for the end of the century and beyond, demand estimates were revised downwards. The question then arose of whether planting rates should be reduced, or whether planting should be aimed at the export market (Hanson, 1980). It has been suggested that the level of uncertainty about future demand for timber has increase in recent years (Haynes and Adams, 10983; Adams and Haynes, 1985). If so, the translation of policy objectives into area requirements may become increasing difficult. In Britain, the advent o nuclear warfare meant that the original objective of building up a strategic reserve of timber for use during ear was becoming irrelevant. This change did not, however, lead to a radical revision of the planting programme. Expansion continued, on the basis of economic strategy. In short, planting programmes can achieve a momentum that outlives the conditions in which they were established, and forest expansion may become a goal in its own right.

While increasing timber production through expansion of the forest area is a major theme in countries such a Britain, Ireland and New Zealand, it may also be an important issue in countries which are well endowed with forest resources. In some developed countries, private forest owners have little interest in maximizing wood production. The needs of forest-products industries have meant that sub-maximal production from private forests has emerged as a particular issue to which policy is directed (for the example of Finland, see Vehkamaki, 1986). In many developing countries, stated or unstated policy has been focused on increasing timber production from natural forests in the hope that socio-economic development would be fostered.

Timber production is therefore usually a central o primary objective in national forestry policies. In addition, social objectives are frequently included, and these may be given special emphasis at certain times and in certain areas. Afforestation has often been viewed as a means of providing employment. A strong regional dimension is sometimes apparent, and forest service in New Zealand has suffered from being used as a social agency in providing employment (Willis and Kirby, 1987). Similar sentiments are widely held, if less directly expressed, elsewhere, and reflect something of the tension that arises when multiple objectives are pursued. The same authors refer to the repeated calls to the forest service in New Zealand to repair the environmental damage caused by burning and over-grazing. This is but one aspect of the environmental objectives that are incorporated in many national policies, especially since the 1970s (*e.g.* Schmithusen, 1986). Similarly, recreation ha been increasingly acknowledged as a policy objective. As various objectives were added incrementally to the management of state forests in New Zealand between the 1950s and the 1980s, including recreation, nature conservation and landscape conservation (Tilling, 1988). Similarly, recreation was added to the management objectives of state forests in the Netherlands in the 1960s, and nature conservation in the 1970s (Grandjean, 1987). In some countries the emergence of new issues such as the environment has resulted in new management agencies and changing relationships between agencies and divisions. Under such circumstances the integrity and coherence of policies becomes increasingly difficult to maintain.

The typical national forest policy has become more complex through time, and simultaneously has often become less precise and specific. Policy formulation has increasingly become a matte of compromise and of reconciling the views of different interest groups, rather than one of devising blueprints or guidelines for the management and use of the forest resource. Policy statements in the developed world frequently refer to multiple use, for example in the cases of the national forest strategy for Australia, as announced in 1986 (Bartiett, 1988) an in the United States. In the latter, the Multiple Use-Sustained Yield Act of 1960 was as much a compromise as a milestone (Cox *et al.*, 1985): whilst it and subsequent legislation such as the National Forest Management Act of 1976 established multiple use as a general principle, many practical problems remained in translating this principle into practice. Multiple use was fined as the management of all the various resources of the national forests, in the combination that would best meet the needs of the American people. Such a definition seems unexceptionable, but it is also vague and lacking in specific criteria. The US Forest Service was left to interpret these definitional terms, and as a guide to decision-making or a standard for measuring performance the act had limited meaning. The bill leading to the act was attacked by the Sierra Club, a major and powerful environmental group. One of the fundamental points of criticism was that foresters were competent to identify resource problems and to propose possible solutions, but not to choose between uses (Steen, 1976). The bill contained a list of multiple uses, arranged in alphabetical order. Despite the fact that the order was held to be insignificant, some judicious manipulation is reported to have been undertaken: 'fish and wildlife' became 'wildlife and fish', for example, and 'forage' was translated into 'range' (Dana and Fairfax, 1980). Whatever the accuracy of the allegations an the significance of the reordering may have been, there is in this issue a reminder that the apparent policy as enshrined in law is but the tip of an iceberg: much lies below the surface.

Nevertheless, such legislation is of at least symbolic significance, indicating that the days of the 'industrial' forest, in which timber production was the sole or overwhelmingly dominant function, were over. The broadening of policies to incorporate environmental and recreational elements is perhaps epitomized by the Multiple Use-Sustained Yield Act in the United States, but many other countries demonstrate similar trends, whether or not they are manifested in major legislation.

When policies do evolve through time, they tend to become more complex and more vague: if they fail to evolve, they become increasingly irrelevant. Inertia may be a powerful factor even when evolution appears to occur: the typical pattern of incremental change may be basically too slow to respond to emerging forest issues (*e.g.* Tikkanen, 1986). Senior officials in forest services may have spent their formative years during periods of different or simpler policies, and unconsciously or otherwise may be strongly influenced by these earlier phases. Furthermore, the addition of new policy objectives (such as environmental and recreational goals), is not always matched by the provision of new instruments. Therefore the effective policy may differ from the stated policy.

In some instances major differences can exist between policy as stated and as practiced. Sometimes there is a simple inability or unwillingness to implement or enforce policy provisions. In Colombia, for example, a permit it required before forest is cleared and there may be a requirement for reforestation, but in practice few landowners observe these regulations and staff and resources are inadequate for enforcement (Green, 1984). Some plans or apparent policies have been dismissed by Westoby (1989) as 'window dressing'. His comments are directed especially towards certain developing countries where forest destruction continues despite the announcement of grandiose plans apparently directed at conserving the forest resource. In his view, such plans and policies are merely designed to mollify international conservationists. Elements of window dressing, however, are not confined to such situations, but extend to developed countries where environmental or recreational objectives are apparently incorporated into policy in order to satisfy interest groups. Whether their incorporation is always followed by changed practice is another matter.

## Policy Limitations

The gap between rhetoric and reality means that although national forest policies may be necessary for effective management of the forest resource, their existence is unlikely to be a sufficient condition for that management. In addition to the points just discussed, three other factors are significant.

First, the recent and continuing 'internationalization' of the forest industry Le Heron, 1988) may mean that national policies become increasingly inadequate in scale and scope, in the same way in which national governments may not always be a match for transnational corporations. Direct foreign involvement in some developing countries may be deceasing, but the international dimension in forestry in some developed counties is increasing. For example, in the United States, Canadian and Scandinavian forest-industry companies have been investing in forests, as have British financial institutions (*e.g.* Yoho, 1985). In response to such tens, some counties may feel a need to devise foreign forest policies, as well a domestic ones. For example, some importing countries such as France may be prompted to devise foreign forest policies as well as domestic ones, as supplies become scarcer and market conditions increasingly favour the seller. Such policies could include elements aimed at encouraging investment in joint ventures in tropical counties (Huguet, 1980). Whether such foreign policies can be effectively harmonized with national domestic policies in the host countries is debatable. A g rowing international or transactional dimension is also reflected by the attempts of the European Community to formulate a common policy. Such attempts encounter great difficulty, because the problems to which policy objectives need to be directed become more numerous as geographical scale increases. In Mediterranean Europe, for example, the problems are related more to fire and environmental protection, whereas production goals and the afforestation of 'surplus' land released from agriculture are relatively more important further north. On the still wider scale, it remains to be seen how successful international policies such as those enshrined in the Tropical Forestry Action Plan are in practice. None of the major international conventions deals directly with forest resources: of the 113 concluded between 1921 and 1983, fewer than ten are related to forest management, and then usually merely tangentially (Mayda, 1986). For example, one tree species (of mahogany) is included in the convention on International Trade in Endangered Species (CITES) concluded in Washington in 1973.

Second, national forest policies do not exist in isolation. Various other agricultural, economic and social policies are likely to be in operation at the same time. And if it has proved difficult to harmonize the various internal elements of forest policy, it is almost impossible to do so on the broader front. Adequate integration is often lacking at both the intra-sectoral and inter-sectoral levels. In Canada, for example, the need to co-ordinate the host of government departments and agencies is seen as even greater than that of establishing a national policy (Wetton, 1978). In practice, non-forest policies may have a far greater effect on the use and management of the forest resource than have forest policies. In relation to forest problems in the developing world, Romm (1986) goes so far as to argue that ' the primary objectives of forest management policy must be outside the forest' (p. 102). I indeed the main threats to the forest are from factors such as population growth and agricultural expansion, the validity of his argument is apparent. And the significance of non-forest policies for the forest resource are not confined to the developing world. Conflicts between sectors are numerous in the developed world also. In Britain, for example, policies of expansion of both agriculture and forestry were pursued after the Second World War. These policies were to a large extent worked out separately: *de facto,* however, agricultural expansion had priority, not least because it was administered by a larger and more powerful government department. Similarly, government positions on forestry and deforestation in developing countries in Africa and elsewhere are often confusing: deforestation may officially be decried, but at the same time land clearance for agriculture encouraged (*e.g.* Hosier, 1988). Numerous calls have been made for integrated land-use policies (*e.g.* Papanastasis, 1986), but these are likely to remain pipedreams. If agreed and meaningful policies cannot be devise for the forest sector, what hope is there for more general land-use policies.

I serious conflicts sometimes arise between forest and agricultural policies, much greater problems exist in the relationship between forest resource use and economic, fiscal and tenurial policies. Such policies may have a far greater impact on the extent and condition of the forest than forestry policies. This is as true of policies pursued at the supra-national and non-governmental levels as well as of those pursued by national governments. The classic example is Brazilian Amazonia, where fiscal incentives, tax holidays, and the nature of land tenure have contributed greatly to pasture-driven deforestation (*e.g.* Hecht, 1989): a scathing indictment of government development policies, sponsored by the World Bank, is presented by Mahar (1988). He proposes five immediate changes in policy: the discontinuation of fiscal incentives for livestock projects, a moratorium on disbursement of fiscal incentive funds for projects in the Carajas area, the modification of policies that recognize deforestation as a form of land improvement, and thus as a mean of establishing tenurial rights. The character of land tenure, and hence of the pattern of forest ownership and management objectives, is largely dependent, for better or worse, on government policy. Laws that assign property rights over public forests to private parties on condition that such lands are 'developed' or 'improved' favour expansion of agriculture at he expense of the forest (*e.g.* Gillis and Repetto, 1988). The significance of tenurial policies and enactments in not new: legislation on forest property an use rights in many European countries in the past facilitated the transition of the forest from the 'pre-industrial' to the 'industrial' stage in the same way as it has done so much more recently in much of the developing world. For example, an act of 1805 abolished all common rights in private forests in Denmark (Sabroe, 1954). Whether intentionally or otherwise, this prepared the way for a fundamental change in the pattern of forest use and management.

Similarly, the framing of tax regulations can wittingly or unwittingly have major effects on how the resource is managed. Tax 'holidays' to foreign corporations, in the hope that industrial development will be fostered, may mean that forest destruction is accelerated. Adjustments to tax regulations can have an almost overnight effect on afforestation rates in countries such as Britain, irrespective of whether they were intended to have any effect on forestry.

Third, planning, or the local implementation of policy, can pose major problems. By their very nature, national forest policies are formulated at the centre. To have effect, they have to flow 'outward' and 'downward'; to reach local inhabitants and forest owners. This problem is reviewed for the case of Japan by Shimotori (1986), but is by no means restricted to that country. In addition, there is the fundamental problem that policies are formulated nationally, but the implications vary regionally. Problems emerge especially in relation to planting targets. In the case of New Zealand, for example, these have been established without significant consultation with regional and local authorities (Noran, 1989). This is also true in other counties such as Britain, and highlights the tensions that exist between sectoral and land-use planning. Different degrees of support or enthusiasm for central government or sectoral polices exist at different levels. In New Zealand, for example, regional planning authorities have been more supportive o central government targets for afforestation than have the local planners (*e.g.* Abbiss, 1986). Local planning authorities have to cope with the local outworkings of these national policies, and often feel frustrated because of their limited or inappropriate powers and instruments. Furthermore, one of the traditional tools of land-use planning is zoning, but this practice of allocating zones of land to specific uses fits uncomfortably with the shift towards multiple use that is characteristic of the 'post-industrial forest'.

There are, therefore, several serious and basic limitations to forestry policy. Numerous prescriptions for more effective policies have been offered: one example is illustrated in Table 15.2. It is easy to identify defects and weaknesses in existing policies and to point to areas where improvement could and should be achieved. It is easy to urge, for example, that forest policy should be treated as a whole, and that forest policies need to be strengthened. It is much more difficult to convert the prescription into reality in the ace of competing issues, changing political agendas and conflicting interests. It is doubtful whether a completely comprehensive and consistent forest policy can ever be formulated. To this extent, the fervent advocacy of national policies, as expressed by World Forestry Congresses and commentators such as Westoby (1985) is futile. To accept that all national forest policies are flawed and incomplete, however is not to suggest that they are pointless and meaningless. Partial though they may be, they can still help to instill principles of management and to establish the broadest of guidelines for the use of the forest resource. In the final analysis, however, they can be but one o the preconditions for forest conservation.

**Table 15.2: Prescriptions for forest policies.**

| |
|---|
| ☆  Forest policies need to be strengthened and implemented more rigorously |
| ☆  Forest policies should give more emphasis to the needs and wishes o people |
| ☆  There must be a close co-ordination between forest policy and other policies |
| ☆  Forest policy must be treated as a coherent whole |
| ☆  National policies should recognize the growing importance of forestry's international dimension |
| ☆  Those responsible for forest policy should do what they can to protect forests against avoidable damage and destruction |
| ☆  Forest administration and individual forest officers should become more outward looking, and become more concerned with people |

*Source*: Compiled from Hummel (1984) (pp. xi-xii).

# Glossary

**Acclimatisation** The physiological adaptation to climatic cariations.

**Adaptation** Adjustment in natural or human systems in response to actual or expected climatic stimuli or their effects, which moderates harm or exploits beneficial opportunities.

**Adaptive Capacity** The ability of a system to adjust to climate change (including climate variability and extremes) to tone down potential damages, to take advantage of opportunities, or to cope with the consequence.

**Aerosols** A collection of air-borne solid or liquid particles, with a typical size between 0.01 and 10mm that reside in the atmosphere for at least several hours. May be of natural or anthropogenic origin. May influence climate directly through scattering and absorbing radiation, and indirectly through acting as condensation nuclei for cloud formation or modifying the optical properties and lifetime of clouds.

**Afforestation** Planting of new forests on lands that historically have not contained forests.

**Aggregate Impacts** Total impacts summed up across sectors and/or regions. Requires knowledge of the relative importance of impacts in different sectors and regions. Measures of aggregate impacts include, for example, the total number of people affected, change in net primary productivity, number of systems undergoing change, or total economic costs.

**Albedo** That fraction of solar radiation which is reflected by a surface or object, often expressed as a percentage. Snow-covered surfaces have a high albedo whereas vegetation-covered surfaces and oceans have a low albedo. The Earth's albedo varies mainly through varying cloudiness, snow, ice, leaf area, and land-cover changes.

**Anadromous Species** A species of fish, such as salmon, that spawn in freshwater and then migrate into the ocean to grow to maturity.

**Annex B countries/Parties** Group of countries included in Annex B in the Kyoto Protocol that have agreed to a target for their greenhouse gas (GHG) emissions, including all the Annex I countries (as amended in 1998) except Turkey and Belarus.

**Annex I countries/Parties** Group of countries included in Annex I (as amended in 1998) to the *United Nations Framework Convention on Climate Change* (UNIFCCC), including all the developed countries in the orgnaisation for the Economic Cooperation and Development (OECD), and economies in transition. By default, the other countries are referred to as non Annex I countries. Annex I countries commit themselves specifically to the aim of returning individually or jointly to their 1990 levels of greenhouse gas (GHG) emissions by the year 2000.

**Annex II countries** Group of countries included in Annex II to the *United Nations Framework Convention on Climate Change* (UNIFCCC), including all developed countries in the Economic Cooperation and Development (OECD). These countries are expected to provide financial resources to assist developing countries to comply with their obligations, such as preparing national reports, and also promote the transfer of environmentally sound technologies to developing countries.

**Antarctic Circumpolar Current** The movement of shallow-to-deep Southern Ocean waters from west to east around the globe, circumnavigating Antarctica, in response to the rotation of the Earth and planetary winds. The current was first reported by James Cook in 1775.

**Antarctic Intermediate** Water Created through large-scale cooling and Ekman convergence in the Southern Ocean.

**Anthropogenic Emissions** Emission of greenhouse gases (GHG's), GHG precursors, and aerosols associated with human activities. Include burning of fossil fuels for energy, deforestation, and land-use changes that result in net increase in emissions.

**Anthropogenic** Resulting from or produced by human beings.

**Anticipatory (proactive) Adaptation** Adaptation that takes place before impacts of climate change are observed.

**Aquaculture** Breeding and rearing fish, shellfish, etc. or growing plants for food in special ponds.

**Aquifer** A stratum of permeable rock that bears water. An unconfined aquifer is recharged directly by local rainfall, rivers, and lakes; the rate of recharge is influenced by the permeability of the overlying rocks and soils. A confined aquifer has an overlying impermeable bed and the local rainfall does not influence the aquifer.

**Arbovirus** Any of various viruses transmitted by arthropods, including the causative agents of dengue fever, yellow fever, and some types of encephalitis.

**Arid Regions** Ecosystems with <250 cm precipitation per year.

**Assigned amounts (AAs)** Under the *Kyoto Protocol*, the total amount of greenhouse gas (GHG) emissions that each Annex B country has agreed that its emissions will not exceed in the first commitment period (2008 to 2012) is the assigned amount. Calculated by multiplying the country's total GHG emissions in 1990 by five (for 5 year commitment period) and then by the percentage it agreed to as listed in Annex B of the Kyoto Protocol (*e.g.* 92% for the European Union, 93% for the USA).

**Atmosphere** The gaseous envelope that surrounds the Earth. The dry atmosphere consists almost entirely of nitrogen (78.1% volume mixing ratio) and oxygen (20.0% volume mixing ratio), plus several trace gases, such as argon (0.93%), helium, and radioactively active greenhouse gases (GHG's) such as carbon dioxide (0.035%) and ozone. The atmosphere also contains water vapour, whose amount is highly variable but typically 1 per cent volume mixing ratio. The atmosphere also contains clouds and aerosols.

**Autonomous (spontaneous) Adaptation** Adaptation that is not a conscious response to climatic stimuli but rather is triggered by ecological changes in natural systems and by market or welfare changes in human systems.

**Base flow** Sustained flow in a river or stream mainly produced by groundwater runoff, delayed subsurface runoff, and/or lake outflow.

**Baseline/Reference** Any datum against which change is measured; might be a 'current baseline', that represents observable, present-day conditions, or a 'future baseline', which is a projected future set of conditions excluding the driving factor of interest. Alternative interpretations of the reference conditions can give rise to multiple baselines.

**Basin** The drainage area of a stream, river, or lake.

**Benthic Organisms** The biota living on, or very near, the bottom of the sea, river, or lake.

**Biodiversity** The numbers and relative abundances of different genes (genetic diversity), species, and ecosystems (communities) in a particular area.

**Biodiversity Hot Spots** Areas with high concentrations of endemic species facing extraordinary habitat destruction.

**Biofuel** A fuel produced from dry organic matter or combustible oils produced by plants. Examples include alcohol (from fermented sugar), black liquor from the paper manufacturing process, wood, and soybean oil.

**Biomass** The total mass of living organisms in a given area or volume; recently dead plant material is often included as dad biomass.

**Biome** A grouping of similar plant and animal communities into broad landscape units found growing under similar environmental conditions.

**Biosphere** The part of the Earth system comprising all ecosystems and living organisms in the atmosphere, on land (terrestrial biosphere), or in the oceans (marine biosphere), including derived dead organic matter such as litter, soil organic matter, and oceanic detritus.

**Black carbon** Operationally defined species based on measurement of light absorption and chemical reactivity and/or thermal stability : consists of soot, charcoal, and/or possible light absorbing refractory organic matter.

**Bog** A poorly drained area rich in accumulated plant materials, frequently surrounding a body of open water and having a characteristic flora (such as sedges, heaths, and sphagnum).

**Bottom-up models** A modeling approach that includes technological and engineering details in the analysis.

**Breakwater** An offshore wall or jueety that, by breaking the force of the wave, protects a harbour, anchorage, beach, or shore area.

**Capacity building** A process of developing the technical skills and institutional capability in developing countries and economics in transition to enable them to participate in all aspects of adaptation to mitigation of and research on climate change, and the implementation of the Kyoto Mechanisms, etc.

**Carbon cycle** The flow of carbon (in various forms such as a carbon dioxide) through the atmosphere, ocean, terrestrial biosphere, and lithosphere.

**Carbon dioxide ($CO_2$)** A naturally occurring gas, and also a by-product of burning fossil fuels and biomass, as well as land-use changes and other industrial processes. It is the main anthropogenic greenhouse gas that affects the Earth's radiate balance, and the reference gas against which other GHG's are measured; has a Global Warming Potential of 1.

**Carbon dioxide fertilization** The stimulation of plant growth as a result of increased atmospheric carbon dioxide concentration. Depending on their mechanism of photosynthesis, certain types of plants are more sensitive to changes in atmospheric carbon dioxide concentration. In particular, plants that produce a three-carbon compound ($C_3$) during photosynthesis- including most trees and such agricultural corps as rice, wheat, soybeans, potatoes and vegetables-generally show a larger response than plants than produce a four-carbon compound ($C_4$) during photosynthesis-mainly of tropical origin, including grasses and the important crops maize, sugar cane, millet and sorghum.

**Carbon Flux** Transfer of carbon from one carbon pool to another in units of measurement of mass per unit area and time (*e.g.* tC).

**Carrying Capacity** The number of individuals in population that the resources of a habitat can support.

**Catchments** An area that collects and drains rainwater.

**Certified Emission Reaction (CER) Unit** Equal to 1 ton (metric ton) of carbon dioxide-equivalent emissions reduced or sequestered through a Clean Development Mechanism (CDM) project, calculated using Global Warming Potentials.

**Changes' Disease** A parasitic disease caused by *Trypanosoma cruzi* and transmitted by triatomine bugs with two clinical periods : acute (fever, swelling of the spleen, edemas) and chronic (digestive syndrome, potentially fatal heart condition).

**Chlorofluorocarbons (CFCs)** Greenhouse gases (GHG's) covered under the 1987 Montreal Protocol and used for refrigeration, air conditioning, packaging, insulation, solvents, or aerosol propellants. They are not destroyed I the lower atmosphere but can drift into the upper atmosphere where, under suitable conditions, they break down ozone. CFCs are being replaced by other compounds, including hydro chlorofluorocarbons and hydro fluorocarbons, which are GHGs covered under the Kyoto Protocol.

**Clean Development Mechanism (CDM)** This is intended to meet two objectives : (i) to assist Parties not included in Annex I in achieving sustainable development and in contributing to the ultimate objective of the convention; and (ii) to assist Parties included in Annex I in achieving compliance with their quantified emission limitation and reduction commitments (CER Units from CDM projects undertaken in non-Annex I countries that limit or reduce GHG emissions, when certified by operational entities designated by Conference of the Parties/Meeting of the Parties, can be accrued to the investing government or industry from Parties in Annex B. A share of the proceeds from the certified project activities is used to meet administrative expenses and to assist those developing country Parties which are specially vulnerable to the negative effects of climate change too meet the costs of adaptation.

**Climate** In a narrow sense, defined as the 'average weather' or more rigorously, as the statistical description in terms of the mean and variability of relevant quantities over a time period ranging from months to centuries or millennia. The classical period is 3 decades, as defined by the World Meteorological Organisation (WMO). These quantities are very often surface variables such as temperature, precipitation and wind.

**Climate Change** (i) Refers to any change in climate over time, whether due to natural variability or as a consequence of human activity (ii) The United Nations Framework Convention on Climate Change (UNFCCC) defines it as a change of climate which is attributed directly or indirectly to human activity that alters the composition of the global atmosphere and which is in addition to natural climate variability observed over comparable time periods.

**Climate Feedback** An interaction mechanism between processes in the climate system, when the result of an initial process triggers changes in a second process that in turn influences the initial one. A positive feedback intensifies the original process and a negative feedback reduces it.

**Climate Impact Assessment** Consequences of climate change on natural and human systems. Depending on adaptation, one can distinguish between potential impacts and residual impacts.

**Climate Impacts** Consequence of climate change on natural and human systems. Depending on adaptation, one can distinguish between potential impacts and residual impacts.

**Climate Impacts** Consequences of climate change on natural and human systems. Depending on the consideration of adaptation, one can distinguish between potential impacts and residual impacts. Potential Impacts–All impacts that may occur given a projected change in climate, without considering adaptation. Residual Impacts- The impacts of climate change that would occur after adaptation.

**Climate Model** (Hierarchy) A numerical representation of the climate system based on the physical, chemical, and biological properties of its components, their interactions and feedback processes, that can account for all or some of its known properties. The climate system is represented by models of varying complexity (*i.e.* for any one component or combination of components, a hierarchy of models can be identified; these models differ in such aspects as the number of spatial dimensions; the extent to which physical, chemical or biological processes are explicitly represented; or the level at which empirical parameterizations are involved. Coupled atmosphere/ ocean/ sea-ice General Circulation Models (AOGCMs) comprehensively represent the climate system. There is an evolution towards more complex models with active chemistry and biology. Climate models are being applied, as an effective research tool, not only to study and simulate the climate, but also for operational purposes, including monthly, seasonal, and interannual climate predictions.

**Climate Prediction** A climate prediction or forecast is attempted to produce a most likely description or estimate of the actual evolution of the climate in the future (*e.g.* at seasonal, interannual, or long-term time scales).

**Climate Projection** A projection of the response of the climate system to emission or concentration scenarios of greenhouse gases and aerosols, or radiative forcing scenarios, usually based upon simulations by climate models. Climate projections may be distinguished from climate predictions so as to emphasis that the projections depend upon the emission/ concentration/ radiative forcing scenario used, which are based on assumptions, concerning, for example, future socioeconomic and technological developments that may or may not be realized and are, therefore, subject to considerable uncertainty.

**Climate Scenario** A plausible, simplified representation of the future climate, based on an internally consistent set of climatological relationships, constructed for clear use in studying the potential consequences of anthropogenic climate change, often serving as input to impact models. Climate projections act as the raw materials for creating climate scenarios, but these scenarios usually also require additional information, *e.g.* about the current climate, A 'climate change scenario' represents the difference between a climate scenario and the current climate.

**Climate sensitivity** Equilibrium climate sensitivity refers to the equilibrium change in global mean surface temperature following a doubling of the atmospheric (equivalent) carbon dioxide concentration. More generally, it also refers to the equilibrium change in surface air temperature following a unit change in radioactive forcing $(C^\circ/Wm^{-2})$.

**Climate System** The highly complex consisting of five major components and their interactions : the atmosphere, the hydrosphere, the cryosphere, the land surface, and the biosphere. It evolves in time under the influence of its own internal dynamics and because of external forcing such as volcanic eruptions, solar variations and human-induced forcings such as the changing composition of the atmosphere and land use.

**Climate Variability** Variations I the mean state and other statistics of the climate on all temporal and spatial scales beyond that of individual weather events. It may be due to natural internal processes within the climate system (internal variability), or to variations in natural or anthropogenic external forcing (external variability).

**Chlorine Equivalents** Ozone recovery depends on how fast stratospheric concentration soft both chlorine and bromine decline. Those concentrations have been calculated into equivalent stratospheric chlorine loading values to provide an estimate for future ozone depletion.

**Coccolithophore** Microscopic single celled marine algae (phytoplankton) which secrete carbonate plates known as coccoliths.

**Conveyor belt** Term used to describe the cycling and recycling of water through the world's oceans.

**Coping Range** The variation in climatic stimuli that a system can absorb without producing significant impacts.

**Coral Bleaching** The paling in colour of corals resulting from a loss of symbiotic algae. Bleaching occurs in response to physiological shocks in response to abrupt changes in temperature, salinity, and turbidity.

**Cost-effective** Manner in which some technology or measure delivers a good or service at equal or lower cost than current practice, or the least-cost alternative for the achievement of some specific target.

**Cryosphere** The component of the climate system consisting of all snow, ice, and permafrost on and beneath the surface of the Earth and ocean.

**Cryptosporidiosis** An opportunistic infection caused by an intestinal parasite common in animals. Transmission occurs through ingestion of food or water contaminated with animal faces. The parasite causes severe chronic diarrhoea, especially in people with HIV.

**Deepwater** It forms when sea water freezes to sea ice. The local release of salt and resulting increase in water density makes coldwater saline and sink to the ocean floor.

**Deep Western Boundary Current** A thermohaline current, steered by the Carioles effect, which flows along the western margin of an ocean basin.

**Deepwater Formation** Occurs when seawater freezes to form sea ice. The local release of salt and the resulting increase in water density leads to the formation of saline coldwater that sinks to the ocean floor.

**Demand-side Management** Policies and programmes designed for a specific purpose, *e.g.*, reducing greenhouse gas (GHG) emissions to influence consumer demand for goods and/or services, *e.g.* to reduce consumer demand for electricity and other energy sources.

**Deposit-refund System** Combines a deposit or fee (tax) on a commodity with a refund or rebate for implementation of a specified action (*e.g.* emission reduction).

**Desert** An ecosystem having less than 100mm precipitation per year.

**Desertification** Land degradation in arid, semi-arid, and dry sub-humid areas resulting room climatic variations and human activities. Land degradation is a reduction or loss in arid, semi-arid, and dry sub-humid areas of the biological or economic productivity and complexity of rain-fed cropland, irrigated cropland, or range, pasture, forest and woodlands resulting from land uses or from processes including those arising from human activities and habitation patterns, such as : (i) soil erosion caused by wind and/or water (ii) deterioration of the physical, chemical and biological or economic properties of soil; and (iii) long-term loss of natural vegetation.

**Detection and Attribution** Climate varies continually on all time scales. Detection of climate change shows that climate has changed in some defined statistical sense, without providing a reason for that change. Attribution of causes of climate change establishes the most likely causes for the detected change.

**Dimethylsulphide (DMS)** The breakdown product of a chemical released from some marine plankton. DMS react so form cloud condensation nuclei (CNN) affect the appearance and radioactive properties of clouds, and also the lifetime and amount of cloud.

**Disturbance Regime** Frequency, intensity, and types of disturbances, such as fires, insect or pest outbreaks, floods and droughts.

**Diurnal Temperature Range** The difference between the maximum and minimum temperature during a day.

**Double Dividend** The effect that revenue-generating instruments, such as carbon taxes or auctioned (tradable)carbon emission permits, can : (*i*) limit or reduce GHG emissions, and (*ii*) offset at least part of the potential welfare losses of climate policies through recycling the revenue in the economy to reduce other potential distortionary taxes.

**Downscaling** Reducing the scale of a model from a global to regional level.

**Drought** The phenomenon of significantly subnormal precipitation, causing serious hydrological imbalances that adversely affect land resource production systems.

**Economic Potential** Economic potential is the portion of technological potential for greenhouse gas (GHG) emission reduction or energy efficiency improvements that might be achieved cost-effectively by creating markets, reducing market failures, or increasing financial and technological transfers.

**Economies in Transition (EITs)** Countries with national economies in the process of changing from a planned economic system to a market economy.

**Ecosystem** A distinct system of interacting living organisms, along with their physical environment. Has somewhat arbitary boundaries; depending on the focus of interest or study. Its extent can range from very small spatial scales to ultimately, the whole Earth.

**Ecosystem** Services Ecological processes or functions of value to individuals or society.

**Ecotone** Transition area between adjacent ecological communities (*e.g.* between forest and grasslands), usually involving competition between organisms common to both.

**Edaphic** Of or relating to the soil; factors inherent in the soil.

**Effective Rainfall** That portion of the total rainfall which is available for plant growth.

*El Nino***-Southern Oscilation (ENSO)** *El Nino*, in its original sense, is a warm water current that periodically flows along the coast of Ecuador and Peru and disrupts the local fishery. It is associated with a fluctuation of the intertropical surface pressure pattern and circulation in the Indian and Pacific Oceans, called the Southern Oscillation. This coupled atmosphere ocean phenomenon is collectively known as *El Nino*-Southern Oscillation, or ENSO. During an *El Nino* event, the prevailing trade winds weaken but the equatorial countercurrent strengthens, causing warm surface waters in the Indonesian area to flow eastward to overlie the cold waters of the Peru current. The ENSO has great impact on the wind, sea surface temperature, and precipitation patterns in the tropical Pacific. It has climatic defects throughout the Pacific region and in many other parts of the world. The opposite of an *El Nino* event in termed *La Nina*.

**Emissions** The release of GHGs and/or their precursors and aerosols into the atmosphere over a specified area and period of time.

**Emission Scenario** A plausible representation of the future development of emissions of substances that are potentially radioactively active (*e.g.*, greenhouse gases, aerosols), based on a coherent and internally consistent set of assumptions about such driving forces as demographic and socioeconomic development, technological change, and their relationships.

**Emissions Reduction Unit (ERU)** Equal to 1 metric ton of carbon dioxide emissions reduced or sequestered arising from a joint Implementation (defined in the Kyoto Protocol) project calculated using Global Warming Potential.

**Emissions Tax** Levy imposed by a government on each unit of carbon dioxide-equivalent emissions by a source subject to the tax. Since virtually all of the cabin in fossil fuels is ultimately emitted as carbon dioxide, a levy on the carbon content of fossil fuels–a carbon tax–is equivalent to an emissions tax for emissions caused by fossil-fuel combustion. An energy tax-a levy on the energy content of fuels is aimed to reduce demand for energy (and hence reduces $CO_2$ emissions) from fossil-fuel use. International emissions/carbon/energy tax is a tax imposed on specified sources in participating countries by an international agency. The revenue is distributed or used as specified by participating countries or the international agency.

**Emissions Trading** A market-based approach to achieve environment objectives. It allows those reducing greenhouse gas (GHG) below what is required, to use or trade the excess reductions to offset emissions at another source inside or outside the country.

**Endemic** Restricted or peculiar to a locality or region.

**Endorheic Lake** A lake with no outflow; a closed take.

**Energy Balance** Averaged over the globe and over longer time periods, the energy budget of the climate system must be in balance. Because the climate system derives all its energy from the stun, this balance implies that, globally, the amount of incoming solar radiation must on average be equal to the sum of the outgoing reflected solar radiation and the outgoing infrared radiation emitted by the climate system. Any perturbation of this global radiation balance, whether human-induced or natural, is termed radioactive forcing.

**Environmentally Sound Technologies(ESTs)** Technologies that protect the environment, are less polluting, use resources sustainably, recycle more of their wastes and products, and handle residual wastes in a more acceptable manner than other technologies for which they were substitutes. The ESTs imply mitigation and adaptation technologies, hard and solve technologies.

**Enzootic** A disease affecting the animals in an area. Corresponds to an endemic disease among humans.

**Epidemic** Occurring suddenly in numbers much more than normal expectancy, said especially of infectious diseases but applied also to any disease, injury, or other health-related event occurring in such outbreaks.

**Equilibrium and Transient Climate Experiment** An 'equlibrium climate experiment' is one in which a climate model is allowed to fully adjust to a change in radiative forcing. These experiments point to the difference between the initial and final states of the model, but not on the time-dependent response. This latter may be analyzed if the forcing is allowed to evolve gradually according to a prescribed emissions scenario. Such an experiment then becomes a 'transient climate experiment'.

**Equivalent Carbon Dioxide** The concentration of carbon dioxide that would cause the same amount of radiative forcing as a given mixture of carbon dioxide and other greenhouse gases.

**Erosion** The process of removal and transport of soil and rock by weathering, mass wasting, and the action of streams, glaciers, waves, winds, and underground water.

**Eutrophication** The process by which a body of water (often shallow) is enriched (either naturally or by pollution) in dissolved nutrients with a seasonal deficiency in dissolved oxygen.

**Evaporation** The process by which a liquid becomes a gas.

**Evapotranspiration** The combined process of evaporation from the Earth's surface and transpiration from vegetation.

**Exorheic Lake** A take drained by out flowing rivers.

**Exposure** The nature and degree to which a system is exposed to significant climatic variations.

**Exposure Unit** An activity, group, region, or resource that is subjected to climatic stimuli.

**Externalities** Byproducts of activities that affect the well-being of people or the environment, where those impacts are not reflected in market prices. The costs or benefits associated with externalities do not enter cost-accounting statements.

**Extinction** The complete disappearance of an entire species.

**Extirpation** The disappearance of a species from a part of its range; local extinction.

**Extreme Weather Event** An event that is rare within its statistical reference distribution at a particular place. An extreme weather event would normally be a rare as or rarer than the 10th or 90th percentile. The characteristics of what is called 'extreme weather' may vary from place to place. An 'extreme climate event is an average of a number of weather events over a certain period of time, an average which is itself extreme (*e.g.* rainfall over a season).

**Feedback** A process that triggers changes in a second process that in turn influences the original one; a positive feedback intensifies the original process, and a negative feedback reduces it.

**Fen** Low land covered wholly or partly with water unless artificially drained.

**Food Insecurity** People lacking secure access to sufficient amounts of safe and nutrition food for normal growth and development and for active and healthy life. Caused by the unavailability of food, insufficient purchasing power, inappropriate distribution, or inadequate use of food at the household level. May be chronic, seasonal, or transitory.

**Foraminifera** Single-celled marine animals (protozoa) which usually secrete a carbonate shell.

**Forest** A vegetation type dominated by trees.

**Fossil Fuels** Carbon-based fuels from fossil carbon deposits such as coal, oil, and natural gas.

**Freshwater Lens** A lenticular fresh groundwater body that underlies an oceanic Island. It is underlain by saline water.

**Fuel Switching** Reducing carbon dioxide emissions by switching to low-carbon fuels, suck as from coal to natural gas.

**Functional Diversity** The number of functionally different organisms in an ecosystem (also referred to as 'functional types' and 'functional groups').

**General Circulation** The large scale motions of the atmosphere and the ocean as a consequence of differential heating on a rotating Earth, aiming to restore the energy balance of the system through transport of heat momentum.

**General Equilibrium Analysis** An approach that considers simultaneously all the markets in an economy, allowing for feedback effects between individual markets.

**Geomorphic** Pertaining to the form of the Earth or its surface features.

**Glacial** The cold part of a fixed cycle of warm and cool periods during a major ice age. The cycle is related to changing heat from the sun due to shifts in the pattern of Earth's orbit and tilt.

**Glacier** A mass of land ice flowing downhill (by internal deformation and sliding at the base) and constrained by the surrounding topography (*e.g.*, the sides of a valley or surrounding peaks); the bedrock topography is the major influence on the dynamics and surface slope of a glacier. Maintained by accumulation of snow at high altitudes; balanced by melting at low altitudes or discharge into the sea.

**Global Surface Temperature** The global surface temperature is the area-weighted global average of : (i) the sea surface temperature over the oceans (*i.e.*, the sub-surface bulk temperature in the first few meters of the ocean), and (ii) the surface air temperature over land at 1.5 m above the ground.

**Greenhouse Effect** Greenhouse gases (GHGs) effectively absorb infrared radiation emitted by the Earth's surface, by the atmosphere itself due to the same gases, and by clouds. Atmospheric radiation is emitted to all sides, including downward to the Earth's surface. Thus GHGs trap heat within the surface troposphere system-called the 'natural greenhouse effect'. Atmospheric radiation is intimately coupled to the temperature of the level at which it is emitted. In the troposphere, the temperature generally decreases with height. Effectively, infrared radiation emitted to space originates from an altitude with a temperature of, on average,– 19°C, in balance with the net incoming solar radiation, whereas the Earth's surface remains at a higher temperature of, on average, 14°C. An increase in the concentration of GFGs increases infrared opacity of the atmosphere, and hence holds to an effective radiation into space from a higher altitude at a lower temperature, generating a radioactive forcing, an imbalance that can only be compensated for by an increase of the temperature of the surface-troposphere system. This is termed the 'enhanced greenhouse effect'.

**Greenhouse Gases (GHGs)** GHGs are those gaseous constituents of the atmosphere, both natural and anthropogenic, that absorb and emit radiation at specific wavelengths within the spectrum of infrared radiation emitted by the Earth's surface, the atmosphere and clouds. This property causes the greenhouse effect. Water vapour ($H_2O$), carbon dioxie ($CO_2$), nitrous oxide ($N_2O$), methane ($CH_4$), and ozone ($O_3$) are the primary GHGs in the Earth's atmosphere. Also, there are several entirely human-made atmospheric GHGs, *e.g.* the halocarbons and other chlorine and bromine containing substances. Beside $CO_2$, $N_2O$ and $CH_4$, the Kyoto Protocol deals with the GHGs sulphur hexafluoride ($SF_6$), hydrofluorocarbons (FHCs) and perfluorocarbons (PFCs).

**Groin** A low, narrow jetty, usually extending roughly perpendicular to the shoreline, designed to protect the shore from erosion by currents, tides, or waves, or to trap sand for the purpose of building up a beach.

**Gross Domestic Product (GDP)** The sum of gross value added, at purchasers' prices, by all resident and non-resident producers in the economy, plus any taxes and minus any subsidies not included in the value of the products in a country or a geographic region for a given period of time, normally 1 year. It is calculated without deducing for depreciation of fabricated assets or depletion and degradation of natural resources. GDP is an often used though incomplete measure of welfare.

**Gross Primary Production (GPP)** The amount of carbon fixed from the atmosphere through photosynthesis.

**Groundwater Recharge** The process by which external water is added to the zone of saturation of an aquifer, either directly into a formation or indirectly by way of another formation.

**Habitat** The particular place where an organism or species lives; a locally circumscribed portion of the total environment.

**Halocarbons** Carbon compounds containing either chlorine, bromine, or fluorine, they act as powerful greenhouse gases in the atmosphere. The chlorine- and bromine0containing halocarbons also deplete the ozone layer.

**Halocline** A layer in the ocean in which the rate of salinity variation with depth is much larger than layers immediately above or below it.

**Heat Island** An area within an urban area characterized by ambient temperatures greater than those of the surrounding area because of the absorption of solar energy by materials, like asphalt.

**Health** Any of the various low-growing shrubby plants of open wastelands, usually growing on acidic, poorly drained soils.

**Hedging** Balancing the risks of acting too slowly (*e.g.* in climate change mitigation) against acting too quickly; depends on society's attitude towards risks.

**Human Settlement** A place or area occupied by settlers.

**Human System** Any system in which human organizations play a major role. Often, the term is synonymous with 'society' or 'social system' (*e.g.*, agricultural system, political system, technological system, economic system).

**Hydrofluorocarbons (HFCs)** Among the six greenhouse gases (GHGs) that have to be curbed under the Kyoto Protocol. Are produced commercially as a substitute for chlorofluorocarbons (CFCs). Hydrofluorocarbons mostly are used in refrigeration and semiconductor manufacturing. Their Global Warming Potentials range from 1,300 to 11,700.

**Hydrosphere** The component of the climate system composed of liquid surface and subterranean water, such as oceans, seas, rivers, freshwater lakes, underground water, etc.

**Hypoolimnion** the part of a lake below the thermocline made up of stagnant water of essentially uniform temperature except during the period of overturn.

**Ice Cap** A dome shaped ice mass coffering a highland area that is considerably smaller in extent than an ice sheet.

**Ice Sheet** A sufficiently deep mass of land ice to cover most of the underlying bedrock topography, so that its shape is determined by its internal dynamics (the flow of the ice as it deforms internally and slides at its base). There are only two large ice sheets in the modern world-on Greenland and Antarctica.

**Ice Shelf** A floating ice sheet of considerable thickness attached to a coast (usually of great horizontal extent with a level or gently undulating surface); often a seaward extension of ice sheets.

**Immunosuppression** Reduced functioning of an individual's immune system.

**Industrial** Revolution a period of rapid industrial growth with important social and economic consequences, beginning in England during the second half of the 18$^{th}$ century; it then spread to Europe and USA. The invention of the steam engine triggered this development. The Revolution marked the beginning of a great increase in the use of fossil fuels and emission of, in particular, fossil carbon dioxide. The terms 'preindustrial' and 'industrial' refer, arbitrarily, to the periods before and after the year 1750, respectively.

**Inertia** Delay or resistance in the response of the climate, biological, or human systems to factors that after their rate of change, including continuation of change in the system after the cause of that change has disappeared.

**Intrared Radiation** Radiation emitted by the Earth's surface, the atmosphere, and clouds. Has a distinctive range of wavelengths ('spectrum') longer than the wavelength of the red colour in the visible part of the spectrum. The spectrum of infrared radiation is practically distinct from that of solar or short-wave radiation because of the difference in temperature between the Sun and the Earth-atmosphere system.

**Infrastructure** The basic equipment, utilities, productive enterprises, installations, and services essential for the development, operation, and growth of an organization, city, or, nation.

**Integrated Assessment** Analysis that combines results and models from the physical, biological, economic, and social sciences, and the interactions between these components, in a consistent framework to evaluate the status and the consequences of environmental change and the policy responses to it.

**Interaction Effect** The result of the interaction of climate change policy instruments with existing tax systems, including both cost-increasing tax interaction and cost-reducing revenue-recycling effect. The former reflects the impact that greenhouse gas (GHG) policies can have on the functioning of labour and capital markets through their effects on real wages and the real return to capital. By restricting the allowable GHG gas emissions, permits, regulations, or a carbon tax raise the cost of production and the prices of output, thus reducing the real return to labour and capital. For policies that raise revenue for the government (carbon taxes and auctioned permits), the revenues may be recycled to reduce existing distortionary taxes.

**Interdecadal Pacific Oscillation (IPO)** A long timescale oscillation in the ocean-atmosphere system that shifts climate in the Pacific region every one to three decades.

**Interglacial** The warm part of the ice age cycle.

**Introduced Species** A species occurring in an area outside its historically known natural range as a result of accidental dispersal by humans (also referred to as 'exotic species' or 'alien species).

**Joint Implementation (JI)** A market based implementation mechanism defined under the Kyoto Protocol that allows Annex I countries or companies from these countries to implement projects jointly that limit or reduce emissions, or enhance sinks, and to share the Emissions Reduction Units. It activity is also permitted by the United Nations Framework Convention on Climate Change (UNFCCC).

**Kyoto Mechanisms** Economic mechanisms based on market principles that Parties to the Kyoto Protocol can use with a view to lessening the potential economic impacts of greenhouse gas emission-reduction requirements.

**Kyoto Protocoal** The Kyoto Protocol was adopted at the Third Session of the Conference of the Parties (COP) to the United Nations Framework Convention on Climate Change (UNFCCC) in 1997 in Kyoto (Japan). Incorporates legally binding commitments, besides those included in the UNFCCC Countries included in Annex B of the Protocol (most OECD countries and EITs) agreed to reduce their anthropogenic GHG emissions by at least 5 per cent below 1990 levels in the commitment period 2008 to 2012. Kyoto Protocol has entered into force recently.

**Land Use** The arrangements, activities, and inputs undertaken in a certain land-cover type (a set of human actions). The social and economic purposes (*e.g.* grazing, timbre extraction, conservation) for which land is managed.

**Landslide** A mass of material that has slipped downhill by gravity, often assisted by water when the material is saturated.

**Large-Scale Singularities :** Abrupt and dramatic changes in systems in response to smooth changes in driving forces. For example, a gradual increase in atmospheric GHG concentrations may lead to such large-scale singularities as slowdown or collapse of the thermohaline circulation or collapse of the West Antarctic Ice Sheet.

**Leaching Removal** of soil elements or applied chemicals through percolation.

**Leakage** The part of emissions reductions in Annex B countries that may be offset by an increase of the emission in the non-constrained countries above their baseline levels, through; (i) relocation of energy-intensive production in non-constrained regions; (ii) increased consumption of fossil fuels in these regions through decline in the international price of oil and gas triggered by lower demand for these energies; and (iii) changes in incomes (hence in energy demand) because of better terms of trade. Also refers to the situation in which a carbon sequestration activity (*e.g.*, tree planting) on one piece of land inadvertently, directly or indirectly, triggers an activity, which in whole or part counteracts the carbon effects of the initial activity.

**Leapfrogging** The opportunities in developing countries to bypass sea earl stages of technology development, historically observed in industrialized countries, and apply the most advanced presently available technologies in the energy and other economic sectors, through investments in technological development and capacity building.

**Limnology** Study of lakes and their biota.

**Lithosphere** The upper layer of the solid Earth, both continental and oceanic, which is composed of all crystal rocks and the cold, mainly elastic, part of the uppermost mantle.

**Littoral Zone** A coastal region; the shore zone between high and low watermarks.

**Local Agenda 21** Local Agenda 21s are the local plans for environment and development that each local authority is expected to develop in consultation with its population, as a means of reorienting their policies, plans, and operations towards the achievement of sustainable development goals.

**Lock-in Technologies and Practices** Technologies and practices that have market advantages arising from existing institutions, services, infrastructure, and available resources and are difficult to change because of their widespread use and the presence of associated infrastructure and socio-cultural patterns.

**Maladaptations** Any changes in natural or human systems that inadvertently increase vulnerability to climatic stimuli; an adaptation that fails in reducing vulnerability but increases it instead.

**Market Impacts** Impacts linked to market transactions, and which directly affect gross domestic product (GDP, a country's national accounts)–for example, changes in the supply and price of agricultural goods.

**Mean Sea Level (MSL)** The average relative sea level over a period, such as a month or a year, long enough to average out transients such as waves.

**Metazoan** An animal whose body consists of many cells. See also protozoan.

**Methylbromide** A widely-used chemical that threatens the ozone layer. It releases bromine, which is 30 to 60 times as destructive to ozone as is chlorine. Methylbromide ($CH_3Br$) is used as a fumigant for soils and commodities, including the quarantine treatment of some products for international trade, and as a transport fuel additive. Total annual anthropogenic release is more than double the 40,000 tons released annually in the early 1980s. Besides, natural sources such as biomass burning contribute another 30,000 to 50,000 tons a year. More than half of all the $CH_3Br$ produced is released into the atmosphere, where concentrations are currently between 8 and 15 parts per thousand million by volume. No single alternative chemical is at present available to replace the various methyl bromide applications.

**Microbial Loop** Complex food web involving bacteria, unicellular animals and plants, viruses, and sissolved and particulate organic material. Dissolved and particulate material, released from organisms, is utilized by bacteria, which are grazed by protozoa which is turn are grazed by metazoan. Around 50 per cent or more of primary production passes through the microbial loop rather than along the classical food chain of phytoplankton to herbivore.

**Microclimate** Local climate at or near the Earth's surface.

**Mitigation** An anthropogenic intervention to reduce the sources or enhance the sinks of greenhouse gases.

**Mixed Layer** The upper region of the ocean well-mixed by interaction with the overlying atmosphere.

**Model** A computer programme that can simulate ozone changes given the changes in the lactors affecting ozone.

**Monsoon** Wind in the general atmospheric regulation typified by a seasonal persistent wind direction and by a pronounced change in direction from one season to the next.

**Montane** The biogeographic zone made up of relatively moist, cool upland slopes below timberline and characterized by the presence of large evergreen trees as a dominant life form.

**Montreal Protocol** The Montreal Protocol on Substances that Deplete the Ozone Layer was adopted in Montreal in 1987, and subsequently adjusted and amended in London (1990), Copenhagen (1992), Vienna (1995), Montreal (1997), and Beijing (1999). It controls the consumption and production of chlorine- and bromine-containing chemicals that destroy stratospheric ozone, such as chlorofluorocarbons (CFCs), methyl chloroform, carbon tetrachloride, and others.

**Morbidity** Rate of occurrence of disease or other health disorder within a population, taking account of the age-specific morbidity rates. Health outcomes include chronic disease incidence/prevalence, rates of hospitalization, primary care consultations, disability-days (*i.e.*, days when absent from work), and prevalence of symptoms.

**Mortality** Rate of occurrence of death within a population within a specified time period; calculation of mortalit6y takes account of age-specific death rates, and can thus yield measures of life expectancy and the extent of premature death.

**Nanoplankton** Phytoplankton whose dimensions range from 10 to 50 mm.

**Net Biome Production (NBP)** Net gain or loss of carbon from a region, NBP is equal to Net Ecosystem Production minus the carbon lost due to a disturbance (*e.g.*, a forest fire or a forest harvest).

**Net Carbon Dioxide Emissions** Difference between sources and sinks of carbon dioxide in a given period and specific area or region.

**Net Ecosystem Production (NEP)** Net gain or loss of carbon from an ecosystem. NEP is equal to Net Primary Production minus the carbon lost through heterotrophic respiration.

**Net Primary Production (NPP)** The increase in plant biomass or carbon of a unit of a landscape. NPP is equal to Gross Primary Production minus carbon lost through autotrophic respiration.

**Nitrogen Fertilisation** Enhanbcement of plant growth through the addition of nitrogen compounds. Typically refers to fertilization from anthropogenic sources of nitrogen such as human made fertilisers and nitrogen oxides released from burning fossil fuels.

**Nitrogen Oxides (NO$_x$)** Any of several oxides of nitrogen.

**Nitrous Oxide (N$_2$O)** A powerful GHG emitted through soil cultivation practices, especially the use of commercial and organic fertilizers, fossil-fuel combustion, nitric acid production, and biomass burning. One of the six greenhouse gases to be curbed under the Kyoto Protocol.

**No Regrets Policy** One that would generate net social benefits whether or not there is anthropogenic climate change.

**Non-Annex I Countries/Parties** The countries that have ratified or acceded to the United Nations Framework Convention on Climate Change (UNFCCC) that are not included in Annex 1 of the Climate Convention.

**Non-Linearity** A process is described as 'nonlinear' when there is no simple proportional relation between cause and effect.

**Non-Market Impacts** Impacts that affect ecosystems or human welfare, but which are not directly linked to market transactions; for example, an increased risk of premature death.

**Non-Point-Source Pollution** Pollution from sources that cannot be defined as discrete points, such as areas of crop production, timber, surface mining, disposal of refuse, and construction.

**No-regrets Policy** One that would generate net social benefits whether or not there is anthropogenic climate change.

**Non-Annex I Countries/Parties** The countries that have ratified or acceded to the United Nations Framework Convention on Climate Change (UNFCCC) that are not included in Annex I of the Climate Convention.

**Non-Linearity** A process is described as 'nonlinear' when there is no simple proportional relation between cause and effect.

**Non-Market Impacts Impacts** that affect ecosystems or human welfare, but which are not directly linked to market transctions; for example; an increased risk of premature death.

**Non-Point-Source Pollution** Pollution from sources that cannot be defined as discrete points, such as areas of crop production, timber, surface mining, disposal of refuse, and construction.

**Non-regrets Policy** A policy that would generate net social benefits regardless of whether or not climate changes. No regrets opportunities for greenhouse gas emissions reduction are those options whose benefits such as reduced energy costs and reduced emissions of pollutants equal or exceed their costs to society, excluding the benefits of avoided climate change. No-regrets potential refers to the gap between the market potential and the socio-economic potential.

**North Atlantic Oscillation (NAO)** The North Atlantic Oscillation consists of opposing variations of barometric pressure near iceland and near the Azores. It is the dominant mode of winter climate variability in the North Atlantic region ranging from central North America to Europe.

**Obligate Species** Species restricted to one particularly characteristic mode of life.

**Ocean Conveyor Belt** The theoretical route by which water circulates around the entire global ocean, driven by wind and the thermohaline circulation.

**Ocean Ventilation** Downswelling of water from near the surface to the deep ocean.

**Oligotrophic** : Relatively unproductive areas of the sea, lakes and rivers with low nutrient content.

**Opportunity Costs** The cost of an economic activity forgone by the choice of another activity.

**Orography** The study of the physical geography of mountains and mountain systems.

**Ozone ($O_3$)** Ozone, the triatomic form of oxygen ($O_3$), is a gaseous atmospheric constituent. In the troposphere it is formed both naturally and by photochemical reactions involving gases resulting from human activities (photochemical 'smog'). In high concentrations, tropospheric ozone can prove harmful to diverse living organisms. Tropospheric ozone is a greenhouse gas. In the stratosphere, ozone is formed by the interaction between solar ultraviolet radiation and molecular oxygen ($O_2$). Stratospheric ozone has critical role in the stratospheric radiative balance. Its concentration is highest in the ozone layer. Depletion of stratospheric ozone, due to chemical reactions that may be enhanced by climate change, results in an in creased ground-level flux of ultraviolet-ß radiation.

**Ozone Layer** A layer in the stratosphere in which the concentration of ozone is greatest. It extends from about 12 to 40 km. This layer is being depleted by human emissions of chlorine and bromine compounds. Every year, during the Southern Hemisphere spring, a very strong depletion of the ozone layer occurs over the Antarctic region, also caused by human-made chlorine and bromine compounds in combination with the specific meterorological conditions of that region. This phenomenon is called the ozone hole.

**Perfluorocarbons (PFCs)** Among the six greenhouse gases to be reduced under the Kyoto Protocol. Are by-products of aluminium smelting and uranium enrichment. They replace chlorofluorocarbons in manufacturing semiconductors. The Global Warming Potential of PFCs is 6500-9200 times that of carbon dioxide.

**Permafrost** Pernnially frozen ground that occurs wherever the temperature remains below $0°C$ for several years.

**Planned Adaptation** is the result of a deliverate policy decision, based on knowledge that conditions have changed or are about to change and that action is required to return to maintain, or achieve a desired state.

**Precursors** Atmospheric compounds which themselves are not greenhouse gases (GHGs) of aerosols, but which have an effect on GHG or aerosol concentrations by taking part in physical or chemical processes regulating their production or destruction rates.

**Primary Energy** Energy embodied in natural resources (*e.g.*, coal, crude oil, sunlight, uranium) that has not undergone any anthropogenic conversion or transformation.

**Primary Production** The conversion of solar energy into chemical energy via photosynthesis by marine plants in the surface waters of the ocean. During photosynthesis dissolved forms of carbon in the water are taken up by marine algae and converted to organic material.

**Profile** A smoothly changing set of concentrations representing a possible pathway towards stabilization. The word 'profile' is used to distinguish such pathways from emissions pathways, which are usually referred to as 'scenarios'.

**Projection (generic)** A projection is a potential future evaluation of a quantity or set of quantities, often computed with the aid of a model. Projections are distringuished from 'prediction' in order to emphasise that projections involve assumptions concerning, for instance, future socioeconomic and technological developments that may or may not be realized, and are therefore subject to substantial uncertainty.

**Proxy** A proxy climate indicator is a local record that is interpreted, using physical and biophysical principles, to represent some combination of climate-related variations back in time. Climate-related data derived in this way are referred to as proxy data. Examples; tree ring records, characteristics of corals, and some data derived from ice cores.

**Quasi-Biennial Oscillation (QBO)** Alternation of easterly and westerly wind regimes in the stratosphere in equatorial latitudes with a periodicity of roughly 24 to 30 months. The alternation has substantial effects on atmospheric transport. When the stratospheric winds are westerly, a 6 to 8 per cent ozone deficiency is observed in mid-polar latitudes. When they are easterly, a similar surplus is usually recorded.

**Radiative Forcing** The change in the net vertical irradiance (expressed in Wm$^{-2}$ at the tropopause due to an internal change or a change in the external forcing of the climate system, such as, a change in the concentration of carbon dioxide or the output of the Sun.

**Rangeland** Unimproved grasslands, shrublands, savannahs, and tundra.

**Reactive Adaptation** that occurs after impacts of climate change have been observed.

**Relative Sea level** Sea level measured by a tide gauge with respect to the land upon which it is situated.

**Renewables** Energy sources that are, within a short time frame relative to the Earth's natural cycles, sustainable, and include non-carbon technologies such as solar energy, hydropower, and wind, as well as carbon-neutral technologies such as biomass.

**Reserves** Refer to those occurrences that are identified and measured as economically and technically recoverable with current technologies and prices.

**Reservoir** A component of the climate system, other than the atmosphere, which can store, accumulate, or release a substance of concern (*e.g.*, carbon a greenhouse gas, or a precursor). Oceans, soils, and forests exemplify reservoirs of carbon. Pool is an equivalent term (but its definition often includes the atmosphere). The absolute quantity of substance of concerns, held within a reservoir at a specified time, is called the stock. The term also means an artificial or natural storage place for water, such as a lake, pond, or aquifer, from which the water may be withdrawn for such purposes as irrigation, water supply, or irrigation.

**Resilience** Amount of change a system can undergo without changing state.

**Resource base** It includes both reserves and resources.

**Resources** Those occurrences with less certain geological and/or economic characteristics, but which may be potentially recoverable with foreseeable technological and economic developments.

**Runoff** That part of precipitation that does not evaporate. In some countries, runoff implies surface runoff only.

**Salinisation** The accumulation of salts in soils.

**Saltwater intrusion/encroachment** Displacement of fresh surface water or groundwater by the advance of saltwater due to its greater density, usually in coastal and estruarine areas.

**Sea-level Rise** An increase in the mean level of the ocean. Eustatic sea-level rise is a change in global average sea level brought about by an alteration to the volume of the world ocean. Relative sea-level rise occurs where there is a net increase in the level of the ocean relative to local land movements. Whereas climate modelers focus on estimating eustatic sea-level change, impact researchers focus on relative sea-level change.

**Seawall** A human made wall or embankment along a shore to prevent wave erosion.

**Semi-arid Regions** Ecosystems that have more than 250 mm precipitation per year but are not highly productive; usually classified as rangelands.

**Sensitivity** The degree to which a system is affected, either adversely or beneficially, by climate-related stimuli. The effect may be direct (*e.g.*, a change in crop yield in response to a change in the mean, range, or variability of temperature) or indirect (*e.g.*, damages caused by an increase in the frequency of coastal flooding due to sea-level rise).

**Sequestration** Increasing the carbon content of a carbon reservoir other than the atmosphere. Biological approaches include direct removal of carbon dioxide from the atmosphere through land-use change, afforestation, reforestation, etc., to enhance soil carbon in agriculture. Physical approaches are separation and disposal of carbon dioxide from flue gases or from processing fossil fuels to produce hydrogen and carbon dioxide-rich fractions and long-term storage in underground in depleted oil and gas reservoirs, coal seams, and saline aquifers.

**Silt** Unconsolidated or loose sedimentary material whose constituent rock particles are finer than grains of sand but larger than clay particles.

**Sink** Any process, activity or mechanism that removes a greenhouse gas (GHG), an aerosol, or a precursor of GHG or aerosol from the atmosphere.

**Socio-economic Potential** The socio-economic potential represents the level of greenhouse gas mitigation that would be approached by a overcoming social and cultural obstacles to the use of cost-effective technologies.

**Soil Carbon Pool** The relevant carbon in the soil, including various forms of soil organic carbon (humus) and inorganic soil carbon and charcoal. Excludes soil biomass (*e.g.* roots, bulbs, etc.) and soil fauna (animals).

**Solar Radiation** Radiation emitted by the Sun, also referred to as short-wave radiation. It has a distinctive range of wavelengths (spectrum) determined by the temperature of the Sun.

**Southern Oscillation** A large-scale atmospheric and hydrospheric fluctuation centered in the equatorial Pacific Ocean, exhibiting a pressure anomaly, alternatively high over the Indian Ocean and high over the South Pacific. Its period is slightly variable, averaging 2.33 years. The variation in pressure is accompanied by variations in wind strengths, ocean currents, sea-surface temperatures, and precipitation in the surrounding areas.

**Spatial and Temporal Scales** Climate can vary on a broad range of spatial and temporal scales. Spatial scales range from local (less than 100,000 sq.km), through regional (100,000 to 10 million sq km) to continental (10 to 100 million sq km). Temporal scales range from seasonal to geological (up to hundreds to millions of years).

**SRES Scenarios** Special Report on Emissions Scenarios (SRES). These scenarios are emissions scenarios used, inter alia, as basis for the climate projections in the Third Assessment Report.

**Stabilisation** The achievement of stabilization of atmospheric concentrations of one or more greenhouse gases (*e.g.*, $CO_2$-equivalent group of GHGs).

**Stakeholders** Person or entity holding grants, concessions, or any other type of value that would be affected by a particular action or policy.

**Standards** Rules or codes mandating or defining product performance (*e.g.*, grades, dimensions, characteristics, test methods, and rules for use). International product and/or technology or performance standards fix minimum requirements for attested products and/or technologies in countries where they are adopted. The standards reduce greenhouse gas emissions associated with the manufacture or use of the products and/or application of the technology.

**Stimuli (Climate-Related)** All the elements of climate change, including mean climate characteristics, climate variability, and the frequency and magnitude or extremes.

**Stochastic Events** Events involving a random variable, chance, or probability.

**Storm Surge** The temporary increase, at some specific location, in the height of the sea due to low atmospheric pressure and/or strong winds. The surge is the excess above the level expected from the tidal variation alone at that time and place.

**Stratification** Generally occurs as surface waters warm in summer-spring and is broken down by mixing processes, such as strong winds and surface cooling, especially during winter.

**Stratified** Arranged in layers where lighter water overlies denser water, especially near the sea surface.

**Stratosphere** Highly stratified region of atmosphere above the troposphere extending from about 10 km (ranging from 9 km in high latitudes to 16 km in the tropics on average) to about 50 km.

**Streamflow** Water within a river channel, usually expressed in $m^3$ $sec^{-1}$.

**Sub-Antarctic Mode Water (SAMW)** A type of water in the Sub-Antarctic Zone of the Southern Ocean. The SAMW is the deep surface layer of water with uniform temperature and salinity created by convective processes in the winter. It can identified by a temperature of around$-1.8°C$ and salinity of around 34.4 PSU, and is separated from the overlying surface water by a halocline at around 50 m in the summer. Although not commonly considered to be a water mass. It contributes to the Central Water of the Southern Hemisphere, and is also responsible for the formation of Antarctic Intermediate Water in the Eastern part of the South Pacific Ocean. Sometimes also known as Winter Water.

**Subantarctic Water** A south-west Pacific Ocean surface water mass, with relatively low salt concentrations, cold temperatures, high concentrations of nutrients, located between the Subtropical and SubAntarctic Fronts. Production in these waters is iron-limited.

**Submergence** A rise in the water level in relation to the land, so that areas of formerly dry land become inundated; it results either from a sinking of the land or from a rise of the water level.

**Subsidence** The sudden sinking or gradual downward settling of the Earth's surface with little or no horixontal motion.

**Subtropical Water** A south-west Pacific Ocean surface water mass, with relatively high salt content concentrations, warm temperatures, low concentrations of nutrients, located north of the Subtropical Front.

**Succession** Transition in the composition of plant communities following disturbance.

**Sulfur Hexafluoride ($SF_6$)** : One of the six greenhouse gases to be curbed under the Kyoto Protocol. Largely used in heavy industry to insulate high-voltage equipment and to help in the manufacturing of cable-cooling systems. Its Global Warming Potential is 23,900.

**Surface Runoff** The water that travels over the soil surface to the nearest surface strem; runoff of a drainage basin that has not passed beneath the surface since precipitation.

**Sustainable Development** Development that meets the needs of the present without compromising the ability of future generations to meet their own needs.

**Synoptic** Relating to or displaying atmospheric and weather conditions as they exist simultaneously over a broad area.

**Taiga** Coniferous forests of northern North America and Eurasia.

**Targets and Time Tables** A target is the reduction of a specific percentage of greenhouse gas (GHG) emissions from a baseline date (*e.g.*, below 1990 levels') to be achieved by a set date or time Table (*e.g.*, 2008 to 2012). These targets and time tables are, in effect, an emissions cap on the total amount of GHG emissions that can be emitted by a country or region in a given time period.

**Technological Potential** The amount by which it is possible to reduce greenhouse gas emissions or improve energy efficiency by implementing some proven technology or practice.

**Technology** A piece of equipment or a technique for performing some particular activity.

**Technology Transfer** Processes that cover the exchange of knowledge, money, and goods among different stakeholders that lead to the spreading of technology for adapting to or mitigating climate change.

**Thermal Erosion** The erosion of ice-rich permafrost by the combined thermal and mechanical action of moving water.

**Thermal Expansion** For sea level, it refers to the increase in volume (and decrease in density) that results from warming water. A warming of the ocean leads to an expansion of the ocean volume and hence an increase in sea level.

**Thermocline** The region in the world's ocean. Typically at a depth of 1 km, where temperature decreases rapidly with depth and which marks the boundary between the surface and the ocean.

**Thermohaline Circulation/currents** The part of the Earth's ocean circulation system that is driven by differences in density of adjacent water masses. The density differences are a result of differences in temperature and salinity.

**Timberline** The upper limit of tree growth in mountains or high latitudes.

**Tolerable-windows Approaches** To analyze GHG emissions as they would be constrained by adopting a long-term climate–rather than GHG concentration stabilization–target (*e.g.*, expressed in terms of temperature or sea level changes or the rate of such changes). These approaches evaluate the implications of such long-term targets for short or medium-term "tolerable" ranges of global GHG emissions.

**Top-down Models** The terms "top" and "bottom" are shorthand for aggregate and disaggregated models. The top-down label drives from how modelers applied macro-economic theory and econometric techniques to historical data on consumption, prices, incomes, and factor costs to model final demand for goods and services, and supply from main sectors, like the energy sector, transportation, agriculture, and industry. Accordingly, top-down models evaluate the system from aggregate economic variables, as compared to bottom-up ones that consider technological options or project specific climate change mitigation policies.

**Total Column Ozone** The total amount of ozone in a column of air reaching from the Earth's surface to Space, measured in Dobson Units (1 DU = $2.673 \times 10^{16}$ molecules/ $cm^2$).

**Transient Climate Response** The globally averaged surface air temperature increase, averaged over a 20-year period, centered at the time of $CO_2$ doubling (*i.e.*, at year 70 in a 1% per year compound $CO_2$ increase experiment with a global coupled climate model).

**Transpiration** The emission of water vapour from the surfaces of leaves or other plant parts.

**Tropopause** The boundary between the troposphere and the stratosphere.

**Tropospehre** The lowest part of the atmosphere from the surface to about 10 km in altitude in mid-latitudes (ranging from 9 km in high latitudes to 16 km in the tropics on average) where clouds and "weather'" phenomena occur. In the troposphere, temperatures generally decrease with height.

**Tsunami** A large tidal wave produced by a submarine, landslide, or volcanic eruption.

**Tundra** A treeless, level, or gently undulating, plain characteristic or arctic and sub arctic regions.

**Ultraviolet (UV)-β Radiation** Solar radiation within a wavelength range of 280-315 nm, most of which is absorbed by stratospheric ozone. Enhanced UV-β radiation suppresses the immune system and can have other adverse effects on living organisms.

**Ungulate** A hoofed, typically herbivorous, quadruped mammal (such as a ruminant, swine, camel, horse, rhinoceros, or elephant).

**United Nations Framework Convention on Climate Change (UNFCCC)** The Convetion was adopted on 9 May 1992 in New York and signed at the 1992 Earth Summit in Rio de Janeiro by more than 150 countries and the European Community. Its ultimate objective is the 'stabilisation of greenhouse gas concentrations in the atmosphere at a level that would prevent dangerous anthropogenic interference with the climate'.

**Uptake** The addition of a substance of concern to a reservoir. The uptake of carbon-containing substances, in particular carbon dioxide, is often called (carbon) sequestration.

**Upwelling** Transport of deeper water to the surface, usually caused by horizontal movements of surface water.

**Urbanisation** The conversion of land from a natural state or managed natural state (such as agriculture) to cities; a process driven by net rural-to-urban migration through which an increasing percentage of the population in any nation or region come to live in settlements that are defined as 'urban centres'.

**Vector** An organism, such as an insect, that transmits a pathogen from one host to another.

**Vernalisation** Hastening the flowering and fruiting of plants by treating seeds, bulbs, or seedlings so as to shorten vegetative period.

**Vulnerability** The degree to which a system is susceptible to, or unable to cope with, adverse effects of climate change, including climate variability and extremes. A function of the character, magnitude, and rate of climate variation to which a system is exposed, its sensitivity, and its adaptive capacity.

**Water Consumption** Amount of extracted water that is irretrievably lost at a given territory during its use (evaporation and goods production). Water consumption is equal to water withdrawal minus return flow.

**Water Mass** A body of water defined by its temperature and salinity characteristics, and created by surface processes at specific locations. Water masses are physically modified as they move long, depending on the rates of mixing with other water masses.

**Water Stress** A country is water stressed if the available freshwater supply relative to water withdrawal acts constrains its development. Withdrawals exceeding 20 per cent of renewable water supply are used as indicative of water stress.

**Water Use Efficiency** Carbon gain in photosynthesis per unit water lost in evapotranspiration. Expressed on a short-term basis as the ratio of photosynthetic carbon gain per unit transpirational water loss, or an a seasonal basis as the ratio of net primary production or agricultural yield to the amount of available water.

**Water Withdrawal** Amount of water extracted from water bodies.

**WRE Profiles** The carbon dioxide concentration profiles leading to stabilization as defined by Wigley, Richels, and Demands (1996) whose initials provide the acronym. For any given stabilization level, these profiles cover a broad range of possibilities.

# Index

www.ingramcontent.com/pod-product-compliance
Lightning Source LLC
Chambersburg PA
CBHW021438180326
41458CB00001B/326